Fundamentals of Premixed Turbulent Combustion

Fundamentals of Premixed Turbulent Combustion

Andrei Lipatnikov

CRC Press
Taylor & Francis Group
Boca Raton London New York

CRC Press is an imprint of the
Taylor & Francis Group, an **informa** business

CRC Press
Taylor & Francis Group
6000 Broken Sound Parkway NW, Suite 300
Boca Raton, FL 33487-2742

First issued in paperback 2017

© 2013 by Taylor & Francis Group, LLC
CRC Press is an imprint of Taylor & Francis Group, an Informa business

No claim to original U.S. Government works

ISBN-13: 978-1-4665-1024-1 (hbk)
ISBN-13: 978-1-138-07441-5 (pbk)

Library of Congress Cataloging-in-Publication Data

Lipatnikov, Andrei.
    Fundamentals of premixed turbulent combustion / Andrei Lipatnikov.
      p. cm.
    Includes bibliographical references and index.
    ISBN 978-1-4665-1024-1 (hardback)
    1. Gas-turbines--Combustion. I. Title.

TJ778.L57 2012
621.43'3--dc23   2012028552

Visit the Taylor & Francis Web site at
http://www.taylorandfrancis.com

and the CRC Press Web site at
http://www.crcpress.com

To my daughter Anna

Contents

Preface

Having learnt to start fire about 30,000 years ago, mankind tried to control and use combustion over millenniums. The Second Industrial (Technological) Revolution drastically accelerated the process and, in particular, begot such a powerful source of energy as burning of turbulent gases in internal combustion engines, both reciprocating and gas turbine engines. Thanks to the great progress made by the industry in the development and production of engines over the past one and a half centuries, turbulent flames dominate in energy conversion, especially in transportation, and the role played by turbulent combustion in sustainable development of the society may be scarcely exaggerated.

Internal combustion engines transform chemical energy into thermal form by burning a fuel in different regimes, ranging from nonpremixed to fully premixed modes. If a fuel has not mixed with an oxidizer, for example, oxygen in air, before ignition, then combustion proceeds in the nonpremixed regime and the burning rate is controlled by the rate of mixing of the fuel and oxidizer. Nonpremixed flames are typical for diesel engines and many gas turbine combustors. If a fuel has mixed with an oxidizer for a sufficiently long time so that the gas mixture has become homogeneous, then combustion proceeds in the premixed regime and the local burning rate is controlled not only by diffusion but also by chemical reactions. Premixed turbulent flames are typical for Otto (spark ignition) engines. Moreover, lean burning of premixed gases is considered to be a promising combustion technology for future, clean and highly efficient gas turbine combustors. In contemporary internal combustion engines, an intermediate, partially premixed burning mode is widely used to reduce fuel consumption and emissions. Partially premixed combustion occurs if the fuel and oxidizer were mixed during a sufficiently short time before ignition so that the composition of the unburned mixture varies in space and time and even pure fuel or pure oxidizer may be observed in some spatial regions at some instants.

For the reasons that will be discussed later, this book is mainly focused on premixed flames, with recent advances in inhomogeneously (partially) premixed turbulent combustion also being addressed. As regards nonpremixed flames, discussion on this burning mode will be very brief and solely restricted to terms, methods, and concepts that are necessary for understanding the peculiarities of partially premixed turbulent combustion.

Against a background of the phenomenal technological and industrial breakthrough in developing and producing highly efficient and powerful internal combustion engines, the progress made by combustion science is much less impressive, especially as far as premixed turbulent flames are concerned. Despite (i) the long history of industrial use, (ii) the knowledge of the underlying physical laws and balance equations, and (iii) the rapid development of advanced laser diagnostic techniques and computer hardware and software, etc., premixed turbulent flame theory should be considered to be in an initial stage of development, close to the 70-year-old concepts by the founding fathers Damköhler and Shchelkin. Although a number of physical mechanisms and local phenomena have been either discovered or hypothesized to act inside a premixed turbulent flame, we still do not know what mechanisms dominate and under what conditions. Accordingly, we are not only unable to quantitatively predict (without tuning!) the basic characteristics of many premixed turbulent flames but also, sometimes, puzzled by the simplest questions. For instance, in certain cases, we cannot confidently answer what mixture, for example, a lean gasoline–air or a very lean hydrogen–air mixture, will burn faster in a particular turbulent flow.

Such a limited success in developing the theory of premixed turbulent combustion is mainly associated with the strong nonlinearity of the problem, as well as with the following three challenges: (i) wide range of relevant scales, (ii) multivariance, and (iii) multidisciplinarity.

Premixed turbulent combustion involves interaction of processes characterized by significantly different length scales and timescales. First, even a laminar flame itself is a multiscale nonlinear phenomenon. Because the rates of different reactions within the flame may differ by several orders of magnitudes and depend strongly nonlinearly (exponentially) on temperature, variations in the concentrations of different species are localized to spatial zones of significantly different widths. For a typical laminar hydrocarbon–air flame under normal conditions, the most important length scales range from 1 mm (for fuel, oxygen, temperature, H_2O, CO_2, etc.) to 0.1 mm (e.g., CH) or even smaller scales. Second, nonreacting constant-density turbulence is another multiscale nonlinear phenomenon, which involves eddies of very different scales, ranging from the scales of the entire flow to much smaller Kolmogorov scales. Third, as far as a premixed turbulent flame is concerned, the problem is associated with a substantially wider range of important scales, because the largest scales characterizing a laminar flame are often of the order of or even less than the smallest scales of turbulence. With the exception of the simplest cases, such a multiscale nonlinear problem is tractable neither analytically nor numerically even using the most powerful computers available today. Accordingly, simplified methods are required to evaluate the basic characteristics of a premixed turbulent flame.

When developing such a method, a researcher is challenged not only by the nonlinear multiscale nature of premixed turbulent combustion but also by a variety of local phenomena and physical mechanisms that affect a turbulent flame. Any simplified approach implies that certain phenomena and physical mechanisms dominate, whereas others play a minor role. How could the governing phenomena be revealed? Certainly, an answer to this question depends substantially on the particular conditions and goals of research, but, moreover, it is still largely the matter of choice, as far as a premixed turbulent flame is concerned.

Furthermore, even if we restrict ourselves to a single "simple" phenomenon, for example, interaction of a laminar flame with a two-dimensional vortex pair that is relevant to premixed turbulent combustion, a study of even such a relatively simple problem requires deep knowledge in several fields of science, such as thermodynamics, chemical kinetics, molecular transport processes, and fluid dynamics.

Thus, a researcher dealing with premixed turbulent combustion has to investigate the nonlinear interaction of various local phenomena characterized by significantly different length scales and timescales, with each particular phenomenon involving interactions of turbulent eddies, chemical reactions, heat and mass transfer, etc.

As far as numerical modeling of combustion in engines is concerned, the task is strongly impeded by the lack of an approach that is well recognized to be a predictive tool for different premixed turbulent flames. Due to the variety of eventually important local phenomena, a number of models compete in the premixed combustion literature, with the validation of the vast majority of them being scarce. Popularity of a model depends, sometimes, on the advertising skills and name(s) of its author(s) much more than on the predictive capabilities of the model or its consistency with the underlying physics. When analyzing the competing models, not only a newcomer but also an experienced researcher runs the risks either (i) to be buried under the avalanche of papers that highlight a number of different trends and phenomena relevant to the subject or (ii) to select a research tool based mainly on occasional factors, such as fashion, advertisement, traditions, etc.

In order to overcome these difficulties, a researcher has not only to be well educated in different relevant disciplines but also to acquire deep general understanding of the physics of premixed turbulent combustion. In particular, he or she has (i) to know the most important peculiarities of various premixed turbulent flames, (ii) to understand the physical mechanisms of premixed turbulent combustion, (iii) to be aware of the approaches to describe these mechanisms, including the domain of validity, advantages and disadvantages of each particular approach, etc. An expert in premixed turbulent combustion should not only know well a few research tools advertised by his/her teacher or by a well-recognized authority but also, and first of all, be able to analyze and compare the different concepts and viewpoints with one another and with the available experimental data; should not only follow the arguments but also assess and, if necessary, criticize them; should not only trust but also check; etc.

Certainly, no lecture course or textbook can be a substitute for years of research that are necessary to acquire such a deep understanding of premixed turbulent combustion, but an advanced graduate text could help the reader to make a few steps in the right direction. By overviewing the physics of premixed turbulent flames, summarizing the acquired knowledge, emphasizing the advantages and disadvantages of the competing approaches, and highlighting the unresolved issues, this book is aimed at helping the reader to make a deliberate choice of his/her research tool. It is more a what-has-been-done and what-should-be-done book than a how-to-do book. It is aimed at stimulating reflection in the first place and at instructing in using particular advanced methods only in the second place.

To reach the claimed goals, this book highlights the phenomenology of premixed turbulent flames and attempts to combine aspects that seem to be incompatible on the face of it, namely, simplicity and topicality. In other words, the book attempts to discuss the state of the art in premixed turbulent combustion in a simple manner. On the one hand, the simplicity of discussion is necessary in order to provide a newcomer, for example, a postgraduate student, with a general overview of the physics of premixed turbulent combustion and to kindle his/her interest in the subject. On the other hand, advanced concepts and tools, as well as unresolved issues, are considered in order to (i) prepare the newcomer to do his/her own research and to orient well in the contemporary literature and (ii) to make the book useful not only for newcomers but also for experienced researchers. In a sense, the book is a hybrid of an advanced graduate text intended primarily for postgraduate students and a scientific monograph intended for experts in the field.

While a typical graduate text on combustion covers a wide range of relevant problems and addresses turbulent flames only briefly by discussing the simplest, but often outdated, concepts, the scope of this book is much more narrow (mainly premixed turbulent combustion), with the aim of overviewing and more deeply considering topical research issues investigated today, as well as advanced research tools used for those purposes. In this respect, the book is more close to a scientific monograph. While some other books intended for researchers in the field of turbulent combustion emphasize the mathematical aspects of the subject, this book highlights the phenomenology and physical mechanisms of premixed turbulent burning and is aimed at simplifying the discussions as much as possible. In this respect, the book is more close to a graduate text. Moreover, while some other books intended for experienced researchers and engineers thoroughly consider advanced methods and concepts developed recently, but are mainly restricted to a few approaches put forward by the respective authors, this book is aimed at overviewing various, often competing, methods and concepts used and developed by the premixed turbulent combustion community today. However, contrary to an imaginary "encyclopedia" on premixed turbulent flames (not yet written), which review all actual methods, concepts, ideas, etc., the list of subjects discussed in the book is far from being complete, with the focus on the physical mechanisms and phenomenology of premixed turbulent combustion, as already noted. Furthermore, while some other books review various advanced methods and concepts but do not emphasize their weak points, this book disputes many widely accepted assumptions and models. In fact, one of its goals is to stimulate the readers to critically perceive any new scientific result or hypothesis by comparing it with other available data and concepts.

One the one hand, critical analysis of certain mainstream approaches may make this book particularly useful for experts in the field. On the other hand, to make the discussion on the state of art in premixed turbulent combustion interesting not only for the experts but also for newcomers, such as postgraduate students, I was sometimes forced either to substantially oversimplify the problems by highlighting the phenomenology to the detriment of mathematical rigorousness or to restrict myself to a very brief description (or even only mentioning) of certain advanced methods and concepts that are widely used. When selecting methods or concepts to be discussed in more detail, I showed preference to tools that aided me personally to better understand the physics of premixed turbulent combustion, whereas certain popular areas of research into premixed turbulent flames, for example, large-eddy simulation or combustion chemistry, are only briefly touched upon, because they appear to be of much more importance for applications than for understanding the basic physics of flame–turbulence interaction. Therefore, the book conveys essentially my (possibly very biased) views of what matters in the field. Nevertheless, if a method, or a concept, or an issue, is only mentioned or addressed very briefly in the book, references to key relevant papers are provided to aid interested readers in independently studying the case more deeply. Furthermore, even if many issues are considered in more detail in the book, the discussion is sometimes restricted to highlighting the phenomenological aspects and summarizing the most important findings and issues, rather than to reviewing the papers that the summary is based on. In such cases, for scrutinizing the issue, interested readers are referred to particular sections of three review papers published by me and Jerzy Chomiak in *Progress in Energy and Combustion Science* in 2002, 2005, and 2010.

Thus, this book is primarily intended (i) as an advanced graduate text for post-graduate students who deal with premixed turbulent flames in their study and (ii) as a scientific monograph for experts in the field. The book may also be valuable for a wider readership, ranging from graduate students in engineering sciences, chemistry, and physics to professional engineers and researchers in combustion and engine fields.

To some degree, this book is conceptually similar to Chapters I–IV in the book *Physics of Gas Combustion* by E.S. Shchetinkov (1965, in Russian) and to the book *Combustion: A Study in Theory, Fact and Application* by J. Chomiak (1990). Certainly, I have an opportunity to extend the discussion by analyzing the results, concepts, and methods obtained and developed in the past decades after the publication of the two cited books.

Because combustion science is based on thermodynamics, chemical kinetics, transport phenomena, and fluid mechanics, a scientific background in these disciplines is surely desirable for the readers. Nevertheless, to aid the readers who do not have a complete command of all of them, Chapters 1–3 are aimed at briefly summarizing the prerequisite knowledge of these disciplines. Subsequently, the main contents of the book will be presented in such a way as to make it accessible to the readers with no advanced knowledge on the aforementioned subjects. Certainly, I assume that the readers have a scientific background in mathematics, general physics, and general chemistry.

The main contents of the book consist of four parts. The first three of them deal with premixed turbulent flames and address (i) the experimental data on the general appearance of such flames, for example, the overall burning rate, flame speed and thickness, spatial profiles of mean temperature (see Chapter 4); (ii) the physical mechanisms that could affect the flame behavior (see Chapter 5 and, in part, Chapter 6); and (iii) the physical and numerical models aimed at predicting the key features of premixed turbulent combustion (see the remainder of Chapter 6 and Chapter 7). The fourth part addresses nonpremixed (see Chapter 8) and partially premixed (see Chapter IX) combustion, with the discussion of the former burning mode in Chapter 8 being very brief and solely restricted to introducing terms, methods, and concepts that are necessary for discussing partially premixed combustion in Chapter 9.

Thus, contrary to a typical combustion textbook, which addresses both premixed and diffusion, that is, nonpremixed, flames, the focus of this book is on premixed combustion for the following two reasons. First, in order not only to introduce a newcomer to the subject but also to discuss the agenda and advanced tools of contemporary research on it in a simple manner, an author should have his/her own long-term experience in studying the subject. Because I never investigated diffusion flames by myself, I do not consider myself to be a proper person for discussing the state of the art in the subject. Second, because the burning rate in diffusion flames is mainly controlled by mixing, whereas the burning rate in premixed flames is controlled by both chemical reactions and mixing, premixed turbulent combustion is substantially a more difficult problem than diffusion burning. Accordingly, the contemporary models of diffusion flames are elaborated much better than the models of premixed turbulent flames. For instance, if the efforts of the diffusion combustion community are aimed at predicting pollutant formation, while the overall burning rate and the mean temperature field can be computed with acceptable accuracy in a statistically stationary diffusion flame, the latter problem still strongly challenges the premixed combustion community. Therefore, the manner of discussion on contemporary research topics relevant to diffusion combustion would likely differ from the manner of discussion adopted in this book. If the latter is based on highlighting the phenomenology and physical mechanisms, a monograph on diffusion flames would likely to deal more with the mathematical and numerical aspects of the subject and to consider them in a more detailed manner.

Andrei N. Lipatnikov
Chalmers University of Technology
Gothenburg, Sweden

Acknowledgments

I am indebted to many people for their help in my research and in the preparation of this book. First of all, I would like to thank Jerzy Chomiak for his long-term cooperation, which is of great importance to me, and for a number of valuable comments regarding the book. Moreover, I am very grateful to Andrei Betev, Alexey Burluka, Vladimir Karpov, Vladimir Sabelnikov, and Vladimir Zimont for the fruitful discussions that substantially broadened my understanding of premixed turbulent combustion. I am thankful to many colleagues from various universities and research centers for sharing their data and/or figures with me, in particular, to Andy Aspden, Fernando Biagioli, Nilanjan Chakraborty, Robert Cheng, Friedrich Dinkelacker, James F. Driscoll, Mathew John Dunn, Simone Hochgreb, Hong G. Im, Satoru Ishizuka, Peter Kalt, Clemens Kaminski, Alexander Konnov, Dimitrios C. Kyritsis, Chang K. Law, Zhongshan Li, Tiangfeng Lu, Shinnosuke Nishiki, Ishwar K. Puri, Bruno Renou, Ramanan Sankaran, Jackie Sung, and Victor Yakhot. The work on this book was started through the encouragement of Ingemar Denbratt, who leads the combustion division of the department I work in. Thanks are due to many of my colleagues from the Department of Applied Mechanics at Chalmers University of Technology. I am grateful to George Soros and his International Science Foundation, Swedish Research Council, Chalmers Combustion Engine Research Center, Swedish Energy Agency, and Swedish Gas Turbine Center for granting my research on premixed turbulent combustion.

My parents, wife Lena, and daughter Anna also deserve many thanks for their patience.

Author

Andrei N. Lipatnikov, PhD, is an associate professor at the Department of Applied Mechanics at Chalmers University of Technology, Gothenburg, Sweden. He graduated from the Moscow Institute of Physics and Technology (MIPT) in 1984 and obtained his PhD degree from the same institute in 1987. From 1987 until 1996, Dr. Lipatnikov had been employed by the MIPT. In 1996, he was invited to join the Internal Combustion Engine Group at Chalmers University of Technology. The academic activities of Dr. Lipatnikov have been concerned with research on turbulent and laminar combustion processes, pollutant formation in flames, autoignition of premixed mixtures, and numerical modeling of combustion in spark ignition engines. He has published more than 180 scientific contributions, including 55 journal papers dealing with the aforementioned subjects. Dr. Lipatnikov is also a member of the Combustion Institute.

1 General Knowledge on Reacting Gas Mixtures

The goal of this chapter is to summarize the general knowledge on gas mixtures that is relevant to turbulent combustion. The summary is very brief because the terms, quantities, and equations addressed in the following sections are discussed in detail in proper general courses, for example, thermodynamics, chemistry, and fluid dynamics.

BASIC CHARACTERISTICS OF GAS MIXTURES

STATE EQUATION

The theory of turbulent combustion considers the reacting mixture to be an ideal gas and uses the well-known ideal gas state equation

$$p = \frac{\rho}{M} R^0 T, \qquad (1.1)$$

where

p is the pressure
ρ is the density
M is the molecular weight of the mixture
T is the temperature
$R^0 = 8.314$ J/mol
K $= 1.986$ cal/mol K is the universal gas constant.

Note that the pressure averaged over a volume filled with a gas mixture is commonly designated as capital P, whereas p is the local pressure in a point. In a typical flame characterized by a low Mach number, the difference between p and P is much less than the value of P. For this reason and because capital P is reserved to designate the probability density function, a single symbol p will be used to denote the two pressures in the following text.

Very often, variations in the molecular weight M in flames are neglected and Equation 1.1 reads

$$\frac{p}{\rho T} = \frac{p_0}{\rho_0 T_0}, \qquad (1.2)$$

where the subscript 0 is associated with the state of the mixture before combustion. Furthermore, if pressure variations within a flame are also neglected, then Equation 1.2 is further simplified:

$$\rho T = \rho_0 T_0. \tag{1.3}$$

Equation 1.3 is widely invoked in studies of turbulent combustion, while Equation 1.1 is commonly used in the numerical simulations of laminar flames.

MIXTURE COMPOSITION

A flame expands in a gas mixture that contains at least two species. In the simplest cases, the unburned gas is a mixture of either two (a fuel and an oxidizer) or three (a fuel, an oxidizer, and an inert diluent) species. Therefore, a proper description of gas mixture composition is the first step toward understanding the combustion process.

In the literature, there are a number of quantities that characterize the composition of a gas mixture. For instance, the partial density, ρ_l, is equal to the mass of the l-th species per unit volume and the molar concentration, c_l, is equal to the number of moles of the l-th species per unit volume. The partial density and the molar concentrations are measured in kilograms per cubic meter (kg/m^3) and moles per cubic meter (mol/m^3), respectively, in SI units. Obviously,

$$\sum_{l=1}^{N} \rho_l = \rho, \tag{1.4}$$

$$\sum_{l=1}^{N} c_l = c = \frac{p}{R^0 T}, \tag{1.5}$$

where

c is the number of moles of the mixture per unit volume
N is the number of different species in the mixture.

Moreover,

$$c_l = \frac{\rho_l}{M_l}, \tag{1.6}$$

where M_l is the molecular weight of the l-th species. In a typical unburned fuel–air mixture, the values of ρ, c, and $M \approx 0.03$ kg/mol are close to those of the corresponding characteristics of the air in the mixture, for example, $\rho \approx 1$ kg/m^3 and $c \approx 30$ mol/m^3 under atmospheric conditions.

The partial pressure p_l of the l-th species, defined as

$$p_l = \frac{\rho_l}{M_l} R^0 T,$$ (1.7)

is another characteristic of mixture composition. Partial pressures are of particular importance when computing the temperature of the adiabatic combustion products. Obviously,

$$\frac{p_l}{p} = \frac{\rho_l}{M_l} \frac{M}{\rho} = c_l \frac{M}{\rho} = \frac{c_l}{c}.$$ (1.8)

In SI units, the partial pressure is measured in pascals (Pa), where 1 Pa is equal to 1 N/m². In the combustion literature, the pressure is often reported in atmospheres (1 atm $\approx$ 105 kPa), MPa (1 atm $\approx$ 0.1 MPa), or bars (1 atm $\approx$ 1 bar). Equations 1.5 and 1.8 show that

$$\sum_{l=1}^{N} p_l = p.$$ (1.9)

The mole fraction X_l is equal to the number of moles of the l-th species per unit mole of the mixture, that is,

$$X_l = \frac{p_l}{p} = \frac{c_l}{c}.$$ (1.10)

Equations 1.5 and 1.10 yield

$$\sum_{l=1}^{N} X_l = 1.$$ (1.11)

Moreover, Equations 1.4, 1.8, and 1.10 yield

$$\sum_{l=1}^{N} M_l X_l = \sum_{l=1}^{N} M_l \frac{p_l}{p} = \sum_{l=1}^{N} M_l \frac{\rho_l}{M_l} \frac{M}{\rho} = M \frac{\sum_{l=1}^{N} \rho_l}{\rho} = M.$$ (1.12)

Finally, the mass fraction Y_l is equal to the mass of the l-th species per unit mass of the mixture, that is,

$$Y_l = X_l \frac{M_l}{M} = \frac{p_l}{p} \frac{M_l}{M} = \frac{\rho_l}{\rho} = c_l \frac{M_l}{\rho}.$$ (1.13)

Equations 1.12 and 1.13 yield

$$\sum_{l=1}^{N} Y_l = 1. \tag{1.14}$$

Both the mole and the mass fractions are nondimensional quantities.

The aforementioned quantities characterize the mixture composition in terms of various species, with the concentrations of most species changing during the combustion process. On the contrary, the number of atoms of a particular chemical element is always conserved in chemical reactions. Accordingly, to take advantage of the element conservation, the mass fractions Z_m of the m-th element are widely used. The mass fraction of the m-th element is equal to the total mass of the atoms of this element per unit mass of the mixture. If a gas contains N species, then

$$Z_m = M_m \sum_{l=1}^{N} b_{ml} \frac{Y_l}{M_l} = \frac{M_m}{M} \sum_{l=1}^{N} b_{ml} X_l, \tag{1.15}$$

where

b_{ml} is the number of atoms of the m-th element in the molecule of the l-th species
M_m and M_l are the molecular weights of the atom and the species, respectively.

Obviously,

$$\sum_{m=1}^{N_Z} Z_m = 1, \tag{1.16}$$

by virtue of the definition of the element mass fraction. Here, N_Z is the number of different elements in the mixture.

If a mixture contains five species (H_2O, CO_2, CO, O_2, and N_2) and four elements (H, O, C, and N), then

$$Z_H = \frac{2}{18} Y_{H_2O},$$

$$Z_O = 16 \left(\frac{Y_{H_2O}}{18} + 2\frac{Y_{CO_2}}{44} + \frac{Y_{CO}}{28} + 2\frac{Y_{O_2}}{32} \right),$$

$$Z_C = 12 \left(\frac{Y_{CO_2}}{44} + \frac{Y_{CO}}{28} \right),$$

$$Z_N = 14 \left(2\frac{Y_{N_2}}{28} \right) = Y_{N_2}.$$

The mass fractions Z_m are nondimensional quantities.

STOICHIOMETRY AND EQUIVALENCE RATIOS

The aforementioned characteristics of mixture composition are used in different fields of science. There are also quantities specific to combustion applications. To introduce such quantities, let us consider the burning of a hydrocarbon fuel, C_xH_y, in the air.

In this case, the products contain many species (e.g., H_2O, CO_2, CO, H_2, O_2, N_2, NO, H, O, OH, H_2O_2, and HO_2), but only water (H_2O) and carbon dioxide (CO_2) are the final combustion products, that is, the burned gas should consist only of H_2O, CO_2, O_2, and N_2, so that 100% of the chemical energy bound in the fuel is converted to thermal energy. This conversion is commonly called heat release. If the products contain, for example, carbon monoxide (CO) under certain conditions, then a part of the chemical energy is still bound in the molecules of CO and may be converted to thermal energy under other conditions.

Thus, the maximum amount of thermal energy is released if all of the carbon contained in a fuel is oxidized to carbon dioxide and all of the hydrogen contained in the fuel is oxidized to water. The released energy is used to heat the gas mixture. Because the molecules that do not react, for example, nitrogen (N_2), just absorb the energy, a higher product temperature is associated with a lower concentration of such inert molecules. However, not only nitrogen but also oxygen molecules may be inert under certain conditions. If a mixture contains too little fuel, only a part of oxygen molecules will take part in chemical reactions and heat release, whereas the rest of the O_2 will behave as inert molecules. Therefore, the highest temperature is associated with the products that contain only H_2O, CO_2, and N_2, with all oxygen molecules consumed in the combustion reactions (note that we do not include the differences in the heat capacities of various molecules in this simplified discussion). In such a case, the relative concentration of inert molecules, which absorb heat, is minimum, whereas the relative concentration of reacting molecules, which contribute to heat release, is maximum.

Thus, an increase in gas temperature within a flame seems to be maximum if all molecules of the fuel and oxygen are converted to water and carbon dioxide. This full conversion of C_xH_y and O_2 to H_2O and CO_2 is possible only if the ratio of the concentrations of the fuel and oxygen is equal to a specific quantity, which is called the stoichiometric ratio and is controlled by the fuel formula. Indeed, let us consider a global reaction,

$$C_xH_y + aO_2 \rightarrow bCO_2 + dH_2O.$$

Since the number of C atoms cannot be changed in the reaction, we may write $b = x$. Similarly, the conservation of H and O atoms yields $2d = y$ and $2a = 2b + d$, respectively. Therefore, the coefficient a is equal to $x + y/4$, that is, in order for both the fuel and the oxidizer to be fully converted to water and carbon dioxide, the unburned mixture should contain $x + y/4$ molecules of oxygen per molecule of the fuel.

The ratio of the number of oxidizer molecules to the number of fuel molecules that is necessary for the full conversion of both the fuel and the oxidizer to the final combustion products is called the molar stoichiometric ratio. For hydrocarbons, this ratio is equal to $x + y/4$, that is, it is controlled by the fuel formula. For instance, it is equal to $1 + 4/4 = 2$ for methane (CH_4), 3.5 for ethane (C_2H_6), 5 for propane (C_3H_8), and 12.5 for *iso*-octane (C_8H_{18}). If the ratio of the number of oxidizer molecules to the number of fuel molecules is equal to the molar stoichiometric ratio, then such a mixture is called the stoichiometric mixture.

In addition to the molar stoichiometric ratio presented earlier, the mass stoichiometric ratio St is widely used in the combustion literature. It is equal to the ratio of oxidizer mass to fuel mass in the case of the full conversion of the fuel and the oxidizer to final combustion products. If molecular oxygen is considered to be the oxidizer, then the mass stoichiometric ratio is equal to

$$St = \frac{M_{O_2}}{M_{C_xH_y}}\left(x + \frac{y}{4}\right). \tag{1.17}$$

Note that the mass stoichiometric ratio St weakly depends on the number x of carbon atoms in the fuel molecule, for example, $St = 4$ and 3.5 for methane (CH_4) and *iso*-octane (C_8H_{18}), respectively.

Alternative definitions to Equation 1.17 are also used in the literature. For instance, the aforementioned global reaction may be rewritten as follows:

$$C_xH_y + (x + 0.25y)(O_2 + 3.7N_2) \rightarrow xCO_2 + 0.5yH_2O + 3.7(x + 0.25y)N_2,$$

and if air is considered to be an oxidizer, then

$$St' = \frac{M_{O_2} + 3.7M_{N_2}}{M_{C_xH_y}}\left(x + \frac{y}{4}\right). \tag{1.18}$$

In the following text, we will use the mass stoichiometric ratio associated with oxygen, that is, Equation 1.17, if the opposite is not specified.

If the ratio of the concentrations of a fuel and oxygen differs from the stoichiometric ratio in a fuel–air mixture, the composition of this mixture is commonly characterized by the equivalence ratio Φ in the combustion literature. The equivalence ratio

$$\Phi = \frac{(X_F/X_O)}{(X_F/X_O)^{st}} = \frac{(Y_F/Y_O)}{(Y_F/Y_O)^{st}} = St\frac{Y_F}{Y_O} \tag{1.19}$$

is the ratio of fuel–air ratios in considered and stoichiometric unburned mixtures. If $\Phi < 1$, the mixture is called lean. If $\Phi = 1$, the mixture is called stoichiometric. If $\Phi > 1$, the mixture is called rich. In a nonstoichiometric C_xH_y–air mixture,

$(M_{O_2} + 3.7M_{N_2})(x + 0.25y)$ grams or $(x + 0.25y)$ molecules of the air are mixed with Φ grams or molecules, respectively, of the fuel. Accordingly, the formula for the aforementioned global reaction is as follows:

$$\Phi C_x H_y + (x + 0.25y)(O_2 + 3.7N_2) \rightarrow \text{Products.}$$

In some papers (especially in the Russian literature), another quantity, $\lambda = 1/\Phi$, is used instead of the equivalence ratio and is called the coefficient of the air abundance. The mixture is called lean (rich) if λ is greater (less) than unity.

The equivalence ratio may easily be related to the other concentrations discussed in the section "Mixture Composition." For instance, in the stoichiometric $C_x H_y$–air mixture,

$$Y_F^{st} = \frac{M_F}{M_F + \left(x + \dfrac{y}{4}\right)(M_{O_2} + 3.7M_{N_2})} = \frac{12x + y}{12x + y + \left(x + \dfrac{y}{4}\right)(32 + 3.7 \times 28)}$$

and

$$Y_{O_2}^{st} = \frac{32\left(x + \dfrac{y}{4}\right)}{12x + y + \left(x + \dfrac{y}{4}\right)(32 + 3.7 \times 28)},$$

where M_F is the fuel molecular weight. If $\Phi \neq 1$, then

$$Y_F = \frac{\Phi(12x + y)}{\Phi(12x + y) + \left(x + \dfrac{y}{4}\right)(32 + 3.7 \times 28)}, \tag{1.20}$$

$Y_{O_2} + Y_{N_2} = 1 - Y_F$, and $Y_{O_2}/Y_{N_2} = 32/(3.7 \times 28) = 0.31$. Equation 1.20 may also be used for calculating the equivalence ratio, if the fuel mass fraction is known.

What products are formed in the case of $\Phi \neq 1$? For a lean ($\Phi < 1$) mixture, the answer is simple: the abundant oxygen behaves like inert nitrogen and the global reaction is as follows:

$$\Phi C_x H_y + (x + 0.25y)(O_2 + 3.7N_2) \rightarrow \Phi(x CO_2 + 0.5y H_2O)$$

$$+ (1 - \Phi)(x + 0.25y)O_2 + 3.7(x + 0.25y)N_2.$$

For a moderately rich ($\Phi > 1$, but the difference between Φ and unity is not large) mixture, one may assume that the abundant fuel is oxidized to carbon monoxide, which is not further oxidized to carbon dioxide due to the lack of oxygen. Accordingly,

$$\Phi C_x H_y + (x + 0.25y) O_2 \rightarrow \left[x + (1 - \Phi)(x + 0.5y) \right] CO_2$$

$$+ 0.5\Phi y H_2 O + 2(\Phi - 1)(x + 0.25y) CO,$$

where the inert nitrogen molecules are skipped for the sake of brevity. Such a global reaction may model the burning of a rich fuel–air mixture only if the number of CO_2 molecules is not negative, that is, $\Phi - 1 \leq (1 + 0.5y/x)^{-1}$. For alkanes, also known as paraffin, $y = 2x + 2$ and the equivalence ratio should be lower than $1 + (2 + 1/x)^{-1}$, for example, 4/3 for methane and 25/17 for *iso*-octane, so that the aforementioned simple reaction may be used. For a larger Φ, a part of the hydrogen cannot be oxidized to water due to the lack of sufficient number of oxygen molecules. In such a case, the products contain not only H_2O, CO_2, and CO, but also H_2. Finally, if $\Phi x > 2(x + 0.25y)$, then the available number of oxygen molecules would not be sufficient for the carbon from the fuel to be completely oxidized to CO and the products would contain unburned hydrocarbons. However, flames do not propagate in such rich mixtures.

As will be discussed later, the aforementioned simple global reactions are idealized models and real combustion products contain a number of various species even in the stoichiometric case.

ENERGY, ENTHALPY, AND HEAT CAPACITIES

For an adiabatic system, the first law of thermodynamics reads

$$dE + pdV = 0, \tag{1.21}$$

where

E is the internal energy of the system
V is its volume.

For a spatially uniform system, the law may be rewritten as follows:

$$de + pd\rho^{-1} = dh - dp/\rho = 0, \tag{1.22}$$

where e and $h = e + p/\rho$ are the specific, that is, per unit mass, internal energy and enthalpy, respectively. Both of these quantities are measured in joules per kilogram (J/kg).

Equation 1.22 shows that both the specific internal energy and the specific enthalpy may equally be used for describing the behavior of an adiabatic system. For some applications, e is a more appropriate quantity. For instance, if the volume of a system is constant, then the first law of thermodynamics simply reads $e = \text{const}$ for a spatially uniform adiabatic system (for a nonuniform system, e may be associated with the specific internal energy averaged over the system). On the contrary, if the pressure is constant, then the use of the enthalpy simplifies the discussion, as Equation 1.22 yields $h = \text{const}$. Because the pressure may be considered to be constant within many flames, we will mainly use h in the following text.

For a single-component ideal gas, the partial derivatives

$$c_V = \left(\frac{\partial e}{\partial T}\right)_V \tag{1.23}$$

and

$$c_P = \left(\frac{\partial h}{\partial T}\right)_P, \tag{1.24}$$

taken at a constant volume and a constant pressure, respectively, are called the specific heat capacities at constant volume and constant pressure, respectively. The mole heat capacities C_V and C_P are equal to Mc_V and Mc_P, respectively. In an ideal gas, $C_P - C_V = R^0T$.

Both c_v and c_P depend on temperature, therefore,

$$e(T) = e(T_0) + \int_{T_0}^{T} c_V dT \tag{1.25}$$

and

$$h(T) = h(T_0) + \int_{T_0}^{T} c_P dT \tag{1.26}$$

for a single-component mixture at a constant volume and a constant pressure, respectively. The reference temperature, T_0, may be set arbitrarily, for example, $T_0 = 0$ K or $T_0 = 298.15$ K. The latter value is widely used in the combustion literature.

For a mixture of N species, the partial specific heat capacities $c_{v,l}$ and $c_{P,l}$, the partial specific internal energies e_l, and the partial specific enthalpies $h_l = e_l + p_l/\rho_l$ may be introduced similarly to the partial densities ρ_l and pressures p_l. Then,

$$c_V = \sum_{l=1}^{N} c_{V,l} Y_l, \tag{1.27}$$

$$e = \sum_{l=1}^{N} e_l Y_l = \sum_{l=1}^{N} Y_l \int_{T_0}^{T} c_{V,l} dT + \sum_{l=1}^{N} e_l(T_0) Y_l = \int_{T_0}^{T} c_V dT + \sum_{l=1}^{N} e_l(T_0) Y_l, \quad (1.28)$$

$$c_P = \sum_{l=1}^{N} c_{P,l} Y_l, \quad (1.29)$$

and

$$h = \sum_{l=1}^{N} h_l Y_l = \sum_{l=1}^{N} Y_l \int_{T_0}^{T} c_{P,l} dT + \sum_{l=1}^{N} \Delta h_l^0 Y_l. \quad (1.30)$$

Note that

$$\sum_{l=1}^{N} Y_l \int_{T_0}^{T} c_{P,l} dT \neq \int_{T_0}^{T} \sum_{l=1}^{N} Y_l c_{P,l} dT = \int_{T_0}^{T} c_P dT,$$

because variations in the temperature and species mass fractions are not independent of each other.

For the reason discussed later, a new symbol Δh_l^0 is introduced to denote the reference enthalpy $h_l(T_0)$ in Equation 1.30.

Using Equations 1.1, 1.8, 1.11, and 1.13, one can easily show that

$$h = \sum_{l=1}^{N} h_l Y_l = \sum_{l=1}^{N} e_l Y_l + \sum_{l=1}^{N} \frac{pM}{\rho M_l} Y_l = e + \sum_{l=1}^{N} \frac{p}{\rho} X_l = e + \frac{p}{\rho} = e + \frac{R^0 T}{M}. \quad (1.31)$$

The partial mole enthalpy H_l is the enthalpy of 1 mol of the l species, that is, $H_l = M_l h_l$. Accordingly,

$$H = \sum_{l=1}^{N} H_l X_l = \sum_{l=1}^{N} h_l M_l X_l = M \sum_{l=1}^{N} h_l Y_l = Mh. \quad (1.32)$$

The first (integral) term on the right-hand side (RHS) of Equation 1.28 is associated with an increase in the thermal energy (i.e., the kinetic energy of the chaotic motion of gas molecules) when the temperature is increased from T_0 to T. The second (sum) term involves the "chemical" energy bound in molecules. This energy may be released, that is, transformed into thermal form, in chemical reactions. Such a transformation of chemical energy into thermal form is called heat release. Due to the heat release, the thermal energy is increased and the temperature of the combustion products is higher than the temperature of the reactants.

In the combustion literature, the ability of different molecules to release (or absorb) heat is commonly quantified using the reference enthalpy $h_l(T_0)$, called the standard specific heat Δh_l^0 of formation of the l-th species at a reference temperature T_0. The standard specific heat of formation of a substance is equal to the enthalpy change that accompanies the formation of 1 kg of this substance in its standard state from its constituent elements in their standard states. The standard state of an element is the most stable form of the element at a reference temperature (typically, 298.15 K) and under pressure equal to 1 bar. Accordingly, $\Delta h_l^0 = 0$ for all elements in their standard states, for example, O_2, H_2, N_2, He, Ar, Xe, and solid carbon in the form of graphite, because the enthalpy is not changed when such an element is "formed from itself." For other species, the value of Δh_l^0 depends both on the species formula and on the reference temperature.

Similarly, the standard mole heat ΔH_l^0 of formation of the species l may be introduced (the word "mole" is often skipped). Obviously, $\Delta H_l^0 = M_l \Delta h_l^0$. The mole enthalpy ΔH_l^0 is reported in the thermodynamic tables.

ADIABATIC COMBUSTION TEMPERATURE

Let us consider the simplest combustion process modeled by a single irreversible reaction

$$F + O \rightarrow P,$$

where F, O, and P designate a fuel, an oxidizer, and products, respectively. Note that any of these three symbols is associated not only with a single chemical component, for example, H_2, O_2, and H_2O, respectively, but also with a mixture of different molecules, for example, the oxidizer may be associated with the air, while P may designate a mixture of CO_2, H_2O, CO, H_2, O_2, and N_2. If the pressure is constant and uniform, then the specific enthalpy is not changed under adiabatic conditions. Accordingly,

$$\int_{T_0}^{T_u} c_{P,u} dT + Y_{F,u} \Delta h_F^0 + Y_{O,u} \Delta h_O^0 = h_u = h_b = \int_{T_0}^{T_b} c_{P,b} dT + Y_{P,b} \Delta h_P^0. \tag{1.33}$$

Let us, for simplicity, assume that the mixture heat capacity, c_P, does not depend on the temperature, that is, $c_{P,u} = c_{P,b} = c_P$. Then, Equation 1.33 reads

$$c_P \left(T_b - T_u \right) = Y_{F,u} \Delta h_F^0 + Y_{O,u} \Delta h_O^0 - Y_{P,b} \Delta h_P^0 \equiv q \tag{1.34}$$

and the adiabatic combustion temperature, T_b, is estimated as follows:

$$T_b = T_u + \frac{Y_{F,u}\Delta h_F^0 + Y_{O,u}\Delta h_O^0 - Y_{P,b}\Delta h_P^0}{c_P} = T_u + \frac{q}{c_P}. \tag{1.35}$$

The quantity q is called the specific (per unit mass) heat of reaction.

Equation 1.35 shows that the combustion temperature is increased when the standard specific heat of formation of the reactants is increased and/or the standard specific heat of formation of the products is decreased. Accordingly, combustion products are commonly associated with the negative values of ΔH_f^0. For instance, $\Delta H_{H_2O}^0 = -241.826$ kJ/mol or $\Delta h_{H_2O}^0 \approx -13.435$ MJ/kg, $\Delta H_{CO_2}^0 = -393.522$ kJ/mol or $\Delta h_{CO_2}^0 = -8.944$ MJ/kg, and $\Delta H_{CO}^0 = -110.529$ kJ/mol or $\Delta h_{CO}^0 \approx -3.947$ MJ/kg. Reactants may be characterized by both positive standard heat of formation, for example, $\Delta H_{C_2H_4}^0 = 52.283$ kJ/mol or $\Delta h_{C_2H_4}^0 \approx 1.867$ MJ/kg, and negative standard heat of formation, for example, $\Delta H_{CH_4}^0 = -74.873$ kJ/mol or $\Delta h_{CH_4}^0 \approx -4.68$ MJ/kg.

Equation 1.35 may be used for a rough estimate of the adiabatic combustion temperature (e.g., see Figure 1.1). For different fuels, the maximum T_b given by the simple Equation 1.35 is associated with the stoichiometric mixtures, because the fraction of inert species, which are heated by the energy released in the combustion reaction, is minimum in this case.

The calculated dependencies of the combustion temperature on the equivalence ratio are linear. For instance, if we consider the combustion reaction

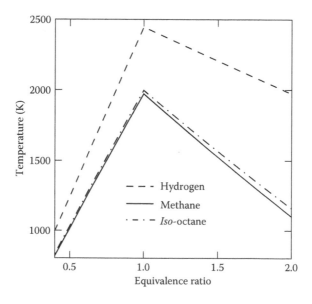

FIGURE 1.1 The adiabatic combustion temperature estimated for various fuel–air mixtures using Equation 1.35 with $c_p = 1.4$ kJ/(kg K) and $T_u = 300$ K.

$$\Phi C_x H_y + (x + 0.25y)(O_2 + 3.7N_2) \rightarrow$$

$$\Phi(xCO_2 + 0.5yH_2O) + (1 - \Phi)(x + 0.25y)O_2 + 3.7(x + 0.25y)N_2$$

for a lean $(\Phi < 1)$ mixture, then $(Y_{l,u} - Y_{l,b})\Delta h_l^0 = -Y_{l,b}\Delta h_l^0 \propto -\Phi\Delta h_l^0 = |\Phi\Delta h_l^0|$ for products containing CO_2 and H_2O (remember that $\Delta h_{O_2}^0 = \Delta h_{N_2}^0 = 0$), that is, the sum on the RHS of Equation 1.35 decreases linearly when the equivalence ratio decreases.

CALCULATION OF TEMPERATURE AND COMPOSITION OF ADIABATIC COMBUSTION PRODUCTS

For many applications, Equation 1.35 is oversimplified because it allows for neither an increase in $c_{P,l}$ by T nor the differences in $c_{P,l}$ for different species. Moreover, real combustion products contain not only H_2O, CO_2, CO, O_2, and N_2, but also H_2, H, O, OH, H_2O_2, HO_2, NO, etc. The mass fractions of the latter species are typically very low, but not zero.

Different methods may be used to improve the accuracy of the computations of the adiabatic combustion temperature and the composition of the equilibrium products. Modern computer programs, for example, a freely available (www.arcl02.dsl.pipex.com) GASEQ code by Morley, do so by minimizing the Gibson energy of a considered system under the constraints of atom conservation. In this chapter, a simpler approach will be considered, as its application to turbulent flames yields sufficiently accurate results.

If burned products contain N species, then we have to calculate N mass (or mole) fractions, temperature (T_b), pressure (p), density (ρ_b), and molecular weight (M_b), that is, $N + 4$ variables. Accordingly, we have to solve $N + 4$ equations.

If N species involve $n < N$ elements (typically, the species involve C, H, O, N, and Ar from the air, i.e., $n = 5$), then we have n equations $Z_m = \text{const}$. Any of these equations may be replaced with the normalizing Equation 1.11 or 1.14, but only n of these $n + 2$ equations are linearly independent from one another.

Since the pressure variations within a typical flame are very low, the pressure is commonly considered to be known when calculating the temperature and the composition of the adiabatic combustion products, that is, we have one more equation $p = \text{const}$. Alternatively, if combustion occurs in a closed constant volume, then the pressure increases, but $\rho = \text{const}$.

The ideal gas state Equation 1.1, the enthalpy (if $p = \text{const}$) or energy (if volume is constant) conservation, and Equation 1.12 are the three additional equations. In total, we have $n + 4$ equations. Finally, $N - n$ equations result from the conditions of chemical equilibrium. As shown in thermodynamics textbooks, the equilibrium condition for a reaction

$$\sum_{l=1}^{N} n_{k,l} S_l \leftrightarrow \sum_{l=1}^{N} m_{k,l} S_l, \tag{1.36}$$

where S_l designates the l-th species, reads

$$\frac{\prod_{l=1}^{N} p_l^{m_l}}{\prod_{l=1}^{N} p_l^{n_l}} = K_P(T), \qquad (1.37)$$

with the equilibrium constant $K_P(T)$ for partial pressures depending solely on the temperature. If concentrations are characterized by other quantities (not partial pressures), then the corresponding equilibrium constants may depend not only on T but also on p. For instance, for reaction $F + O \rightarrow P$, we have

$$K_X(T,p) = \frac{X_P}{X_F X_O} = \frac{p_P p}{p_F p_O} = p K_P(T). \qquad (1.38)$$

To illustrate the calculation of the adiabatic combustion temperature for a $C_x H_y$–air flame under a constant pressure, let us summarize the necessary equations for the products that contain 12 species; H_2O, CO_2, CO, O_2, N_2, H_2, H, O, OH, NO, N, and Ar. We need 16 equations as follows:

1. $p = \text{const.}$

2. The ideal gas state equation—$p = \dfrac{\rho}{M} R^0 T$. (This equation may be solved for the density after the integration of the other 15 equations.)

3. Conservation of C atoms—$Z_{C,u} = \dfrac{12}{M_b}\left(X_{CO_2} + X_{CO}\right)$, where M_b is given in grams per mol (g/mol).

4. Conservation of H atoms—$Z_{H,u} = \dfrac{1}{M_b}\left(2X_{H_2O} + 2X_{H_2} + X_H + X_{OH}\right)$.

5. Conservation of N atoms—$Z_{N,u} = \dfrac{14}{M_b}\left(2X_{N_2} + X_{NO} + X_N\right)$.

6. Conservation of Ar atoms—$Z_{Ar,u} = \dfrac{40}{M_b} X_{Ar}$.

7. Normalizing condition—$\displaystyle\sum_{l=1}^{N} X_l = 1$. (Conservation of O atoms may be used alternatively.)

8. Conservation of the specific enthalpy—$h_u = \displaystyle\sum_{l=1}^{N} \frac{M_l}{M_b} X_l \left(\int_{T_0}^{T_b} c_{P,l} dT + \Delta h_l^0\right)$.

9. Equilibrium in the reaction $H_2O = 2H + O$—$\dfrac{X_H^2 X_O}{X_{H_2O}} = p^{-2} K_{P,H_2O}(T)$.

10. Equilibrium in the reaction $H_2 = 2H — \dfrac{X_H^2}{X_{H_2}} = p^{-1} K_{P,H_2}(T)$.

11. Equilibrium in the reaction $OH = H + O — \dfrac{X_H X_O}{X_{OH}} = p^{-1} K_{P,OH}(T)$.

12. Equilibrium in the reaction $O_2 = 2O — \dfrac{X_O^2}{X_{O_2}} = p^{-1} K_{P,O_2}(T)$.

13. Equilibrium in the reaction $CO_2 = CO + O — \dfrac{X_{CO} X_O}{X_{CO_2}} = p^{-1} K_{P,CO_2}(T)$.

14. Equilibrium in the reaction $N_2 = 2N — \dfrac{X_N^2}{X_{N_2}} = p^{-1} K_{P,N_2}(T)$.

15. Equilibrium in the reaction $NO = N + O — \dfrac{X_N X_O}{X_{NO}} = p^{-1} K_{P,NO}(T)$.

16. Molecular weight of the combustion products— $M_b = \displaystyle\sum_{l=1}^{N} M_l X_l$.

Because Equations 8 through 15 are nonlinear, the system is commonly solved using an iteration method. The dependencies of $K_P(T)$ and $c_{P,l}(T)$, as well as the heats of formation Δh_l^0, may be found in *NIST-JANAF Thermochemical Tables* (4th edn, 1998). The quantities with subscript u characterize the unburned gas and are controlled by the initial conditions. For instance, if the fuel formula $C_x H_y$, the air composition $(O_2 + aN_2 + bAr)$, the initial temperature T_u, and the pressure p are known, then

$$Z_{C,u} = \frac{12x\Phi}{(12x+y)\Phi + \left(x+\dfrac{y}{4}\right)(32+28a+40b)},$$

$$Z_{H,u} = \frac{y\Phi}{(12x+y)\Phi + \left(x+\dfrac{y}{4}\right)(32+28a+40b)},$$

$$Z_{O,u} = \frac{32\left(x+\dfrac{y}{4}\right)}{(12x+y)\Phi + \left(x+\dfrac{y}{4}\right)(32+28a+40b)},$$

$$Z_{N,u} = \frac{28a\left(x+\dfrac{y}{4}\right)}{(12x+y)\Phi + \left(x+\dfrac{y}{4}\right)(32+28a+40b)},$$

$$Z_{\mathrm{Ar},u} = \frac{40b\left(x+\dfrac{y}{4}\right)}{(12x+y)\Phi+\left(x+\dfrac{y}{4}\right)(32+28a+40b)},$$

$$h_u = \int_{T_0}^{T_u}\left(c_{P,\mathrm{C}_x\mathrm{H}_y}Y_{\mathrm{C}_x\mathrm{H}_y,u}+c_{P,\mathrm{O}_2}Y_{\mathrm{O}2,u}+c_{P,\mathrm{N}_2}Y_{\mathrm{N}2,u}+c_{P,\mathrm{Ar}}Y_{\mathrm{Ar},u}\right)dT+Y_{\mathrm{C}_x\mathrm{H}_y,u}\Delta h^0_{\mathrm{C}_x\mathrm{H}_y},$$

$$M_u = \frac{(12x+y)\Phi+\left(x+\dfrac{y}{4}\right)(32+28a+40b)}{\Phi+\left(x+\dfrac{y}{4}\right)(1+a+b)},$$

where $Y_{\mathrm{C}_x\mathrm{H}_y,u}=Z_{\mathrm{C},u}+Z_{\mathrm{H},u}$, $Y_{\mathrm{O}2,u}=Z_{\mathrm{O},u}$, $Y_{\mathrm{N}2,u}=Z_{\mathrm{N},u}$, and $Y_{\mathrm{Ar},u}=Z_{\mathrm{Ar},u}$.

The results of the computations of the adiabatic combustion temperatures for hydrogen–air, methane–air, and *iso*-octane–air mixtures are shown in Figure 1.2 (cf. these more accurate data with the results of rough estimates of the combustion temperature, reported in Figure 1.1).

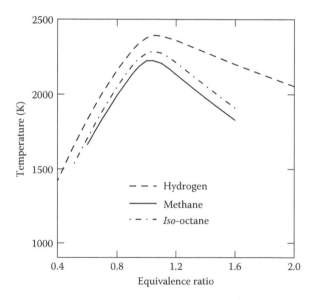

FIGURE 1.2 The adiabatic combustion temperatures computed for various fuel–air mixtures at $T_u = 300$ K and $p = 0.1$ MPa.

The curves shown in Figure 1.2 indicate that the maximum adiabatic combustion temperatures are reached in near-stoichiometric, slightly rich mixtures. For instance, the maximum temperatures are associated with $\Phi = 1.06$ for hydrogen, $\Phi = 1.03$ for methane, and $\Phi = 1.04$ for *iso*-octane. When the equivalence ratio decreases markedly from $\Phi = 1$, the temperature decreases, because the same amount (per one mole of the fuel) of energy released in chemical reactions is consumed to heat a larger amount of inert diluents (nitrogen and oxygen in lean mixtures). When the equivalence ratio increases markedly from $\Phi = 1$, the temperature still decreases, because incompletely burned fuel acts as an inert diluent. For rich hydrogen–air mixtures, a decrease in T_b with an increasing Φ is substantially less pronounced than for methane–air, *iso*-octane–air, or lean H_2–air mixtures, because, due to a low molecular weight, the addition of n moles of H_2 to a rich hydrogen–air mixture leads to an increase in the heat capacity of the mixture that is much less than the increase by the addition of n moles of the air to a lean mixture or n moles of CH_4 to a rich methane–air mixture.

The mole fractions of certain combustion products, calculated for *iso*-octane–air mixtures under different conditions, are reported in Table 1.1. In lean and stoichiometric mixtures, the products contain mainly water and carbon dioxide. In a rich mixture ($\Phi = 1.6$), carbon is oxidized mainly to CO, rather than CO_2. An increase in the unburned gas temperature slightly shifts the equilibrium from the final combustion products (H_2O and CO_2) to carbon oxide and radicals (cf. data computed for $T_u = 300$ and 600 K at $\Phi = 1.0$). An increase in the pressure acts in the opposite direction (cf. data computed for $p = 1$ and 40 atm at $\Phi = 1.0$). The combustion temperature is increased by T_u, for example, for the stoichiometric *iso*-octane–air mixture, $T_b = 2273$ and 2415 K for $T_u = 300$ and 600 K, respectively. It is worth emphasizing that an increase in T_u by two times or 300 K results in increasing T_b by 6% or just 142 K. One may use the simple rule, $T_{b,2} - T_{b,1} \approx (T_{u,2} - T_{u,1})/2$, for rough estimates. The combustion temperature is also increased by the pressure, but the

TABLE 1.1

Mole Fractions of Certain Products of Combustion of *Iso*-Octane–Air Mixtures under Different Conditions

	$\Phi = 1.0$ $T_u = 300$ K $p = 1$ atm	$\Phi = 0.6$ $T_u = 300$ K $p = 1$ atm	$\Phi = 1.6$ $T_u = 300$ K $p = 1$ atm	$\Phi = 1.0$ $T_u = 600$ K $p = 1$ atm	$\Phi = 1.0$ $T_u = 300$ K $p = 40$ atm
X_{H_2O}	0.134	0.0863	0.118	0.129	0.138
X_{CO_2}	0.110	0.0768	0.0460	0.0995	0.118
X_{CO}	0.0136	0.198×10^{-4}	0.126	0.0229	0.596×10^{-2}
X_{O_2}	0.623×10^{-2}	0.0791	0.477×10^{-7}	0.0102	0.236×10^{-2}
X_O	0.335×10^{-3}	0.133×10^{-4}	0.707×10^{-7}	0.953×10^{-3}	0.501×10^{-4}
X_H	0.461×10^{-3}	0.394×10^{-6}	0.235×10^{-3}	0.121×10^{-2}	0.676×10^{-4}
X_{OH}	0.318×10^{-2}	0.274×10^{-3}	0.305×10^{-4}	0.587×10^{-2}	0.132×10^{-2}

effect is less pronounced, for example, $T_b = 2273$ and 2349 K for $p = 1$ and 40 atm, respectively.

DENSITY RATIO

The density ratio $\sigma = \rho_u/\rho_b$ is a very important combustion characteristic, because flame-induced flow perturbations are controlled by the density gradient and hence by the density ratio.

When modeling turbulent combustion, the simplified state Equation 1.3 is often invoked and, accordingly, the density ratio is equal to T_b/T_u. However, due to the change in the molecular weight of a reacting mixture in a flame, this simple equality is just the first approximation. Figure 1.3 shows that the difference in ρ_u/ρ_b and T_b/T_u may be significant, especially in hydrogen–air and rich paraffin–air mixtures. It is also worth noting that the maximum density ratio is commonly associated with a richer mixture as compared with the mixture characterized by the maximum combustion temperature. For instance, the maxima T_b/T_u and ρ_u/ρ_b are reached at $\Phi = 1.03$ and 1.06, respectively, for methane–air flames and at $\Phi = 1.06$ and 1.11, respectively, for *iso*-octane–air flames.

In many models of premixed turbulent combustion, the so-called heat release factor, $\tau = \sigma - 1$, is used instead of the density ratio, as terms proportional to τ vanish in the constant-density case. In some theoretical papers, another quantity, $1 - \sigma^{-1} = \tau/\sigma$, is used to characterize the heat release. The computed data shown in Figure 1.3 indicate that $\sigma = 5-8$, $\tau = 4-7$, and $1 - \sigma^{-1} = 0.8-0.875$ under typical conditions, with the extreme values of these quantities associated with slightly rich mixtures. In the constant-density case, $\sigma = 1$, while $\tau = 0$.

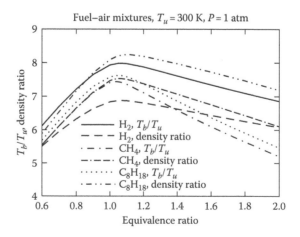

FIGURE 1.3 Density and temperature ratios versus the equivalence ratio.

CHEMICAL REACTIONS IN FLAMES

In flames, a huge number of chemical reactions may occur and the real chemistry of combustion is very different from the global reactions that are used to indicate the main reactants and products. For instance, in the simplest case of the stoichiometric hydrogen–oxygen mixture, the global reaction is as follows:

$$2H_2 + O_2 \rightarrow H_2O.$$

However, the rate of this reaction is too slow under typical flame conditions and there are much more effective reaction paths. Table 1.2 shows the H_2/O_2 part of a chemical mechanism of the $CO/H_2/O_2$ system developed by Maas and Warnatz (1988). Here, numbers I and I' are relevant to the forward and backward reactions, that is, the rates of reactions I and I' are equal to one another in the equilibrium state. Symbols B, β, and E_A will be explained later.

As compared with the aforementioned single reaction, this mechanism is associated with a much higher rate of conversion of hydrogen and oxygen to water, because it involves chain reactions. For instance, in reaction III, which is an example of a chain-propagation reaction, one H_2 molecule is converted to one H_2O molecule and the hydroxyl radical OH is converted to the atom H. Then, the OH radical reappears in reaction I and may convert one more H_2 molecule to one more H_2O molecule. Thus, even low concentrations of the OH radical can cause rapid conversion of hydrogen and oxygen to water. Because the rates of reactions that involve such active particles as atoms (with the exception of inert gases) and radicals (both the atoms and radicals are called chain carriers) are typically much higher than the rates of reactions between stable molecules, for example, H_2 and O_2, the rate of the water production via the chain mechanism shown in Table 1.2 is much higher than the rate of the aforementioned single reaction. Moreover, in reaction I, one stable O_2 molecule and one H atom are converted to one O atom and one OH molecule, that is, the number of chain carriers is increased. This process, called chain branching, substantially increases the net rate of the conversion of hydrogen and oxygen to water. Chain carriers disappear in termination reactions such as termolecular recombination reactions V, VI, and VII. To start up the chain reactions by forming chain carriers, initiation reactions are necessary, such as reactions V' and VII', in which the radicals H and O, respectively, appear. Even if the rates of these initiation reactions are low, the overall rate of the chain process may be high, as one chain carrier formed in an initiation reaction may cause the conversion of a large number of reactant molecules to products. This number depends on the ratio of the rates of the chain-branching and termination reactions.

In general, combustion occurs via very complicated chain mechanisms that involve a huge number, N_R, of chemical reactions. For instance, Figure 1.4 summarizes the various chemical mechanisms that have been developed for modeling hydrocarbon–air flames. Even the smallest mechanisms addressed in Figure 1.4 involve about 200 reactions and 30 species.

TABLE 1.2
Mechanism of H_2/O_2 System

	Reaction			B (cm, mol, s)	B	E_A (kJ/mol)
I	$O_2 + H$	$\rightarrow$	$OH + O$	2.00×10^{14}	0.00	70.30
I'	$OH + O$	$\rightarrow$	$O_2 + H$	1.46×10^{13}	0.00	2.08
II	$H_2 + O$	$\rightarrow$	$OH + H$	5.06×10^{4}	2.67	26.30
II'	$OH + H$	$\rightarrow$	$H_2 + O$	2.24×10^{4}	2.67	18.40
III	$H_2 + OH$	$\rightarrow$	$H_2O + H$	1.00×10^{8}	1.60	13.80
III'	$H_2O + H$	$\rightarrow$	$H_2 + OH$	4.45×10^{8}	1.60	77.13
IV	$OH + OH$	$\rightarrow$	$H_2O + O$	1.50×10^{9}	1.14	0.42
IV'	$H_2O + O$	$\rightarrow$	$OH + OH$	1.51×10^{10}	1.14	71.64
V	$H + H + M$	$\rightarrow$	$H_2 + M$	1.80×10^{18}	-1.00	0.00
V'	$H_2 + M$	$\rightarrow$	$H + H + M$	6.99×10^{18}	-1.00	436.08
VI	$H + OH + M$	$\rightarrow$	$H_2O + M$	2.20×10^{22}	-2.00	0.00
VI'	$H_2O + M$	$\rightarrow$	$H + OH + M$	3.80×10^{23}	-2.00	499.41
VII	$O + O + M$	$\rightarrow$	$O_2 + M$	2.90×10^{17}	-1.00	0.00
VII'	$O_2 + M$	$\rightarrow$	$O + O + M$	6.81×10^{18}	-1.00	496.41
VIII	$H + O_2 + M$	$\rightarrow$	$HO_2 + M$	2.30×10^{18}	-0.80	0.00
VIII'	$HO_2 + M$	$\rightarrow$	$H + O_2 + M$	3.26×10^{18}	-0.80	195.88
IX	$HO_2 + H$	$\rightarrow$	$OH + OH$	1.50×10^{14}	0.00	4.20
IX'	$OH + OH$	$\rightarrow$	$HO_2 + H$	1.33×10^{13}	0.00	168.30
X	$HO_2 + H$	$\rightarrow$	$H_2 + O_2$	2.50×10^{13}	0.00	2.90
X'	$H_2 + O_2$	$\rightarrow$	$HO_2 + H$	6.84×10^{13}	0.00	243.10
XI	$HO_2 + H$	$\rightarrow$	$H_2O + O$	3.00×10^{13}	0.00	7.20
XI'	$H_2O + O$	$\rightarrow$	$HO_2 + H$	2.67×10^{13}	0.00	242.52
XII	$HO_2 + O$	$\rightarrow$	$OH + O_2$	1.80×10^{13}	0.00	-1.70
XII'	$OH + O_2$	$\rightarrow$	$HO_2 + O$	2.18×10^{13}	0.00	230.61
XIII	$HO_2 + OH$	$\rightarrow$	$H_2O + O_2$	6.00×10^{13}	0.00	0.00
XIII'	$H_2O + O_2$	$\rightarrow$	$HO_2 + OH$	7.31×10^{14}	0.00	303.53
XIV	$HO_2 + HO_2$	$\rightarrow$	$H_2O_2 + O_2$	2.50×10^{11}	0.00	-5.20
XV	$OH + OH + M$	$\rightarrow$	$H_2O_2 + M$	3.25×10^{22}	-2.00	0.00
XV'	$H_2O_2 + M$	$\rightarrow$	$OH + OH + M$	2.10×10^{24}	-2.00	206.80
XVI	$H_2O_2 + H$	$\rightarrow$	$H_2 + HO_2$	1.70×10^{12}	0.00	15.70
XVI'	$H_2 + HO_2$	$\rightarrow$	$H_2O_2 + H$	1.15×10^{12}	0.00	80.88
XVII	$H_2O_2 + H$	$\rightarrow$	$H_2O + OH$	1.00×10^{13}	0.00	15.00
XVII'	$H_2O + OH$	$\rightarrow$	$H_2O_2 + H$	2.67×10^{12}	0.00	307.51
XVIII	$H_2O_2 + O$	$\rightarrow$	$OH + HO_2$	2.80×10^{13}	0.00	26.80
XVIII'	$OH + HO_2$	$\rightarrow$	$H_2O_2 + O$	8.40×10^{12}	0.00	84.09
XIX	$H_2O_2 + OH$	$\rightarrow$	$H_2O + HO_2$	5.40×10^{12}	0.00	4.20
XIX'	$H_2O + HO_2$	$\rightarrow$	$H_2O_2 + OH$	1.63×10^{13}	0.00	132.71

Source: Reprinted from Maas, U. and Warnatz, J., *Proc. Combust. Inst.*, 22, 1695–1704, 1988. With permission.

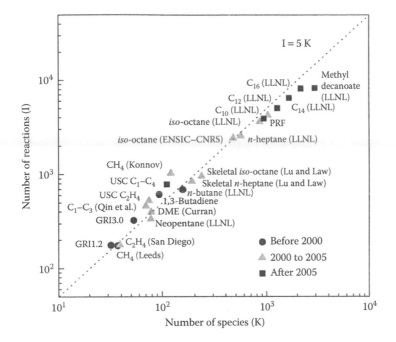

FIGURE 1.4 The number of reactions versus the number of species for various chemical mechanisms for hydrocarbon fuels. (Reprinted from *Progress in Energy and Combustion Science*, 35, Lu, T. and Law, C.K., Toward accommodating realistic fuel chemistry in large-scale computations, 192–215, Copyright 2009, with permission from Elsevier.)

Any chemical mechanism may be generalized by Equation 1.36 written for $k = 1, \ldots, N_R$ reactions. A single k-th reaction in Equation 1.36 is associated with two reactions, one forward (K) and one backward (K'), reactions in Table 1.2.

If $l = 1, 2, 3, 4, 5, 6, 7,$ and 8 are associated with H, O, OH, H_2O_2, HO_2, H_2, O_2, and H_2O, respectively, then for the mechanism shown in Table 1.2, we have

$$n_{1,1} = n_{1,7} = m_{1,2} = m_{1,3} = 1,$$

$$n_{1,2} = n_{1,3} = n_{1,4} = n_{1,5} = n_{1,6} = n_{1,8} = m_{1,1} = m_{1,4} = m_{1,5} = m_{1,6} = m_{1,7} = m_{1,8} = 0,$$

$$n_{2,2} = n_{2,6} = m_{2,1} = m_{2,3} = 1,$$

$$n_{2,1} = n_{2,3} = n_{2,4} = n_{2,5} = n_{2,7} = n_{2,8} = m_{2,2} = m_{2,4} = m_{2,5} = m_{2,6} = m_{2,7} = m_{2,8} = 0,$$

$$n_{3,3} = n_{3,6} = m_{3,1} = m_{3,8} = 1,$$

$$n_{3,1} = n_{3,2} = n_{3,4} = n_{3,5} = n_{3,7} = n_{3,8} = m_{3,2} = m_{3,3} = m_{3,4} = m_{3,5} = m_{3,6} = m_{3,7} = 0,$$

$$n_{4,3} = 2,$$

$$m_{4,2} = m_{4,8} = 1,$$

$$n_{4,1} = n_{4,2} = n_{4,4} = n_{4,5} = n_{4,6} = n_{4,7} = n_{4,8} = m_{4,1} = m_{4,3}$$

$$= m_{4,4} = m_{4,5} = m_{4,6} = m_{4,7} = 0,$$

etc.

Symbol M in reactions V through VIII and XV designates a third body, that is, any species, rather than a species that is not involved in other reactions. Thus, if the H_2/O_2 system is considered, then $N = 8$ (H, O, OH, H_2O_2, HO_2, H_2, O_2, and H_2O) in Table 1.2, while $N = 9$ in Equation 1.36. For the $H_2/O_2/N_2$ system, $N = 9$ and 10, respectively, if the oxidation of nitrogen is disregarded. The third body is not converted in reactions V through VIII and XV. Typically, a third body either removes (in the form of its kinetic energy) a bond energy liberated in a recombination reaction, for example, the forward reactions V through VIII and XV, or provides kinetic energy to excite and then destroy a stable molecule, for example, the backward reactions V' through VIII' and XV'.

If, in Equation 1.36, $l = 9$ is associated with a third body, then

$$n_{5,1} = 2,$$

$$n_{5,9} = m_{5,6} = m_{5,9} = 1,$$

$$n_{5,2} = n_{5,3} = n_{5,4} = n_{5,5} = n_{5,6} = n_{5,7} = n_{5,8} = m_{5,1} = m_{5,2} = m_{5,3} = m_{5,4} = m_{5,5}$$

$$= m_{5,7} = m_{5,8} = 0.$$

The net rate of production (consumption) of the l-th species in a chemical mechanism generalized by Equation 1.36 is equal to

$$\frac{dc_l}{dt} = W_l = \sum_{k=1}^{N_R} (m_{k,l} - n_{k,l})(w_{f,k} - w_{b,k}), \tag{1.39}$$

where the subscripts f and b designate the forward and backward reactions, respectively. For instance, for the mechanism mentioned in Table 1.2,

$$W_{HO_2} = w_{f,8} - w_{b,8} - \left(w_{f,9} - w_{b,9}\right) - \left(w_{f,10} - w_{b,10}\right) - \left(w_{f,11} - w_{b,11}\right) - \left(w_{f,12} - w_{b,12}\right)$$

$$- \left(w_{f,13} - w_{b,13}\right) - 2\left(w_{f,14} - w_{b,14}\right) + w_{f,16} - w_{b,16} + w_{f,18} - w_{b,18} + w_{f,19} - w_{b,19}.$$

Note that the integer numbers $n_{k,l}$ and $m_{k,l}$ are always equal to one another for a third body, as it is inert, that is, it is not converted to another species in chemical reactions. Therefore, the symbol M may be ignored when considering Equation 1.39.

For reactions that do not involve third bodies, the forward and backward reaction rates are modeled as follows:

$$
w_{f,k} = k_{f,k} \prod_{l=1}^{N-1} c_l^{n_{kl}} = k_{f,k} \rho^{\sum_{l=1}^{N-1} n_{kl}} \prod_{l=1}^{N-1} \left(\frac{Y_l}{M_l} \right)^{n_{kl}},
$$

$$
w_{b,k} = k_{b,k} \prod_{l=1}^{N-1} c_l^{m_{kl}} = k_{b,k} \rho^{\sum_{l=1}^{N-1} m_{kl}} \prod_{l=1}^{N-1} \left(\frac{Y_l}{M_l} \right)^{m_{kl}}.
$$

(1.40)

Here, the number $N - 1$ is used, because the N-th species in Equation 1.36 is associated with a third body. For instance,

$$
w_{f,1} = k_{f,1} c_H c_{O_2},
$$

$$
w_{b,1} = k_{b,1} c_O c_{OH},
$$

$$
w_{f,9} = k_{f,9} c_H c_{HO_2},
$$

$$
w_{b,9} = k_{b,9} c_{OH}^2.
$$

Because chemical reactions between the molecules occur when the molecules collide, the reaction rate is proportional to the probability of the collision, that is, to the concentrations of the reacting molecules. The coefficients $k_{f,k}$ and $k_{b,k}$ are the reaction rate coefficients for the forward and backward k-th reactions.

Because the symbol M in reactions V through VIII and XV designates any molecule or atom, a single reaction V means, in fact, eight reactions:

(V-1) $H + H + H \rightarrow H_2 + H,$
(V-2) $H + H + O \rightarrow H_2 + O,$
(V-3) $H + H + OH \rightarrow H_2 + OH,$
(V-4) $H + H + H_2O_2 \rightarrow H_2 + H_2O_2,$
(V-5) $H + H + HO_2 \rightarrow H_2 + HO_2,$
(V-6) $H + H + H_2 \rightarrow H_2 + H_2,$
(V-7) $H + H + O_2 \rightarrow H_2 + O_2,$
(V-8) $H + H + H_2O \rightarrow H_2 + H_2O.$

If the reaction rate coefficients were the same for all of these reactions, then the rate of the forward reaction V would be equal to

$$
w_{f,5} = k_{f,5} c_H^2 c.
$$

However, the reaction rate coefficients for the aforementioned eight reactions are different. This difference is commonly characterized by the catalytic efficiency z_l of the l-th species. If, for example, $z_{H_2} = 1$ and $z_{O_2} = 0.35$, then the rate of reaction V-7 is lower than the rate of reaction V-6 by a factor of 0.35. In the general case,

$$k_{f,k-l} = z_l k_{f,k},$$ (1.41)

where $k_{f,k-l}$ is the rate coefficient for the k-th reaction with $M = S_l$, for example, $k_{f,5-6}$ is the rate coefficient for reaction V-6. Therefore, for reactions involving a third body (see reactions V through VIII and XV in Table 1.2)

$$w_{f,k} = k_{f,k} \prod_{l=1}^{N} c_l^{n_{kl}},$$

$$w_{b,k} = k_{b,k} \prod_{l=1}^{N} c_l^{m_{kl}},$$ (1.42)

where

$$c_N = \sum_{l=1}^{N-1} z_l c_l$$ (1.43)

and $n_{k,N} = m_{k,N} = 1$. Maas and Warnatz (1988) have reported the following catalytic efficiencies: $z_{H_2} = 1$, $z_{O_2} = 0.35$, and $z_{H_2O} = 6.5$. The catalytic efficiencies for the other five species included in the discussed H_2/O_2 mechanism are of minor importance, because the concentrations of these five species are very low and, hence, their contribution to the sums on the RHS of Equation 1.43 is negligible. If a hydrogen–air flame is considered, then $z_{N_2} = 0.5$ (Maas and Warnatz, 1988).

Certainly, Equations 1.42 and 1.43 are applicable for any reaction, provided that $n_{k,N} = m_{k,N} = 0$ for reactions that do not involve a third body.

The reaction rate coefficients are commonly approximated as follows:

$$k_{f,k} = B_{f,k} T^{\beta_{fk}} \exp\left(-\frac{\Theta_{f,k}}{T}\right),$$

$$k_{b,k} = B_{b,k} T^{\beta_{bk}} \exp\left(-\frac{\Theta_{b,k}}{T}\right),$$ (1.44)

where

$B_{f,k}$ and $B_{b,k}$ are the pre-exponential factors
$\Theta_{f,k}$ and $\Theta_{b,k}$ are the activation temperatures of the forward and backward reactions, respectively.

These reaction constants, as well as the power exponents $\beta_{f,k}$ and $\beta_{b,k}$, are reported in Table 1.2 (after Maas and Warnatz, 1988), where $E_A = R^0\Theta$.

In the SI system, the units of $k_{f,k}$ and $k_{b,k}$ are $(\text{mol/m}^3)^{1-n_k}/s$ and $(\text{mol/m}^3)^{1-m_k}/s$, respectively, where $n_k = \sum_{l=1}^{N} n_{k,l}$ and $m_k = \sum_{l=1}^{N} m_{k,l}$. The units of $B_{f,k}$ and $B_{b,k}$ are those of $k_{f,k}$ and $k_{b,k}$ divided by $K^{\beta_{f,k}}$ and $K^{\beta_{b,k}}$, respectively.

The key peculiarity of flames is the high activation temperatures required for many important combustion reactions. For instance, for the initiation reactions V′ and VII′, the activation temperatures $\Theta = E_A/R^0$ shown in Table 1.2 are as high as 52,500 and 59,800 K, respectively. These values are more than 20 times higher than the adiabatic combustion temperature for the stoichiometric hydrogen–air mixture under normal conditions.

Because the rates of the forward and backward reactions are equal to one another in the equilibrium state, the ratio of the forward and backward reaction rate coefficients may be expressed via the equilibrium constant for this reaction. Indeed, in the equilibrium state, Equation 1.42 results in

$$\frac{k_{f,k}}{k_{b,k}} = \prod_{l=1}^{N} c_l^{m_{kl}-n_{kl}} = \left(\frac{c}{p}\right)^{-\Delta n_k} \prod_{l=1}^{N} p_l^{m_{kl}-n_{kl}} = \left(R^0 T\right)^{\Delta n_k} K_{P,k}(T), \qquad (1.45)$$

using Equations 1.5 and 1.8. Here,

$$\Delta n_k = \sum_{l=1}^{N-1} \left(n_{k,l} - m_{k,l}\right), \qquad (1.46)$$

with the equality of $n_{k,N} = m_{k,N} = 1$ taken into account for a third body. Using Equation 1.46, Equations 1.39 and 1.42 may be rewritten as

$$W_l = \sum_{k=1}^{N_R} (m_{k,l} - n_{k,l})k_{f,k}\left[\prod_{j=1}^{N} c_j^{n_{k,j}} - \frac{1}{K_{P,k}\left(R^0 T\right)^{\Delta n_k}}\prod_{j=1}^{N} c_j^{m_{k,j}}\right]. \qquad (1.47)$$

For bimolecular reactions, in which two molecules (or atoms) of the reactants are converted into two molecules (or atoms) of the products, for example, see reactions I through IV,

$$\sum_{l=1}^{N} n_{k,l} = 2, \quad \sum_{l=1}^{N} m_{k,l} = 2, \quad \text{and} \quad \Delta n_k = 0,$$

that is, $k_{f,k} = k_{b,k}K_{P,k}$. For monomolecular reactions, in which one molecule of a reactant is decomposed into two molecules (or atoms) of the products, for example, $C_3H_6 \rightarrow C_2H_3 + CH_3$,

$$\sum_{l=1}^{N} n_{k,l} = 1, \quad \sum_{l=1}^{N} m_{k,l} = 2, \quad \Delta n_k = -1, \quad \text{and} \quad \frac{k_{f,k}}{k_{b,k}} = \frac{K_{P,k}}{R^0 T}.$$

For termolecular reactions, in which three molecules (or atoms) of the reactants are converted into two molecules (or atoms) of the products, for example, see reactions V through VIII,

$$\sum_{l=1}^{N} n_{k,l} = 3, \quad \sum_{l=1}^{N} m_{k,l} = 2, \quad \Delta n_k = 1, \quad \text{and} \quad \frac{k_{f,k}}{k_{b,k}} = R^0 T K_{P,k}.$$

The term in square brackets on the RHS of Equation 1.47 may be rewritten as

$$\prod_{j=1}^{N} c_j^{n_{k,j}} - \frac{1}{K_{P,k}\left(R^0 T\right)^{\Delta n_k}} \prod_{j=1}^{N} c_j^{m_{k,j}} = \prod_{j=1}^{N} \left(\frac{\rho}{M} X_j\right)^{n_{k,j}} - \frac{1}{K_{P,k}\left(R^0 T\right)^{\Delta n_k}} \prod_{j=1}^{N} \left(\frac{\rho}{M} X_j\right)^{m_{k,j}}$$

$$= \left(\frac{\rho}{M}\right)^{\sum_{j=1}^{N} n_{k,j}} \left[\prod_{j=1}^{N} X_j^{n_{k,j}} - \frac{1}{K_{P,k} p^{\Delta n_k}} \prod_{j=1}^{N} X_j^{m_{k,j}} \right]$$

using Equations 1.1, 1.5, and 1.10. For a termolecular reaction, $\Delta n_k = 1$ and the second term in the square brackets is decreased when the pressure is increased. At the equilibrium state, which is characterized by $W_l = 0$, this decrease in the factor of $(K_{P,k} p)^{-1}$ should be compensated by an increase in the concentrations of the equilibrium products of the termolecular reaction (e.g., H_2, H_2O, and O_2 in reactions V, VI, and VII, respectively). This effect explains an increase in the equilibrium X_{H_2O} and X_{CO_2} and a decrease in the equilibrium X_H, X_O, and X_{OH} by the pressure, as shown in Table 1.1.

The heat of the reaction, which is defined as

$$Q_k = -\sum_{l=1}^{N} (m_{kl} - n_{kl})\Delta H_l^0, \tag{1.48}$$

is another important characteristic of a chemical reaction. This quantity is related separately to the forward and backward reactions, with Equation 1.48 showing that $Q_{f,k} = -Q_{b,k}$.

If $Q > 0$, the reaction is called exothermic, that is, the chemical energy is released in the reaction and is transformed into thermal energy. Accordingly, the gas temperature increases in the course of the reaction, see Equation 1.35 and bear in mind that $Y_l \Delta h_l^0 = X_l \Delta H_l^0 / M$. For instance, reaction V is characterized by $Q_V = 2\Delta H_H^0 = 435.972$ kJ/mol, that is, this is an exothermic reaction. If $Q < 0$, the reaction is called endothermic, that is, the thermal energy is consumed to destroy the reactant molecule, for example, see the backward reaction V'.

Figure 1.4 shows that the number of species and the number of reactions, included in a detailed chemical mechanism, rapidly grow as the number of carbon atoms in a fuel molecule increases. Certainly, chemical mechanisms that involve a thousand species and ten thousand reactions are not manageable in unsteady three-dimensional (3D) numerical simulations of combustion in an engine or in a laboratory burner. Therefore, the development of reduced chemical mechanisms was the focus of the combustion community over the past decades.

There are two basic methods for reducing the number of reactions in a chemical mechanism. First, steady-state approximations are widely invoked for this purpose. To explain this method, let us consider the well-known Zel'dovich mechanism of NO formation in flames (see a review paper by Miller and Bowman, 1989)

$$XX \quad O + N_2 \rightarrow N + NO,$$
$$XXI \quad N + O_2 \rightarrow O + NO.$$

For simplicity, we have neglected the backward reactions here. This is a classical chain mechanism: a chain carrier, the O atom, reacts with nitrogen to form a product, NO, and another chain carrier, the atom N. Subsequently, the latter chain carrier reacts with oxygen to form one more molecule of the product and to reproduce the former chain carrier. Accordingly, the formation of a single O atom would cause the formation of an infinitely large number of NO molecules if there were no termination reactions.

Under typical conditions, the rate coefficient $k_{f,21}$ for reaction XXI is much higher than the rate coefficient $k_{f,20}$ for reaction XX. Accordingly, even if the concentration of the nitrogen atoms is much less than the concentration of the oxygen atoms, the rates of the two reactions may be equal. In other words, reaction XX is a "narrow throat" as all the nitrogen atoms formed in this, relatively slow, reaction are immediately consumed by a much faster reaction XXI. For such a process, the steady-state approximation works well.

In the particular case considered, the approximation consists of an assumption that the concentration of the N atoms is steady, that is, the rates of the two reactions are equal to one another. From the steady-state condition of

$$\frac{dc_N}{dt} = k_{20}c_Oc_{N_2} - k_{21}c_Nc_{O_2} = 0$$

we can determine the mole concentration of the N atoms

$$c_N = \frac{k_{20}}{k_{21}} \frac{c_Oc_{N_w}}{c_{O_2}}.$$

Here, for the sake of brevity, we skip subscript f, which is associated with the forward reaction rate coefficients. Therefore,

$$\frac{dc_{NO}}{dt} = k_{20}c_Oc_{N_2} + k_{21}c_Nc_{O_2} = 2k_{20}c_Oc_{N_2}.$$

Thus, the steady-state approximation has allowed us to reduce the two equations for dc_N/dt and dc_{NO}/dt, associated with reactions XX and XXI, to a single equation for dc_{NO}/dt.

Second, the partial-equilibrium approximations are also aimed at reducing the number of reactions and species by obtaining algebraic relations between the concentrations of some species. For instance, Table 1.2 shows that the activation temperatures of the termolecular recombination reactions V through VIII and XV are zero, while the bimolecular chain-branching reaction I is characterized by a high activation energy. Accordingly, the rate of the latter reaction depends strongly on the temperature and is very high when T is close to the adiabatic combustion temperature, while the rate of the backward reaction I′ and the rates of the recombination reactions V through VIII and XV depend weakly on T.

For the forward and backward reactions I and I′, respectively, the situation is basically similar to that of reactions XX and XXI. Because one reaction rate coefficient is much higher than the other, the former reaction will dominate until the concentration of its products, which take part in the latter reaction, is so large that the rate of the latter reaction is equal to the former reaction rate. For reactions I and I′, the aforementioned scenario means that a partial, that is, for a single particular reaction, equilibrium between the rates of the forward and backward reactions is reached when $T \approx T_b$. The condition of the partial equilibrium

$$K_{p,1}(T) = \frac{p_{OH}p_O}{p_{O_2}p_H} = \frac{X_{OH}X_O}{X_{O_2}X_H}$$

may be used in order to express one concentration, for example, the concentration of the oxygen atoms, through the concentrations of other species. Accordingly, both the number of reactions and the number of differential equations for the species concentrations, which are similar to Equation 1.39, are reduced by one.

Note that the temperature in the aforementioned equation is lower than T_b, as far as the partial equilibrium is concerned. The temperature slowly increases due to the heat release in the recombination reactions V through VIII and XV, the rates of which depend weakly on T. Thus, the equilibrium is just partial, that is, it is restricted to some, but not all, reactions. The full equilibrium is reached as $T = T_b$.

In addition to the similarity between the steady-state and the partial-equilibrium approximations, emphasized earlier, there are important differences. First, the steady-state approximation is applied to different reactions, while the partial-equilibrium approximation is relevant to the forward and backward steps of the same reaction. Second, the steady-state approximation may involve not only two reactions, as in the Zel'dovich mechanism, but more reactions, that is, all reactions in which the steady-state element is converted. For instance, the reaction

XXII   OH + N → NO + H

may substantially contribute to the NO formation in flames (Miller and Bowman, 1989). For reactions XX through XXII (these three reactions and the corresponding backward reactions are often called "the extended Zel'dovich mechanism"), the steady-state approximation results in

$$\frac{dc_N}{dt} = k_{20}c_Oc_{N_2} - k_{21}c_Nc_{O_2} - k_{22}c_Nc_{OH} = 0$$

and

$$c_N = \frac{k_{20}c_Oc_{N_2}}{k_{21}c_{O_2} + k_{22}c_{OH}}$$

if we neglect the backward reactions for the sake of simplicity.

By using the steady-state and the partial-equilibrium approximations, reduced chemical mechanisms that contain a few reactions have been developed for various fuels. For instance, Peters and Williams (1987) and Seshadri and Peters (1990) have developed the following four-step reduced mechanism of methane combustion:

(RR-I) $O_2 + 3H_2 \leftrightarrows 2H + 2H_2O$
(RR-II) $H + H + M \leftrightarrows H_2 + M$
(RR-III) $CH_4 + 2H + H_2O \leftrightarrows CO + 4H_2$
(RR-IV) $CO + H_2O \leftrightarrows CO_2 + H_2$

with reactions RR-I and RR-II constituting a reduced mechanism of hydrogen combustion. It is worth emphasizing that these four reactions are the so-called overall reactions, rather than the real reactions listed in Table 1.2. For instance, the forward overall reaction RR-II is the sum of two real forward reactions VIII and XIII and one real backward reaction III', rather than the real forward reaction V. Accordingly, the rate of the overall reaction is not equal to the rate of the real reaction V, but it is controlled by the rate of reaction VIII under typical conditions. Similarly, for reactions XX and XXI, associated with the Zel'dovich mechanism, the overall reaction, that is, the sum of reactions XX and XXI, is as follows:

(RR-V) $N_2 + O_2 \rightarrow 2NO$

with the rate of this overall reaction equal to $2w_{20}$, as previously discussed.

Reduced chemical mechanisms have been developed even for heavy paraffin (alkane) fuels such as heptane (Seshadri et al., 1997) and iso-octane (Pitsch et al., 1996). Despite the substantial progress made in this subject, reduced chemical mechanisms are rarely used in the studies of turbulent combustion. The contemporary theory deals mainly with a hypothetical turbulent flame, where the combustion chemistry is reduced to a single global irreversible reaction:

$$(GR-I) \quad \Phi F + (St\,M_F/M_O)O \rightarrow P + \max\{\Phi - 1, 0\}\,F + \max\{1 - \Phi, 0\}(St\,M_F/M_O)O.$$

In numerical simulations, models that deal with the global reaction (GR-I) may invoke input parameters, for example, the laminar flame speed discussed in Chapter 2, which have been computed using a detailed chemical mechanism. A similar approach is used in this book. The physics and the numerical modeling of premixed turbulent combustion are mainly discussed in the simplified case of a single global reaction GR-I, while a detailed combustion chemistry is considered to be of importance when evaluating the laminar flame characteristics that are input parameters for modeling turbulent combustion.

BALANCE EQUATIONS

A general set of balance equations that models reacting flows is discussed in detail in a book by Williams (1985a). Here, we restrict ourselves to a less general set of equations that is commonly used for modeling turbulent flames. The latter equations (Libby and Williams, 1994) can be obtained from the former equations by assuming that

- The mixture is an ideal gas.
- The Mach number is much less than unity.
- There are no heat losses (adiabatic problem).
- There are no body forces.
- The Soret and Dufour effects, pressure-gradient diffusion, and bulk viscosity are negligible.
- The molecular mass and heat fluxes are approximated by Fick's and Fourier's laws, respectively.

In contemporary numerical simulations of laminar premixed flames, more advanced models based on the kinetic theory of gases are often invoked (Giovangigli, 1999), but the use of Fick's and Fourier's laws is sufficient for a basic understanding of the state of the art in turbulent combustion.

Under the aforementioned assumptions, premixed combustion is modeled by the following set of balance equations.

Mass conservation (or continuity) equation:

$$\frac{\partial \rho}{\partial t} + \frac{\partial}{\partial x_k}(\rho u_k) = 0, \tag{1.49}$$

where t is the time and $\mathbf{x} = \{x_1, x_2, x_3\} = \{x, y, z\}$ and $\mathbf{u} = \{u_1, u_2, u_3\} = \{u, v, w\}$ are the spatial coordinates and the flow velocity vector, respectively, and the summation convention applies for the repeated index k. In a constant-density case, the equation reads

$$\nabla \cdot \mathbf{u} \equiv \frac{\partial u_k}{\partial x_k} = 0. \tag{1.50}$$

The divergence $\nabla \cdot \mathbf{u}$ of the velocity vector u is called dilatation.
Momentum conservation (Navier–Stokes) equation:

$$\frac{\partial}{\partial t}(\rho u_i) + \frac{\partial}{\partial x_k}(\rho u_k u_i) = \frac{\partial \tau_{ik}}{\partial x_k} - \frac{\partial p}{\partial x_i}, \tag{1.51}$$

where

$$\tau_{ij} = \mu\left(\frac{\partial u_i}{\partial x_j} + \frac{\partial u_j}{\partial x_i} - \frac{2}{3}\frac{\partial u_k}{\partial x_k}\delta_{ij}\right) \tag{1.52}$$

is the viscous stress tensor, δ_{ij} is the Kronecker delta, and μ is the dynamic molecular viscosity of the mixture. The two terms on the RHS of Equation 1.51 are associated with the viscous and the pressure forces, respectively.

By taking the curl of the Navier–Stokes equation and assuming that μ is constant, the balance equation

$$\frac{\partial \omega_i}{\partial t} + u_k\frac{\partial \omega_i}{\partial x_k} = \omega_k\frac{\partial u_i}{\partial x_k} + \nu\frac{\partial^2 \omega_i}{\partial x_k\partial x_k} - \omega_i\frac{\partial u_k}{\partial x_k} + \frac{1}{\rho^2}\varepsilon_{ijk}\frac{\partial \rho}{\partial x_j}\frac{\partial p}{\partial x_k} \tag{1.53}$$

for the vorticity vector $\boldsymbol{\omega} = \{\omega_1, \omega_2, \omega_3\} = \{\omega_x, \omega_y, \omega_z\}$, defined as

$$\omega_i = \varepsilon_{ijk}\frac{\partial u_j}{\partial x_k}, \tag{1.54}$$

can be obtained. Here, $\nu = \mu/\rho$ is the molecular kinematic viscosity of the mixture, and the alternating (or Levi-Civita) symbol ε_{ijk} means 1 ($\varepsilon_{123} = \varepsilon_{312} = \varepsilon_{231}$), or -1 ($\varepsilon_{132} = \varepsilon_{213} = \varepsilon_{321}$), or 0 otherwise. In vector form, Equations 1.53 and 1.54 read

$$\frac{\partial \boldsymbol{\omega}}{\partial t} + (\mathbf{u}\cdot\nabla)\boldsymbol{\omega} = (\boldsymbol{\omega}\cdot\nabla)\mathbf{u} + \nu\nabla^2\boldsymbol{\omega} - \boldsymbol{\omega}(\nabla\cdot\mathbf{u}) + \frac{1}{\rho^2}\nabla\rho\times\nabla p \tag{1.55}$$

and

$$\boldsymbol{\omega} = \nabla\times\mathbf{u}, \tag{1.56}$$

respectively.

The use of vorticity (Equation 1.53 or 1.55) for studying turbulent flames is beneficial for two reasons. First, because turbulent flows are always rotational, vorticity characterizes turbulence better than velocity. Second, in a constant-density case, the last two terms on the RHS of Equation 1.53 or 1.55 vanish and the vorticity equation reads

$$\frac{\partial \boldsymbol{\omega}}{\partial t} + (\mathbf{u} \cdot \nabla)\boldsymbol{\omega} = (\boldsymbol{\omega} \cdot \nabla)\mathbf{u} + \nu \nabla^2 \boldsymbol{\omega}. \tag{1.57}$$

The first term on the RHS of this equation is associated with vorticity generation (or damping) due to the stretching (or compressing) of the vortexes by the flow. In turbulent flows, the former process dominates and this term is generally a production term.

The last two terms on the RHS of Equation 1.55 are specific to variable-density flows and, in particular, flames. These two terms are associated with a decrease in vorticity when the vortex size increases due to the expansion of hot gas (the conservation of the angular momentum) and vorticity generation by pressure forces (the so-called baroclinic torque), respectively.

Species balance equations:

$$\frac{\partial}{\partial t}(\rho Y_l) + \frac{\partial}{\partial x_k}(\rho u_k Y_l) = \frac{\partial}{\partial x_k}\left(\rho D_l \frac{\partial Y_l}{\partial x_k}\right) + M_l W_l, \tag{1.58}$$

where D_l is the molecular diffusivity of the l-th species in the mixture and the reaction rate, W_l, is modeled by Equation 1.47. The last term on the RHS of Equation 1.58 is multiplied by M_l, because $d\rho Y_l/dt = M_l dc_l/dt = M_l W_l$ by virtue of Equations 1.13 and 1.39.

The two terms on the RHS of Equation 1.58 describe the flux of the l-th species due to the molecular diffusion and production (or consumption) of the l-th species in the chemical reactions, respectively.

Summing up Equation 1.58 written for $l = 1, \ldots, N$ and using Equation 1.14 and the continuity Equation 1.49, we obtain

$$\frac{\partial}{\partial x_k}\left(\rho \sum_{l=1}^{N} D_l \frac{\partial Y_l}{\partial x_k}\right) = 0, \tag{1.59}$$

because

$$\sum_{l=1}^{N} M_l W_l = \sum_{l=1}^{N} M_l \frac{dc_l}{dt} = \sum_{l=1}^{N} \frac{d\rho_l}{dt} = \frac{d\rho}{dt} = 0 \tag{1.60}$$

by virtue of mass conservation in the chemical reactions. Because diffusion fluxes vanish in a spatially uniform unburned fuel–air mixture, Equation 1.59 results in

$$\sum_{l=1}^{N} D_l \frac{\partial Y_l}{\partial x_k} = 0 \tag{1.61}$$

in a flame. Note that some simple expressions used to calculate diffusivities D_l in a multi-component mixture may be inconsistent with Equation 1.61.

Using Equation 1.15 and summing Equation 1.58 written for various species, one can easily obtain the following element conservation equation:

$$\frac{\partial}{\partial t}(\rho Z_m) + \frac{\partial}{\partial x_k}(\rho u_k Z_m) = \frac{\partial}{\partial x_k}\left(\rho \sum_{l=1}^{N} \alpha_{m,l} D_l \frac{\partial Y_l}{\partial x_k}\right), \qquad (1.62)$$

where $\alpha_{m,l} = M_m b_{m,l}/M_l$ (see Equation 1.15). The source term vanishes on the RHS of Equation 1.62, because the total mass of the m-th element is not changed in the chemical reactions. Accordingly, the constraint

$$\sum_{l=1}^{N} \alpha_{m,l} M_l W_l = M_m \sum_{l=1}^{N} b_{m,l} W_l = 0 \qquad (1.63)$$

always holds.

Enthalpy conservation:

$$\frac{\partial}{\partial t}(\rho h) + \frac{\partial}{\partial x_k}(\rho u_k h) = \frac{\partial p}{\partial t} + \frac{\partial}{\partial x_k}\left[\frac{\mu}{Pr}\frac{\partial h}{\partial x_k} + \mu \sum_{l=1}^{N}\left(\frac{1}{Sc_l} - \frac{1}{Pr}\right)h_l \frac{\partial Y_l}{\partial x_k}\right], \quad (1.64)$$

where $Pr = \mu/(\rho a) = \nu/a$ and $Sc_l = \mu/(\rho D_l) = \nu/D_l$ are the Prandtl and Schmidt numbers, respectively, $a = \lambda/(\rho c_p)$ is the molecular heat diffusivity of the mixture, and λ is the thermal conductivity. The ratio of a/D_l is called the Lewis number Le_l for the l-th species. Accordingly, $Le_l = Sc_l/Pr$.

The methods for evaluating the molecular viscosities ν or μ and the molecular transport coefficients λ, a, and D_l are discussed in detail elsewhere (Hirschfelder et al., 1954; Giovangigli, 1999; TRANSPORT, 2000). In this book, these coefficients are considered to be known functions of temperature and pressure for a particular mixture. For a flammable hydrocarbon–air mixture under normal conditions, typical values of the kinematic and dynamic viscosities are 1.5×10^{-5} m²/s and 1.5×10^{-5} kg/ms, respectively. A typical value of the heat diffusivity is 2×10^{-5} m²/s. Molecular mass diffusivities vary substantially for different species. The highest diffusivities are associated with the hydrogen atom and molecular hydrogen, while the coefficient D_l for heavy fuels may be lower by several times in the same mixture. For instance, the Lewis number may be as small as 0.25 for molecular hydrogen and as large as 4 for iso-octane in air. The dynamic viscosity and the thermal conductivity are independent of the pressure. The dynamic viscosity is proportional to T^q with $q < 1$, for example, $q = 0.6$ is widely used in such estimates. Accordingly, ν, a, and D_l are proportional to $p^{-1}T^{q+1}$.

If $Le_l = 1$ for all species and the pressure is constant, then Equation 1.64 is simplified as follows:

$$\frac{\partial}{\partial t}(\rho h) + \frac{\partial}{\partial x_k}(\rho u_k h) = \frac{\partial}{\partial x_k}\left(\frac{\mu}{\text{Pr}}\frac{\partial h}{\partial x_k}\right). \tag{1.65}$$

In the adiabatic case, the enthalpy fluxes vanish at the boundaries and Equation 1.65 has a trivial solution $h = \text{const}$. Thus, the assumption of $\text{Le}_l = 1$ offers an opportunity to substantially simplify the problem and reduce the number of balance equations in the case of constant pressure and spatially uniform initial enthalpy field.

Coming back to the general case (different $\text{Le}_l \neq 1$ and variable pressure), Equation 1.64 may be rewritten as follows:

$$\frac{\partial}{\partial t}(\rho h) + \frac{\partial}{\partial x_k}(\rho u_k h) = \frac{\partial p}{\partial t} + \frac{\partial}{\partial x_k}\left(\lambda\frac{\partial T}{\partial x_k}\right) + \frac{\partial}{\partial x_k}\left(\rho\sum_{l=1}^{N} h_l D_l \frac{\partial Y_l}{\partial x_k}\right), \tag{1.66}$$

because

$$\frac{\partial h}{\partial x_k} - \sum_{l=1}^{N} h_l\frac{\partial Y_l}{\partial x_k} = \frac{\partial}{\partial x_k}\sum_{l=1}^{N} h_l Y_l - \sum_{l=1}^{N} h_l\frac{\partial Y_l}{\partial x_k} = \sum_{l=1}^{N} Y_l\frac{\partial h_l}{\partial x_k}$$

$$= \sum_{l=1}^{N} Y_l\frac{\partial}{\partial x_k}\int_{T_0}^{T} c_{p,l}dT = \sum_{l=1}^{N} Y_l c_{p,l}\frac{\partial T}{\partial x_k} = c_p\frac{\partial T}{\partial x_k}.$$

The last two terms on the RHS of Equation 1.66 are associated with the enthalpy fluxes due to heat conductivity and molecular diffusivity, respectively. Substituting the enthalpy h, determined by Equation 1.30, on the left-hand side (LHS) of Equation 1.66, we have

$$\frac{\partial}{\partial t}(\rho h) + \frac{\partial}{\partial x_k}(\rho u_k h) = \sum_{l=1}^{N} h_l\left(\rho\frac{\partial Y_l}{\partial t} + \rho u_k\frac{\partial Y_l}{\partial x_k}\right) + \sum_{l=1}^{N} Y_l\left(\rho\frac{\partial h_l}{\partial t} + \rho u_k\frac{\partial h_l}{\partial x_k}\right)$$

$$= \sum_{l=1}^{N} h_l\left[\frac{\partial}{\partial x_k}\left(\rho D_l\frac{\partial Y_l}{\partial x_k}\right) + M_l W_l\right] + \sum_{l=1}^{N} Y_l c_{P,l}\left(\rho\frac{\partial T}{\partial t} + \rho u_k\frac{\partial T}{\partial x_k}\right)$$

$$= \sum_{l=1}^{N} h_l\left[\frac{\partial}{\partial x_k}\left(\rho D_l\frac{\partial Y_l}{\partial x_k}\right) + M_l W_l\right] + c_p\left[\frac{\partial}{\partial t}(\rho T) + \frac{\partial}{\partial x_k}(\rho u_k T)\right],$$

using the continuity Equation 1.49 and the species balance Equation 1.58. Therefore, Equation 1.66 results in the following temperature balance equation:

$$c_p \left[\frac{\partial}{\partial t}(\rho T) + \frac{\partial}{\partial x_k}(\rho u_k T) \right] = \frac{\partial p}{\partial t} + \frac{\partial}{\partial x_k}\left(\lambda \frac{\partial T}{\partial x_k} \right) + \rho \sum_{l=1}^{N} c_{P,l} D_l \frac{\partial Y_l}{\partial x_k} \frac{\partial T}{\partial x_k} + W_T, \quad (1.67)$$

where

$$W_T = -\sum_{l=1}^{N} h_l M_l W_l = -\sum_{l=1}^{N} M_l W_l \int_{T_0}^{T} c_{P,l} dT - \sum_{l=1}^{N} \Delta h_l^0 M_l W_l. \quad (1.68)$$

If the heat capacities $c_{P,l}$ are assumed to be equal to c_P, then the third term on the RHS of Equation 1.67 and the first term on the RHS of Equation 1.68 vanish by virtue of Equations 1.61 and 1.60, respectively. Therefore, Equation 1.67 takes the following well-known form:

$$c_p \left[\frac{\partial}{\partial t}(\rho T) + \frac{\partial}{\partial x_k}(\rho u_k T) \right] = \frac{\partial p}{\partial t} + \frac{\partial}{\partial x_k}\left(\lambda \frac{\partial T}{\partial x_k} \right) - \sum_{l=1}^{N} \Delta h_l^0 M_l W_l. \quad (1.69)$$

The balance Equations 1.49, 1.51, 1.58, and 1.64 or 1.67, supplemented with a detailed combustion chemistry, are typically solved numerically to simulate a planar laminar flame. However, this set of balance equations is too large to be theoretically studied and much simpler sets of balance equations are commonly invoked when discussing the physics of turbulent combustion. In the rest of this section, simplified sets of balance equations relevant to turbulent flames will be reported.

First, the contemporary theory of turbulent combustion often deals with a single global irreversible reaction GR-I and assumes that unburned gas, reacting gas, and equilibrium combustion products are mixtures of only three species; a fuel F, an oxidizer O, and a product P. For instance, if $\Phi > 1$ (a rich flame), then a burned mixture contains a fuel and a product with $Y_{F,b} = (\Phi - 1)/(\Phi + St)$ and $Y_{P,b} = (1 + St)/(\Phi + St)$, while $Y_{P,u} = Y_{O,b} = 0$, $Y_{F,u} = \Phi/(\Phi + St)$, and $Y_{O,u} = St/(\Phi + St)$.

Second, the unsteady pressure term on the RHS of Equation 1.64 is commonly disregarded when developing models of a premixed turbulent flame. Note that models developed by invoking such a simplification may subsequently be used to solve the averaged Equation 1.64, with the unsteady pressure term being retained on the RHS, for example, in multidimensional numerical simulations of reciprocating internal combustion engines.

Under the two aforementioned simplifications, Equations 1.58 and 1.64 read

$$\frac{\partial}{\partial t}(\rho Y_F) + \frac{\partial}{\partial x_k}(\rho u_k Y_F) = \frac{\partial}{\partial x_k}\left(\rho D_F \frac{\partial Y_F}{\partial x_k} \right) - M_F W, \quad (1.70)$$

$$\frac{\partial}{\partial t}(\rho Y_O) + \frac{\partial}{\partial x_k}(\rho u_k Y_O) = \frac{\partial}{\partial x_k}\left(\rho D_O \frac{\partial Y_O}{\partial x_k} \right) - St M_F W, \quad (1.71)$$

$$\frac{\partial}{\partial t}(\rho h) + \frac{\partial}{\partial x_k}(\rho u_k h) = \frac{\partial}{\partial x_k}\left[\frac{\mu}{\text{Pr}}\frac{\partial h}{\partial x_k} + \frac{\mu}{\text{Pr}}\sum_{l=1}^{3}\left(\frac{1}{\text{Le}_l}-1\right)h_l\frac{\partial Y_l}{\partial x_k}\right]. \tag{1.72}$$

The product mass fraction, Y_P, is calculated using the normalizing constraint $Y_F + Y_O + Y_P = 1$. The mass stoichiometric ratio, St, is inserted into Equation 1.71, because reaction GR-I consumes St g of the oxidizer per 1 g of the fuel. In order for such a simplified model to yield a reasonable dependence of the laminar flame speed on the equivalence ratio, one has to invoke the following reaction rate expression:

$$W = BT^{\beta}\exp\left(-\frac{\Theta}{T}\right)\left(\frac{\rho Y_F}{M_F}\right)^{n_F}\left(\frac{\rho Y_O}{M_O}\right)^{n_O} \tag{1.73}$$

with $n_F \neq 1$ and $n_O \neq 1$, contrary to the reaction GR-I. The coefficients B, β, Θ, n_F, and n_O tuned for various fuels were reported by Westbrook and Dryer (1984), see Table 4 in the cited review paper.

Equations 1.49, 1.51, and 1.70 through 1.72 are addressed when differences both in the mass diffusivities D_F and D_O and in the mass D_F and heat a diffusivities play a substantial role. These differences are characterized by the ratio of D_O/D_F and by the Lewis number Le of the deficient reactant, respectively. In lean mixtures, $\text{Le} = \text{Le}_F$, while $\text{Le} = \text{Le}_O$ in rich mixtures.

If the ratio of D_O/D_F is close to unity, then Equation 1.71 may be reduced to Equation 1.70 by properly normalizing the former equation. For instance, in a lean mixture, Equation 1.71 may be rewritten for $(Y_O - Y_{O,b})$ and multiplied by St^{-1} in order to obtain Equation 1.70 with Y_F replaced by $(Y_O - Y_{O,b})/St$. The boundary conditions for Y_F and $(Y_O - Y_{O,b})/St$ are also the same, because $Y_{F,b} = 0$ in a lean mixture and

$$St^{-1}\left(Y_{O,u} - Y_{O,b}\right) = St^{-1}\left(\frac{St}{\Phi + St} - \frac{St(1-\Phi)}{\Phi + St}\right) = \frac{\Phi}{\Phi + St} = Y_{F,u}.$$

Therefore, Equation 1.71 may be skipped if $D_O/D_F = 1$.

Either Equation 1.70 or 1.71 may also be skipped if the mass fraction of the abundant reactant is assumed to be much larger than the mass fraction Y_d of the deficient reactant and, therefore, variations in the former mass fraction in a flame are neglected. Such a simplification is invoked in many theoretical studies, especially as far as the burning of a lean fuel–air mixture is concerned.

If not only $D_F = D_O$, but also the specific heat capacities are assumed to be constant, then the temperature Equation 1.67 reads

$$\frac{\partial}{\partial t}(\rho T) + \frac{\partial}{\partial x_k}(\rho u_k T) = \frac{\partial}{\partial x_k}\left(\rho a\frac{\partial T}{\partial x_k}\right) - \frac{M_P W \Delta h_P^0 - M_F W \Delta h_F^0}{c_p}, \tag{1.74}$$

provided that $\Delta h^0 = 0$ for the oxidizer, for example, oxygen or air. Because

$$c_P \left(T_b - T_u \right) = \left(Y_{F,u} - Y_{F,b} \right) \Delta h_F^0 - Y_{P,b} \Delta h_P^0$$

and $M_P/M_F = 1 + St = Y_{P,b}/(Y_{F,u} - Y_{F,b})$ in the considered case, we arrive at

$$\frac{\partial}{\partial t}(\rho T) + \frac{\partial}{\partial x_k}(\rho u_k T) = \frac{\partial}{\partial x_k}\left(\rho a \frac{\partial T}{\partial x_k} \right) + \frac{T_b - T_u}{Y_{F,u} - Y_{F,b}} M_F W. \tag{1.75}$$

When applying Equation 1.75 to model a real flame, the combustion temperature T_b may be preliminarily calculated using the method discussed in the section "Calculation of Temperature and Composition of Adiabatic Combustion Products."

Finally, Equations 1.70 and 1.75 may be written in the same form:

$$\frac{\partial}{\partial t}(\rho \theta) + \frac{\partial}{\partial x_k}(\rho u_k \theta) = \frac{\partial}{\partial x_k}\left(\rho a \frac{\partial \theta}{\partial x_k} \right) + \frac{M_F W}{Y_{F,u} - Y_{F,b}}, \tag{1.76}$$

$$\frac{\partial}{\partial t}(\rho c) + \frac{\partial}{\partial x_k}(\rho u_k c) = \frac{\partial}{\partial x_k}\left(\rho D \frac{\partial c}{\partial x_k} \right) + \frac{M_F W}{Y_{F,u} - Y_{F,b}}, \tag{1.77}$$

by introducing the normalized temperature $\theta = (T - T_u)/(T_b - T_u)$ and the so-called combustion progress variable c defined as follows:

$$c = \frac{Y_{F,u} - Y_F}{Y_{F,u} - Y_{F,b}}. \tag{1.78}$$

Note that $c = 0$ and 1 in unburned and burned mixtures, respectively.

Because the boundary conditions for the normalized temperature and the combustion progress variable are identical ($\theta_u = c_u = 0$ and $\theta_b = c_b = 1$), the solutions to Equations 1.76 and 1.77 may be different either due to differences in a and D or due to different initial conditions.

If the molecular heat diffusivity of the mixture and the molecular mass diffusivities of the reactants are equal to one another (Le = 1), then a single equation (Equation 1.76 or 1.77) is sufficient to describe the variations in the mixture composition and the temperature within the flame, provided that the initial fields of θ and c are identical.

Thus, a simple model that resolves the structure of a laminar premixed flame involves the continuity Equation 1.49, the Navier–Stokes Equation 1.51, and the following balance equation:

$$\frac{\partial}{\partial t}(\rho c) + \frac{\partial}{\partial x_k}(\rho u_k c) = \frac{\partial}{\partial x_k}\left(\rho D \frac{\partial c}{\partial x_k}\right) + W \qquad (1.79)$$

for the combustion progress variable. Here,

$$W = \frac{\rho(1-c)}{t_r}\exp\left(-\frac{\Theta}{T}\right), \qquad (1.80)$$

$$T = T_u + (T_b - T_u)c, \qquad (1.81)$$

where t_r designates a reaction timescale, which depends on the mixture composition. Strictly speaking, it may also depend on c, cf. Equations 1.73 and 1.80, but we shall disregard this dependence for simplicity in the following text. Despite eventual dependence of t_r on the mixture composition, a factor $(1 - c)$ is retained in Equation 1.80 in order for the reaction rate to vanish in the products.

This model results from the following additional (with respect to the assumptions listed in the beginning of the section "Balance Equations") simplifications:

- Combustion chemistry is reduced to a single irreversible global reaction GR-I.
- The pressure does not depend on time.
- $Le_l = 1$ (therefore, $D_l = a$) for all the species.
- $c_{P,l} = $ const and $c_P = $ const.

Note that the last assumption is not necessary for describing the flame with the balance Equation 1.79, because the enthalpy balance Equation 1.65, the solution to which may be trivial ($h = $ const), holds if $\partial p/\partial t = 0$ and $Le_l = 1$, independently of the behavior of the heat capacities. However, if the heat capacities depend on temperature, then Equation 1.81 may not hold in a general case and the calculation of the temperature in Equation 1.80 would require solving the nonlinear (with respect to T) Equation 1.30.

In the following, the discussion of premixed turbulent flames will be mainly based on Equations 1.79 through 1.81, although the more general Equations 1.70 through 1.73 will sometimes be invoked.

Finally, it is worth stressing again that the key peculiarity of flames is a large ratio of Θ/T in Equation 1.80.

2 Unperturbed Laminar Premixed Flame

A premixed flame is a deflagration wave that propagates in a mixture comprising a fuel and an oxidant, for example, gasoline and oxygen in the air, respectively, and leaves behind hot, nonreacting combustion products, for example, a mixture of water vapor, carbon dioxide, and nitrogen at a temperature of about 2000 K. The pressure p decreases across the flame from the unburned to the burned side, but the decrease in p is very weak and is commonly neglected when calculating the temperature T and the density ρ using the ideal gas law. The density drops across the flame and the density ratio, $\sigma = \rho_u/\rho_b$, is about 6–8 for a typical burning mixture. Here, the subscripts u and b designate unburned (or fresh) mixture and burned mixture (or products), respectively. The term "burned mixture" means combustion products in the state of chemical equilibrium.

Very briefly, the physical mechanism of the propagation of a laminar premixed flame consists of the following steps. An energy released in the chemical reactions heats the reaction products. A part of this energy is transported to the fresh mixture due to molecular heat conductivity, increasing the temperature of the nonreacting gas in the vicinity of the zone where the chemical reactions occur. The increase in T triggers chemical reactions just ahead of the aforementioned zone, which thus moves toward the fresh mixture, and the flame propagates. Therefore, the flame propagation is controlled by the molecular transport and chemical reactions.

This chapter is restricted to the study of unperturbed laminar premixed flame, that is, a planar, one-dimensional (1D), adiabatic flame that has a stationary structure and propagates at a steady speed, S_L^0, against a 1D laminar flow of a spatially uniform fuel–air mixture. The problem is stationary and 1D in the coordinate framework attached to the flame. As the flame is adiabatic, the combustion products are also spatially uniform.

The aforementioned unperturbed flame is an idealized model and a practical laminar premixed flame is always perturbed, for example, by heat losses, flow nonuniformities, unburned mixture nonuniformities, and flame curvature. Due to such perturbations, addressed further in Chapter 5, the speed, S_L, of a practical flame differs from S_L^0, but this difference is commonly significantly smaller than S_L^0. Therefore, a study of the unperturbed flame is not only of basic importance but also of practical importance, and the unperturbed laminar flame speed is the key physical–chemical characteristic of a combustible mixture. For brevity, the symbol S_L will henceforth designate the unperturbed laminar flame speed in this chapter.

The contemporary theoretical studies of an unperturbed laminar premixed flame are focused on predicting the flame structure and speed by allowing for a reduced but still multistep chemistry and using the so-called rate-ratio asymptotic (RRA) approach; for example, see the review papers by Seshadri (1996), Williams (2000), and Buckmaster et al. (2005). In this book, however, the focus of discussion will be on the classical activation-energy asymptotic (AEA) theory developed by Zel'dovich and Frank-Kamenetskii (1938) in the simplest case of a single-step chemistry and discussed in many textbooks on combustion (Zel'dovich et al., 1985; Chomiak, 1990; Warnatz et al., 1996). This theory is sometimes called the thermal theory of laminar premixed flames. The rationale for highlighting the AEA theory is not only its simplicity and physical clarity but also the fact that the theory is still the cornerstone for contemporary models of premixed turbulent combustion.

It is also worth noting that the AEA approach put forward by Zel'dovich and Frank-Kamenetskii (1938) laid the foundations for more rigorous methods of matched asymptotic development, which are widely used in the contemporary theory of unperturbed (Bush and Fendell, 1970) and perturbed (Clavin and Williams, 1982; Frankel and Sivashinsky, 1982, 1983; Matalon and Matkowsky, 1982; Chung and Law, 1988; de Goey and ten Thije Boonkkamp, 1999; Class et al., 2003a,b; Matalon et al., 2003) premixed flames. These theories will not be considered in this book, and interested readers are referred to the cited papers, as well as to books by Buckmaster and Ludford (1982) and Zel'dovich et al. (1985) and to review papers by Clavin (1985), Law and Sung (2000), Sivashinsky (2002), and de Goey et al. (2011).

To highlight the basic physics and avoid unnecessary mathematical complications, we will consider the simplest case of (i) a single irreversible global combustion reaction, the rate of which is controlled by the temperature and concentration of the deficient reactant; (ii) unity Lewis number; and (iii) $c_{P,l} = c_P = $ const. Accordingly, the problem is modeled by the stationary (in the coordinate framework attached to the flame), 1D Equations 1.49, 1.51, and 1.79 through 1.81. It is worth stressing again that this book is focused on turbulent burning and, accordingly, the following consideration of laminar flames is mainly aimed at discussing those points that are of importance for understanding the basic physics of turbulent combustion.

FLOW IN A LAMINAR PREMIXED FLAME

In the coordinate framework attached to an unperturbed laminar premixed flame, the continuity Equation 1.49 and the Navier–Stokes Equation 1.51 read

$$\frac{d}{dx}(\rho u) = 0 \qquad (2.1)$$

and

$$\frac{d}{dx}(\rho u^2) = \frac{d}{dx}\left(\frac{4}{3}\mu \frac{du}{dx}\right) - \frac{dp}{dx}, \qquad (2.2)$$

respectively, using Equation 1.52 for the viscous stress tensor.

FLOW VELOCITY

Let us direct the x-axis from the unburned to the burned mixture. The integration of Equation 2.1 from $-\infty$ to x yields

$$\rho u = \rho_u U_u, \tag{2.3}$$

where $U_u = u(x \to -\infty)$ is the velocity of the unburned mixture. Because the flame is steady in the selected coordinate framework, the magnitude, $S_{L,u}$, of the flame speed with respect to the unburned mixture is equal to U_u. Note that the signs of the x-components of the flow velocity and flame speed vectors are opposite in the selected coordinate framework and depend on the direction of the x-axis, but $S_{L,u}$ is always positive because it designates the magnitude of the latter vector. For brevity, the subscript u is commonly skipped and the symbol S_L is used to designate the magnitude of the flame speed with respect to the unburned mixture. Thus, Equation 2.3 reads

$$\rho u = \rho_u S_L. \tag{2.4}$$

If the simplest version, $\rho T = \rho_u T_u$, of the ideal gas state equation is invoked, then Equation 2.4 results in

$$u = S_L \frac{T}{T_u}. \tag{2.5}$$

Using Equation 1.81, we finally obtain

$$u = S_L (1 + \tau c), \tag{2.6}$$

where $\tau = \sigma - 1 = \rho_u/\rho_b - 1$ is the heat release factor. Thus, the flow velocity depends linearly on the combustion progress variable in the simplest case considered.

If we apply Equation 2.4 to $x \to \infty$, then

$$\rho_b U_b = \rho_u S_L, \tag{2.7}$$

where $U_b = u(x \to \infty)$ is the velocity of the burned mixture. Because the flame is steady in the selected coordinate framework, the flame speed, $S_{L,b}$, with respect to the burned mixture is equal to U_b. Accordingly, Equation 2.7 reads

$$\rho_b S_{L,b} = \rho_u S_L \tag{2.8}$$

or

$$S_{L,b} = \sigma S_L \tag{2.9}$$

that is, the flame speed with respect to the products is higher than the flame speed with respect to the fresh mixture by a factor of σ, which is equal to the density ratio.

Although Equations 2.8 and 2.9 are equivalent to one another, there is an important difference in using them. If the flame thickness is much less than all the other characteristic length scales of a problem, then the flame may be treated as an infinitely thin, self-propagating interface, which separates the unburned and burned mixtures. In such a widely used asymptotic case, Equation 2.9 characterizes the velocity jump at the flame surface and shows that the flame speed is not a continuous quantity because $S_{L,b} \neq S_L$. On the contrary, Equations 2.4 and 2.8 indicate that the mass flux, ρu, is a continuous quantity. This basic difference between the flame speeds S_L and $S_{L,b}$ and the mass fluxes $\rho_u S_L$ and $\rho_b S_{L,b}$, respectively, should be borne in mind, and the use of the mass fluxes, rather than the flame speeds, is strongly recommended when considering the asymptotic case of an infinitely thin laminar premixed flame.

The physical meaning of the flux $\rho_u S_L = \rho_b S_{L,b}$ is clear. It is the mass rate of consumption of the unburned mixture (or the mass rate of formation of the burned mixture) per unit area of a flame surface.

The aforementioned results have been obtained in the coordinate framework attached to the flame. In an arbitrary coordinate framework, the mass flux $\rho_u S_L = \rho_b S_{L,b}$ is invariant with respect to the Galilean transformation, while the observed flame propagation speed depends on the choice of coordinate framework. However, the difference in the observed flame speed and the flow velocity at the flame surface is invariant and is equal to either S_L or $S_{L,b}$.

Because (i) the motion of a flame surface may be considered to be either the motion of the flame with respect to the unburned mixture or the motion of the flame with respect to the burned mixture and (ii) the observed flame speed should be the same in the two cases, we have

$$u_u - S_L = u_b - S_{L,b} \tag{2.10}$$

in an arbitrary coordinate framework. Here, u_u and u_b are the flow velocities just ahead and behind the flame, respectively, with the flow velocity and flame speed vectors pointing in opposite directions. Equation 2.10 is invariant with respect to the Galilean transformation. In the coordinate framework attached to an unperturbed laminar flame, $u_u = U_u$, $u_b = U_b$, and Equation 2.10 results straightforwardly from Equations 2.6 and 2.8.

Pressure Drop

In the coordinate framework attached to a flame, the integration of Equation 2.2 from $-\infty$ to ∞ yields

$$\rho_b U_b^2 - \rho_u U_u^2 = p_u - p_b \tag{2.11}$$

or

$$\Delta p \equiv p_u - p_b = \tau \rho_u S_L^2, \tag{2.12}$$

using Equations 2.4 and 2.7. Thus, in an unperturbed laminar flame, the pressure is decreased from the unburned to the burned side and this pressure drop accelerates the reacting mixture. However, the pressure drop is much less than the pressure itself. Indeed, because the speed of sound, c_s, is of the order of $\sqrt{p/\rho}$, Equation 2.12 shows that

$$\frac{\Delta p}{p} \equiv O\left(\tau \text{Ma}^2\right) \ll 1, \tag{2.13}$$

where $\text{Ma} = S_L/c_s$ is the Mach number in the unburned mixture. Because $c_s \approx 370$ m/s and S_L is less than 0.4 m/s in a typical hydrocarbon–air mixture under room conditions, the ratio of $\Delta p/p$ is of the order of 10^{-5} or even less, and the pressure variations in a typical laminar premixed flame are negligible. This fact justifies the simplification of $p = $ const in open flames.

It is worth stressing, however, that the earlier discussion does not mean that the pressure gradient term may be omitted in the Navier–Stokes equation when simulating laminar combustion. Even if the pressure drop, Δp, is much smaller than p in a laminar flame, the pressure gradient induced by combustion may be substantial due to the small thickness of the flame. Accordingly, the pressure gradient term should always be retained if the Navier–Stokes equation is solved within a laminar premixed flame.

AEA THEORY OF UNPERTURBED LAMINAR PREMIXED FLAME

The earlier analysis shows that the continuity Equation 1.49 and the Navier–Stokes Equation 1.51 may be analyzed independently of the combustion progress variable balance Equation 1.79. The Navier–Stokes Equation 1.51 indicates that the pressure drop across a flame is weak and the continuity Equation 1.49 yields the flow velocity as a function of the density or combustion progress variable; see Equations 2.4 and 2.6, respectively. The goal of this section is to consider the combustion progress variable balance Equation 1.79 for determining the laminar flame speed, following the work of Zel'dovich and Frank-Kamenetskii (1938).

Using Equation 2.4, the combustion progress variable balance Equation 1.79 reads

$$\rho_u S_L \frac{dc}{dx} = \frac{d}{dx}\left(\rho D \frac{dc}{dx}\right) + W, \tag{2.14}$$

where

$$W = \frac{\rho(1-c)}{t_r} \exp\left(-\frac{\Theta}{T}\right), \tag{2.15}$$

$$T = T_u + (T_b - T_u)c, \tag{2.16}$$

and

$$\rho T = \rho_u T_u, \tag{2.17}$$

In order for such a single-reaction model to describe the basic features of a typical laminar premixed flame, the activation temperature Θ of the reaction should be sufficiently large, for example, $\Theta = 15,000–20,000$ K for hydrocarbon–air mixtures. Then, the rate W of the reaction depends strongly on the temperature. For instance, an increase in T by 5% (from 1900 to 2000 K) results in increasing W by a factor of 1.7 if $\Theta = 20,000$ K. The AEA theory addresses the asymptotic case of $\Theta/T_b \rightarrow \infty$. The assumption of a high activation temperature is the key point of the thermal theory of premixed laminar flames.

Structure of Unperturbed Laminar Flame: Reaction Zone

Figure 2.1 shows the dependencies of the normalized reaction rate $W(T)/W_m$ on the normalized temperature T/T_u, calculated using Equations 2.15 through 2.17 for the density ratio $\sigma = 7$ and the four normalized activation temperatures Θ/T_b specified in the legend. Here, W_m is the maximum reaction rate reached at certain $T_m < T_b$. If $T_m < T < T_b$, then the reaction rate decreases when the temperature increases, because either the fuel or the oxidizer is rapidly consumed, that is, $(1 - c)$ drops in Equation 2.15. If $T < T_m$, the reaction rate is increased by an increase in temperature due to the exponential term in Equation 2.15.

When the activation temperature is increased, the role played by the strongly nonlinear exponential term is more significant than that of the term $\rho(1 - c)$ in

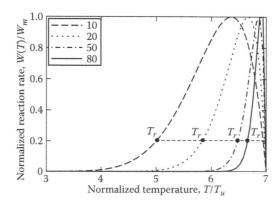

FIGURE 2.1 Dependencies of the normalized reaction rate $W(T)/W_m$ on the normalized temperature T/T_u, calculated using Equations 2.15 through 2.17 for the density ratio $\sigma = 7$ and four normalized activation temperatures Θ/T_b specified in the legends.

Equation 2.15. Accordingly, the decrease in $W(T)$ due to a decrease in temperature at $T < T_m$ becomes more pronounced and the temperature zone, characterized by a substantial reaction rate, becomes narrower. If we associate the low-temperature boundary T_r of the considered zone with a condition of $W(T_r)/W_m = \varepsilon$, where $\varepsilon \ll 1$ is a constant, then T_r is increased by the activation temperature and tends to the adiabatic combustion temperature as $\Theta/T_b \to \infty$. This trend is shown in Figure 2.1 (see circles) for T_r associated with $\varepsilon \approx 0.2$.

Thus, if $\Theta/T_b \gg 1$, the reaction rate depends very strongly on the temperature. Accordingly, the ratio of $W(T)/W_m$ is very low at all points within the flame, with the exception of a narrow temperature interval $T_r < T < T_b$, where T_r is slightly lower than T_b. A part of the flame, characterized by a substantial reaction rate, that is, $T_r < T < T_b$, is called the reaction zone or reactive–diffusive zone. As will be discussed later, the structure of this zone is controlled by the chemical reaction and molecular transport of heat (from the reaction zone to the unburned mixture) and reactants (in the opposite direction).

STRUCTURE OF UNPERTURBED LAMINAR FLAME: PREHEAT ZONE

Outside the reaction zone, that is, at $T_u \le T < T_r$, the reaction rate is negligible and Equation 2.14 is simplified as follows:

$$\rho_u S_L \frac{dc}{dx} = \frac{d}{dx}\left(\rho D \frac{dc}{dx}\right). \tag{2.18}$$

The integration of this equation from $-\infty$ to x (for a flame that propagates from right to left) yields

$$c = \frac{\rho D}{\rho_u S_L} \frac{dc}{dx}. \tag{2.19}$$

The integration of Equation 2.19 from x to the boundary x_r of the reaction zone yields

$$\int_x^{x_r} \frac{\rho_u S_L}{\rho D} dx = \ln c_r - \ln c. \tag{2.20}$$

In the asymptotic case of $\Theta/T_b \to \infty$, the reaction zone is infinitely thin with $T_r \to T_b$ and $c_r \to 1$. Accordingly, Equation 2.20 reads

$$c = \exp\left(-\int_x^{x_r} \frac{\rho_u S_L}{\rho D} dx\right). \tag{2.21}$$

In the simplest case of $\rho D = \rho_u D_u$ we have

$$c = \exp\left(\frac{S_L \xi}{D_u}\right),$$ (2.22)

where $\xi < 0$ is the distance measured from the boundary of the reaction zone.

Thus, within the framework of the AEA theory, a laminar premixed flame is divided into two zones: (i) a preheat zone characterized by a negligible reaction rate and (ii) a much thinner (see the solid curve in Figure 2.1, calculated for $\Theta/T_b = 80$) reaction zone characterized by a substantial reaction rate. The assumption of a negligible reaction rate allows us to easily determine the distribution of the combustion progress variable within the preheat zone (see Equation 2.21). However, the problem of determining the laminar flame speed is still not resolved. To do so, the burning rate within the reaction zone should be considered.

HEAT FLUX FROM REACTION ZONE

In the previous section, the distribution of the combustion progress variable in a preheat zone was determined by neglecting the reaction rate. This assumption allowed us to substantially simplify Equation 2.14 and to make the problem analytically solvable. It is tempting to invoke a similar method, that is, to neglect a term in Equation 2.14, to study a reaction zone. Which term could be neglected? Because the reaction rate is high in a reaction zone, this term is of substantial importance and should be retained. Therefore, let us compare the magnitudes of the convection and diffusion terms within the reaction zone.

The simplest estimate yields

$$\rho_u S_L \frac{dc}{dx} \approx \frac{\rho_u S_L}{\delta_r}$$ (2.23)

and

$$\frac{d}{dx}\left(\rho D \frac{dc}{dx}\right) \approx \frac{(\rho D)_b}{\delta_r^2},$$ (2.24)

where δ_r is the thickness of the reaction zone. Therefore, the ratio of the convection and diffusion terms is of the order of

$$\frac{\rho_u S_L \dfrac{dc}{dx}}{\dfrac{d}{dx}\left(\rho D \dfrac{dc}{dx}\right)} \approx \frac{\rho_u S_L}{(\rho D)_b} \delta_r.$$ (2.25)

The earlier analysis of a preheat zone, for example, see Equation 2.18, shows that

$$\frac{\rho_u S_L}{\Delta_L} \approx \frac{\rho D}{\Delta_L^2},$$

(2.26)

where Δ_L is the thickness of the preheat zone. Because the term ρD depends relatively weakly on the temperature and may be considered roughly constant in a flame, Equations 2.25 and 2.26 indicate that, in a thin reaction zone,

$$\frac{\rho_u S_L \dfrac{dc}{dx}}{\dfrac{d}{dx}\left(\rho D \dfrac{dc}{dx}\right)} \approx \frac{\delta_r}{\Delta_L},$$

(2.27)

that is, the ratio of the convection and diffusion terms is low therein. Therefore, the former term may be neglected and Equation 2.14 reads

$$0 = \frac{d}{dx}\left(\rho D \frac{dc}{dx}\right) + W$$

(2.28)

in the reaction zone.

To solve this equation, let us introduce a new variable,

$$y = \rho D \frac{dc}{dx}.$$

(2.29)

Then,

$$\frac{d}{dx}\left(\rho D \frac{\partial c}{\partial x}\right) = \frac{dy}{dx} = \frac{dy}{dc}\frac{dc}{dx} = \frac{y}{\rho D}\frac{dy}{dc}$$

(2.30)

and Equation 2.28 reads

$$\frac{dy^2}{dc} = -2\rho DW.$$

(2.31)

Because $y(c = 1) = 0$ (see Equation 2.29), the integration of Equation 2.31 within the reaction zone yields

$$\rho D \frac{dc}{dx}\bigg|_r = y_r = \left(\int_{c_r}^{1} 2\rho DW dc\right)^{1/2},$$

(2.32)

using Equation 2.29. Because c_r is very close to unity within the framework of the AEA theory, variations in ρD within the reaction zone may be neglected and Equation 2.32 reads

$$\rho D \frac{dc}{dx}\bigg|_r = \left[2(\rho D)_b \int_{c_r}^{1} W dc \right]^{1/2}.$$ (2.33)

Finally, because the reaction rate drops sharply when c decreases below c_r, integration on the right-hand side (RHS) of Equation 2.33 may be expanded from c_r to zero, that is,

$$\rho D \frac{dc}{dx}\bigg|_r = \left[2(\rho D)_b \int_{0}^{1} W dc \right]^{1/2}.$$ (2.34)

Equation 2.34 determines the flux of heat, released in the chemical processes within the reaction zone, toward the preheat zone.

LAMINAR FLAME SPEED

Obviously, in the stationary case considered, the heat flux that leaves the reaction zone should be equal to the heat flux that enters the preheat zone. The latter flux is determined by Equation 2.19:

$$\rho D \frac{dc}{dx}\bigg|_r = \rho_u S_L c_r \approx \rho_u S_L.$$ (2.35)

A comparison between Equations 2.34 and 2.35 shows that

$$\rho_u S_L = \left[2(\rho D)_b \int_{0}^{1} W dc \right]^{1/2}.$$ (2.36)

If the dependence of ρD on T is neglected, Equation 2.36 is further simplified as follows:

$$S_L = \left[2 D_u \int_{0}^{1} \frac{W}{\rho_u} dc \right]^{1/2}.$$ (2.37)

This is the main result of the AEA theory by Zel'dovich and Frank-Kamenetskii (1938).

It is worth noting that a basically similar equation could be obtained from simple physical reasoning. At the start of this chapter, it was noted that the propagation of a laminar premixed flame is controlled by the molecular transport and chemical reactions. Therefore, from dimension reasoning, one could obtain

$$S_L \propto \sqrt{\frac{D_u}{\tau_c}}, \tag{2.38}$$

where τ_c is a timescale that characterizes the reaction rate. The AEA theory yields

$$\tau_c^{-1} = 2 \int_0^1 \frac{W}{\rho_u} dc. \tag{2.39}$$

Laminar Flame Thickness

As we already discussed, a laminar premixed flame is divided into preheat and reaction zones within the framework of the AEA theory. Because the reaction zone is much thinner than the preheat zone, the flame thickness is approximately equal to the thickness of the preheat zone.

In Figure 2.2, a thick solid curve shows the distribution of the combustion progress variable in a preheat zone, calculated using Equation 2.22 in the simplest case of $\rho D = $ const. Various methods could be invoked to evaluate the thickness of this distribution, but the maximum gradient method is commonly used for this purpose, that is, the flame thickness, Δ_L, is defined as follows:

$$\Delta_L = \frac{1}{\max \left| \dfrac{dc}{dx} \right|} \tag{2.40}$$

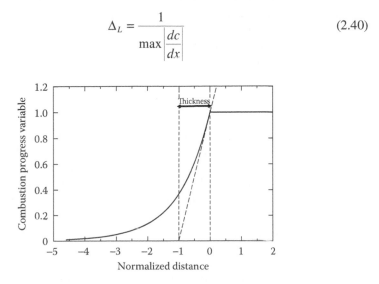

FIGURE 2.2 Determination of the flame thickness using the maximum gradient method. The distance is normalized using a_u/S_L.

or

$$\Delta_L = \frac{T_b - T_u}{\max \left| \dfrac{dT}{dx} \right|}. \tag{2.41}$$

As shown in Figure 2.2, the thickness is equal to the distance between the two points where the tangent to the $c(x)$ profile with the maximum slope (see long dashed line) intersects the lines $c = 0$ and $c = 1$.

Within the framework of the AEA theory, the gradient of the combustion progress variable is highest at the boundary between the preheat zone and the reaction zone (see Equation 2.19). Therefore,

$$\Delta_L = \frac{(\rho D)_b}{\rho_u S_L}. \tag{2.42}$$

If ρD is assumed to be constant, then

$$\Delta_L = \frac{a_u}{S_L} \equiv \delta_L. \tag{2.43}$$

Here, the diffusivity, D_u, of the deficient reactant in the unburned mixture is replaced with the heat diffusivity of the mixture, because we consider the case of unity Lewis number. The thickness δ_L, defined by the RHS of Equation 2.43, is widely used to characterize various possible regimes of premixed turbulent combustion, discussed in the section "Regimes of Premixed Turbulent Combustion" in Chapter 5. It is worth noting, however, that the quantity ρD is increased by the temperature in a typical flame and the ratio of $(\rho D)_b/(\rho D)_u$ is often assumed to scale as $\sigma^{0.6}$. Then, Equations 2.42 and 2.43 result in

$$\Delta_L = \sigma^{0.6} \delta_L. \tag{2.44}$$

Moreover, Figure 2.2 shows that variations in the combustion progress variable in a laminar premixed flame are well pronounced at distances substantially longer than Δ_L. Thus, the thickness δ_L, defined by the RHS of Equation 2.43 and widely used when discussing premixed turbulent combustion underestimates the width of the zone of substantial temperature variations in a laminar premixed flame.

REACTION ZONE THICKNESS

As we already discussed, the reaction zone is very thin in the case of $\Theta/T_b \gg 1$, addressed by the AEA theory. In the temperature space, the reaction zone thickness may be evaluated using the Frank-Kamenetskii expansion of the exponential term on the RHS of Equation 2.15:

$$\exp\left(-\frac{\Theta}{T}\right) = \exp\left(-\frac{\Theta}{T_b}\frac{T_b}{T}\right) = \exp\left[-\frac{\Theta}{T_b}\frac{1}{1-\Delta T/T_b}\right] \approx \exp\left[-\frac{\Theta}{T_b}\left(1+\frac{\Delta T}{T_b}\right)\right]$$

$$= \exp\left(-\frac{\Theta}{T_b}\right)\exp\left(-\frac{\Theta\Delta T}{T_b^2}\right). \tag{2.45}$$

Equations 2.15 and 2.45 show that the reaction rate drops exponentially when the ratio of $\Theta\Delta T/T_b^2 = \Theta(T_b - T)/T_b^2$ increases. Therefore, in the temperature space, the thickness of the reaction zone is T_b^2/Θ, whereas the thickness of the preheat zone is $T_b - T_u$. Accordingly, we may estimate the reaction zone thickness as follows:

$$\delta_r = \Delta_L \frac{T_b^2}{\Theta(T_b - T_u)} = \frac{\Delta_L}{\text{Ze}}, \tag{2.46}$$

where $\text{Ze} \equiv \Theta(T_b - T_u)/T_b^2$ is the Zel'dovich number, which is considered to be much larger than unity within the framework of the AEA theory.

SUMMARY OF AEA THEORY

The AEA theory is aimed at determining the speed of an unperturbed laminar premixed flame. The theory reduces combustion chemistry to a single, global, irreversible reaction with a high activation theory. In the asymptotic case of $\Theta/T_b \gg 1$ (or $\text{Ze} \gg 1$), the theory divides the flame into a preheat zone and a much thinner reaction zone (see Figure 2.3). The thicknesses of the two zones scale as a_u/S_L and $\text{Ze}^{-1}a_u/S_L$, respectively. In the preheat (or convection–diffusion) zone, the reaction rate is negligible and Equation 2.14 reduces to Equation 2.18, that is,

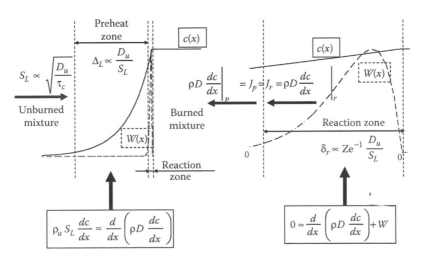

FIGURE 2.3 A summary of the AEA theory.

the structure of this zone is controlled by convection and molecular transport, for example, diffusion. In the reaction (or reaction–diffusion) zone, convection plays a minor role and Equation 2.14 reduces to Equation 2.28, that is, the structure of this zone is controlled by the reaction rate and molecular transport, for example, diffusion. On the boundary between the two zones, the heat flux that leaves the reaction zone, where heat is released in the reaction, is equal to the heat flux that enters the preheat zone and heats the nonreacting mixture. This matching condition (the balance of the two fluxes) controls the laminar flame speed, that is, the speed at which the fresh mixture flows toward the flame in the coordinate framework attached to the flame. Therefore, the laminar flame speed is controlled by heat release in the reaction zone and molecular transport of the released heat (or reactants) from (toward) the reaction zone. Consequently, the laminar flame speed scales as $\sqrt{a_u/\tau_c}$, where the diffusivity, a_u, and the chemical timescale, τ_c, characterize the magnitude of the molecular transport and the rate of the chemical reactions, respectively.

Effect of Lewis Number on Unperturbed Laminar Flame Speed

The previous discussion addressed the case of unity Lewis number. In this case, the distributions of the temperature, density, mass fractions of fuel, oxidizer, and products are similar to one another. In particular, the combustion progress variable c, defined by Equation 1.78, is equal to the normalized temperature $\theta = (T - T_u)/(T_b - T_u)$. The two solid lines in Figure 2.4 sketch the distributions of the normalized temperature and the normalized mass fraction, $g = Y_F/Y_{F,u} = 1 - c$, of the deficient reactant, for example, fuel, in the case of Le = 1. Obviously, $g + \theta = 1$ if Le = 1.

What happens in the case of Le $\neq$ 1? Let us assume that the diffusivity of the deficient reactant, for example, fuel, is increased, that is, the Lewis number becomes lower than unity. If the laminar flame speed were not changed, then the temperature profile in the preheat zone would not change, but the mass fraction profile would become wider, as shown by the dashed curve in Figure 2.4. Therefore, because the temperature

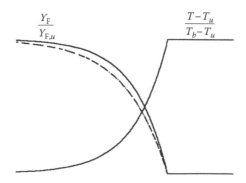

FIGURE 2.4 The effect of the Lewis number on the structure of an unperturbed laminar premixed flame. The solid and dashed curves are associated with Le = 1 and Le < 1, respectively.

reaches T_b, that is, $\theta = 1$, and the deficient reactant vanishes at the same point x in the flame, the mass fraction of the deficient reactant would decrease in the reaction zone; cf. the solid and dashed curves in Figure 2.4. Consequently, the reaction rate would decrease and, hence, the laminar flame speed would decrease.

This simple physical reasoning shows that a decrease in the Lewis number should reduce the unperturbed laminar flame speed. A more sophisticated mathematical analysis discussed elsewhere (Zel'dovich et al., 1985) supports this qualitative conclusion.

LIMITATIONS OF AEA THEORY

The AEA theory is based on two key simplifications: single-step chemistry and high-activation-temperature asymptotic. Both these simplifications may be put into question.

First, as already noted, for a typical hydrocarbon fuel, the activation temperature should be about 20,000 K in order for the AEA theory to match with the experimental data. Thus, the ratio of Θ/T_b and the Zel'dovich number, Ze, are of the order of 10 for a near-stoichiometric hydrocarbon–air mixture under atmospheric conditions. Even if $\Theta/T_b = O(10)$ is high, it is not sufficiently high in order for the difference in the thicknesses of the preheat and reaction zones to be so well pronounced as assumed by the AEA theory. For a typical hydrocarbon–air mixture, the ratio of δ_r/Δ_L is larger than the asymptotic value of Ze^{-1} yielded by the AEA theory at $Ze \rightarrow \infty$. Figure 2.5 shows the profiles of the normalized temperature, T/T_b, and reaction rate, W/W_m, computed for a typical hydrocarbon–air flame using three different values of the activation temperature for a single, global reaction. In each case, $x = 0$ is associated with the point where the reaction rate reaches its maximum

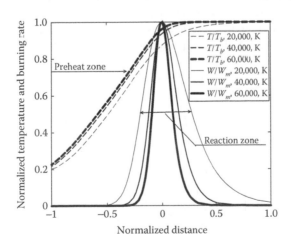

FIGURE 2.5 Profiles of the normalized temperature T/T_b (dashed curves) and normalized reaction rate $W(T)/W_m$ (solid curves), calculated for three activation temperatures Θ specified in the legends. The spatial distance is normalized with the thickness Δ_L evaluated using Equation 2.41.

value, W_m, and the spatial distance is normalized with the thermal flame thickness calculated using Equation 2.41. If $\Theta = 20{,}000$ K, the reaction zone thickness is of the order of unity, that is, Δ_L and δ_r are of the same order. For instance, if the latter thickness is determined to be equal to the distance between the two points where $W/W_m = 0.5$, then the ratio of δ_r/Δ_L is equal to 0.5, 0.3, and 0.2 for $\Theta = 20{,}000$, $40{,}000$, and $60{,}000$ K, respectively. Thus, for $\Theta = 20{,}000$ K associated with typical hydrocarbon–air mixtures, an estimate of $\delta_r/\Delta_L \ll 1$ does not hold even for single-step combustion chemistry.

It is also worth noting that the temperature profiles shown in Figure 2.5 and computed by solving Equation 2.14 differ substantially from the temperature profile shown in Figure 2.2 and calculated by solving the simplified Equation 2.18, which is associated with a preheat zone within the framework of the AEA theory. This difference is also a limitation of the theory.

Second, complex chemistry makes the structure of a real laminar premixed flame different from the structure of the flame addressed by the AEA theory. For instance, the structure of a typical laminar premixed flame is shown in Figure 2.6. In line with the AEA theory, there is a preheat zone ($-0.1 < x < 0.3$ mm), where the temperature increases due to the heat flux from the reaction zone and the mass fraction of oxygen decreases due to the diffusion flux toward the reaction zone, whereas the mass fractions of the radicals are very low, thus indicating negligible reaction rates. Although the preheat zone is thin, its thickness of about 0.36 mm, calculated using Equation 2.41, is larger by a factor of 6 than $\delta_L = 0.06$ mm yielded by Equation 2.43 for this flame ($a_u = 0.22 \times 10^{-4}$ m^2/s, $S_L = 0.38$ m/s) or by a factor of 2 than $\Delta L = 0.2$ mm yielded by Equation 2.44. Moreover, the profile of the normalized temperature shown in Figure 2.6 differs substantially from the profile of the normalized temperature plotted in Figure 2.2.

The most pronounced difference between single-reaction and complex-chemistry flames is the structure of the trailing edges of the flames. Within the framework of the AEA theory, the temperature reaches its equilibrium value, T_b, within a thin reaction zone, outside which the reaction rate is negligible. In a real flame, however,

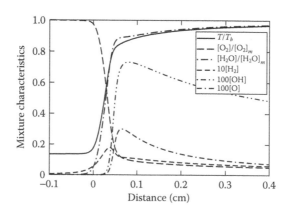

FIGURE 2.6 Spatial profiles of the normalized temperature T/T_b and mass fractions of certain reactants within stoichiometric methane–air flames under atmospheric conditions.

there is a long (about 1.5 cm for the flame shown in Figure 2.6) tail where the temperature is still lower than T_b but slowly increases, whereas the concentrations of the radicals decrease. Such a tail exists because termolecular recombination reactions V through VIII and XV (see Table 1.2) are relatively slow and their rates do not increase when T increases, because the activation temperatures vanish for these reactions. Therefore, even if the rates of the bimolecular reactions characterized by high Θ are strongly increased by the temperature and these reactions rapidly reach equilibrium states in a thin reaction zone, where the largest part of the heat release occurs, a part of the heat release is controlled by slow termolecular reactions that need a much longer recombination zone to reach equilibrium states. Accordingly, division of a laminar flame into a preheat zone and a much thinner reaction zone does not involve the termolecular reactions and the recombination zone.

The aforementioned limitations of the AEA theory have been overcome by the RRA theory, which deals with reduced chemical mechanisms of combustion, for example, see reactions RR-I, RR-II, RR-III, and RR-IV in the section "Chemical Reactions in Flames" in Chapter 1, and better predicts the structure of a real flame, as reviewed elsewhere (Peters, 1994; Seshadri, 1996; Williams, 2000; Buckmaster et al., 2005). Nevertheless, the RRA theory is beyond the scope of this book, which is aimed at discussing the physics of turbulent combustion. As far as this subject is concerned, the AEA theory appears to be a proper research tool because, on the one hand, it is sufficiently simple and clear, but, on the other hand, it offers an opportunity to reveal the basic physical mechanisms of turbulent burning. While the contemporary theory of laminar flames is based on the RRA approach, the contemporary theory of premixed turbulent combustion is still based on the AEA concept, which is also capable of explaining certain important features of laminar flames, as discussed in the next section.

As regards the evaluation of laminar flame speed and thickness as input parameters for modeling turbulent combustion, a more precise RRA theory is not necessary, because the aforementioned input parameters can be either measured or determined by numerically simulating laminar premixed flames with complex chemistry, as discussed in the next section.

LAMINAR FLAME SPEED AND THICKNESS

MEASUREMENTS AND COMPUTATIONS OF LAMINAR FLAME SPEEDS

For practical applications and premixed turbulent combustion modeling, in particular, laminar flame speed and thickness are the most important physical–chemical characteristics of a burning mixture. As discussed earlier, these two quantities may be estimated by invoking the AEA theory. However, due to the aforementioned limitations of the theory, more accurate methods for evaluating S_L and Δ_L are often required.

Two such methods are widely used today. First, the speed of a laminar flame can be measured. When experimentally evaluating S_L, the main challenge consists of making the studied flame as close as possible to the unperturbed one, that is, reducing the influence of various perturbations on the flame. Flame curvature, flow and mixture nonuniformities, heat losses, and transient effects are typical examples of perturbations that should be reduced when measuring laminar flame speeds. Experimental techniques developed for these purposes are beyond the scope of this book. Here, we restrict ourselves to claim that the speed of a typical near-stoichiometric hydrocarbon–air flame can be determined with an accuracy of about 5% under room conditions. A wide database of the values of S_L measured for various fuels, equivalence ratios, initial temperatures, and pressures is already encompassed in the combustion literature. It is worth noting, however, that the vast majority of the experimental studies on laminar premixed flames were focused on evaluating S_L, whereas the available data on Δ_L are much poorer. Moreover, for rich mixtures of heavy hydrocarbons and air, measuring S_L is still difficult, because such laminar flames are particularly susceptible to the instabilities discussed in the section "Flame Instabilities" in Chapter 5. Accordingly, a few sets of experimental data have yet to be reported for such mixtures, with the scatter of the data increased by the equivalence ratio. For instance, Figure 2.7 shows the laminar flame speeds of *iso*-octane–air mixtures measured by several research groups under room conditions and the data from different sources agree reasonably well at $\Phi \leq 1.3$ as far as recent experiments 3–8 are concerned (the earlier measurements made in experiment 1 overestimated the flame speeds, because the influence of flame perturbations on S_L was not addressed in this work, as well as in experiment 2, due to the lack of proper techniques at that time). However, the scatter of data 3–7 is substantially larger at $\Phi = 1.4$ and not much data have been reported for richer mixtures.

Second, substantial progress made in computer hardware and software over the past decades has made possible the determination of S_L and Δ_L by numerically simulating unperturbed adiabatic laminar flames with detailed chemistry, that is, by numerically solving Equations 1.49, 1.51, 1.58, and 1.64 in the planar 1D case. Widely used CFD packages (e.g., CHEMKIN, see http://www.reactiondesign.com; COSILAB, see http://www.rotexo.com; or Cantera, see dgoodwin@caltech.edu) and

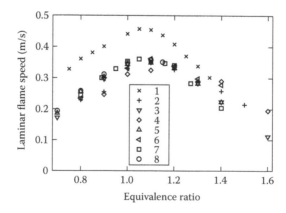

FIGURE 2.7 Laminar flame speed versus equivalence ratio for *iso*-octane–air flames measured under room conditions. 1, Gülder (1982); 2, Metghalchi and Keck (1982); 3, Davis and Law (1998); 4, Kwon et al. (2000); 5, Huang et al. (2004); 6, Kumar et al. (2007); 7, Kelley et al. (2011); 8, Lipzig et al. (2011).

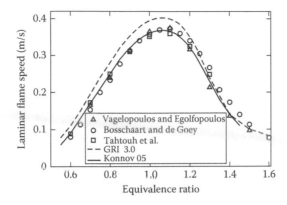

FIGURE 2.8 Laminar flame speed versus equivalence ratio for methane–air flames under room conditions. The symbols show the experimental data by Vagelopoulos and Egolfopoulos (1998), Bosschaart and de Goey (2004), and Tahtouh et al. (2009). The curves have been computed using GRI 3.0 (http://www.me.berkeley.edu/gri_mech/) and Konnov 05 (http://homepages.vub.ac.be/~akonnov/) chemical mechanisms.

various chemical mechanisms reviewed elsewhere (Miller et al., 2005; Lu and Law, 2009) have been developed, and these packages and chemical mechanisms are sufficiently mature to numerically determine the laminar flame speeds for mixtures of hydrogen or light alkane fuels and air under room conditions (e.g., see Figure 2.8). Moreover, such simulations yield not only S_L but also Δ_L.

As far as burning in internal combustion engines is concerned, more work is required in order to accurately determine the laminar flame speeds under such conditions. First, the deficiency and scatter (see Figure 2.7) of the experimental data on $S_L(\Phi)$ do not allow researchers to thoroughly validate the chemical mechanisms for rich mixtures of heavy hydrocarbon fuels used, for example, in gasoline direct injection (GDI) spark ignition (SI) engines. Moreover, detailed chemical mechanisms are too large for such fuels (see Figure 1.4), and practical applications of these mechanisms are still difficult. Furthermore, increasing interest in the burning of alternative fuels in engines demands experimental data and chemical schemes for a number of new alternative fuels, for example, gasoline–ethanol blends and biofuels.

Second, evaluation of S_L under elevated pressures and temperatures associated with the conditions in internal combustion engines is another important unresolved problem. Most of the experimental techniques used for measuring laminar flame speeds are poorly suited to substantially vary the initial temperature and pressure. Moreover, experimental studies on high-pressure laminar premixed flames are restricted for safety reasons. Furthermore, high-pressure laminar premixed flames are subject to hydrodynamic instability, as discussed in the section "Flame Instabilities" in Chapter 5, and this instability impedes accurate evaluation of S_L. As a result, the available experimental data are scanty and contradictory. For instance, Figure 2.9 shows that a few available approximations of the dependence of $S_L(T_u,p)$ for the stoichiometric iso-octane–air mixture yield substantially different laminar flame speeds under elevated pressures and temperatures associated with the conditions in SI engines.

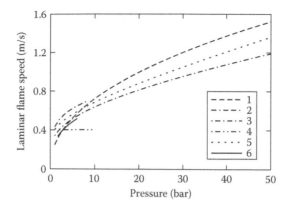

FIGURE 2.9 Laminar flame speed versus pressure for stoichiometric *iso*-octane–air mixtures during adiabatic compression starting from atmospheric conditions. The curves have been calculated using the following approximations: 1, Metghalchi and Keck (1982); 2, Gülder (1982); 3, Müller et al. (1997); 4, Bradley et al. (1998b); 5, Wallesten et al. (2002a,b); 6, Marshall et al. (2011). Curves 3, 4, and 6 are restricted to the conditions for which the approximations were obtained.

On the face of it, the aforementioned problems (scanty experimental data on the laminar flame speeds of heavy hydrocarbons under elevated pressures and temperatures) could be resolved by numerically simulating laminar flames with detailed chemistry. However, the predictive capabilities of the chemical mechanisms of the burning of hydrocarbon–air mixtures may be disputed, because the transport properties and reaction rate constants are still not well known. For many elementary reactions that occur in hydrocarbon–air flames, the rate constants "are only guesses" and "uncertainties in excess of a factor 2 remain the norm" (Williams, 2000). Even for such a simple and widely studied reaction as reaction I in Table 1.2, which plays a very important role in flames, uncertainties in the rate constants are about 10% at 1500 K, but 30% at 800 K (Miller et al., 2005). For hundreds (or even thousands) of other reactions invoked by various detailed chemical mechanisms, uncertainties in the rate constants are much higher. Therefore, to compute S_L using a detailed mechanism that contains several hundred elementary steps, the reaction rates are often tuned on the basis of the experimental data. Typically, the measured values of laminar flame speeds and ignition delay times are utilized for this purpose. However, it is not clear whether or not such a tuned chemical model is capable of predicting laminar flame speeds under conditions that have not yet been investigated experimentally.

Nevertheless, computation of S_L invoking a detailed chemical mechanism is a rapidly growing and well-supported branch of contemporary combustion science and technology. Chemical schemes are becoming more and more mature and could be a predictive tool in the foreseeable future. For instance, recently, Huang et al. (2010) tuned a semidetailed chemical mechanism for gasoline–air mixtures using the experimental data obtained from flames at atmospheric pressure (see Figure 2.10a). After the mechanism tuning was completed, Jerzembeck et al. (2009) reported the experimental data obtained at higher pressures. As shown in Figure 2.10b, the laminar flame speeds computed using

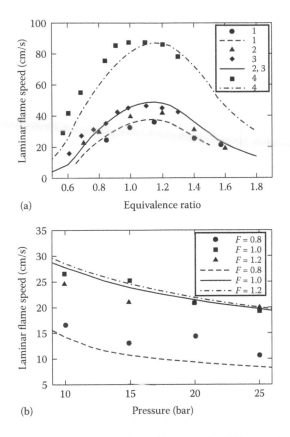

(a)

(b)

FIGURE 2.10 Laminar flame speeds measured (symbols) by different groups and computed (lines) by Huang et al. (2010) for gasoline–air mixtures. (a) Experiments 1–2 by Metghalchi and Keck (1982) and 3–4 by Zhao et al. (2003) performed at $p = 1$ bar and various temperatures of unburned mixtures; 1, $T_u = 298$ K, 2, $T_u = 350$ K, 3, $T_u = 353$ K, and 4, $T_u = 500$ K. (b) Experiments performed by Jerzembeck et al. (2009) at $T_u = 373$ K, elevated pressures, and three different equivalence ratios specified in the legends.

the same mechanism without extra tuning agree well with the high-pressure experimental data, thereby indicating the predictive capabilities of the mechanism.

DEPENDENCE OF LAMINAR FLAME SPEED AND THICKNESS ON MIXTURE COMPOSITION, PRESSURE, AND TEMPERATURE

The laminar flame speeds measured and computed for methane–air and *iso*-octane–air mixtures under room conditions are shown in Figures 2.7 and 2.8, respectively. The speeds of ethane–air, ethylene–air, acetylene–air, and hydrogen–air flames under room conditions are reported in Figure 2.11. The following conclusions could be drawn by analyzing these data:

First, for various near-stoichiometric paraffin–air mixtures, the values of S_L are close to 0.4 m/s under room conditions (see Figures 2.7 and 2.8 and the squares and

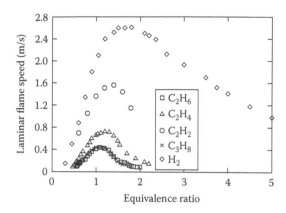

FIGURE 2.11 Laminar flame speeds obtained experimentally by Egolfopoulos et al. (1990) and Aung et al. (1997, 1998) for hydrocarbon–air and hydrogen–air flames, respectively, under room conditions.

crosses in Figure 2.11). Significantly higher values of S_L characterize hydrogen–air mixtures (see diamonds in Figure 2.11).

Second, for hydrocarbon–air mixtures, the highest flame speeds are associated with slightly rich ($\Phi \approx 1.1$ for paraffins) flames and the dependence of S_L on the equivalence ratio, in general, follows the dependence of $T_b(\Phi)$ (cf. Figures 2.7 and 2.8 with Figures 1.2 and 1.3), with the former dependence being much more pronounced. For instance, for lean methane–air mixtures, S_L drops by four times at $\Phi = 0.6$ as compared with stoichiometric flames, whereas the ratio $T_b(\Phi = 0.6)/T_b(\Phi = 1)$ is about 0.8. For hydrogen–air mixtures, the highest S_L is associated with rich flames ($\Phi \approx 1.6$), whereas the highest combustion temperature is reached in near-stoichiometric ($\Phi = 1.06$) mixtures.

The aforementioned observations are well explained by the AEA theory. Indeed, using Equation 2.16, Equation 2.45 reads

$$\exp\left(-\frac{\Theta}{T}\right) \approx \exp\left(-\frac{\Theta}{T_b}\right)\exp\left(-\frac{\Theta(T_b-T_u)}{T_b^2}\frac{T_b-T}{T_b-T_u}\right) = \exp\left(-\frac{\Theta}{T_b}\right)\exp\left(-Ze(1-c)\right).$$

(2.47)

Therefore, the integration of Equation 2.15 yields

$$\int_0^1 \frac{W}{\rho_u}dc \approx \frac{1}{\sigma t_r}\exp\left(-\frac{\Theta}{T_b}\right)\int_0^1 (1-c)\exp\left(-Ze(1-c)\right)dc$$

$$= \frac{1}{\sigma t_r}\exp\left(-\frac{\Theta}{T_b}\right)\left[\frac{1}{Ze^2}(1-e^{-Ze}) - \frac{e^{-Ze}}{Ze}\right] \approx \frac{1}{Ze^2\sigma t_r}\exp\left(-\frac{\Theta}{T_b}\right) \quad (2.48)$$

if Ze ≫ 1 and the laminar flame speed, given by Equation 2.37, may be estimated as follows:

$$S_L \approx \frac{1}{Ze}\left(\frac{2D_u}{\sigma t_r}\right)^{1/2} \exp\left(-\frac{\Theta}{2T_b}\right). \qquad (2.49)$$

This theoretical estimate shows that the dependence of the laminar flame speed on the mixture characteristics is mainly controlled by the adiabatic combustion temperature, T_b, and is also affected by the molecular transport coefficients (it is worth remembering that the heat diffusivity a of the mixture is equal to the mass diffusivity D of the deficient reactant in the case of Le = 1 addressed here).

Therefore, first, for hydrocarbon–air mixtures, which are characterized by high activation temperatures and a weak dependence of the molecular transport coefficients on the equivalence ratio, the dependence of $S_L(\Phi)$ follows the dependence of $T_b(\Phi)$. Second, the former dependence is much more pronounced due to the strongly nonlinear dependence of S_L on T_b, given by Equation 2.49. Third, for rich ($\Phi > 1$) hydrogen–air mixtures, which are characterized by (i) a lower Θ, (ii) a stronger dependence of a_u on Φ, and (iii) a slower decrease in T_b with Φ (see Figure 1.3); the dependence of S_L on Φ is affected not only by a decrease in T_b with Φ but also by an increase in a_u with Φ. As a result, the highest flame speed is associated with $\Phi \approx 1.6$. Fourth, the speeds of hydrogen–air flames are significantly higher than the speeds of paraffin–air flames, because (i) the activation temperature and the Zel'dovich number are lower for the former flames, (ii) the combustion temperature is higher for the former flames, and (iii) the diffusivity of hydrogen is high. Fifth, for acetylene–air and ethylene–air mixtures (see the circles and triangles, respectively, in Figure 2.11), S_L is higher than for paraffin–air mixtures (squares and crosses), because the combustion temperature is higher for the former flames. For instance, if $\Phi = 0.9$, then T_b is equal to 2169, 2289, and 2470 K for C_2H_6, C_2H_4, and C_2H_2, respectively, under room conditions.

This discussion shows that, despite the substantial simplifications discussed in the section "Limitations of AEA Theory," the AEA theory is able to explain the dependence of the laminar flame speed on the mixture composition. This and other achievements of the theory, which will be discussed later, further justify using it in contemporary combustion studies, even if a more sophisticated RRA approach was already developed and the latter approach describes the structure of a laminar premixed flame much better than the AEA theory.

The data on the laminar flame speeds, discussed earlier, allow us to estimate differently defined laminar flame thicknesses. For near-stoichiometric hydrocarbon–air flames, $S_L \approx 0.4$ m/s and $a_u \approx 0.2 \times 10^{-4}$ m²/s. Therefore, the thickness δ_L, defined by Equation 2.43, is approximately equal to 0.05 mm and the thickness Δ_L, defined by Equation 2.44, is approximately equal to 0.17 mm. Due to complex-chemistry effects, the thickness of a real near-stoichiometric hydrocarbon–air laminar flame is larger (about 0.4 mm if determined using the maximum gradient method, see Figure 2.6). For near-stoichiometric hydrogen–air flames, $S_L \approx 2$ m/s and $a_u \approx 0.5 \times 10^{-4}$ m²/s. Therefore, the thickness δ_L, defined by Equation 2.43, is approximately equal to

0.025 mm and the thickness Δ_L, defined by Equation 2.44, is approximately equal to 0.08 mm. Again, due to complex-chemistry effects, the thickness of a real near-stoichiometric hydrogen–air laminar flame is larger.

The dependence of the laminar flame thickness on the equivalence ratio is mainly controlled by the dependence of $S_L(\Phi)$. Accordingly, the maximum flame speed and the minimum thickness are associated with roughly the same equivalence ratio, Φ_m, and the thickness increases as $|\Phi - \Phi_m|$ increases.

Equation 2.49 predicts that the laminar flame speed is increased by the initial temperature due to an increase in T_b by T_u. The experimental data support this theoretical prediction, with the effect being substantially weaker than the strongly nonlinear dependence of S_L on T_b, associated with the exponential term on the RHS of Equation 2.49. The point is that an increase in T_u results in a substantially lower increase in T_b and the timescale t_r may depend on the temperature, cf. the simple Equation 2.15 invoked in the current discussion on the AEA theory with Equation 1.73, which yields more accurate values of the laminar flame speed. The empirical formula

$$S_L \propto T_u^q \tag{2.50}$$

is commonly invoked to approximate the experimental data with q ranging from 1 (Bradley et al., 1998b) to 2.2 (Metghalchi and Keck, 1982). For estimates, the value of $q = 1.6$ is often recommended. Because the heat diffusivity, a_u, is also roughly proportional to T_u^q with $q \approx 1.6$, the laminar flame thickness depends weakly on the initial temperature.

Measured (or computed using a detailed chemical mechanism) dependencies of S_L on the product temperature are commonly used to determine the activation temperature Θ for a single reaction that models combustion chemistry. In the limit case of $\Theta/T_b \to \infty$ addressed by the AEA theory, Equation 2.49 yields

$$\Theta = 2T_b^2 \frac{d \ln S_L}{dT_b}. \tag{2.51}$$

Therefore, by measuring the dependence of S_L on T_b, for example, by varying the volume percentage of the inert species in burning mixtures, one may evaluate the activation temperature using Equation 2.51. The values of Θ determined for various mixtures were reported by, for example, Sun et al. (1999).

Within the framework of the AEA theory, the dependence of the laminar flame speed on pressure is controlled by an equation invoked to approximate the reaction rate W. Equation 1.40 shows that the forward reaction rate is proportional to p, p^2, and p^3 for monomolecular, bimolecular, and termolecular reactions, respectively. Because the diffusivity D is proportional to p^{-1}, Equation 2.37 implies that S_L is proportional to p^q with $q = n/2 - 1$, where n is the reaction order. Thus, according to the theory, the laminar flame speed is increased by pressure if combustion chemistry is reduced to a termolecular reaction, but the opposite effect is observed if a monomolecular single-step chemical reaction is invoked. Both a decrease and an increase in S_L with increasing pressure have been documented. In particular, the experimental data and the results of numerical simulations indicate that the laminar flame speed decreases when the pressure increases for hydrocarbon–air mixtures. For instance, $q = -0.5$ for methane (Kobayashi et al., 1996); $q = -0.25$ for ethane and propane (Kobayashi et al., 1998); and $q = -0.16$ (Metghalchi and Keck, 1982), -0.22 (Gülder, 1982), or -0.28 (Bradley et al., 1998b) for stoichiometric iso-octane–air mixtures. On the contrary, the speeds of rapidly burning hydrogen flames are increased by pressure (Zel'dovich et al., 1985). Thus, hydrogen–air flames are associated with a higher reaction order n than hydrocarbon–air flames. The numerical data reported by Sun et al. (1999) support this trend.

Because the diffusivity D is proportional to p^{-1}, the laminar flame thickness is decreased when the pressure increases, with δ_L proportional to $p^{-n/2}$ within the framework of the AEA theory. The decrease in laminar flame thickness with increasing pressure is well documented.

FLAMMABILITY LIMITS

Figures 2.7, 2.8, and 2.10a indicate that the laminar flame speed rapidly decreases in lean and rich mixtures, that is, when $|\Phi - \Phi_m|$ increases. Numerous experiments show that such a decrease in S_L is limited. For instance, when a hydrocarbon–air mixture becomes so lean that $S_L \approx 0.05$ m/s, a flame cannot sustain itself in such a lean mixture (Peters, 1994). In rich propane–air mixtures, Jarosinski et al. (2002a) reported as low a value of S_L as 0.02 m/s.

The lowest (highest) equivalence ratio that a laminar premixed flame can sustain itself at is called the lean (rich) flammability limit. Contrary to near-stoichiometric flames, it is very difficult to experimentally obtain a weakly perturbed flame if the mixture composition is close to lean or rich flammability limit. Accordingly, measured flammability limits are sensitive to various perturbations (heat losses, buoyancy, flow nonuniformities, flame curvature, etc.) and the available experimental data on flammability limits and flame speeds near those limits are scattered (Jarosinski, 1986). For instance, Jarosinski et al. (2002b) reported $\Phi = 1.6$ and 2.42 for propane–air flames that propagate downward and upward, respectively. The lean flammability limit for typical hydrocarbon–air mixtures is associated with $\Phi = 0.5–0.6$, for example, $\Phi = 0.54$ for propane–air flames under room conditions (Jarosinski et al., 2002b).

In order to explain flammability limits, the thermal theory of flame propagation highlights heat losses (Zel'dovich et al., 1985). The RRA theory associates flammability limits with the competition between radical production in chain-branching reactions, which dominate at high temperatures, and binding radicals in chain-breaking reactions, which dominate at low temperatures (Peters, 1994; Seshadri, 1996). Further discussion on flammability limits is beyond the scope of this book.

SUMMARY

Because the focus of this book is placed on turbulent combustion, it is worth summarizing the features of laminar premixed flames that are of paramount importance for understanding the physics of turbulent burning.

First, a laminar premixed flame may be considered to be a thin interface that separates an unburned, heavy, and cold mixture and burned, light, and hot products. Under room conditions, the flame thickness is less than 1 mm, but the thickness decreases when the pressure increases.

Second, heat release is mainly confined to a reaction zone, which is substantially thinner than the flame.

Third, the flame propagates at a speed S_L with respect to the unburned mixture, and the mass rate of reactant consumption (and product formation) per unit flame surface area is equal to $\rho_u S_L$. The flame speed and mass burning rate are controlled by heat release in the chemical reactions and heat (reactant) transport to the preheat (reaction) zone. The flame speed and mass burning rate depend on the mixture composition, pressure, and temperature. Under room conditions, $S_L = 0.05 - 0.4$ m/s for typical hydrocarbon–air mixtures. Laminar flame speed is increased by an increase in temperature, and, for most hydrocarbon–air mixtures, S_L is decreased when the pressure increases.

Fourth, for studies of turbulent combustion, laminar flame speed and thickness are key input physical–chemical characteristics of a burning mixture. They can be either determined experimentally or computed by simulating a planar laminar premixed flame with complex chemistry. In the subsequent chapters, we will assume that laminar flame speed and thickness are known input parameters for studying turbulent combustion.

3 A Brief Introduction to Turbulence

Although the readers of this text may intuitionally understand what the term "turbulence" means, and everybody can see turbulent flows every day by observing the smoke from a chimney, the water in a river, or a milk drop falling into a glass of water, it is very difficult to define turbulence precisely and concisely. For instance, Hinze (1975) used the following definition:

> Turbulent fluid motion is an irregular condition of flow in which the various quantities show a random variation with time and space coordinates, so that statistically distinct average values can be discerned.

Bradshaw (1971) suggested a more sophisticated definition to emphasize the structure of turbulence:

> Turbulence is a three dimensional time dependent motion in which vortex stretching causes velocity fluctuations to spread to all wavelengths between a maximum determined by the boundary conditions of the flow and minimum determined by viscous forces. It is the usual state of fluid motion except at low Reynolds numbers.

Many other formulations can be found in the literature, but the two aforementioned definitions reflect the features of turbulence on which the mainstream models of premixed combustion are based. These are the randomness and unsteadiness of the flow, the existence of eddies of widely varying sizes, and the important role played by vortex stretching. It is worth stressing, however, that the randomness of a turbulent flow should not be overestimated. Turbulence differs from a totally random process, for example, a turbulent velocity field should satisfy the continuity and Navier–Stokes equations.

This chapter is aimed at briefly introducing the readers to the physics of turbulence and some basic terms. Because the book addresses flames, the following introduction to turbulence will focus on phenomena that are of importance for understanding turbulent combustion. Accordingly, some outstanding results discussed in the courses in turbulence will be beyond the scope of this chapter, whereas other results skipped in the courses in turbulence will be discussed here due to their importance in the study of combustion physics. Interested readers are referred to books by Brodkey (1967), Bradshaw (1971), Monin and Yaglom (1971, 1975), Tennekes and Lumley (1972), Hinze (1975), Townsend (1976), Frisch (1995), Pope (2000), etc., for a deeper understanding of turbulence. The *Album of Fluid Motion* collected by van Dyke (1982) illustrates many important features of turbulent motion well.

CHARACTERISTICS OF AN INCOMPRESSIBLE TURBULENT FLOW

As with any flow, a turbulent flow may be characterized by unsteady three-dimensional (3D) velocity $\mathbf{u}(\mathbf{x},t)$ and pressure $p(\mathbf{x},t)$ fields. However, due to small-scale variations of turbulent quantities in time and space, it is difficult to measure or compute these fields in many practical flows. Even if the rapid development of computer hardware and software has allowed researchers to evaluate $\mathbf{u}(\mathbf{x},t)$ and $p(\mathbf{x},t)$ in direct numerical simulations (DNS) of certain turbulent flows, such simulations are still unpractical in many cases, especially in flames. Most engineering studies on turbulence deal with averaged velocity and pressure fields.

MEAN (AVERAGE) QUANTITIES

The averaged characteristics of a turbulent flow may be determined using different methods. For instance, let us assume that the velocity vector $\mathbf{u}$ is measured in a point $\mathbf{x}$ during a time interval from instant t to instant $t + \tau$. Each component of the vector $\mathbf{u}$ varies in time (see Figure 3.1). Then, the components of the mean velocity vector $\bar{\mathbf{u}}$ may be evaluated as follows:

$$\bar{u}_i(\mathbf{x},t) = \frac{1}{\tau}\int_0^\tau u_i(\mathbf{x},t+\vartheta)d\vartheta, \qquad (3.1)$$

provided that the result of the integration does not depend on τ during a sufficiently long time interval $\tau_0 \ll \tau_1 < \tau < \tau_m$ with $\tau_m - \tau_1 \gg \tau_0$. Here, τ_0 and τ_m are associated with the longest timescale of turbulent motion and the flow lifetime (any flow dries up at some instant), respectively. If the duration τ of the averaging interval is too short, the integral on the right-hand side (RHS) of Equation 3.1 depends on τ due to velocity fluctuations of a longer timescale. Obviously, in a stationary mean flow, the mean velocity does not depend on time; therefore, $\bar{\mathbf{u}}(\mathbf{x}) = \bar{\mathbf{u}}(\mathbf{x})$. The x-component $\bar{u}$ of the mean flow velocity vector $\bar{\mathbf{u}}$ is shown with a dashed straight line in Figure 3.1.

A flow such that the RHS of Equation 3.1 does not depend on τ at $\tau_0 \ll \tau_1 < \tau < \tau_m$ is called a stationary mean flow, and the time-averaging method described by

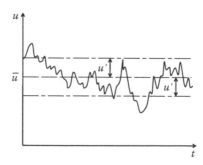

FIGURE 3.1 A sketch of velocity variations at a point.

Equation 3.1 is well suited to determine the mean characteristics of the flow. Jets and wakes and flows behind turbulence-generating grids are examples of stationary mean flows.

If a grid oscillates with a sufficiently low frequency, then the mean flow behind it will be quasi-stationary, that is,

$$0 < \left| \frac{\partial \bar{u}_i(\mathbf{x},t)}{\partial t} \right| \ll \frac{\left| \bar{u}_i(\mathbf{x},t) \right|}{\tau_0}, \tag{3.2}$$

and Equation 3.1 is well suited to determine the mean flow velocity even in this case. The physical meaning of Equation 3.2 is that the variations in the mean flow take a much longer time as compared with the variations in the instantaneous flow; therefore, the former variations may be neglected. Accordingly, the equality of $\overline{\bar{\mathbf{u}}(\mathbf{x})} = \bar{\mathbf{u}}(\mathbf{x})$ is assumed to hold in a quasi-stationary mean flow.

Differentiation of Equation 3.1 yields

$$\frac{\partial \bar{u}_i}{\partial x_j} = \frac{\partial}{\partial x_j} \left[\frac{1}{\tau} \int_0^\tau u_i(\mathbf{x}, t+\vartheta) d\vartheta \right] = \frac{1}{\tau} \int_0^\tau \frac{\partial u_i}{\partial x_j} d\vartheta = \overline{\frac{\partial u_i}{\partial x_j}} \tag{3.3}$$

and

$$\frac{\partial \bar{u}_i}{\partial t} = \frac{\partial}{\partial t} \left[\frac{1}{\tau} \int_0^\tau u_i(\mathbf{x}, t+\vartheta) d\vartheta \right] = \frac{1}{\tau} \int_0^\tau \frac{\partial u_i}{\partial t} d\vartheta = \overline{\frac{\partial u_i}{\partial t}}, \tag{3.4}$$

that is, time averaging commutes with differentiation.

In some cases, the mean characteristics of a turbulent flow may be determined using spatial averaging technique, that is,

$$\bar{u}_i(\mathbf{x}, t) = \frac{1}{V} \int u_i(\mathbf{x}+\mathbf{r}, t) d\mathbf{r}, \tag{3.5}$$

where the integration is performed either along an interval of a 1D line or over a segment of a 2D plane or over a 3D volume. Accordingly, V is either the length of the 1D interval, or the area of the 2D segment, or the 3D volume, respectively. One can easily show that spatial averaging commutes with differentiation.

If the integral on the RHS of Equation 3.5 depends neither on the dimension of the integration domain (1D interval or 2D segment or 3D volume) nor on a length scale l of the domain V, provided that $L \ll l_1 < l < l_m$ and $l_m - l_1 \gg L$, then the mean flow velocity does not depend on x, the flow is called a spatially uniform mean flow, and $\overline{\bar{\mathbf{u}}(t)} = \bar{\mathbf{u}}(t)$. Here, L and l_m are associated with the length scales of the largest turbulent eddies and the spatial variations in the mean flow, for example, due to boundaries, respectively.

In practice, the spatial averaging technique is also applied to two other kinds of flows. First, it is applied to quasi-uniform mean flows determined by the following inequality:

$$\left| \frac{\partial \bar{u}_i(\mathbf{x}, t)}{\partial x_j} \right| \ll \frac{\left| \bar{u}_i(\mathbf{x}, t) \right|}{L}. \tag{3.6}$$

The use of Equation 3.5 in such a case is basically equivalent to the application of time-averaging techniques to quasi-stationary mean flows.

Second, Equation 3.5 may be used for determining spatially averaged velocity even if Equation 3.6 does not hold for one direction but holds for another. For instance, if the velocity $\bar{u}$, evaluated using Equation 3.5 and $V = \Delta x \Delta y \Delta z$, varies rapidly along the x-axis due to a strong pressure gradient in this direction, the velocity $\bar{u}$, evaluated using Equation 3.5 and $V = \Delta y \Delta z$, may be a meaningful characteristic of the mean flow, provided that Equation 3.6 holds for $j = 2$ and 3. In this case, the mean flow velocity depends neither on y nor on z, and $\overline{\mathbf{u}}(x, t) = \overline{\mathbf{u}}(x, t)$. Such a flow is called a statistically 1D flow.

Such an application of the spatial averaging technique is typical for DNS of statistically planar flames, that is, flames where the mean density and temperature vary along the x-axis but do not vary in the yz-plane. If the flame speed varies rapidly with time, Equation 3.1 cannot be used to determine the mean velocity in the flame. Equation 3.5 with a 3D volume V cannot be used for this purpose either due to variations in the mean flow velocity in the x-direction. However, the mean flow velocity can easily be determined if the integration on the RHS of Equation 3.5 is performed over a sufficiently large (as compared with the size of the largest turbulent eddies) segment of the yz-plane.

In some cases, neither Equation 3.1 nor Equation 3.5 characterizes the mean flow. For instance, in the chamber of a reciprocating internal combustion engine, $\overline{\mathbf{u}}(\mathbf{x}, t)$ determined using Equation 3.1 or 3.5 rapidly varies in time and space. To characterize such turbulent flows, the ensemble averaging technique may be used.

If a process is repeated many times, for example, the burning cycles in the chamber of a reciprocating internal combustion engine, one can evaluate

$$\bar{u}_i(\mathbf{x}, t) = \frac{1}{N} \sum_{n=1}^{N} u_{i,n}(\mathbf{x}, t_n), \tag{3.7}$$

provided that the left-hand side (LHS) of Equation 3.7 does not depend on N if $N \gg N_0$. Here, $u_{i,n}(\mathbf{x}, t_n)$ is the i-th component of the velocity vector measured during the n-th realization of the process in point x at instant t_n that is counted from the start of each realization. Such an averaging method is called averaging over an ensemble of random realizations of statistically the same process, or ensemble averaging, for the sake of brevity. The ensemble-average technique is seldom used in experimental studies of turbulent flows, but it is commonly invoked to theoretically substantiate the characterization of unsteady and spatially nonuniform mean flows with average quantities.

Discussion on the relation between time-averaged, spatially averaged, and ensemble-averaged quantities is beyond the scope of this book. In engineering studies, it is commonly assumed that time-averaged, spatially averaged, and ensemble-averaged quantities are equal to one another. Accordingly, if any of the three aforementioned methods of averaging is not valid in a flow, two other methods of averaging may be applied to characterize the flow. If any two of the three aforementioned methods of averaging are not valid in a flow, the remaining third method of averaging may be applied to characterize the flow.

It is worth remembering, however, that, from the basic viewpoint, the problem of averaging in a turbulent flow is not so simple. For instance, some fundamental difficulties in applying the results of the theory of stochastic processes to the description of turbulence are discussed by Burluka (2010, Section 3.4.1).

Although the previous discussion dealt solely with velocity, expressions similar to Equations 3.1 through 3.7 may be applied to evaluate the mean value $\bar{q}$ of any scalar, vector, or tensor quantity q in a turbulent flow, for example, $q = u_i u_j$ or $q = u_i u_j u_k$ or $q = p_{u_i}$ or $q = (u_i - \bar{u}_i)(u_j - \bar{u}_j)$ or $q = (u_i - \bar{u}_i)(u_j - \bar{u}_j)(u_k - \bar{u}_k)$ or $q = (u_i - \bar{u}_i)(p - \bar{p})$.

CONDITIONED QUANTITIES

There are methods of averaging that differ substantially from the aforementioned techniques. Often, researchers are interested not only in the ordinary mean quantities that are determined using equations similar to Equation 3.1 or 3.5 or 3.7, but also in the mean values $\langle q | a = A \rangle$ of a quantity q conditioned on a particular value A (or interval of values) of another quantity a. For instance, in many free flows such as jets and wakes, there is a sharp interface called a viscous superlayer or an entrainment interface that separates a turbulent flow characterized by a large vorticity from the ambient essentially irrotational flow (Townsend, 1976; Kuznetsov and Sabel'nikov, 1990; Pope, 2000). This phenomenon is called intermittency. In such a case, the quantities $\langle q | |\omega| > \omega_0 \rangle$ conditionally averaged, provided that the magnitude $|\omega|$ of the vorticity exceeds a threshold value ω_0, are commonly used to characterize a turbulent flow. The quantities $\langle q | Y > Y_0 \rangle$, where Y_0 is the threshold value of the mass fraction of an admixture, are useful when studying turbulent mixing. The quantities $q_u \equiv \langle q | c = 0 \rangle$ and $q_b \equiv \langle q | c = 1 \rangle$ conditioned on unburned ($c = 0$) and burned ($c = 1$) mixtures, respectively, are widely used in studies of premixed turbulent flames (e.g., see Equations 5.36 and 5.37). Here, c is the combustion progress variable defined by Equation 1.78.

The difference between ordinary averaging and conditional averaging is illustrated in the case of a statistically 1D turbulent flow in Figure 3.2. In this case, the mean value $\bar{q}(x,t)$ of a quantity q may be evaluated using the spatial averaging technique (see Equation 3.5), that is, by integrating $q(x,y,t)$ along an interval AB of length L of a solid straight line parallel to the y-axis (see Figure 3.2a). Certainly, the length L should be much larger than the largest length scales of the turbulent motion. If, in this flow, there is a wrinkled interface (see thick solid curve) that separates two different fluids (e.g., turbulent and nonrotating fluids in a free jet, or clean and polluted fluids in the case of turbulent mixing, or unburned and burned mixtures in the case of a premixed flame), then one can (i) introduce an indicator function $I(x,y,t)$ that is

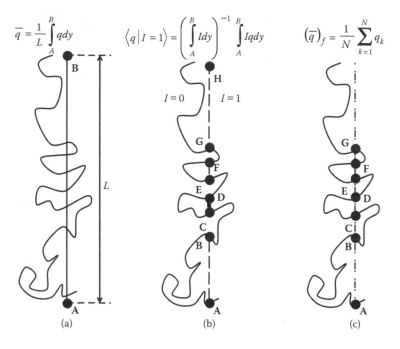

FIGURE 3.2   (a–c) Ordinary and conditional averaging.

equal to zero in one fluid and unity in another fluid ($I = c$ in the case of premixed combustion) and (ii) determine the mean value of the same quantity q conditioned on the latter fluid as follows:

$$\langle q | I = 1 \rangle (x,t) \equiv \frac{\int q(x,y,t) I(x,y,t) dy}{\int I(x,y,t) dy}. \tag{3.8}$$

In fact, this method consists of averaging $q(x,y,t)$ along the parts of the interval AB, in Figure 3.2a, that lie in the mixture characterized by $I = 1$, that is, along the intervals AB, CD, EF, and GH shown with dashed straight lines in Figure 3.2b. In addition to the quantities conditioned on a particular fluid, the quantities conditioned on an interface that separates two different fluids are also widely used, for example, the quantities conditioned on the instantaneous flame front in a turbulent flow of premixed reactants. In Figure 3.2c, such quantities are determined by summing the values of q at intersection points A, B, C, D, E, F, and G that belong to the interface and are characterized by the same x. If the wrinkled curve in Figure 3.2c is associated with an isoscalar surface characterized by $a = A$, then the sum $(q_A + q_B + q_C + q_D + q_E + q_F + q_G)/7$ yields $\langle q | a = A \rangle$ conditioned on the value A of the quantity a. Certainly, more than seven values of q_n should be averaged in order to obtain statistically reliable data.

Because a surface that bounds a domain of conditional averaging moves and changes its shape in a turbulent flow, the spatial and time variations in the magnitude of a conditioned quantity depend not only on the behavior of the quantity within the domain of averaging but also on the surface motion. Accordingly, conditional averaging does not commute with differentiation in a general case. For instance, Libby (1975) has shown that $\nabla \cdot \langle \mathbf{u}|a = A \rangle \neq \langle \nabla \cdot \mathbf{u}|a = A \rangle$, that is, the divergence of a conditioned velocity vector does not vanish in a constant-density flow characterized by $\nabla \cdot \mathbf{u} = 0$ and $\nabla \cdot \bar{\mathbf{u}} = 0$.

REYNOLDS AND FAVRE DECOMPOSITIONS

If the mean flow velocity vector $\bar{\mathbf{u}}(\mathbf{x}, t)$ is determined using either the time-average or spatial-average or ensemble-average technique, then the instantaneous flow velocity vector may be decomposed as follows:

$$\mathbf{u}(\mathbf{x}, t) = \bar{\mathbf{u}}(\mathbf{x}, t) + \mathbf{u}'(\mathbf{x}, t),$$ (3.9)

where $\mathbf{u}' = \{u', v', w'\}$ is called either the velocity pulsation or velocity oscillation or velocity fluctuation vector. In Figure 3.1, the pulsation velocity u' is associated with the difference between the solid curve, which shows the instantaneous velocity $u(t)$, and the dashed straight line, which shows $\bar{u}$. Equation 3.9 is called the Reynolds decomposition.

When modeling variable-density turbulent flows, for example, flames, the mass-weighted (or Favre) decomposition

$$q(\mathbf{x}, t) = \tilde{q}(\mathbf{x}, t) + q''(\mathbf{x}, t)$$ (3.10)

is widely used, where q is an arbitrary scalar or vector or tensor quantity, and

$$\tilde{q} \equiv \frac{\overline{\rho q}}{\bar{\rho}}$$ (3.11)

is the mass-weighted (or Favre) average value of the quantity q. Obviously, $\tilde{q} = \bar{q}$ in the constant-density case.

ROOT-MEAN-SQUARE TURBULENT VELOCITY

The mean quantities introduced in the previous section characterize the mean flow but provide no information about turbulence. A scale of velocity pulsations appears to be a natural characteristic of turbulence. Such a scale cannot be introduced by averaging Equation 3.9, as

$$\overline{u_i'} = \overline{u_i - \bar{u}_i} = \bar{u}_i - \bar{u}_i = 0$$ (3.12)

for any component of the flow velocity vector, because positive and negative pulsations u'_i cancel one another after averaging. Such a cancellation does not occur for u'^2_i, because this quantity is never negative. Accordingly, the magnitude of the flow velocity pulsations in the i-th direction is commonly characterized by

$$\overline{u'^2_i} \equiv \overline{\left(u_i - \bar{u}_i\right)^2}. \tag{3.13}$$

The sum of

$$k = \frac{\overline{u'_k u'_k}}{2} \equiv \frac{1}{2}\left(\overline{u'^2} + \overline{v'^2} + \overline{w'^2}\right) \tag{3.14}$$

is called the turbulent kinetic energy.

In isotropic turbulence, $\overline{u'^2} = \overline{v'^2} = \overline{w'^2}$ and the magnitude of the velocity pulsations may be characterized by a single root-mean-square (rms) turbulent velocity $u' = \sqrt{\overline{u'^2}}$. In Figure 3.1, the rms turbulent velocity is shown with arrows. In anisotropic turbulence, the rms turbulent velocity may be introduced as follows:

$$u' = \sqrt{\frac{2}{3}\bar{k}} = \sqrt{\frac{\overline{u'^2} + \overline{v'^2} + \overline{w'^2}}{3}}. \tag{3.15}$$

When discussing the physics of turbulent combustion, the magnitude of the velocity pulsations is commonly characterized by a single rms velocity u'. However, it is worth remembering that this single quantity is not sufficient to completely characterize the magnitudes of the velocity pulsations in different directions in an anisotropic turbulence.

By comparing Equations 3.9 and 3.15, one can see that the same symbol u' designates two quantities: (i) the difference in $u(\mathbf{x},t)$ and $\bar{u}(\mathbf{x},t)$ in the Reynolds decomposition of the x-component of the instantaneous velocity vector and (ii) the rms turbulent velocity. These two quantities are different, for example, $u'(\mathbf{x},t)$, defined by Equation 3.9, may be negative, while the rms turbulent velocity is positive by definition. To distinguish these two quantities, henceforth, the symbol $u'(\mathbf{x},t)$ will be used when discussing the Reynolds decomposition, while the rms turbulence velocity will simply be denoted as u', without writing the spatial coordinates and time. Moreover, the symbol u' written in an expression that is averaged, for example, $\overline{u'_i u'_j}$ or $\overline{u'_i u'_j u'_k}$ or $\overline{u'_i p'}$, will always be associated with the Reynolds decomposition, even if the spatial coordinates and time are skipped.

PROBABILITY DENSITY FUNCTION

Although the rms turbulent velocity is an important characteristic of the velocity pulsations in a single point (if time-average and ensemble-average techniques are used) or at a single instant (if spatial-average and ensemble-average techniques are used), there are many other average turbulent quantities that provide additional information about turbulence in a single point (for the sake of brevity, we will use the term "single point" when discussing issues relevant to the turbulence characteristics determined either in a single point or at a single instant, henceforth). For instance, knowledge of $\overline{u'^2}$, $\overline{v'^2}$, and $\overline{w'^2}$ is not sufficient for determining the third $\overline{u_i'u_j'u_k'}$ and fourth $\overline{u_i'u_j'u_k'u_l'}$ moments in a general case. The knowledge of all moments, starting from the first ($\overline{u_i}$) to the n-th, is not sufficient for determining the $(n + 1)$-th moments. Thus, even for very large n, the n-th moments contain information that is not contained by all the lower moments together. Accordingly, to fully characterize turbulent pulsations in a single point, one has to know an infinite number of moments. Obviously, the moment approach is unpractical for this purpose.

Probability density function (PDF) is another technique that offers an opportunity to more fully characterize turbulence in a single point. Let us briefly introduce this technique by considering a single component u of a flow velocity vector, for simplicity.

If, for instance, the velocity u measured at a point during a sufficiently long time interval τ (see Figure 3.1) varies between u_{min} and u_{max}, then we can divide the interval $u_{min} \leq u \leq u_{max}$ into N subintervals $\Delta u_1, \Delta u_2, \ldots, \Delta u_N$ such that $u_1 = u_{min}$, $u_{N+1} = u_{max}$, and

$$u_l = u_{min} + \sum_{m=1}^{l-1} \Delta u_m. \tag{3.16}$$

Then, we can measure the total duration Δt_l of the time intervals during which $u_l \leq u \leq u_{l+1}$. The probability of finding $u_l \leq u \leq u_{l+1}$ is equal to

$$\Delta P_l \equiv P\left(u_l < u < u_{l+1}\right) = \frac{\Delta t_l}{\tau}. \tag{3.17}$$

Obviously, $\Delta P_l \geq 0$ and

$$\sum_{l=1}^{N} \Delta P_l = \frac{1}{\tau} \sum_{l=1}^{N} \Delta t_l = 1. \tag{3.18}$$

If $N \to \infty$, then Δu_l, Δt_l, and ΔP_l tend to zero, while the ratio of $\Delta P_l/\Delta u_l$ remains finite. Accordingly, the PDF $P(u)$ is defined as follows:

$$P\left(u\right) = \frac{dP}{du}. \tag{3.19}$$

Because an extension of the interval Δu_l in that velocity is measured cannot reduce the probability ΔP_l of recording $u_l \leq u \leq u_{l+1}$, the PDF $P(u)$ given by Equation 3.19 may not be negative. Equations 3.18 and 3.19 show that

$$\int_{-\infty}^{\infty} Pdu = \int dP = 1. \tag{3.20}$$

For any continuous function $f(u)$, the mean value

$$\overline{f(u)} = \int_{-\infty}^{\infty} fP du \qquad (3.21)$$

can easily be determined if the PDF $P(u)$ is known. Application of Equation 3.21 to $f = u$ and $f = (u - \overline{u})^2$ yields

$$\overline{u} = \int_{-\infty}^{\infty} uP du \qquad (3.22)$$

and

$$\overline{u'^2} = \overline{(u - \overline{u})^2} = \int_{-\infty}^{\infty} (u - \overline{u})^2 P du. \qquad (3.23)$$

The statistics of different turbulent quantities in different turbulent flows is described by different PDFs, but the Gaussian (or normal) PDF,

$$P(u) = \frac{1}{u'\sqrt{2\pi}} \exp\left[-\frac{(u - \overline{u})^2}{2u'^2} \right], \qquad (3.24)$$

is most widely used to characterize the velocity fluctuations. Here, u' is the rms turbulent velocity. A typical Gaussian PDF is shown in Figure 3.3.

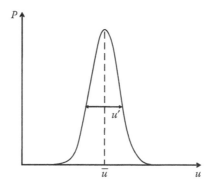

FIGURE 3.3 A typical Gaussian PDF for a flow velocity vector component.

HOMOGENEOUS ISOTROPIC TURBULENCE

A typical natural or industrial turbulent flow is a complicated motion. All the components of the mean velocity vector depend on time and spatial coordinates and are not equal to one another. Similarly, all turbulence characteristics, for example, the Reynolds stress tensor $\overline{u_i' u_j'}$, also depend on time and spatial distance and are different for different directions (different indexes i and j). Obviously, a physical understanding of such a flow is extremely difficult, and a simplified model is required to clarify the basic physics of turbulence.

For this purpose, the concept of statistically stationary, homogeneous, and isotropic turbulence is very useful. Statistical stationarity means that all the average characteristics of turbulence, for example, $\overline{u_i' u_j'}$ or $\overline{u_i' u_j' u_k'}$ or $\overline{u_i' p'}$, are invariants under a shift in time, (i.e., do not depend on time), while the instantaneous velocity pulsations $u_i'(\mathbf{x},t)$ depend on t. Statistical homogeneity means that all the average characteristics of turbulence are invariants under translations (i.e., do not depend on spatial coordinates), while $u_i'(\mathbf{x},t)$ depend on $\mathbf{x}$. Statistical isotropy means that all the average vector and tensor characteristics of turbulence are invariants under rotations and reflections of the coordinate framework. In particular, $\overline{u'^2} = \overline{v'^2} = \overline{w'^2}$, while $u'(\mathbf{x},t) \neq v'(\mathbf{x},t) \neq w'(\mathbf{x},t)$. In statistically isotropic turbulence, the Reynolds stress tensor $\overline{u_i' u_j'}$ is simply equal to $\overline{u'^2}\delta_{ij}$, where δ_{ij} is the Kronecker delta.

The contemporary turbulence theory and the theory of turbulent combustion are based on an analysis of statistically stationary, homogeneous, and isotropic turbulence. This simplification is widely invoked when discussing combustion physics or when developing engineering models of turbulent reacting flows. When applying such a model to the simulation of a laboratory, industrial, or natural flow, it is commonly assumed (often implicitly) that turbulence is isotropic and adjusts itself in an infinitely fast manner to the mean flow, that is, the temporal and spatial variations in the averaged turbulence characteristics are hypothesized to be solely controlled by the temporal and spatial variations in the mean flow. Certainly, the validity of such an approach is limited, for example, it may be oversimplified when studying thermoacoustic instabilities in gas turbine combustors. Nevertheless, the concept of statistically stationary, homogeneous, and isotropic turbulence is very useful for understanding the basic physics of turbulent motion.

CORRELATIONS, LENGTH SCALES, AND TIMESCALES

The aforementioned turbulence characteristics (rms velocity and PDF) are relevant to the flow velocity oscillations in a single point. In addition to such oscillations, another important feature of turbulence is that these quantities, measured in two different points at the same instant (or in the same point at two different instants), correlate with one another, provided that the distance between the two points (or the time interval between the two instants) is sufficiently small. From the physical viewpoint, this feature reflects that a turbulent flow is an ensemble of eddies of different velocity and length scales. Because the velocities measured in two different points on the same eddy depend on one another, the velocities measured in two different points in a turbulent flow statistically correlate with one another if the distance between the points is sufficiently small as compared with the length scale of the largest eddies.

Mathematically, the discussed feature is characterized by introducing correlation tensors, autocorrelation functions, and length scales and timescales. The correlation tensor is defined as follows:

$$R_{ij}(\mathbf{x},\mathbf{r},t) \equiv \overline{u_i'(\mathbf{x},t)u_j'(\mathbf{x}+\mathbf{r},t)}. \tag{3.25}$$

In statistically stationary, homogeneous, and isotropic turbulence, the correlation tensor $R_{ij}(\mathbf{x},\mathbf{r},t)$ may depend solely on the distance $r = |\mathbf{r}|$. It can be shown (Pope, 2000) that

$$R_{ij}(r) = u'^2 \left\{ g(r)\delta_{ij} + \left[f(r) - g(r) \right] \frac{r_i r_j}{r^2} \right\} \tag{3.26}$$

in this case. The functions $f(r)$ and $g(r)$ are called the longitudinal and transverse autocorrelation functions, respectively.

To clarify the physical meaning of these two functions, let us consider Figure 3.4, where $\mathbf{r} = r\mathbf{e}_x$ and $\mathbf{e}_x = \{1,0,0\}$ is the unity vector directed along the x-axis. For the x-component of the flow velocity vector, $R_{11}(r) = \overline{u_1'u_1'} = u'^2 f(r)$, while $R_{22}(r) = \overline{v_1'v_1'} = u'^2 g(r)$ for the y-component of the velocity vector. Thus, the longitudinal autocorrelation function $f(r)$ is associated with the normalized correlation between the same components of the flow velocity vector, measured in two different points located on a line that is parallel to the aforementioned component. Similarly, the transverse autocorrelation function $g(r)$ is associated with the normalized correlation between the same components of the flow velocity vector, measured in two different points located on a line that is perpendicular to the aforementioned component. It is also worth noting that, in isotropic turbulence, $R_{ij} = 0$ if $i \neq j$.

It can be shown (Pope, 2000) that

$$g(r) = f(r) + \frac{r}{2}\frac{\partial f}{\partial r}. \tag{3.27}$$

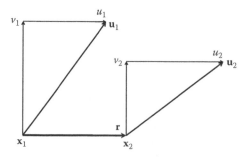

FIGURE 3.4 Velocity directions for a double-velocity correlation.

Here and henceforth, a partial derivative is used instead of an ordinary derivative in order to stress that Equation 3.27 is valid even in the case of statistically homogeneous and isotropic, but unstationary turbulence, that is, if $f = f(r,t)$ and $g = g(r,t)$. Equations 3.26 and 3.27 show that the longitudinal autocorrelation function completely determines the correlation tensor $R_{ij}(\mathbf{x},\mathbf{r},t)$ in statistically homogeneous and isotropic turbulence.

Obviously, the autocorrelation functions tend to unity as $r \to 0$. Accordingly, $R_{11}(r \to 0) \to u'^2$ and $R_{22}(r \to 0) \to u'^2$. Because the velocities measured in two points are statistically independent of one another if the distance between the points is much larger than the length scale of the largest turbulent eddies, $f(r \to \infty) \to 0$ and $g(r \to \infty) \to 0$. Because the correlation between the velocities measured in two points decreases as the distance r between the points increases, the autocorrelation functions reach their maximum values at $r = 0$, that is, $f(r) \leq 1$ and $g(r) \leq 1$. Typical dependencies of $f(r)$ and $g(r)$ are shown in Figure 3.5.

To characterize a mean distance at which the motions in two different points correlate, longitudinal and transverse integral length scales are introduced as follows:

$$L_{\parallel} = \int_0^\infty f(r)\,dr \qquad (3.28)$$

and

$$L_{\perp} = \int_0^\infty g(r)\,dr, \qquad (3.29)$$

respectively. These length scales are simply equal to the areas under the $f(r)$ and $g(r)$ curves, respectively. As shown in Figure 3.5, the two length scales are of the order of but smaller than the length scale of the largest turbulent eddies, that is, the largest

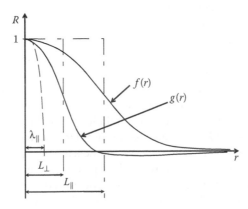

FIGURE 3.5 The Eulerian longitudinal, $f(r)$, and transverse, $g(r)$, autocorrelation functions versus the distance between the two points.

distance r at which both $f(r)$ and $g(r)$ are finite. Using Equation 3.27, one can easily show that

$$L_\perp = \int_0^\infty g\,dr = \frac{1}{2}\int_0^\infty \left[f + \frac{\partial}{\partial r}(rf) \right]dr = \frac{1}{2}\int_0^\infty f\,dr = \frac{L_\parallel}{2} \qquad (3.30)$$

in statistically homogeneous and isotropic turbulence.

The aforementioned autocorrelation functions allow us to introduce two more length scales:

$$\lambda_\parallel = \left(-\frac{1}{2}\frac{\partial^2 f}{\partial r^2} \right)_{r\to 0}^{-1/2} \qquad (3.31)$$

and

$$\lambda_\perp = \left(-\frac{1}{2}\frac{\partial^2 g}{\partial r^2} \right)_{r\to 0}^{-1/2}. \qquad (3.32)$$

The two scales are called the longitudinal and transverse Taylor length scales, respectively. If the behavior of the autocorrelation functions $f(r)$ and $g(r)$ at $r \to 0$ is approximated by parabolas, then these parabolas are described by the following equations: $1 - (r/\lambda_\parallel)^2$ and $1 - (r/\lambda_\perp)^2$, respectively. Therefore, the Taylor length scales are equal to the distance r at which these parabolas intersect the axis. Using Equation 3.27, it can be shown that $\lambda_\perp = \lambda_\parallel / \sqrt{2}$ in the discussed case. As illustrated in Figure 3.5, the Taylor length scales are markedly smaller than the corresponding integral length scales, that is, the Taylor length scales characterize small turbulent eddies (but not the smallest ones, as will be discussed later).

In the turbulent combustion literature, the difference between the longitudinal and transverse length scales is seldom emphasized and a single integral length scale L and a single Taylor length scale λ are commonly used to characterize turbulence. The point is that (i) the turbulent burning rate depends weakly on L than on u', as will be discussed in the next chapter, and (ii) the capabilities of the contemporary models of premixed turbulent combustion for predicting the variations in an integral turbulence length scale within a flame have not yet been demonstrated. Therefore, a factor of the order of 2 may be disregarded when specifying length scales. Following the common practice and for the sake of simplicity, we will use a single length scale L (or a single length scale λ) when discussing turbulent combustion.

The aforementioned correlation tensor, autocorrelation functions, and length scales characterize spatial correlations. Similar quantities may be introduced to characterize the correlation between the velocities measured at two different instants. Such time correlations may be introduced in two different ways. First, one can consider the correlation between the velocities measured in the same point, that is, in the Eulerian framework, at two different instants. Then,

$$T_{E,ij}(\mathbf{x},t,\vartheta) \equiv \overline{u_i'(\mathbf{x},t)u_j'(\mathbf{x},t+\vartheta)} = u'^2 f_E(\vartheta)\delta_{ij} \tag{3.33}$$

in statistically stationary, homogeneous, isotropic turbulence and

$$\tau_E = \int_0^\infty f_E(\vartheta)d\vartheta \tag{3.34}$$

is the Eulerian timescale.

Second, one can consider the correlation between the velocities measured for the same fluid particles, that is, in the Lagrangian framework, at two different instants. Then,

$$T_{L,ij}(\mathbf{x},t,\vartheta) \equiv \overline{u_i'(\mathbf{x},t)u_j'(\mathbf{x}_1(\mathbf{x},\vartheta),t+\vartheta)} = u'^2 f_L(\vartheta)\delta_{ij} \tag{3.35}$$

in statistically stationary, homogeneous, isotropic turbulence and

$$\tau_L = \int_0^\infty f_L(\vartheta)d\vartheta \tag{3.36}$$

is the Lagrangian timescale.

Based on the rms turbulent velocity and the aforementioned timescales, one can introduce the length scales $l_E = u'\tau_E$ and $L_L = u'\tau_L$. In a typical turbulent flow, the integral length scales $L_\parallel$ and $L_\perp$ and the Lagrangian length scale L_L are of the same order. These three scales characterize large turbulent eddies.

The length scale l_E is typically smaller. Indeed, in many cases, for example, flows in channels, jets, and wakes, the mean flow velocity $\bar{u}$ is much larger than the rms turbulent velocity u'. Here, the x-axis is directed along the mean flow line, for example, along the axis of a channel. Accordingly, $u'(x,y,z,t+\vartheta) \approx u'(x-r,y,z,t)$, provided that $r = \bar{u}\vartheta$ and the distance r and time ϑ are sufficiently small in order for $\bar{u}(x,y,z,t) \approx \bar{u}(x,y,z,t+\vartheta) \approx \bar{u}(x-r,y,z,t)$. The latter approximate equalities hold for distances r and times ϑ associated with notable values of the autocorrelation functions $f(r)$ and $f_E(\vartheta)$ because the decomposition of a flow into mean and turbulent motions implies that the length scales and timescales of turbulent eddies are smaller than the length scales and timescales of the mean flow variations. Therefore,

$$L_\parallel = \int_0^\infty f(r)dr = \int_0^\infty \frac{\overline{u'(0)u'(r)}}{u'^2}dr \approx \bar{u}\int_0^\infty \frac{\overline{u'(0)u'(\vartheta)}}{u'^2}d\vartheta = \bar{u}\int_0^\infty f_E(\vartheta)d\vartheta = \bar{u}\tau_E. \tag{3.37}$$

Accordingly, $l_E = u'\tau_E \approx L_\parallel(u'/\bar{u}) \ll L_\parallel$. The length scale l_E is seldom used in the literature due to the lack of clear physical meaning.

Taylor's Hypothesis

For flows characterized by a low magnitude of velocity pulsations as compared with the magnitude of the mean flow velocity, Taylor has hypothesized that the advection of the turbulent field is controlled by the mean flow, that is, $q(x, y, z, t + \vartheta) \approx q(x - \bar{u}\vartheta, y - \bar{v}\vartheta, z - \bar{w}\vartheta, t)$.

According to Taylor's hypothesis, the evolution of turbulence characteristics along the mean flow direction is the manifestation of the development of a turbulent field convected by the mean flow. For instance, when a flow passes a grid, turbulence is generated by the grid. Behind the grid, the turbulence decays due to the transformation of the turbulent energy into heat by viscous forces. The turbulence decay results in a decrease in the rms turbulent velocity by $\Delta u'$ during the time interval Δt. Because a fluid element that passes the grid at instant t will be convected by the mean flow at distance $\Delta x = \bar{u}\Delta t$ during the time interval Δt, the aforementioned decrease in the rms turbulent velocity will be documented at the distance Δx from the grid. All the characteristics of such a flow may be statistically stationary, despite the decay of the turbulence. If the mean flow velocity $\bar{u}$ is constant, the considered statistically stationary process becomes unsteady in a coordinate framework where the mean flow velocity vanishes, but the grid moves at speed $\bar{u}$. Irrespective of whether a laminar flow passes through a fastened grid with a mean velocity $\bar{u}$ or a similar grid moves with the same velocity $\bar{u}$ through a motionless gas, the decay of a grid-generated turbulence is the same transient phenomenon in the two cases even if it looks like a statistically stationary process in the former one.

Thus, according to Taylor's hypothesis, due to the advection of a turbulent field by the mean flow, such an inherently transient process as turbulence development (decay in the considered example) may manifest itself as statistically stationary spatial variations in the turbulence characteristics along the mean flow line. This well-known feature of turbulent flows is of paramount importance for understanding the phenomenology of turbulent combustion. Unfortunately, although Taylor's hypothesis is a classical contribution well recognized by the turbulence community, this hypothesis is not discussed in many combustion papers, in which the basic difference between a statistically stationary turbulent flame and a fully developed turbulent flame is neglected. Similar to the decaying turbulence behind a grid, a developing (or transient) turbulent flame may be statistically stationary!

Energy-Spectrum Function

As we already discussed, a turbulent flow consists of eddies of substantially different sizes. Some scales of turbulent motion are characterized by the length scales and timescales that were introduced in the section "Correlations, Length Scales and Timescales." To describe the distribution of the energy of a turbulent flow over all turbulent eddies, the following Fourier transform

$$\Psi_{ij}(\boldsymbol{\kappa}, \mathbf{x}, t) = \frac{1}{(2\pi)^3} \iiint R_{ij}(\mathbf{x}, \mathbf{r}, t) e^{-i\boldsymbol{\kappa}\cdot\mathbf{r}} d\mathbf{r} \qquad (3.38)$$

of the correlation tensor $R_{ij}(\mathbf{x},\mathbf{r},t)$ determined by Equation 3.25, is commonly introduced. By virtue of the general property of a Fourier transform

$$R_{ij}\left(\mathbf{x},\mathbf{r},t\right) = \iiint \Psi_{ij}\left(\mathbf{x},\mathbf{\kappa},t\right)e^{i\mathbf{\kappa}\cdot\mathbf{r}}d\mathbf{\kappa} \tag{3.39}$$

and, therefore,

$$\overline{u_i'u_j'}\left(\mathbf{x},t\right) = R_{ij}\left(\mathbf{x},0,t\right) = \iiint \Psi_{ij}\left(\mathbf{x},\mathbf{\kappa},t\right)d\mathbf{\kappa}. \tag{3.40}$$

Subsequently, the turbulent kinetic energy is equal to

$$\bar{k}\left(\mathbf{x},t\right) = \frac{1}{2}\overline{u_j'u_j'}\left(\mathbf{x},t\right) = \frac{1}{2}\iiint \Psi_{jj}\left(\mathbf{x},\mathbf{\kappa},t\right)d\mathbf{\kappa} = \int_0^\infty E\left(\kappa,\mathbf{x},t\right)d\kappa, \tag{3.41}$$

where $\kappa = |\mathbf{\kappa}|$,

$$E\left(\kappa,\mathbf{x},t\right) = \frac{1}{2}\oint \Psi_{jj}\left(\kappa,\mathbf{x},t\right)dS\left(\kappa\right) = \frac{1}{2}\iiint \Psi_{jj}\left(\mathbf{x},\mathbf{\kappa},t\right)\delta\left(|\mathbf{\kappa}|-\kappa\right)d\mathbf{\kappa} \tag{3.42}$$

is the energy-spectrum function, $S(\kappa)$ is the sphere with radius κ and centered at the origin in the wavevector space, and $\delta(|\mathbf{\kappa}| - \kappa)$ is the Dirac delta function. The quantity $E(\kappa)d\kappa$ is the contribution to the turbulent kinetic energy from all wavevectors $\mathbf{\kappa}$ such that $\kappa \leq |\mathbf{\kappa}| \leq \kappa + d\kappa$. Accordingly, the energy-spectrum function has the dimension cubic meter per second squared (m³/s²).

In a typical turbulent flow, the energy-spectrum function $E(\kappa)$ is positive in a wide range of wave numbers, such that the two values of κ that bound this range differ by several orders of magnitude. The lowest bounding value of κ is of the order of $2\pi/L$, because the length scale of the largest turbulent eddies is of the order of L. The energy-spectrum function $E(\kappa)$ contains information about the distribution of the turbulent energy over eddies of various length scales $l \propto \kappa^{-1}$, but, contrary to the velocity-spectrum tensor $\Psi_{ij}(\mathbf{x},\mathbf{\kappa},t)$, $E(\kappa)$ does not keep information about the direction of the velocity pulsations or information about the direction of the Fourier modes.

SUMMARY

Turbulent fluctuations in a single point are characterized either by an infinite set of moments (the second moments $\overline{u_i'u_j'}$, the third moments $\overline{u_i'u_j'u_k'}$, the fourth moments $\overline{u_i'u_j'u_k'u_l'}$, etc.) or by a PDF. Within the framework of the moment approach, the rms turbulent velocity u' (or turbulent kinetic energy $\bar{k} = 1.5u'^2$) is the most widely used single-point turbulence characteristic.

The two-point behavior of turbulence is characterized by the spatial-correlation tensor $R_{ij}(\mathbf{x},\mathbf{r},t)$ defined by Equation 3.25 and the Eulerian $T_{E,ij}(\mathbf{x},t,\tau)$ and Lagrangian $T_{L,ij}(\mathbf{x},t,\tau)$ time-correlation tensors defined by Equations 3.33 and 3.35, respectively. For statistically stationary, homogeneous, isotropic turbulence, the spatial-correlation tensor $R_{ij}(\mathbf{x},\mathbf{r},t)$ is fully determined by the longitudinal autocorrelation function $f(r)$. Similarly, the time-correlation tensors $T_{E,ij}(\mathbf{x},t,\tau)$ and $T_{L,ij}(\mathbf{x},t,\tau)$ are fully determined by the Eulerian and Lagrangian autocorrelation functions, respectively. Based on the aforementioned autocorrelation functions and the transverse autocorrelation function $g(r)$, the following scales are commonly introduced:

- The longitudinal and transverse integral length scales $L_{\parallel}$ and $L_{\perp}$, respectively, which characterize large turbulent eddies
- The longitudinal and transverse Taylor length scales $\lambda_{\parallel}$ and $\lambda_{\perp}$, respectively, which characterize small turbulent eddies
- The Eulerian timescale τ_E, which is approximately equal to $L_{\parallel}/\bar{u}$ in many flows
- The Lagrangian timescale τ_L and length $L_L = u'\tau_L$ scale, which characterize large turbulent eddies

The distribution of the turbulent energy over eddies of various length scales is characterized by the energy-spectrum function $E(\kappa)$.

THEORY OF HOMOGENEOUS ISOTROPIC TURBULENCE

Viscous Dissipation

A very important feature of a turbulent flow is that the viscous forces play a substantial role only in small-scale spatial regions characterized by large velocity gradients. To show this, let us compare the magnitudes of the convection (the second term on the LHS) and viscous (the first term on the RHS) terms in the Navier–Stokes equation (Equation 1.51). The simplest estimate yields

$$\frac{\partial}{\partial x_k}\left(\rho u_k u_i\right) \propto \frac{\rho U^2}{l},$$

$$\frac{\partial \tau_{ik}}{\partial x_k} \propto \frac{\mu U}{l^2}.$$

(3.43)

Accordingly, the ratio of the two terms scales is Ul/v, where U and l are the velocity and length scales, respectively, of the local velocity variations. If an eddy is sufficiently small so that $Ul/v \ll 1$, then the viscous term dominates in the Navier–Stokes equation by smoothing out the flow nonuniformities, that is, the kinetic energy of the eddy motion is rapidly dissipated by the viscous forces (the energy is transformed to heat). Accordingly, such very small eddies cannot exist in a turbulent flow and the velocity-spectrum function $E(\kappa)$ vanishes at sufficiently large wave numbers κ. On the contrary, if an eddy is sufficiently large such that $Ul/v \gg 1$, then the eddy motion is weakly affected by the viscous forces, because the corresponding term in the Navier–Stokes equation is much smaller than the convection term.

Thus, when considering the velocity-spectrum function $E(\kappa)$, we may assume that the viscous dissipation of the turbulent energy plays a substantial role only in the interval $(\kappa, \kappa + \Delta\kappa)$ of wave numbers such that

$$\frac{\sqrt{E(\kappa)\Delta\kappa}}{\kappa v} = O(1). \tag{3.44}$$

Here, the numerator of the fraction on the LHS estimates the velocity scale associated with the eddies of a length scale $l \approx \kappa^{-1}$. The behavior of substantially larger eddies characterized by lower wave numbers κ, is weakly affected by the viscous dissipation, because the viscous term in the Navier–Stokes equation is much less than the convection term associated with the large-scale eddies. Substantially smaller eddies, characterized by higher wave numbers κ, do not exist in the flow, as they are rapidly dissipated by the viscous forces. The range of wave numbers associated with substantial dissipation of the turbulent kinetic energy by the viscous forces is called the dissipation (or viscous) range of turbulence spectrum.

The dissipation rate ε of the turbulent kinetic energy plays an important role in the turbulence theory. As previously discussed, the dissipation rate is controlled by processes localized to the small-scale eddies. It is defined as follows (see also the sections "Kolmogorov's Theory of Turbulence" and "The $k - \varepsilon$ Model"):

$$\varepsilon \equiv 2v s_{jk} s_{jk}, \tag{3.45}$$

where

$$s_{jk} \equiv \frac{1}{2}\left(\frac{\partial u'_j}{\partial x_k} + \frac{\partial u'_k}{\partial x_j}\right) \tag{3.46}$$

is the fluctuating rate-of-strain tensor. The dissipation rate has the dimension cubic meter per second squared (m^2/s^3). It can be shown (Pope, 2000) that the mean dissipation rate

$$\bar{\varepsilon} = 2v\int_0^\infty \kappa^2 E(\kappa)d\kappa. \tag{3.47}$$

A comparison of Equations 3.41 and 3.47 indicates that the mean turbulent kinetic energy is controlled by significantly larger eddies as compared with eddies that control the mean dissipation rate. Indeed, due to the factor κ^2 in the integral on the RHS of Equation 3.47, the term $\kappa^2 E(\kappa)$ reaches a peak value at a substantially larger wave number as compared with the term $E(\kappa)$ integrated on the RHS of Equation 3.41.

Energy Cascade and Turbulent Stretching

If turbulence is statistically stationary, then small-scale eddies should receive energy at a rate that is equal to $\bar{\varepsilon}$ in order to compensate for the viscous dissipation. The only source of this energy is larger-scale turbulent eddies, which, in turn, withdraw energy from even larger-scale eddies. For the largest turbulent eddies, the only source of energy is the mean flow (see also the section "The $k - \varepsilon$ Model").

Based on this simple physical reasoning, Richardson (1926) proposed a hypothesis about turbulent cascade, which became the basis of the well-known Kolmogorov's theory of turbulence. According to Richardson (1926) and Kolmogorov (1941a,b, 1942), the large-scale turbulent eddies withdraw energy from the mean flow and transfer it to eddies of a smaller scale. The latter eddies further transfer the energy to eddies of even smaller scales, etc. Finally, the energy is dissipated, that is, transformed into thermal energy by the molecular viscosity, within the smallest eddies, the length of which is of the order of the Kolmogorov length scale η, which will be defined later (see Equations 3.56 and 3.64). Thus, the rate $\bar{\varepsilon}$ of the viscous dissipation of the turbulent energy is equal to the rate of the energy withdrawal from the mean flow; therefore, $\bar{\varepsilon}$ is a basic characteristic of turbulence at all scales.

The physical mechanism of the transfer of energy from larger to smaller turbulent eddies is commonly associated with turbulent stretching (Bradshaw, 1971). To illustrate this physical mechanism, let us consider the evolution of two vortexes, $\boldsymbol{\omega}_1 = \{0,0,\omega_1\}$ and $\boldsymbol{\omega}_2 = \{0,-\omega_2,0\}$, perpendicular to each other (see Figure 3.6) in the following flow:

$$\mathbf{u} = \{0, -\dot{s}y, \dot{s}z\}, \tag{3.48}$$

which is the simplest model of stretching and satisfies the continuity equation $\nabla \mathbf{u} = 0$. The quantity $\dot{s}$ is called the stretch (or strain) rate.

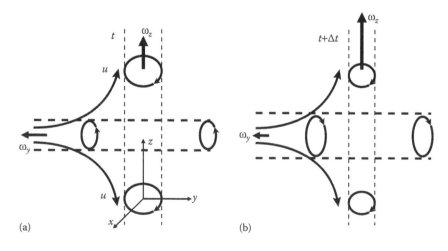

(a)

(b)

FIGURE 3.6 (a,b) Energy transfer via vortex stretching.

Let us assume for simplicity that the vortexes do not interact with one another. If the radii of the vortexes are sufficiently large, then the evolution of the vortexes is weakly affected by the viscous forces (as discussed in the section "Viscous Dissipation"), and the vorticity Equation 1.57 reads

$$\frac{D\boldsymbol{\omega}}{Dt} = (\boldsymbol{\omega} \cdot \nabla)\mathbf{u}, \tag{3.49}$$

where

$$\frac{D}{Dt} = \frac{\partial}{\partial t} + \mathbf{u} \cdot \nabla \tag{3.50}$$

is the time derivative taken along a flow line. For vortexes $\boldsymbol{\omega}_1 = \{0,0,\omega_1\}$ and $\boldsymbol{\omega}_2 = \{0,-\omega_2,0\}$, Equation 3.49 reads

$$\frac{D\omega_1}{Dt} = \dot{s}\omega_1 \tag{3.51}$$

and

$$\frac{D\omega_2}{Dt} = -\dot{s}\omega_2, \tag{3.52}$$

respectively. Thus, the flow field modeled by Equation 3.48 increases the vorticity in vortex 1 but decreases the vorticity in vortex 2. Because an increase (decrease) in the vorticity is associated with a decrease (increase) in the vortex radius, the flow field stretches vortex 1 but compresses vortex 2 (see Figure 3.6). When the vortex radius decreases, the energy of the rotational motion is transferred to smaller scales under the influence of stretching.

Although the same flow field may stretch one vortex and compress another, as previously shown, the former process statistically dominates in a turbulent flow, that is, in the absence of the viscous forces, the magnitude of the vorticity increases with time. For instance, in the aforementioned example, the magnitude of the vorticity vector $\boldsymbol{\omega} = \boldsymbol{\omega}_1 + \boldsymbol{\omega}_2$ increases with time, as

$$|\omega| = \sqrt{\omega_1^2 + \omega_2^2} = \omega_0\sqrt{\exp(2\dot{s}t) + \exp(-2\dot{s}t)} = \omega_0\sqrt{2\cosh(2\dot{s}t)} \geq \omega_0 \tag{3.53}$$

if $\omega_1(0) = \omega_2(0) = \omega_0$.

It is worth stressing that stretching of the turbulent eddies is an essentially 3D phenomenon, and the key role played by vortex stretching in the turbulent cascade highlights that turbulence is also an essentially 3D phenomenon. Therefore, 1D or 2D models of turbulence substantially oversimplify the problem, even if such models are able to yield some valuable knowledge.

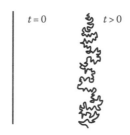

FIGURE 3.7   Growth of the area of a material surface due to turbulent stretching.

Stretching of the turbulent eddies not only plays an important role in the energy transfer from larger to smaller eddies but also controls the behavior of the material surfaces in a turbulent flow. For example, if at $t = 0$, an admixture is uniformly seeded in a half space ($x < 0$) and the molecular diffusivity of the admixture is negligible, then a surface that separates the flow with the admixture from the clean flow will be wrinkled by turbulent eddies of various sizes (see Figure 3.7) and the increase in the surface area will be controlled by the mean stretch rate $\bar{s}$.

Batchelor (1952) has shown that the area of a material surface grows exponentially with time,

$$A = A_0 \exp\left(\bar{s}t\right), \tag{3.54}$$

in statistically stationary, homogeneous, isotropic turbulence. This feature of turbulence is of paramount importance for understanding the physics of turbulent combustion.

KOLMOGOROV'S THEORY OF TURBULENCE

As previously discussed, the energy withdrawn by the large-scale turbulent eddies from the mean flow is transferred to smaller and smaller eddies and is then dissipated due to the viscosity in the smallest-scale eddies. Moreover, the viscous dissipation of the turbulent energy is only localized to the smallest-scale eddies, whereas the larger-scale eddies are not straightforwardly affected by the viscous forces. On the basis of this physical scenario, Kolmogorov (1941a) formulated two similarity hypotheses, which became the cornerstone of the theory of turbulence.

According to the first hypothesis of similarity, the statistics of the smallest-scale turbulent eddies is universal and is uniquely determined by the viscosity ν and mean dissipation rate $\bar{\varepsilon}$.

This hypothesis allows us to easily estimate the key characteristics of the smallest-scale turbulent eddies. For dimensional reasons, we obtain

$$u'_\eta = \left(\nu\bar{\varepsilon}\right)^{1/4}, \tag{3.55}$$

$$\eta = v^{3/4}\bar{\varepsilon}^{-1/4}, \tag{3.56}$$

and

$$\tau_\eta = \frac{\eta}{u'_\eta} = v^{1/2}\bar{\varepsilon}^{-1/2}. \tag{3.57}$$

These three scales are called the Kolmogorov velocity scale, length scale, and timescale, respectively. Equations 3.55 and 3.57 may be rewritten in the following forms:

$$u'_\eta = (\bar{\varepsilon}\eta)^{1/3} \tag{3.58}$$

and

$$\tau_\eta = \eta^{2/3}\bar{\varepsilon}^{-1/3}. \tag{3.59}$$

Substitution of $U = u'_\eta$ and $l = \eta$ into the estimate given by Equation 3.43 shows that the convection and viscous terms are of the same order if

$$\mathrm{Re}_\eta \equiv \frac{u'_\eta \eta}{v} = 1, \tag{3.60}$$

in line with the physical reasoning discussed in the section "Viscous Dissipation." The ratio Re_η is called the turbulent Reynolds number based on the Kolmogorov scales.

Equation 3.57 may be rewritten as follows:

$$\bar{\varepsilon} = \frac{v}{\tau_\eta^2} = v\left(\frac{u'_\eta}{\eta}\right)^2. \tag{3.61}$$

A comparison of this equation with the definition of the dissipation rate given by Equations 3.45 through 3.46 shows the velocity gradient associated with the smallest turbulent eddies scales as $u'_\eta/\eta = \tau_\eta^{-1}$.

Because all of the energy dissipated in the smallest eddies was withdrawn early by the largest turbulent eddies from the mean flow, the dissipation rate should be controlled by the characteristics of the largest eddies, while the viscosity controls the scales of the eddies where the energy is dissipated. For dimensional reasons,

$$\bar{\varepsilon} \propto \frac{u'^3}{L}. \tag{3.62}$$

Then, Equations 3.55 through 3.57, 3.59, and 3.62 yield

$$u'_\eta \propto u' \, \text{Re}_t^{-1/4}, \tag{3.63}$$

$$\eta \propto L \, \text{Re}_t^{-3/4}, \tag{3.64}$$

and

$$\tau_\eta \propto \tau_t \, \text{Re}_t^{-1/2}, \tag{3.65}$$

where

$$\text{Re}_t = \frac{u'L}{\nu} \tag{3.66}$$

is the turbulent Reynolds number. Obviously, to distinguish the dissipation (viscous) range of the turbulence spectrum from the larger-scale turbulent eddies, the ratio of the length scale L of the large eddies to the length scale η of the smallest eddies should be sufficiently high. Equation 3.64 shows that it is possible only if the turbulent Reynolds number is sufficiently large, that is,

$$\text{Re}_t \gg 1. \tag{3.67}$$

Thus, the discussed theory is asymptotically valid at $\text{Re}_t \to \infty$, as clearly stated by Kolmogorov (1941a), who formulated his famous similarity hypotheses "for sufficiently large Reynolds number."

According to the second hypothesis of similarity, at sufficiently large Reynolds numbers, there is $\eta \ll l \ll L$ range of scales of turbulent motions, such that the statistics of the eddies of scale l is uniquely determined by the mean dissipation rate $\bar\varepsilon$ but is independent of the viscosity. This range of scales is called the inertial range of the turbulence spectrum.

The second hypothesis of similarity allows us to estimate the key characteristics of the turbulent eddies of a length scale l such that $\eta \ll l \ll L$. For dimensional reasoning, we have

$$u'_l \propto \left(\bar\varepsilon l\right)^{1/3}, \tag{3.68}$$

$$\tau_l \propto l^{2/3}\bar\varepsilon^{-1/3}, \tag{3.69}$$

and

$$\dot{s}_l = \frac{u'_l}{l} = \tau_l^{-1} \propto l^{-2/3}\bar\varepsilon^{1/3}. \tag{3.70}$$

Moreover, in the inertial range, the dependence of the energy-spectrum function on the wave number may also be obtained from dimensional reasoning:

$$E(\kappa) \propto \bar{\varepsilon}^{2/3} \kappa^{-5/3} .$$ (3.71)

This is the famous $-5/3$ spectrum predicted theoretically by Kolmogorov and documented in numerous measurements and DNSs discussed in any textbook on turbulence (e.g., see Pope, 2000).

It is worth stressing two features of the Kolmogorov turbulence that are of paramount importance for understanding the physics of turbulent combustion. First, a comparison of Equation 3.68 with Equation 3.58 and Equation 3.69 with Equation 3.59 shows that the Kolmogorov scaling given by Equations 3.68 and 3.69 for the inertial range of the turbulence spectrum is also valid for the dissipation range. Therefore, in both ranges, the velocity scales and timescales of the turbulent eddies are increased by their length scales. On the contrary, Equation 3.70 indicates that the stretch-rate scale is increased by the wave number. Consequently, the highest instantaneous stretch rates are associated with the smallest turbulent eddies of the Kolmogorov scale. Bearing in mind that stretching increases the area A of a material surface in a turbulent flow (as discussed in the section "Energy Cascade and Turbulent Stretching"), we may conclude that the increase in $A(t)$ is mainly controlled by the small-scale turbulent eddies. Indeed, DNS studies on turbulent flows have shown that (Yeung et al., 1990; Bray and Cant, 1991; Girimaji and Pope, 1992)

$$\frac{dA}{dt} \propto \frac{A}{\tau_\eta} .$$ (3.72)

Second, Equations 3.63 through 3.65 show that $\eta/L \ll \tau_\eta/\tau_t \ll u'_\eta/u'$ at large Reynolds numbers. In other words, the difference between the Kolmogorov velocity u'_η and the rms velocity u', which characterize the small- and large-scale turbulent eddies, respectively, is much less than the difference in the length scales of the small- and large-scale eddies. For instance, if $Re_t = 10,000$, then $\eta/L \approx 0.001$, whereas $u'_\eta/u' \approx 0.1$. Certainly, a ratio of the order of 0.1 is sufficiently low for many applications, but $u'_\eta/u' \approx 1/3$ if $Re_t = 100$, as occurs in many laboratory premixed turbulent flames. Thus, even if the Kolmogorov velocity is substantially lower than the rms velocity, u'_η is not negligible as compared with S_L in a typical premixed flame. The influence of small-scale eddies on a flame should not be ignored, especially as the increase in the area of a material surface is controlled by the small-scale eddies, as previously discussed.

In numerical simulations of turbulent combustion, the important role played by small-scale turbulent eddies is sometimes disregarded by arguing that the energy-spectrum function decreases rapidly with increasing wavenumber, for example, see Equation 3.71. However, it is worth remembering that the energy-spectrum function is the energy per unit wave number, rather than the energy itself. Integration of

Equation 3.71 over an interval $d\kappa$ results in a substantially weaker dependence of the energy on the wave number, that is, $u_\kappa'^2 \propto \kappa^{-2/3}$ in line with Equation 3.68.

The previous discussion dealt with homogeneous isotropic turbulence. However, the Kolmogorov theory is valid in a more general case, provided that the turbulent Reynolds number is sufficiently large. As argued by Kolmogorov (1941a), if $Re_t \gg 1$, the statistics of the turbulent eddies of scales $l \ll L$ is homogeneous and isotropic even if the mean flow and the large-scale turbulent eddies are statistically nonuniform and anisotropic. In such a flow, the turbulence spectrum may be divided into two ranges: (i) energy-containing eddies, which are statistically nonuniform and anisotropic, and (ii) universal equilibrium eddies, which are statistically uniform and isotropic. The universal equilibrium range may also be divided into the inertial and dissipation subranges previously discussed. The ensemble of eddies within the universal equilibrium range is called locally isotropic turbulence.

Thus, in a turbulent flow, there are large energy-containing eddies of scale L and smaller eddies up to the Kolmogorov scale η. The Taylor microscales are associated with small-scale turbulent eddies, which, however, are substantially larger than the eddies of the Kolmogorov scale at high Reynolds numbers. Indeed, it can be shown (Pope, 2000) that

$$\bar{\varepsilon} = 15\nu \frac{u'^2}{\lambda_\perp^2} \tag{3.73}$$

in isotropic homogeneous turbulence. Equations 3.56, 3.62, and 3.73 yield

$$\lambda \propto \eta \, Re_t^{1/4} \propto L \, Re_t^{-1/2} \tag{3.74}$$

and

$$Re_\lambda \equiv \frac{u'\lambda}{\nu} \propto Re_t^{1/2}. \tag{3.75}$$

INTERNAL INTERMITTENCY AND VORTEX FILAMENTS

The previous discussion dealt mainly with the so-called background (or Kolmogorov) turbulence, which contains most of the energy of turbulent motion. However, the real isotropic turbulence is not reduced to the self-similar cascade of turbulent eddies of different sizes, described well by the classical theory by Kolmogorov (1941a,b, 1942). Various measurements and DNSs discussed in detail elsewhere (Frisch, 1995; Lipatnikov and Chomiak, 2005a) indicate the existence of spotty regions characterized by high magnitudes of either the vorticity or the local dissipation rate in turbulent flows. This phenomenon is called internal intermittency.

These spotty regions are sometimes associated with elongated, coherent structures called "vortex filaments" or "worms" (e.g., see Figure 3.8). The length of the vortex filaments can be as large as L, but their width scales either as η or λ, according

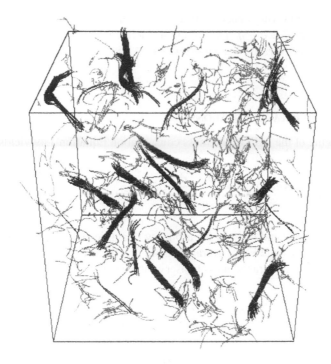

FIGURE 3.8   Vortex lines in a homogeneous turbulent flow with $Re_\lambda \approx 77$ obtained by DNS. The local vorticity intensity is keyed to shading, ranging from light gray (low) to black (high). (Reprinted from She, Z.-S., Jackson, E., and Orszag, A., *Proc. R. Soc. Lond. A*, 434, 101–124, 1991. With permission.)

to different data. See Section 7.3.2 in a review paper by Lipatnikov and Chomiak (2005a) for other data on vortex filaments.

The volume occupied by the vortex filaments is very small at high Re_t and their energy is much less than the total energy of the turbulent motion. Owing to the latter fact, vortex filaments are often considered to be an interesting, but minor feature of turbulence, with discussion on them commonly skipped in textbooks on fluid mechanics. However, vortex filaments may play a very important role in the propagation of premixed turbulent flames, as discussed in the section "Flame Propagation along Vortex Tubes" in Chapter 5.

REYNOLDS-AVERAGED NAVIER–STOKES APPROACH TO NUMERICAL SIMULATIONS OF TURBULENT FLOWS

The most popular engineering approach to simulations of turbulent flows consists of numerically solving the Navier–Stokes Equation 1.51 that is averaged using a method discussed in the section "Mean (Average) Quantities." In this section, an averaged Navier–Stokes equation and an averaged balance equation for the second moments (Reynolds stress tensor) will be derived and discussed. However, bearing in mind the subsequent use of the equations for modeling turbulent flames, the derivation will be performed by allowing density variations and using the Favre-averaged quantities $\tilde{q} = \overline{\rho q}/\overline{\rho}$ and $q'' \equiv q - \tilde{q}$. The following derived equations are reduced to the standard Reynolds-averaged Navier–Stokes (RANS) equations if $\rho = \text{const}$, $\tilde{q} = \overline{q}$, and $q'' = q'$.

FAVRE-AVERAGED BALANCE EQUATIONS

Favre-averaging of the LHS of the mass conservation Equation 1.49 yields

$$\overline{\frac{\partial}{\partial t}(\overline{\rho}+\rho')} + \overline{\frac{\partial}{\partial x_k}\left[\rho(\tilde{u}_k + u_k'')\right]} = \frac{\partial \overline{\rho}}{\partial t} + \frac{\partial}{\partial x_k}(\overline{\rho}\tilde{u}_k) + \frac{\partial}{\partial x_k}\overline{\rho u_k''} = \frac{\partial \overline{\rho}}{\partial t} + \frac{\partial}{\partial x_k}(\overline{\rho}\tilde{u}_k), \quad (3.76)$$

because $\overline{q'} = \overline{\rho q''} = 0$ and $\overline{b\tilde{q}} = \overline{b}\tilde{q}$ for arbitrary quantities b and q due to the definitions of the Reynolds and Favre averages. Note that the use of the original Reynolds decomposition in the averaged mass conservation equation results in

$$\overline{\frac{\partial}{\partial t}(\overline{\rho}+\rho')} + \overline{\frac{\partial}{\partial x_k}\left[(\overline{\rho}+\rho')(\overline{u}_k + u_k')\right]} = \frac{\partial \overline{\rho}}{\partial t} + \frac{\partial}{\partial x_k}(\overline{\rho}\cdot\overline{u}_k) + \frac{\partial}{\partial x_k}\overline{\rho' u_k'}, \quad (3.77)$$

because $\overline{\overline{b}q'} = \overline{b}\,\overline{q'} = 0$ for arbitrary quantities b and q. A comparison of Equations 3.76 and 3.77 clearly shows that the use of Favre decomposition offers an opportunity to write the averaged balance equations in a more compact form in the case of variable density.

Thus, the Favre-averaged mass conservation (continuity) reads

$$\frac{\partial \overline{\rho}}{\partial t} + \frac{\partial}{\partial x_k}(\overline{\rho}\tilde{u}_k) = 0. \quad (3.78)$$

By using the method previously discussed, one can average the Navier–Stokes Equation 1.51:

$$\frac{\partial}{\partial t}(\overline{\rho}\tilde{u}_i) + \frac{\partial}{\partial x_k}(\overline{\rho}\tilde{u}_k\tilde{u}_i) = -\frac{\partial}{\partial x_k}\overline{\rho u_k'' u_i''} + \frac{\overline{\partial \tau_{ik}}}{\partial x_k} - \frac{\partial \overline{p}}{\partial x_k}, \quad (3.79)$$

because

$$\overline{\rho ab} = \overline{\rho(\tilde{a}+a'')(\tilde{b}+b'')} = \overline{\rho}\tilde{a}\tilde{b} + \tilde{a}\overline{\rho b''} + \tilde{b}\overline{\rho a''} + \overline{\rho a'' b''} = \overline{\rho}\tilde{a}\tilde{b} + \overline{\rho a'' b''} \quad (3.80)$$

for arbitrary quantities a and b. In Equation 3.79, the ratio of the convection term (the second term on the LHS) to the viscous term (the second term on the RHS) scales as

$$\frac{\overline{\rho \tilde{u}_k \tilde{u}_i}}{\overline{\tau}_{ik}} \propto \frac{U^2}{\nu(U/L)}. \tag{3.81}$$

At high Reynolds numbers, this ratio is low and the mean viscous term is often neglected as compared with the mean convection term.

Contrary to the original Navier–Stokes equation, the averaged Equation 3.79 is not closed, that is, the Reynolds stresses $\overline{\rho u_k'' u_i''}$ cannot be determined by solving Equations 3.78 and 3.79. One method of closing Equation 3.79 consists of invoking the second-order balance equations.

Such equations can be derived using the following method. Let us (i) multiply the Navier–Stokes equations for the velocity components u_i and u_j with the velocity components u_j and u_i, respectively; (ii) sum up the two obtained equations; and (iii) average the result. We obtain

$$\frac{\partial}{\partial t}\overline{\rho u_i u_j} + \frac{\partial}{\partial x_k}\overline{\rho u_i u_j u_k} = \overline{u_i \frac{\partial \tau_{ik}}{\partial x_k}} + \overline{u_j \frac{\partial \tau_{ik}}{\partial x_k}} - \overline{u_j \frac{\partial p}{\partial x_i}} - \overline{u_i \frac{\partial p}{\partial x_j}}, \tag{3.82}$$

because

$$\frac{\partial}{\partial t}(\rho q) + \frac{\partial}{\partial x_k}(\rho u_k q) = \rho \frac{\partial q}{\partial t} + \rho u_k \frac{\partial q}{\partial x_k} \tag{3.83}$$

for any quantity q due to the mass conservation Equation 1.49. Because

$$\overline{ab} = \overline{(\tilde{a} + a'')b} = \overline{\tilde{a}b} + \overline{a''b} = \tilde{a}\overline{b} + \overline{a''} \cdot \overline{b} + \overline{a''b'} \tag{3.84}$$

and

$$\overline{\rho abq} = \overline{\rho(\tilde{a} + a'')(\tilde{b} + b'')(\tilde{q} + q'')}$$

$$= \overline{\rho}\tilde{a}\tilde{b}\tilde{q} + \tilde{a}\overline{\rho b'' q''} + \tilde{b}\overline{\rho a'' q''} + \tilde{q}\overline{\rho a'' b''} + \overline{\rho a'' b'' q''} \tag{3.85}$$

for arbitrary quantities a, b, and q, Equation 3.82 reads

$$\frac{\partial}{\partial t}\left(\overline{\rho}\tilde{u}_i\tilde{u}_j\right) + \frac{\partial}{\partial t}\overline{\rho u_i'' u_j''} + \frac{\partial}{\partial x_k}\left(\overline{\rho}\tilde{u}_i\tilde{u}_j\tilde{u}_k\right) + \frac{\partial}{\partial x_k}\left(\tilde{u}_i\overline{\rho u_j'' u_k''}\right)$$

$$+ \frac{\partial}{\partial x_k}\left(\tilde{u}_j\overline{\rho u_i'' u_k''}\right) + \frac{\partial}{\partial x_k}\left(\tilde{u}_k\overline{\rho u_i'' u_j''}\right) + \frac{\partial}{\partial x_k}\overline{\rho u_i'' u_j'' u_k''}$$

$$= \tilde{u}_j\frac{\partial \overline{\tau}_{ik}}{\partial x_k} + \tilde{u}_i\frac{\partial \overline{\tau}_{jk}}{\partial x_k} + \overline{u_j''\frac{\partial \tau_{ik}}{\partial x_k}} + \overline{u_i''\frac{\partial \tau_{jk}}{\partial x_k}} - \tilde{u}_j\frac{\partial \overline{p}}{\partial x_i} - \tilde{u}_i\frac{\partial \overline{p}}{\partial x_j} - \overline{u_j''\frac{\partial p}{\partial x_i}} - \overline{u_i''\frac{\partial p}{\partial x_j}}. \tag{3.86}$$

Let us (i) multiply the averaged Navier–Stokes Equation 3.79 for the mean velocity components $\tilde{u}_i$ and $\tilde{u}_j$ with the velocity components $\tilde{u}_j$ and $\tilde{u}_i$, respectively, and (ii) sum up the two equations. We obtain

$$\frac{\partial}{\partial t}\left(\bar{\rho}\tilde{u}_i\tilde{u}_j\right)+\frac{\partial}{\partial x_k}\left(\bar{\rho}\tilde{u}_i\tilde{u}_j\tilde{u}_k\right)=-\tilde{u}_j\frac{\partial}{\partial x_k}\overline{\rho u_i''u_k''}-\tilde{u}_i\frac{\partial}{\partial x_k}\overline{\rho u_j''u_k''}+\tilde{u}_j\frac{\partial\overline{\tau}_{ik}}{\partial x_k}$$

$$+\tilde{u}_i\frac{\partial\overline{\tau}_{jk}}{\partial x_k}-\tilde{u}_j\frac{\partial\overline{p}}{\partial x_i}-\tilde{u}_i\frac{\partial\overline{p}}{\partial x_j}. \tag{3.87}$$

Subtraction of Equation 3.87 from Equation 3.86 yields the following balance equation:

$$\frac{\partial}{\partial t}\overline{\rho u_i''u_j''}+\frac{\partial}{\partial x_k}\left(\tilde{u}_k\overline{\rho u_i''u_j''}\right)=-\overline{\rho u_i''u_k''}\frac{\partial\tilde{u}_j}{\partial x_k}-\overline{\rho u_j''u_k''}\frac{\partial\tilde{u}_i}{\partial x_k}-\frac{\partial}{\partial x_k}\overline{\rho u_i''u_j''u_k''}$$

$$+\overline{u_j''\frac{\partial\tau_{ik}}{\partial x_k}}+\overline{u_i''\frac{\partial\tau_{jk}}{\partial x_k}}-\overline{u_j''\frac{\partial p}{\partial x_i}}-\overline{u_i''\frac{\partial p}{\partial x_j}} \tag{3.88}$$

for the Favre-averaged Reynolds stresses $\overline{\rho u_i''u_j''}$.

However, the obtained balance equation involves terms, for example, the third moments $\overline{\rho u_i''u_j''u_k''}$, that cannot be determined by solving Equations 3.78, 3.79, and 3.88. An attempt to invoke a balance equation for $\overline{\rho u_i''u_j''u_k''}$ does not seem to be meaningful, because such an equation will involve new unclosed terms such as the fourth moments $\overline{\rho u_i''u_j''u_k''u_m''}$. Therefore, the unclosed terms (either the Reynolds stresses $\overline{\rho u_k''u_i''}$ in Equation 3.79 or the last five terms on the RHS of Equation 3.88) need modeling. A relation that expresses such a term through known quantities ($\bar{\rho}$, $\tilde{u}_i$, $\tilde{c}$, and, if Equation 3.88 is considered, $\overline{\rho u_k''u_i''}$) is called a closure of the term. There are various approaches to closing the previously derived averaged balance equations. Here, we restrict ourselves to the well-known $k-\varepsilon$ model (Jones and Launder, 1972; Launder and Spalding, 1972), which still dominates in engineering applications. In the rest of this chapter, we will consider an incompressible (or constant-density) turbulent flow where $\nabla\cdot\mathbf{u}=\nabla\cdot\overline{\mathbf{u}}=0$.

$k-\varepsilon$ MODEL

Within the framework of the $k-\varepsilon$ model, the Reynolds stresses are closed as follows:

$$\overline{u_i'u_j'}=-\nu_t\left(\frac{\partial\overline{u}_i}{\partial x_j}+\frac{\partial\overline{u}_j}{\partial x_i}\right)+\frac{2}{3}\overline{k}\delta_{ij}, \tag{3.89}$$

where

$$\nu_t=C_\mu\frac{\overline{k}^2}{\varepsilon} \tag{3.90}$$

is the turbulent viscosity and $C_\mu = 0.09$ is a constant. The last term on the RHS of Equation 3.89 is necessary in order for the trace $\overline{u'_k u'_k} = \overline{u'_1 u'_1} + \overline{u'_2 u'_2} + \overline{u'_3 u'_3}$ to be equal to $2\overline{k}$ in an incompressible flow. Equations 3.14, 3.62, 3.66, and 3.90 show that the ratio of the turbulent viscosity, v_t, to the kinematic viscosity, v, is of the order of the turbulent Reynolds number, Re_t. This estimate justifies the neglect of the viscous term on the RHS of Equation 3.79 when modeling turbulent flows characterized by large Reynolds numbers.

In order for the model to be closed, the mean turbulent kinetic energy $\overline{k}$ and the dissipation rate $\overline{\varepsilon}$ should be determined. The former quantity may be evaluated by solving a balance equation, which results from Equation 3.88. In the case of $\rho - \mathrm{const}$, the trace of the tensor Equation 3.88 is as follows:

$$\frac{\partial \overline{k}}{\partial t} + \frac{\partial}{\partial x_k}\left(\overline{u}_k \overline{k}\right) = -\overline{u'_j u'_k}\frac{\partial \overline{u}_j}{\partial x_k} - \frac{\partial}{\partial x_k}\overline{u'_k k} + \frac{1}{\rho}\overline{u'_j \frac{\partial \tau_{jk}}{\partial x_k}} - \frac{1}{\rho}\overline{u'_k \frac{\partial p}{\partial x_k}}. \tag{3.91}$$

Because, using Equations 1.52 and 3.46, we obtain

$$\frac{1}{\rho}\overline{u'_j \frac{\partial \tau_{jk}}{\partial x_k}} = \overline{v u'_j \frac{\partial}{\partial x_k}\left(\frac{\partial u_j}{\partial x_k} + \frac{\partial u_k}{\partial x_j}\right)} = \overline{v u'_j \frac{\partial^2 u_j}{\partial x_k^2}} + \overline{v u'_j \frac{\partial^2 u_k}{\partial x_j \partial x_k}} = \overline{v u'_j \frac{\partial^2 u_j}{\partial x_k^2}}, \tag{3.92}$$

$$2\frac{\partial}{\partial x_j}\left(u'_k s_{jk}\right) - 2s_{jk}s_{jk} = \frac{\partial}{\partial x_j}\left[u'_k\left(\frac{\partial u'_j}{\partial x_k} + \frac{\partial u'_k}{\partial x_j}\right)\right] - \frac{1}{2}\left(\frac{\partial u'_j}{\partial x_k} + \frac{\partial u'_k}{\partial x_j}\right)\left(\frac{\partial u'_j}{\partial x_k} + \frac{\partial u'_k}{\partial x_j}\right)$$

$$= \frac{\partial u'_k}{\partial x_j}\left(\frac{\partial u'_j}{\partial x_k} + \frac{\partial u'_k}{\partial x_j}\right) + u'_k \frac{\partial^2 u'_k}{\partial x_j^2} - \frac{\partial u'_j}{\partial x_k}\frac{\partial u'_j}{\partial x_k} - \frac{\partial u'_j}{\partial x_k}\frac{\partial u'_k}{\partial x_j}$$

$$= u'_k \frac{\partial^2 u'_k}{\partial x_j^2}, \tag{3.93}$$

and

$$\overline{u'_k \frac{\partial p}{\partial x_k}} = \frac{\partial}{\partial x_k}\overline{u'_k p}, \tag{3.94}$$

in an incompressible flow, Equations 3.45 and 3.91 read

$$\frac{\partial \overline{k}}{\partial t} + \frac{\partial}{\partial x_k}\left(\overline{u}_k \overline{k}\right) = \frac{\partial T_k}{\partial x_k} + \mathrm{P} - \overline{\varepsilon}, \tag{3.95}$$

where

$$T_k = -\overline{u'_k k} - \frac{\overline{u'_k p}}{\rho} + 2v\overline{u'_j s_{jk}}, \tag{3.96}$$

$$P = -\overline{u'_j u'_k} S_{jk},$$ (3.97)

and

$$S_{jk} = \frac{1}{2}\left(\frac{\partial \overline{u}_j}{\partial x_k} + \frac{\partial \overline{u}_k}{\partial x_j}\right)$$ (3.98)

is the mean rate-of-strain tensor.

The first term on the RHS of Equation 3.95 is associated with the redistribution of the turbulent energy in space in an inhomogeneous flow. This term vanishes after integration over a volume bounded by a surface where $T_k = 0$, that is, the total turbulent energy in a flow is not affected by this term. Accordingly, it is called the transport term and is modeled by analogy with Equation 3.89:

$$T_j = \frac{\nu_t}{\sigma_k}\frac{\partial \overline{k}}{\partial x_j},$$ (3.99)

where $\sigma_k = 0.9$ is a constant.

The last term on the RHS of Equation 3.95 is the mean dissipation rate, and it always reduces the turbulent energy. In order for turbulence to be statistically stationary, the term P should compensate for the dissipation. Accordingly, this term is called the generation term. It is positive in many turbulent flows, for example, in a 2D boundary layer, $\overline{u}_1 \gg \overline{u}_2$, $\partial \overline{q}/\partial x_1 \ll \partial \overline{q}/\partial x_2$, and

$$P \approx \nu_t \left(\frac{\partial \overline{u}_1}{\partial x_2}\right)^2 > 0.$$ (3.100)

It is worth stressing that the generation term is controlled by the gradient of the mean flow velocity (see Equations 3.89, 3.97, and 3.98). Thus, this mathematical analysis supports the physical hypothesis that turbulence withdraws energy from the mean flow.

In principle, a balance equation for the dissipation rate can be obtained by invoking the same method as the one used to derive Equation 3.91, that is, by differentiating and summing up the Navier–Stokes equation in order to obtain a balance equation for $s_{jk}s_{jk}$. However, at large Reynolds numbers, the $\overline{\varepsilon}$-equation derived in this way involves an unclosed source term that scales as $Re_t^{1/2}$ and an unclosed sink term that also scales as $Re_t^{1/2}$, while the unsteady and convection terms are finite. Therefore, in order to close the equation in the asymptotic case of $Re_t \to \infty$, addressed by the turbulence theory, one has to model a finite difference in two infinitely large terms. Such a problem is ill-posed.

A breakthrough was provided by Kolmogorov (1942). Based on the concept of turbulent cascade, he pointed out that the dissipation rate characterizes not only the small-scale eddies but also the large-scale eddies. Therefore, the evolution of the dissipation rate may be modeled using the balance equations for turbulent kinetic energy and one more large-scale quantity, with the structure of the latter equation being similar to the structure of the $\bar{k}$-equation, because both equations address large-scale flow characteristics. Kolmogorov (1942) developed a $\bar{k} - \bar{\omega}$ model with $\bar{\omega} \propto u'/L \propto \bar{\varepsilon}/\bar{k}$. In the literature, balance equations for other large-scale quantities may be found. Models of that type may be called $\bar{k} - u'^m L^n$ models. The $k - \varepsilon$ model is associated with $m = 3$ and $n = -1$, and the $\bar{k} - \bar{\omega}$ model corresponds to $m = 1$ and $n = -1$. Other examples are $\bar{k} - L$ ($m = 0, n = 1$), $\bar{k} - \tau_t$ ($m = -1, n = 1$), and $\bar{k} - \nu_t$ ($m = n = 1$) models. After numerous tests, the $k - \varepsilon$ model became the most popular one.

As previously discussed, the $\bar{\varepsilon}$-equation is similar to the $\bar{k}$-equation, that is,

$$\frac{\partial \bar{\varepsilon}}{\partial t} + \frac{\partial}{\partial x_k}\left(\bar{u}_k \bar{\varepsilon}\right) = \frac{\partial}{\partial x_k}\left(\frac{\nu_t}{\sigma_\varepsilon}\frac{\partial \bar{\varepsilon}}{\partial x_k}\right) + C_{\varepsilon 1}\frac{\bar{\varepsilon}}{\bar{k}}P - C_{\varepsilon 2}\frac{\bar{\varepsilon}^2}{\bar{k}}, \tag{3.101}$$

where $\sigma_\varepsilon = 1.3$, $C_{\varepsilon 1} = 1.44$, and $C_{\varepsilon 2} = 1.92$ are empirical constants.

It is worth stressing again that both $\bar{k}$-Equation 3.95 and ε-Equation 3.101 are balance equations for large-scale characteristics of turbulence, that is, the $k - \varepsilon$ model is strongly based on the concept of turbulent cascade.

TURBULENT DIFFUSION

GRADIENT DIFFUSION APPROXIMATION

The balance equation for the mean concentration of an admixture in a turbulent flow may be obtained using the method discussed in the section "Favre-Averaged Balance Equations." In particular, averaging the species conservation Equation 1.58 yields

$$\frac{\partial \bar{Y}}{\partial t} + \frac{\partial}{\partial x_k}\left(\bar{u}_k \bar{Y}\right) = -\frac{\partial}{\partial x_k}\overline{u'_k Y'} + \frac{\partial}{\partial x_k}\left(\overline{\rho D \frac{\partial Y}{\partial x_k}}\right) \tag{3.102}$$

in a nonreacting ($W = 0$) constant-density turbulent flow. Here, the turbulent scalar flux $\overline{u'_k Y'}$ needs a closure.

If the turbulent-viscosity concept, for example, the $k - \varepsilon$ model, is used to simulate turbulence, the flux is commonly closed, invoking the gradient diffusion approximation:

$$\overline{u'_k Y'} = -D_t \frac{\partial \bar{Y}}{\partial x_k}, \tag{3.103}$$

where

$$D_t = \frac{\nu_t}{\sigma_c} \tag{3.104}$$

is the turbulent diffusivity and different values (from 0.3 to 1) of the constant σ_c may be found in the literature. If $Pr = O(1)$, the ratio of the turbulent diffusivity to the molecular diffusivity scales as the turbulent Reynolds number Re_t (see Equations 3.14, 3.62, 3.66, and 3.90), that is, $D_t/D \gg 1$ in a typical turbulent flow. Based on such an estimate, the molecular transport terms are often neglected as compared with the turbulent transport terms in the balance equations for Reynolds- or Favre-averaged species mass fractions. Nevertheless, as will be discussed in the subsequent chapters, premixed combustion may be strongly affected by the molecular transport properties, even if $Re_t \gg 1$.

If the Reynolds number is sufficiently high in order for the molecular diffusion term (the last term on the RHS of Equation 3.102) to be neglected, then Equations 3.102 through 3.104 have the following particular solution:

$$\bar{Y} = Y_0 \sqrt{\frac{t_0}{t}} \exp\left(-\frac{\xi^2}{2\delta_m^2}\right) \tag{3.105}$$

in the case of $\bar{Y} = \bar{Y}(x,t)$, $\bar{\mathbf{u}} = \{\bar{u}, 0, 0\}$, $\bar{u} = $ const, and $D_t = $ const. Here, $\xi = x - \bar{u}t$ and $\delta_m = \sqrt{2D_t t}$ is the thickness of the turbulent mixing layer. Note that Equation 3.105 is not the unique solution. For instance, the complementary error function distribution:

$$\bar{Y} = 1 - \frac{1}{2}\text{erfc}(\zeta) = 1 - \frac{1}{\sqrt{\pi}} \int_\zeta^\infty e^{-\eta^2} d\eta, \tag{3.106}$$

where $\zeta = \xi/\sqrt{2}\,\delta_m$ also satisfies Equations 3.102 through 3.104 under the aforementioned conditions. Note that Equations 3.105 and 3.106 satisfy Equations 3.102 through 3.104 even if D_t depends on time; but $\delta_m^2 = 2\int_0^t D\,dt$ in this case.

Depending on the boundary conditions, both Equations 3.105 and 3.106 may describe turbulent mixing. For instance, two examples of turbulent diffusion are shown in Figure 3.9. Sketch (a) is associated with the supply of a pollutant through a small pipe into a clean ($Y = 0$) turbulent flow in the hypothetical case of zero molecular diffusivity. In this case, the thin layer filled with the pollutant ($Y = 1$) and surrounded by the clean flow is convected by the mean flow along the x-axis and is randomly moved by the turbulent eddies in the transverse y-direction. An instantaneous image of this layer is shown with a bold wrinkled line. Owing to turbulent fluctuations, this layer oscillates in space with time, and as a result, the transverse distribution $\bar{Y}(y)$ of the time-averaged mass fraction of the pollutant is described by the Gaussian function. One can easily check that Equation 3.105 with $\xi = y$ and $t = x/\bar{u}$ satisfies the statistically stationary 2D Equation 3.102, provided that $\bar{v} = 0$, and the turbulent diffusion in the x-direction is negligible as compared with the convection of the pollutant by the mean flow. Two Gaussian distributions of $\bar{Y}(y)$, associated with two different x, are shown in Figure 3.9a.

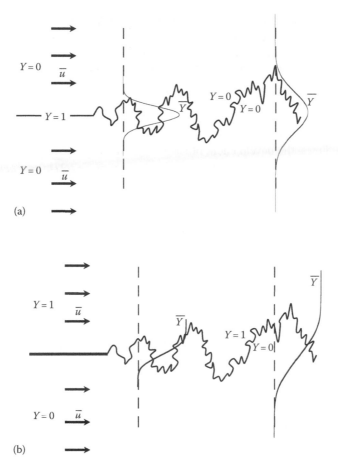

FIGURE 3.9 Two examples of turbulent mixing: (a) diffusion from a point source of an admixture and (b) mixing of two gases. Molecular diffusion is neglected in both cases.

According to Equation 3.105, the thickness of the mixing layer, that is, the length of the y-interval where $\overline{Y}$ is notable due to the oscillations of the instantaneous pollutant layer, increases with the distance from the pipe exit and scales as $\sqrt{2D_t t} = \sqrt{2D_t\, x/\overline{u}}$. Accordingly, the maximum (vs. y, i.e., along a line $x = $ const) mean mass fraction $\overline{Y}_{\max}(x) = \overline{Y}(x, y = 0)$ decreases with an increase in the distance x in order for the integral of $\int_{-\infty}^{\infty} \overline{Y}(x, y)\, dy$ to be constant, because the pollutant is neither consumed nor produced during the mixing process. It is worth stressing that the decrease in $\overline{Y}_{\max}(x) = \overline{Y}(x, y = 0)$ does not mean that the instantaneous mass fraction Y of the pollutant is lower than unity. If the molecular diffusivity $D = 0$, then Y may be equal to either 0 or 1. The decrease in the average value $\overline{Y}(x, y)$ with the axial distance x means a decrease in the probability of finding the pollutant layer at a particular transverse distance y, because the y-range where the layer may be found increases with x.

Figure 3.9b looks similar to Figure 3.9a, but two different gases, for example, Ar and He, separated by a wall at $x < 0$ mix in this case. On account of the change in the boundary conditions (from $\bar{Y}(y \to -\infty) = \bar{Y}(y \to \infty) = 0$ to $\bar{Y}(y \to -\infty) = 0$ and $\bar{Y}(y \to \infty) = 1$), the latter problem is described by Equation 3.106 with $\xi = y$, $t = x/\bar{u}$, and $\zeta = y/2\sqrt{D_t t}$.

It is worth stressing that the two simple problems previously discussed and described by the transient Equations 3.105 and 3.106, respectively, are examples of statistically stationary but developing turbulent flows. Despite the statistical stationarity of the flows, the development of such flows manifests itself in the increase in thickness of the mixing layer, with the thickness scaling as $\sqrt{2D_t t} = \sqrt{2D_t\, x/\bar{u}}$. The transient nature of the flow is masked due to the axial convection of the growing (i.e., developing) mixing layer by the mean flow.

The previous discussion dealt with the case of a developing mixing layer, but a fully developed turbulent diffusivity given by Equations 3.90 and 3.104 was used in the discussion. However, turbulent diffusivity D_t can also develop (Taylor, 1921, 1935) in a mixing layer and the development of D_t is of paramount importance for understanding the phenomenology of premixed turbulent combustion.

Taylor's Theory

Let us consider the problem shown in Figure 3.9a. Let us also assume that the turbulence is statistically stationary, homogeneous, and isotropic. For simplicity, we will address a 2D case.

The pollutant layer oscillates solely because it is moved by the turbulent eddies. Let us consider a volume of the pollutant. If this volume is observed in a point $\mathbf{x} = \{x,y\}$ at an instant $t = 0$, then the subsequent motion of the volume is described by the following kinematic equation:

$$\frac{d\mathbf{x}}{dt} = \mathbf{u}\left(\mathbf{x},t\right). \tag{3.107}$$

In particular,

$$\frac{dy}{dt} = v'\left(\mathbf{x},t\right), \tag{3.108}$$

because $\bar{v} = 0$ in the considered case. Therefore,

$$y\left[\mathbf{x}(t = 0),t\right] = \int_0^t v'\left[\mathbf{x}(\eta),\eta\right]d\eta, \tag{3.109}$$

where integration is performed in the Lagrangian framework, that is, along the trajectory of the selected pollutant volume. Multiplying the LHS (RHS) of Equation 3.108 with the LHS (RHS) of Equation 3.109, we obtain

$$\frac{1}{2}\frac{d}{dt}\overline{y^2}\left[\mathbf{x}(t=0),t\right] = \int_0^t v'(\mathbf{x},t)v'\left[\mathbf{x}(\eta),\eta\right]d\eta = \int_0^t v'(\mathbf{x},t)v'\left[\mathbf{x}(t-\vartheta),t-\vartheta\right]d\vartheta.$$

(3.110)

By averaging this expression and using the Lagrangian autocorrelation function $f_L(\vartheta)$, defined by Equation 3.35 in statistically stationary, homogeneous, and isotropic turbulence, we arrive at

$$\frac{1}{2}\frac{d}{dt}\overline{y^2} = u'^2\int_0^t f_L(\vartheta)d\vartheta.$$

(3.111)

At far distances x from the pipe exit, associated with the long development of the turbulent mixing layer, the integral on the RHS of Equation 3.111 tends to the Lagrangian timescale τ_L; therefore,

$$\frac{1}{2}\frac{d}{dt}\overline{y^2} = u'^2\tau_L,$$

(3.112)

provided that $\tau_L \ll t = x/\overline{u}$.

In the opposite limit case of $\tau_L \gg t = x/\overline{u}$, the autocorrelation function $f_L(\vartheta)$ tends to unity, that is, the turbulent motion is fully correlated during a short time interval. Accordingly, Equation 3.111 reads

$$\frac{1}{2}\frac{d}{dt}\overline{y^2} = u'^2 t,$$

(3.113)

provided that $\tau_L \gg t = x/\overline{u}$.

Equations 3.112 and 3.113 are exact theoretical results obtained in the two opposite limit cases. In a general case, the RHS of Equation 3.111 may be integrated, invoking a proper empirical approximation of the autocorrelation function. In particular, the exponential parameterization

$$f_L(t) = \exp\left(-\frac{t}{\tau_L}\right)$$

(3.114)

is commonly used for that purpose. Substitution of Equation 3.114 into Equation 3.111 yields

$$\frac{1}{2}\frac{d}{dt}\overline{y^2} = u'^2\tau_L\left[1-\exp\left(-\frac{t}{\tau_L}\right)\right],$$

(3.115)

which reduces to Equations 3.112 and 3.113 in the proper limit cases. If the growing turbulent mixing layer is convected by the mean flow, that is, $t = x/\bar{u}$, then we have

$$\frac{\bar{u}}{2}\frac{d}{dx}\overline{y^2} = u'^2\tau_L\left[1-\exp\left(-\frac{x}{\bar{u}\tau_L}\right)\right]. \tag{3.116}$$

Equations 3.111 through 3.116 discussed in this section describe the increase in the mean displacement $\overline{y^2}$ of the instantaneous pollutant layer, while Equation 3.105 obtained in the section "Gradient Diffusion Approximation" deals with the spatial distribution of the mean mass fraction $\bar{Y}$ of the pollutant. To bridge the two approaches, let us discuss the following three questions: Is there any relation between $\bar{Y}(t, y)$ and $\overline{y^2}(t)$? Is Equation 3.112, obtained in the limit case of $\tau_L \ll t = x/\bar{u}$, consistent with Equation 3.104, which holds in the case of a constant diffusivity? What is the turbulent diffusivity within the framework of Taylor's theory?

First, in the studied case ($Y = 1$ in the thin pollutant layer and $Y = 0$ outside the layer), the mean mass fraction of the pollutant in point x at instant t is simply proportional to the probability P of finding the layer in this point at that instant. Therefore, using Equation 3.105 and the normalizing constraint given by Equation 3.20, one can easily show that

$$P(x, y) = \frac{1}{\sqrt{2\pi}\delta_m}\exp\left(-\frac{y^2}{2\delta_m^2}\right), \tag{3.117}$$

where the thickness δ_m depends on x in the case considered (see Figure 3.9a). Therefore,

$$\overline{y^2} = \int_{-\infty}^{\infty} y^2 P\, dy = \frac{1}{\sqrt{2\pi}\delta_m}\int_{-\infty}^{\infty} y^2 \exp\left(-\frac{y^2}{2\delta_m^2}\right)dy = \delta_m^2 = 2D_t\frac{x}{\bar{u}} = 2D_t t \tag{3.118}$$

if $D_t = \text{const}$. Thus, the thickness $\delta_m(x)$ of the turbulent mixing layer in the spatial profile of $\bar{Y}(y)$, given by Equation 3.105 with $\xi = y$, is equal to the mean displacement $\overline{y^2}(x)$ of the instantaneous pollutant layer.

The aforementioned analysis and, in particular, Equation 3.117, shows that the gradient diffusion closure given by Equation 3.103 is, in fact, based on a hypothesis that the probability of finding the instantaneous mixing layer is the Gaussian function. A similar conclusion may be drawn by analyzing the case sketched in Figure 3.9b.

Second, differentiation of Equation 3.118 yields

$$\frac{1}{2}\frac{d\overline{y^2}}{dt} = D_t. \tag{3.119}$$

Subsequently, Equations 3.112 and 3.119 result in

$$D_t = u'^2 \tau_L \tag{3.120}$$

if $\tau_L \gg t = x/\bar{u}$. Similar to Equation 3.104, which is commonly used in conjunction with the $k - \varepsilon$ model, Equation 3.120 yields the diffusivity D_t controlled solely by the turbulence characteristics. The two equations are consistent with each other, provided that

$$\tau_L = \frac{C_\mu}{\sigma_c} \frac{\bar{k}^2}{u'^2 \bar{\varepsilon}} = \frac{3C_\mu}{2\sigma_c} \frac{\bar{k}}{\bar{\varepsilon}} \propto \frac{L}{u'}. \tag{3.121}$$

Finally, Equation 3.119 may be considered to be the definition of diffusivity in a general case, for example, if the inequality of $\tau_L \gg t = x/\bar{u}$ does not hold. Then, Equations 3.115 and 3.119 yield

$$D_t = u'^2 \tau_L \left[1 - \exp\left(-\frac{t}{\tau_L} \right) \right], \tag{3.122}$$

which reduces to $D_t = u'^2 t$ and $D_t = u'^2 \tau_L$ in the limit cases of $t \ll \tau_L$ and $t \gg \tau_L$, respectively. If Equation 3.122 is substituted into Equation 3.103, then Equations 3.105 and 3.106 will satisfy the diffusion Equation 3.102 with $\delta_m^2 = 2\int_0^t D_t(\vartheta)d\vartheta$. Accordingly, $\delta_m^2 = u'^2 t^2$ and $\delta_m^2 = 2u'^2 \tau_L t$ at $t \ll \tau_L$ and $t \gg \tau_L$, respectively.

Equation 3.120 shows that the fully developed turbulent diffusivity is controlled by the rms turbulent velocity and Lagrangian timescale, with the two quantities characterizing large-scale turbulent eddies. This result implies that the growth of the turbulent mixing layer, that is, turbulent diffusion, is controlled by the large-scale eddies. Indeed, within the framework of the Kolmogorov theory of turbulence, diffusivity D_l associated with eddies of a length scale l is of the order of $\bar{\varepsilon}^{1/3} l^{4/3}$, that is, the largest diffusivity is associated with the largest eddies.

Molecular Diffusion in Turbulent Flows

In the previous discussion, the molecular diffusion term was neglected in the balance Equation 3.102 in the case of $Re_t \gg 1$. However, the neglect of this term does not mean that molecular mixing is of minor importance in intense turbulence. Turbulent eddies only advect and wrinkle the molecular mixing layer, but cannot make the mixture spatially uniform at scales smaller than the Kolmogorov length scale η. Indeed, the time required in order for molecular diffusion to mix nonuniformities of a length scale l is of the order of l^2/D, while the time required in order for "thin convoluted sheets of thickness η separating" the two mixing fluids to be formed due to turbulent stretching is of the order of $(l^2/\bar{\varepsilon})^{1/3}$ (Yakhot, 2008). If the Schmidt number $Sc = \nu/D \approx 1$, then the ratio of the two timescales is of the order of $(l/L)^{4/3}\mathrm{Re}_t \propto (l/\eta)^{4/3}$ and is much larger than unity only if $\eta \ll l$. At scales smaller than η, the former timescale l^2/D is less than $(l^2/\overline{\varepsilon})^{1/3}$ and mixing is controlled by molecular diffusion. Therefore, if $\mathrm{Sc} \approx 1$ as occurs in a typical combustible gas mixture, then "the mixing process in turbulent flows involves three main steps: (i) entrainment, creating pockets of material B in a turbulent flow enriched by a substance A; (ii) advection and stretching leading to formation of thin convoluted sheets of thickness η separating the reactants; ... (iii) molecular diffusion across" the aforementioned sheets (Yakhot, 2008). Thus, at any Reynolds number, mixing is completed by molecular transport within the aforementioned "thin convoluted sheets of thickness η separating" the two substances. An increase in the Reynolds number results in a decrease in the thickness of the sheets (molecular mixing layers) but does not make molecular mixing a minor process. Only molecular transport can make the mixture composition spatially uniform at scales smaller than η. Turbulence accelerates mixing by producing the aforementioned sheets, thereby increasing the local concentration gradients.

SUMMARY

In constant-density flows, a turbulent scalar flux $\overline{u'_k Y}$ is commonly modeled, invoking the gradient diffusion closure given by Equation 3.103. This closure is, in fact, based on a hypothesis that the probability of finding the instantaneous mixing layer is the Gaussian function.

Turbulent diffusivity is defined by Equation 3.119, where $\sqrt{\overline{y^2}}$ is the thickness of the mixing layer.

Turbulent diffusivity is a developing quantity with $D_t = u'^2 t$ and $D_t = u'^2 \tau_L$ in the limit cases of $t \ll \tau_L$ and $t \gg \tau_L$, respectively. The semiempirical Equation 3.104, which is commonly used in conjunction with the $k - \varepsilon$ model, is only valid in the latter limit case.

The increase in the thickness of a turbulent mixing layer with the distance from an admixture source in a statistically stationary case is the manifestation of the development of the turbulent diffusion process, with this transient phenomenon masked by the convection of the mixing layer by the mean flow.

NOTES

First, although the following three features of turbulent motions:

- Exponential increase in the area of a material surface due to turbulent stretching (see Equation 3.54 and Figure 3.7)
- Internal intermittency (see Figure 3.8)
- Development of turbulence in statistically stationary flows and development of turbulent diffusivity in particular

are not highlighted in a typical textbook on fluid dynamics and turbulence, these features are of paramount importance for understanding the physics of turbulent combustion.

Second, it is worth remembering that the mean dissipation rate $\bar{\varepsilon}$, rather than the rms turbulent velocity u', is the key characteristic of the Kolmogorov turbulence. Therefore, if a nondimensional criterion is invoked to model the influence of turbulence on a flame, the use of $\bar{\varepsilon}$ in such a criterion appears to be a more justified choice than the use of u'. In particular, numerous attempts to parameterize the aforementioned influence invoking solely the ratio of u'/S_L do not seem to be consistent with the underlying physics of turbulence.

4 Phenomenology of Premixed Turbulent Combustion

This chapter introduces the main theme of the book, which focuses on the physics of premixed turbulent combustion. The goals of this chapter are to highlight the key global effects of turbulence on premixed flames and to discuss the following issues:

- What happens to a laminar premixed flame when it enters a turbulent flow?
- How does the influence of turbulence on premixed combustion appear generally?
- What differences between laminar and turbulent flames are best pronounced?
- What are the key general features of premixed turbulent flames?

The most important effects of turbulence on premixed combustion may be seen with the naked eye by comparing the images of laminar and turbulent Bunsen flames (see Figure 4.1). When a fuel–air mixture is drawn through a channel to the open atmosphere and is ignited near the channel exit, a conical, statistically stationary flame may be obtained under certain conditions. Such a flame is widely used in the experimental studies of premixed combustion and is called a Bunsen flame. It is stabilized near the channel rim, and the conical shape of the flame results from (i) the flame propagation from the stabilization zone into the unburned mixture, that is, to the axis of the flow and (ii) the axial convection of the propagating flame by the mean flow.

Typical images of laminar (left) and turbulent (right) Bunsen flames are shown in Figure 4.1. These photographs taken by Summerfield et al. (1955) clearly reveal the effect of turbulence on premixed combustion, because the mixture and the mean flow rate are the same in each case and a striking difference between the two flames is caused by inserting a grid into the flow (right image). The grid produces turbulence, with all other things being equal, and the grid-generated turbulence strongly changes the general appearance of the flame. Based on the discussed images, the effect of turbulence on the general appearance of a Bunsen flame is shown in Figure 4.2, where the left and right sketches are associated with the laminar and turbulent premixed flames, with all other things being equal.

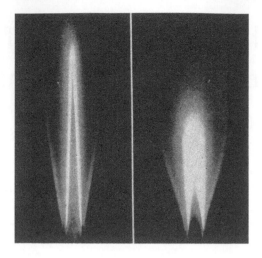

FIGURE 4.1 Photographs of laminar (*left*) and turbulent (*right*) Bunsen flames. (Reprinted from Summerfield, M., Reiter, S.H., Kebely, V., and Mascolo, R.W., *Jet Propul.* 25, 377–384, 1955. With permission.)

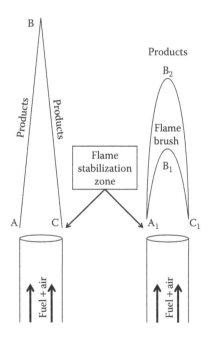

FIGURE 4.2 Sketches of laminar (*left*) and turbulent (*right*) Bunsen premixed flames.

MEAN FLAME BRUSH THICKNESS

First, Figures 4.1 and 4.2 show that a turbulent flame is much thicker than a laminar flame. This is a very important and well-documented quantitative difference between laminar and turbulent premixed combustion. This difference is well pronounced not only for Bunsen flames but also for all other known types of premixed flames.

Second, there is also a qualitative difference between laminar and turbulent Bunsen flames. The thickness of the laminar flame on the left image in Figure 4.1 is roughly the same at various distances from the burner exit, with the exception of the flame tip and stabilization zone. On the contrary, the thickness of the turbulent flame on the right image increases with the distance from the burner exit.

The increase in the mean thickness, δ_t, of the flame brush with distance X from the flame-stabilization zone is a typical feature of premixed turbulent combustion. For instance, this phenomenon is well pronounced not only in Bunsen flames but also in V-shaped flames (see Figure 4.3). If a rod is fixed at the exit of a Bunsen burner and the mixture is ignited near the rod, then an inverse conical (V-shaped) flame may be stabilized by a recirculation zone behind the rod. The conical shape of the flame results from (i) the flame propagation from the stabilization zone into the unburned mixture, that is, from the axis to the periphery of the flow and (ii) the axial convection of the propagating flame by the mean flow. Therefore, the burned (unburned) mixture occupies the central part (the periphery) of the flow in a V-shaped flame, while the opposite occurs in a Bunsen flame.

Even if a statistically stationary flame is considered, for example, the Bunsen flame in Figure 4.1, the V-shaped flame in Figure 4.3, or the confined flame in Figure 4.4, the increase in $\delta_t(X)$ is a manifestation of the development of a turbulent

FIGURE 4.3 A typical photograph of a premixed turbulent V-shaped flame. (Reprinted from Bill, R.G., Naimer, I., Talbot, L., Cheng, R.K., and Robben, F., *Combust. Flame*, 43, 229–242, 1981. With permission.)

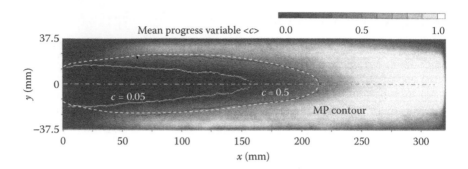

FIGURE 4.4 A confined, premixed turbulent flame stabilized by the recirculation zones caused by an abrupt channel expansion. The contours marked with $c = 0.05$ and $c = 0.5$ (these are the reference values of the mean combustion progress variable) show the leading edge of the flame brush and the mean flame surface, respectively. (Reproduced from Griebel, P., Siewert, P., and Jansohn, P., *Proc. Combust. Inst.*, 31, 3083–3090, 2007. With permission.)

flame as it is advected by the mean flow from the flame-stabilization zone. Similarly, the growth of the turbulent mixing layer is a manifestation of the development of turbulent diffusion, as discussed in the section "Taylor's Theory" in Chapter 3.

The development of premixed turbulent combustion is especially pronounced if an expanding flame is considered. For instance, Figure 4.5a shows the spatial profiles $\bar{c}(r)$ of the Reynolds-averaged combustion progress variable, obtained by Renou et al. (2002) from a statistically spherical, premixed flame that expanded in an unconfined turbulent flow after spark ignition. The maximum slope of the profile decreases as the flame kernel expands, thereby indicating an increase in the mean flame brush thickness. Indeed, the thickness (see triangles in Figure 4.5b), determined using the maximum gradient method,

$$\delta_t = \max^{-1} \left| \frac{d\bar{c}}{dr} \right|, \qquad (4.1)$$

as well as the visible flame speed $S_t = dR_f/dt$ (circles), increases with time after spark ignition. Here, r is the radial coordinate, and the mean flame radius, R_f, is associated with the isoscalar surface characterized by $\bar{c}(r = R_f) = 0.5$. The increase in $\delta_t(t)$ and $S_t(t)$ is a clear manifestation of the premixed turbulent flame development.

The simplest way to explain the increase in the mean turbulent flame brush thickness is to consider the effect to be basically similar to the growth of the turbulent mixing layer. The two phenomena are caused by the random advection of a thin layer that separates two fluids by turbulent eddies. In the former case, the unburned and burned mixtures are separated by a thin flame front (see Figures 5.17 and 5.18 in the section "Development of Premixed Turbulent Flame in the Flamelet Regime" in Chapter 5), which retains an inherently laminar structure if its thickness is much less than the Kolmogorov length scale. In the latter case, clean and polluted flows are separated by an infinitely thin (if the molecular diffusivity $D = 0$) surface (see the wrinkled bold curves in Figure 3.9). If the laminar flame speed, S_L, is substantially

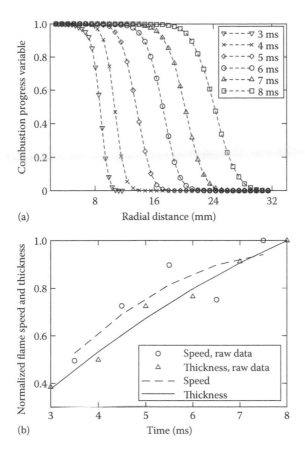

(a)

(b)

FIGURE 4.5 The evolution of the Reynolds-averaged combustion progress variable in an expanding, statistically spherical flame. (a) The radial profiles of $\bar{c}(r)$ measured by Renou et al. (2002) at different times (see legends) after spark ignition. (b) The mean flame brush thickness (*triangles*) and the observed flame speed dR_f/dt (*circles*). The curves show the second-order polynomial fit to the measured data.

less than u', then the influence of the self-propagation of the flame front on its large-scale topology is sometimes assumed to be of minor importance as compared with the role played by the random advection of the flame front by large-scale turbulent eddies. Within the framework of such a concept, Taylor's theory of turbulent diffusion, discussed in the section "Taylor's Theory" in Chapter 3, may provide a reasonable estimate of the increase in the mean turbulent flame brush thickness.

Indeed, the experimental data (see Figure 4.6 and Figures 21 and 22 in a review paper by Lipatnikov and Chomiak [2002a]) show that the increase in δ_t is described well by the equation

$$\delta_t^2 = 4\pi u'^2 \tau_L t \left\{ 1 - \frac{\tau_L}{t} \left[1 - \exp\left(-\frac{t}{\tau_L} \right) \right] \right\}$$

(4.2)

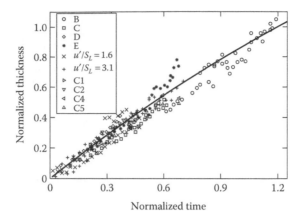

FIGURE 4.6 The normalized mean flame brush thickness, $\delta_t/(\sqrt{2\pi}L)$, versus the normalized time, t/τ_L. The curve has been calculated using Equation 4.2. The symbols show the experimental data with the experimental conditions labeled following the original papers. C1, C2, C4, and C5—data obtained by Renou et al. (2002) from expanding, spherical, unconfined, stoichiometric methane–air and propane–air flames. *Crosses and pluses*—data obtained by Atashkari et al. (1999) from expanding, spherical, confined, stoichiometric methane–air flames at two different u' specified in the legends. B, C, D, and E—data obtained by Goix et al. (1990) from stationary, V-shaped, lean hydrogen–air flames behind various turbulence-generating grids.

in many flames. Equation 4.2 results from Taylor's theory (see Equations 3.119 and 3.122), provided that $\delta_t^2 = 2\pi y^2$. Here, the factor 2π results from the difference between the definition of the thickness, δ_m, of the turbulent mixing layer in the section "Turbulent Diffusion" in Chapter 3 and the definition of the flame brush thickness by Equation 4.1. For example, Equation 3.106 yields $\max^{-1}\left|d\bar{Y}/d\xi\right| = \sqrt{2\pi}\delta_m$.

However, it would be too optimistic to consider Equation 4.2 to be a universal solution. For instance, Figure 4.7 shows the mean flame brush thicknesses of some V-shaped flames and Bunsen flames, evaluated using the maximum gradient method (see Equation 4.1). Here, the flame development time has been estimated as X/U, where X is the distance from the flame holder and U is the mean flow velocity at the burner exit. First, it is worth stressing again the increase in $\delta_t(X)$ for all these flames. Second, in some cases (see symbols 3, 6, and 10) Equation 4.2 describes the increase in the mean flame brush thickness well. For flames 1 and 2, the equation overestimates the increase in δ_t, and this trend may be associated with the countergradient scalar transport discussed in the section "Countergradient Scalar Transport in Turbulent Premixed Flames" in Chapter 6. For flames 4, 5, 7–9, and 11, Equation 4.2 underestimates the thickness. Certainly, such a simple test is not sufficient to claim that the equation is wrong, because the disagreement between the measured data and Equation 4.2 may be associated with a number of reasons: for example, (i) the aforementioned estimate of the flame development time is oversimplified; (ii) the Lagrangian length scale and timescale were not evaluated in the analyzed experiments; (iii) the mean thickness obtained by processing a profile of $\bar{c}(y)$ is greater

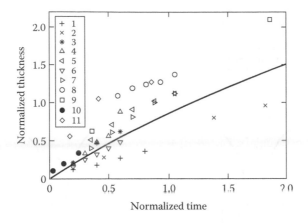

FIGURE 4.7 The normalized mean flame brush thickness, $\delta_t/(\sqrt{2\pi}L)$, versus the normalized time, $X/(U\tau_L)$. The curve has been calculated using Equation 4.2. Symbols 1–3 show the data obtained by Namazian et al. (1986) from V-shaped flames 1–3, respectively. Symbols 4–7 show the data obtained by Cheng et al. (1988) from V-shaped flames 1–4, respectively. Symbol 8 shows the data obtained by Shepherd (1996) from a V-shaped flame. Symbols 9–11 show the data obtained by Filatyev et al. (2005) from Bunsen flames 3c, 8b, and 8c, respectively.

than the actual δ_t, because the y-axis is not normal to the mean flame brush; (iv) the turbulence characteristics depend on X in the discussed flows.

The increase in the mean flame brush thickness is a very important feature of premixed turbulent combustion, which should be borne in mind. The available data allow us to claim that a typical premixed turbulent flame is a developing flame characterized by an increasing mean thickness. In other words, the development of a premixed turbulent flame is a long process, whereas the development of a laminar premixed flame takes a very short time.

Nevertheless, there are premixed turbulent flames that do not show an increase in δ_t. These are (a) twin counterflow flames stabilized in the vicinity of the stagnation plane resulting from the collision of two identical turbulent flows of premixed reactants, (b) a flame stabilized in a jet impinging at a wall, and (c) a flame stabilized by a low swirl. Sketches of these three flames are shown in Figure 4.8. The mean thickness of any of these flames is much greater than the laminar flame thickness and is statistically stationary. However, the physical mechanisms that control the mean thickness of the three aforementioned flames differ substantially from the physical mechanisms that control the mean thicknesses of other premixed turbulent flames.

The point is that the mean flow is two- or three-dimensional (2D or 3D) and divergent in the three cases sketched in Figure 4.8. For instance, the mean flow of an unburned mixture ahead of an impinging jet flame (or half an axially symmetrical twin counterflow flame) is well approximated as follows: $\bar{u} = U - a_r x$ and $\bar{v} = a_r r/2$, where $\bar{u}$ and $\bar{v}$ are the axial and radial components, respectively, of the flow velocity vector, r is the radial distance, and the x-axis is normal to the mean flame brush and points to the products with $x = 0$ at the leading edge of the flame brush. The

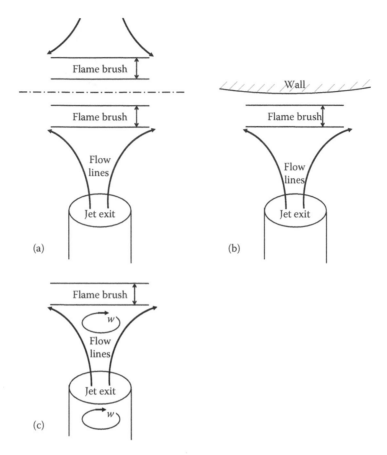

FIGURE 4.8 A sketch of (a) twin counterflow flames, (b) an impinging jet flame, and (c) a swirl-stabilized flame.

same approximation held within the flame if the density were constant. The axial gradient, a_t, of the mean flow velocity, called the mean strain rate, bounds the random advection of the thin flame front. For instance, as turbulent eddies convect the flame front upstream in an impinging jet (sketched in Figure 4.8b), the flame front is counterconvected by a stronger mean axial flow due to an increase in $\bar{u}$ with the distance from the wall. This mechanism restricts fluctuations of the axial position of the flame front. Therefore, the mean flame brush thickness δ_t, that is, the length of the x-interval where the thin flame front may be observed, depends on the mean strain rate (see Section 4.5 in a review paper by Lipatnikov and Chomiak [2002a], as well as papers by Lipatnikov and Chomiak [2002b] and Biagioli [2004]). More specifically, δ_t is decreased when a_t is increased.

If $\rho = $ const for simplicity and, based on Taylor's theory discussed in section "Taylor's Theory" in Chapter 3, we assume that the rate of increase of the mean flame brush thickness due to random advection of the thin flame front by turbulent eddies is of the order of u' or less, then δ_t should not increase when the difference, $\Delta\bar{u}$ in the mean axial velocities at the leading and trailing edges of the flame brush is of the order of u'. Accordingly, the maximum thickness is estimated as $\Delta \bar{u} \approx a_t \delta_{t,\max} \approx u'$, that is, it depends straightforwardly on the mean strain rate. Just after ignition, $\delta_t \approx \delta_L$, the variation $\Delta \bar{u} \approx a_t \delta_L \ll u'$, in the mean axial flow velocity across the flame brush is much less than the rate of increase of δ_t due to random advection of the instantaneous flame front by turbulent eddies, and the mean flame brush thickness increases. However, as soon as $\delta_t \approx \delta_{t,\max} \propto u'/a_t$, the mean velocity variation becomes so large that it cannot be overcome by random advection, and the thickness becomes statistically stationary.

Thus, although the nonuniformity of the mean flow of the unburned mixture may limit the development of a premixed turbulent flame under certain conditions (e.g., see the flames sketched in Figure 4.8), the vast majority of the other premixed turbulent flames are developing flames, with the flame development manifesting itself in the increase in the mean flame brush thickness. In particular, the existence of a fully developed premixed turbulent flame has not yet been shown in the simplest case of an unconfined unperturbed flame, provided that $\rho_b < \rho_u$. Here, the term "unperturbed" means that the flame is statistically planar and 1D and propagates against a statistically stationary, planar, and 1D turbulent flow of a homogeneous unburned mixture. Such a case is the standard case for the theoretical studies, as well as for the development of various models, of premixed turbulent combustion.

Can an unperturbed premixed turbulent flame be fully developed in principle? At least in the case of a constant density, this seems to be possible, but it requires a long time, for example, of the order of $\tau_t (u'/S_L)^2$, as argued by Zimont (2000) and discussed in the section "Modeling of Effects of Turbulence on Premixed Combustion in RANS Simulations" in Chapter 7 (see Equation 7.19 through 7.21). This time is much longer than the turbulence timescale, τ_t, if the rms velocity u' is significantly greater than S_L, as occurs in a typical turbulent flame. On the contrary, the burning time is comparable with τ_t in a typical combustor. This estimate explains why fully developed premixed turbulent flames are not observed under typical conditions. Moreover, as discussed at the end of the section "Instabilities of Laminar Flamelets in Turbulent Flows" in Chapter 5, the influence of heat release on the flow does not allow an unperturbed premixed turbulent flame to fully develop in the case of a variable density.

TURBULENT FLAME SPEED AND BURNING VELOCITY

An inspection of the images shown in Figure 4.1 or the sketches drawn in Figure 4.2 reveals one more very important difference between the general appearances of the laminar and turbulent flames, with all other things being equal. The length of a laminar Bunsen flame is substantially greater than that of a turbulent flame. Accordingly, the angle φ between the flame surface and the main flow direction is markedly smaller in the laminar case. If, for simplicity, we assume that the mean flow velocity $\bar{u}_n$, normal to such a flame, is equal to $U \sin \varphi$, where U is the mean axial flow velocity in the nozzle, then the flame speed is equal to

$$S_t = \bar{u}_n = U \sin \varphi \qquad (4.3)$$

due to the stationarity of the mean flame surface. Equation 4.3 implies that the laminar flame speed is significantly less than the speed S_t of the self-propagation of the turbulent flame, because the angle φ is smaller in the former case (see Figures 4.1 and 4.2). Certainly, it is difficult to define the mean surface of the turbulent flame due to its thickness, but Equation 4.3 yields $S_t > S_L$ for any isoscalar surface (e.g., a surface characterized by a reference value $T_u < T_0 < T_b$ of the mean temperature) within the flame brush, that is, for any isoscalar surface bounded by the lines $A_1B_1C_1$ and $A_1B_2C_1$ in Figure 4.2.

Alternatively, we can compare the burning rates of the laminar and turbulent Bunsen flames shown in Figure 4.1. Because the flow rates were the same in both cases and the fuel–air mixture was fully burned, the total burning rates were also the same. However, the specific burning rate $\rho_u U_t$, evaluated per unit flame surface area, was higher in the turbulent flow, because the area of any mean flame surface on the right side in Figure 4.2 is smaller than the area of the laminar flame surface on the left side in Figure 4.2. Therefore, the burning velocity U_t, that is, the specific burning rate normalized with the unburned gas density ρ_u, was higher in the turbulent flow.

Thus, turbulence increases both the flame speed and the burning velocity as compared with the laminar flame, with all other things being equal. An increase in S_t and U_t by u' is often considered to be the most important global effect of turbulence on premixed combustion. However, the substantial thickness and growth of the flame brush appear to be equally important phenomena.

DETERMINATION OF TURBULENT FLAME SPEED AND BURNING VELOCITY

Before further considering the influence of turbulence on the flame speed and burning velocity, it is worth discussing the problem of determining these quantities. Intuitively, the flame speed, S_t, is the speed of a mean flame surface with respect to the mean flow, while the burning velocity, U_t, is the burning rate per unit area of a mean flame surface, normalized using ρ_u. From the theoretical viewpoint, the turbulent burning velocity is a well-defined quantity in the case of an unperturbed premixed flame. To determine U_t in this case, one may average Equation 1.79 and integrate the mean rate $\overline{W}/\rho_u$ across the flame brush, that is,

$$U_t = \frac{1}{\rho_u} \int_{-\infty}^{\infty} \overline{W}(x)\,dx. \tag{4.4}$$

However, even in this simplest case, a physically meaningful definition of S_t is an issue (Lipatnikov and Chomiak, 2002c), because different isoscalar surfaces within a growing mean flame brush are characterized by different speeds with respect to the mean flow. In the case of a laboratory premixed turbulent flame, which may be curved and/or may propagate in a statistically 2D or 3D flow, the problem of determining U_t and S_t is much more difficult, as discussed in this section.

In a general case, the turbulent burning velocity and flame speed may be measured either locally or globally. The local burning velocity is determined by Equation 4.4, where integration is performed along the local normal to the mean flame brush. The global burning velocity is equal to

$$U_t = \frac{Q}{\rho_u A_t},$$ (4.5)

where

Q is the total oncoming mass flux of an unburned mixture
A_t is the area of a mean flame surface

The local flame speed is equal to the magnitude of the difference between the observed speed of an infinitesimal segment of a reference surface chosen to characterize the mean flame position and the mean flow velocity at this segment. In a statistically stationary case, the local flame speed is simply equal to the magnitude $|\bar{u}_n(\mathbf{x})|$ of the local mean flow velocity normal to the mean flame position, that is, $S_t(\mathbf{x}) = |\bar{u}_n(\mathbf{x})|$. Finally, the global flame speed characterizes the entire flame brush. For instance, if the mean surface of a Bunsen flame is assumed to have a conical shape with an angle of 2φ at the tip, then the global flame speed may be evaluated using Equation 4.3. If the same conical mean flame surface is used for evaluating the global burning velocity and flame speed, then these two quantities are equal to each other in the aforementioned Bunsen flame. However, in a general case, the values of the local and global turbulent flame speeds and that of the local and global turbulent burning velocities differ from each other and are sensitive to the choice of a reference surface associated with the mean flame position.

The turbulent flame speed is sometimes called the displacement speed, S_d, while the burning velocity is called the consumption speed, S_c, or the consumption velocity, U_c. The use of such terms appears to be acceptable as far as the global flame (displacement) speed and burning (consumption) velocity are concerned. This is not so for the local displacement speed, s_d, and the local consumption speed, s_c, or velocity, u_c, because these terms and symbols are commonly used for characterizing the local burning rate in thin, inherently laminar, instantaneous flame fronts that separate the unburned and burned mixtures in a turbulent flow, as will be discussed in the next chapter (e.g., see Equation 5.84). In the latter case, s_c or u_c is not integrated across the turbulent flame brush, and it fluctuates in space and time, contrary to the local burning velocity evaluated using Equation 4.4. In this book, the terms local displacement speed, s_d, and local consumption speed, s_c, or velocity, u_c, will be used in the latter sense, that is, s_d is not a local turbulent flame speed and s_c or u_c is not a local turbulent burning velocity.

In a laminar flow, due to the stationarity and smallness of the flame thickness, both the burning velocity and the flame speed are equal to the same S_L in the first approximation, with the flame speeds determined for various isoscalar surfaces being approximately equal to one another. If a laminar flame propagates normal to itself at a distance $\Delta x = S_L \Delta t$ during the time interval Δt, then $\rho_u A \Delta x$ g of the reactants is burned, where A is the flame surface area. Accordingly, the burning velocity is equal to S_L. This equality is exact either for a planar, 1D laminar flame propagating in a spatially uniform mixture or for an infinitesimally thin laminar flame. For a typical curved or strained laminar flame characterized by a finite thickness, δ_L, the difference between the burning velocity and the flame speed is much less than S_L, because δ_L is much less than the length scales of the flow variations.

In a turbulent flow, the definition of U_t and S_t is a complicated issue, because the choice of the mean flame surface is ambiguous due to the substantial thickness and development of a turbulent flame brush. First, due to the increase in the turbulent flame brush thickness, the speeds of the different isoscalar surfaces are different and, hence, the definition of S_t is ambiguous. For instance, Equation 4.2 indicates that the difference in the speeds of the leading and trailing edges of the turbulent flame brush scales as follows $d\delta_t/dt \propto u'$ if the flame development time t is substantially less than the Lagrangian timescale, τ_L, of turbulence.

Second, due to the large thickness of a turbulent flame brush, the mean flow velocity varies substantially within it. In the statistically stationary, planar, 1D case, the velocity variations could be taken into account by integrating the averaged continuity Equation 3.78, which yields $\overline{\rho}\tilde{u} = \text{const}$. However, such a solution works neither in developing flames nor in 2D (or 3D) mean flows. For instance, Figure 4.9 clearly shows that the mass flux $\overline{\rho}\tilde{u}$ normal to the mean turbulent V-flame brush drops by almost a factor of 4 from the leading edge to the trailing edge, whereas the counterpart flux is almost constant in a typical laminar flame the thickness of which is much less than the length scales of the flow variations.

Accordingly, the speed of an isoscalar surface within a thick turbulent flame brush with respect to the local flow (i.e., the difference between the observed speed of the surface and the mean flow velocity on it) depends on the choice of the surface, thereby contributing to the ambiguity of the definition of a turbulent flame speed. For instance, for the twin counterflow flames sketched in Figure 4.8a, the flow field is commonly modeled as $|u| = |a_r x|$, $v = a_r r/2$ in the vicinity of the symmetry axis $r = 0$, outside the nozzle, and before the flame brush. Here, $x = 0$ at the stagnation

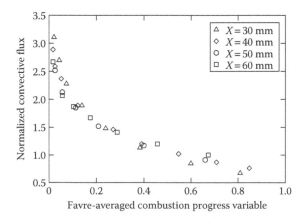

FIGURE 4.9 Mean normal convective fluxes measured by Gouldin (1996) across the V-flame brush at four downstream distances from the nozzle, specified in the legends. The fluxes are normalized using ρ_u and S_L.

plane and a cylindrically symmetrical case is considered. Even if we restrict ourselves to a constant-density "flame" for simplicity, the difference in the speeds of the self-propagation of the leading and trailing edges, which are equal to the local mean axial flow velocities normal to the mean flame brush in the statistically stationary case considered, is of the order of $a_t \delta_t$, that is, comparable with S_t under typical conditions.

As regards the turbulent burning velocity, its definition is unambiguous at least for a statistically planar flame, because the areas of the various isoscalar surfaces are equal to one another and the x-axis in Equation 4.4 is a well-defined straight line in this case. Such a planar flame could develop and/or could propagate in a 2D or 3D mean flow (statistically planar flames stabilized in 2D or 3D mean flows are sketched in Figure 4.8). For these reasons, the turbulent burning velocity is considered to be a better-defined quantity than the turbulent flame speed and U_t will be preferably used henceforth in this book. In particular, any model of a developing unperturbed premixed turbulent flame addresses U_t, rather than S_t, which is sensitive to the choice of the reference surface due to the growth of the flame brush. When discussing various combustion models, henceforth in this book, U_t will designate the local burning velocity determined using Equation 4.4. It is worth remembering, however, that experimenters often measure the global turbulent burning velocity. Accordingly, when discussing the experimental data, U_t may designate the global burning velocity.

If a turbulent flame brush is curved, then the definition of the turbulent burning velocity is also an issue, because, due to the significant thickness of the flame brush, (i) the normals to different reference surfaces may not be parallel, for instance, the x-coordinate in Equation 4.4 may be a curvilinear coordinate and (ii) the areas of the different reference surfaces may be substantially different. The latter effect may be of substantial importance in experimental investigations. For instance, Figure 4.10 clearly shows that the values of a global turbulent burning velocity, obtained from

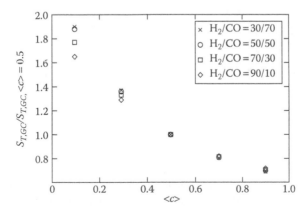

FIGURE 4.10 The dependence of the normalized value of the global turbulent burning velocity on the mean combustion progress variable used to set a reference isoscalar surface for evaluating U_t. The normalization was done using the burning velocity measured at $\langle c \rangle = 0.5$. (Adapted from Venkateswaran, P., Marshall, A., Shin, D.H., Noble, D., Seitzman, J., and Lieuwen, T., *Combust. Flame*, 158, 1602–1614, 2011. With permission.)

H_2/CO–air Bunsen flames, depend substantially on the mean combustion progress variable that is used to set a reference surface for evaluating U_t. If the reference surface is associated with the leading (trailing) edge of the flame brush, the surface area is smaller (larger) (see on the right side of Figure 4.2) and the measured global burning velocity is higher (lower).

Due to the previously discussed effects (the increase in the mean flame brush thickness and the nonuniformity of the mean velocity field within the thick turbulent flame brush), differences as large as 400% (i.e., $S_t/U_t \approx 4$) between turbulent flame speeds and burning velocities, measured using different methods, have been reported in the literature, as discussed in detail elsewhere (see Section 5.2 in a review paper by Lipatnikov and Chomiak [2010]). Such effects will be considered in the section "Burning Velocity and Speed of an Expanding Spherical Flame" in Chapter 7, for an expanding, statistically spherical, turbulent premixed flame.

Therefore, when analyzing the experimental or numerical data, particular attention should be paid to the method invoked to evaluate U_t or S_t. Even if different methods yield qualitatively similar dependencies of U_t (or S_t) on the basic mixture and flow characteristics, as discussed in detail in Sections 3.3 and 3.4 of a review paper by Lipatnikov and Chomiak (2002a), the values of U_t (or S_t) obtained using different methods may be substantially different from a quantitative viewpoint. A significant scatter of the values of U_t and S_t, measured by different groups using different methods for approximately the same mixture and turbulence characteristics, is shown in Figure 4.11.

Nevertheless, it is worth stressing once more that, despite the aforementioned difficulties in determining U_t or S_t in a general case and the significant quantitative

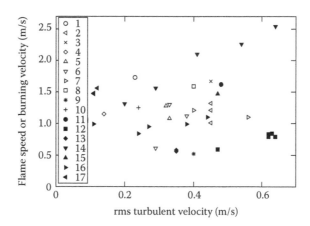

FIGURE 4.11 The turbulent flame speeds and burning velocities obtained by different groups from stoichiometric methane–air turbulent flames under room conditions. 1: Ho et al. (1976); 2: Smith and Gouldin (1979); 3: Gouldin and Dandekar (1984); 4: Gulati and Driscoll (1986a,b); 5 and 6: Cheng et al. (1988), V-shaped and impinging jet flames, respectively; 7, 8, and 9: Cheng and Shepherd (1991), V-shaped, Bunsen, and impinging jet flames, respectively; 10: Chan et al. (1992); 11: Cheng (1995); 12: Bourguignon et al. (1996); 13: Shepherd (1996); 14: Aldredge et al. (1998); 15: Shy et al. (2000a); 16: Filatyev et al. (2005); 17: Savarianandam and Lawn (2006).

scatter of the experimental data, at least the local turbulent burning velocity is well defined in the case of a statistically planar, 1D, premixed flame. Moreover, if the mean flow is also statistically planar and 1D, then the local and global burning velocities are equal to one another. Accordingly, U_t is a theoretically meaningful quantity and will be widely used to discuss various models of the influence of turbulence on premixed combustion in Chapter 7.

To minimize the scatter in the values of the turbulent burning velocities and flame speeds due to the differences in the methods of measuring them, the following discussion will be based mainly on a single experimental database obtained using a single, well-defined method. We will analyze the behavior of the turbulent burning velocity, invoking the experimental database encompassed by Karpov and Severin (1978, 1980). The database appears to suit the goals of the book for a number of reasons. First, it is representative, that is, it offers an opportunity to reveal the key trends in the behavior of U_t, which were also documented by many other groups using different methods, as reviewed in Sections 3.3 and 3.4 of a paper by Lipatnikov and Chomiak (2002a). Second, it is extensive, for example, it contains data of a large set of substantially different mixtures. Third, it is reliable and useful for model validation, because the data have been obtained utilizing a well-defined method under well-defined conditions. Because these pioneering experiments are rarely addressed in the combustion literature in English language, it is worth describing them in detail.

Karpov and Severin (1978) measured the global turbulent burning velocity as follows. A premixed gas was ignited by a spark in the center of a constant-volume bomb equipped with eight fans that generated homogeneous, isotropic turbulence in the central part of the bomb. After spark ignition, a statistically spherical flame kernel expanded toward the walls of the bomb. As the flame kernel grew, the pressure in the bomb increased. The pressure increase, $p(t)$, was recorded under laminar and turbulent conditions. In the laminar case, images of the expanding flames were simultaneously obtained using high-speed Schlieren cinematography and the radii, R_f, of the flames were evaluated by processing these images. Thus, for a particular combustible mixture, the raw experimental data consisted of (i) the radii $R_f(t, u' = 0)$ of the laminar flame measured at different instants after spark ignition, (ii) the pressure diagram $p(t, u' = 0)$ obtained from the same laminar flame, and (iii) the pressure diagrams $p(t, u')$ obtained from various turbulent flames characterized by different rms velocities u'. In order to reduce the influence of the variations in p and T_u due to the mixture compression on the burning rate, the measurements were restricted to the so-called prepressure period characterized by $p(t) - p_0 \ll p_0$ and $T_u(t) - T_0 \ll T_0$, where the subscript 0 designates the conditions at the ignition instant.

As far as the aforementioned raw data are concerned, the effect of turbulence on premixed combustion manifested itself in a more rapid increase in $p(t, u')$ as compared with $p(t, u' = 0)$, that is, turbulence increased the burning rate in these measurements.

In such an experiment, the pressure in the bomb increases due to the transformation of the chemical energy bound in the reactants into the thermal form. Accordingly, the pressure increase is straightforwardly controlled by the mass of burned mixture (Lewis and von Elbe, 1961). Therefore, the measured pressure diagrams allow one to evaluate the total burning rate. To determine the global turbulent burning velocity, that is, the normalized (with ρ_u) burning rate per unit flame surface area, a mean flame surface should be introduced.

To do so, first, Karpov and Severin (1978) assumed that the volume occupied by the intermediate states (between unburned and burned) of the mixture was much less than the volumes V_u and V_b occupied by the unburned gas and the adiabatic combustion products, respectively, in both the laminar and the turbulent cases. In the former case, this assumption is justified by the small thickness of the laminar premixed flame. The validity of this assumption in the latter case will be discussed in the section "Experimental Data" in Chapter 5. Second, Karpov and Severin (1978) associated the mean surface of a turbulent flame with an equivalent spherical surface that enveloped the spherical volume filled solely by the products. Third, due to the straightforward link between the pressure increase and the mass of burned gas, the radius of this equivalent spherical surface is directly related to the pressure. For each particular mixture, such a relation, $R_f(p)$, was determined by comparing $R_f(t,u' = 0)$ and $p(t,u' = 0)$, obtained from the laminar flame, which had a spherical shape and a small thickness $\delta_L \ll R_f$ in the measurements discussed earlier. Fourth, the radius of the equivalent spherical surface in a turbulent flow was evaluated by substituting the measured $p(t,u')$ into the relation $R_f(p)$, obtained from the counterpart laminar flame. Finally, the global burning velocity was determined using the expression

$$U_t = \frac{\rho_b}{\rho_u} \frac{dR_f}{dt} \tag{4.6}$$

and approximating the obtained $R_f(t)$ curves with straight lines in the range of 15 mm $< R_f < $ 40 mm. Equation 4.6 is valid for the equivalent spherical surface (or for an infinitely thin laminar spherical flame). Indeed, because the mass of burned gas in the considered spherical volume

$$m_b = \frac{4\pi}{3} \rho_b R_f^3 \tag{4.7}$$

increases with the rate

$$\frac{dm_b}{dt} = 4\pi \rho_b R_f^2 \frac{dR_f}{dt}, \tag{4.8}$$

and the area of the equivalent spherical (or the laminar flame) surface is equal to $4\pi R_f^2$, we have

$$U_t = \frac{1}{4\pi \rho_u R_f^2} \frac{dm_b}{dt}. \tag{4.9}$$

Substitution of Equation 4.8 into Equation 4.9 yields Equation 4.6.

It is worth stressing that in addition to the method for measuring the global turbulent burning velocity, developed by Karpov and Severin (1978) and discussed earlier, other methods for evaluating the global and local S_t and U_t were also elaborated. For further discussion on the problem of measuring the turbulent burning velocity and flame speed, interested readers are referred to Sections 3.2 and 5.2 of review papers by Lipatnikov and Chomiak (2002a and 2010), respectively, and to the references cited therein, in particular to the contributions by Shepherd and Kostiuk (1994), Shepherd (1995, 1996), Kostiuk and Shepherd (1996), and Gouldin (1996).

DEPENDENCE OF TURBULENT BURNING RATE ON RMS VELOCITY AND LAMINAR FLAME SPEED

A typical dependence of the burning velocity, U_t, on the rms turbulent velocity, u', is shown in Figure 4.12. It is worth stressing that the following discussion will be limited to highlighting the key features in the behavior of the turbulent burning velocity, whereas the physical mechanisms that control the behavior of U_t will be discussed in Chapter 5.

The most known effect of turbulence on premixed combustion consists of an increase in U_t by u'. Inspection of Figure 4.12 and many other similar figures reported in the literature indicates several ranges of variations in u', characterized by different responses of the burning velocity to an increase in the rms velocity.

First, if $u' < u'_l$, the dependence of U_t on u' appears to be well described by a straight line:

$$U_t = S_L + Cu', \qquad (4.10)$$

where C is independent of u' but may depend on the mixture composition.

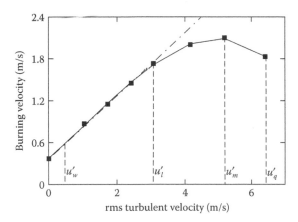

FIGURE 4.12 A typical dependence of the burning velocity on the rms turbulent velocity. The symbols show the experimental data obtained by Karpov and Severin (1978) from a propane–air mixture with $\Phi = 1.1$. The solid curve approximates the data with a third-order polynomial. The dotted-dashed line shows the linear approximation of the data obtained at $u' < u'_l$.

Second, if $u'_l < u' < u'_m$, the burning velocity is still increased by u', but the slope dU_t/du' is decreased when u' is increased. This phenomenon is called bending or the bending effect.

Third, if $u'_m < u' < u'_q$, the burning velocity depends weakly on the rms velocity or U_t is decreased when u' is increased. Note that the former trend (weak dependence) is well documented, whereas the latter trend (decrease), shown in Figure 4.12, has been observed in a few experimental studies of expanding, statistically spherical, premixed turbulent flames.

Finally, combustion is quenched by strong turbulence and the flame cannot expand after spark ignition if $u'_q < u'$.

It is worth stressing that the quantities u'_l, u'_m, and u'_q are not universal and depend substantially on the mixture composition. For instance, Figure 4.13 shows that for a richer ($\Phi = 1.67$) propane–air mixture, u'_m is substantially lower than for another propane–air mixture ($\Phi = 1.25$) that is closer to the stoichiometric mixture. A higher velocity u'_m is commonly associated with a higher laminar flame speed, with all other things being equal.

The aforementioned four ranges ($u' < u'_l$, $u'_l < u' < u'_m$, $u'_m < u' < u'_q$, and $u'_q < u'$) of variations in the rms velocity are still insufficient to fully describe the response of the burning velocity to turbulence. The point is that the data shown in Figure 4.12 have been obtained either from a laminar flame or from turbulent flames characterized by $u' > S_L$. In the literature, very little data obtained from the flames characterized by $u' < S_L$ can be found, and these data are controversial. Therefore, the use of Equation 4.10 to evaluate the burning velocity in weakly turbulent (i.e., $u' < S_L$) flames does not seem to be well justified. Certainly, the parameterization of the entire range of $u' < u'_l$ with a single linear relation is very attractive due to its simplicity and the long-term tradition reflected in the vast majority of the textbooks on combustion. However, alternative standpoints should not be rejected until weakly turbulent premixed flames are well studied.

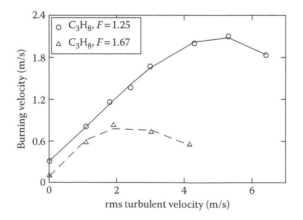

FIGURE 4.13 The burning velocity versus the rms velocity for two rich ($\Phi = 1.25$ and 1.67) C_3H_8–air mixtures. The symbols show the data measured by Karpov and Severin (1978). The curves approximate the data with third-order polynomials.

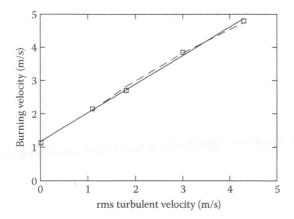

FIGURE 4.14 The burning velocity versus the rms turbulent velocity. The symbols show the data obtained by Karpov and Severin (1980) from lean ($\Phi = 0.71$) hydrogen–air flames. The straight line linearly approximates the entire data set. The dashed curve shows the power-law approximation $U_t \propto u'^{0.6}$ of the data associated with $S_L < u'$.

As regards the "linear" increase in $U_t(u')$ reported in many papers, it is worth remembering that the burning velocities measured in a limited range of rms velocities ($S_L < u' < u'_t$) may be approximated by various dependencies $U_t(u')$. For instance, Figure 4.14 shows the experimental data (symbols) by Karpov and Severin (1980), which are very well approximated with a straight solid line if the entire range of $0 \le u' < u'_t$ is considered. However, due to the lack of burning velocities measured at $0 \le u' < S_L$, an analysis of a narrower range of $S_L < u' < u'_t$ is equally justified. If the latter range is considered, then the same data are equally well approximated (dashed curve) by a substantially nonlinear expression $U_t \propto u'^{0.6}$.

Moreover, the use of Equation 4.10 may result in difficulties when describing the dependence of the burning velocity on the turbulence length scale L and mixture characteristics, such as the laminar flame speed S_L and thickness $\delta_L = a_u/S_L$, where a_u is the heat diffusivity of the unburned gas.

For instance, Figure 4.15 shows the dependencies of U_t on u' obtained from various mixtures. Each dependence of $U_t(u')$ considered separately shows the trends already emphasized when discussing Figure 4.12. However, the data considered all together indicate two more trends. First, the turbulent burning velocity is increased by the laminar flame speed. Second, the slope dU_t/du' is also increased by S_L in the range of $u' < u'_t$.

To describe the latter trend, the coefficient C in Equation 4.10 should be increased by S_L. At the same time, this coefficient must be independent of u' if U_t is assumed to depend linearly on u' at $u' < u'_t$. Therefore, if, following common practice, we assume that U_t depends on u', L, S_L, and δ_L, then for dimensional reasons, we have $C = C(L/\delta_L) = C(LS_L/a_u)$ and C should be an increasing function of its argument for dU_t/du' to be increased by S_L. On the face of it, such a parameterization could be acceptable, but there is a weak point. The use of an increasing $C(L/\delta_L)$ implies that both the laminar flame speed and an integral length scale of turbulence affect the turbulent

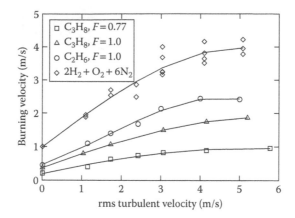

FIGURE 4.15 The burning velocity versus the rms turbulent velocity for various mixtures specified in the legends, where F designates the equivalence ratio. The symbols show the data measured by Karpov and Severin (1978, 1980). The curves approximate the data with third-order polynomials.

burning velocity equally strongly. However, the available experimental data indicate that U_t depends on L substantially weakly than on S_L (see Equations 4.11 through 4.25 and Section 3.3 in a review paper by Lipatnikov and Chomiak [2002a]).

Therefore, although the parameterization of the burning velocity by a single linear Equation 4.10 in the range of $0 \leq u' < u'_i$ is very attractive for simplicity reasons, it seems to be more consistent to consider the ranges of weak ($0 \leq u' < S_L$) turbulence and moderate ($S_L < u' < u'_m$) turbulence separately from one another. The following discussion will mainly focus on moderately ($S_L < u' < u'_m$) turbulent flames, because (i) they have been studied more extensively than weakly ($0 \leq u' < S_L$) or strongly ($u'_m < u'$) turbulent flames and (ii) they are characterized by the highest burning rate and, hence, are of the most practical importance.

The aforementioned hypothesis that the dependencies of U_t on u' are different at weak turbulence and moderate turbulence is further supported by processing the most extensive experimental databases on turbulent burning velocities and flame speeds.

Empirical Parameterizations of Turbulent Burning Velocity and Flame Speed

When parameterizing the turbulent burning velocity and flame speed, turbulence is commonly characterized by the rms velocity, u', and an integral length scale, L, while the mixture is commonly characterized by the laminar flame speed, S_L, and either the thickness, $\delta_L = a_u/S_L$, or the heat diffusivity, a_u. Definitely, this list of controlling parameters is not complete, and U_t (or S_t) may also depend on the density ratio; Prandtl, Schmidt, and Lewis numbers; flame development time; external pressure gradient or acceleration (Veynante and Poinsot, 1997); mean flow velocity

(Filatyev et al., 2005); etc. In the remainder of this section, the latter parameters are not taken into account due to the scarcity of target-directed experimental research into the effects of most of the parameters on the turbulent burning velocity (or flame speed) and for the following reasoning:

First, when processing the experimental data, the eventual dependence of U_t (or S_t) on the density ratio is not addressed because it is masked by the dependence of U_t (or S_t) on S_L. Indeed, the variations in the density ratio are accompanied by the variations in the laminar flame speed in a typical experiment, with the former variations being much less (in percentage) than the latter. For instance, for lean ($\Phi = 0.6$) and stoichiometric methane–air mixtures, the laminar flame speeds differ by 3.5 times (0.8 and 0.36 m/s, respectively), whereas the difference in the density ratios is less than 35% (5.6 and 7.5, respectively).

Second, because the Prandtl number is of the unity order and roughly the same for various hydrocarbon–air mixtures, the difference in the heat diffusivity, a_u, and the kinematic viscosity, v_u, of the mixture is commonly neglected, that is, Pr = 1 in the following if the opposite is not specified. As regards the influence of the differences in the molecular transport coefficients of the main reactants and heat, that is, Sc and Le, on the turbulent burning velocity and flame speed, this very important and extensively studied issue will be discussed in the section "Dependence of Turbulent Burning Velocity on Differences in Molecular Transport Coefficients."

Third, the dependence of U_t on the flame development time will be addressed using the physical arguments in the sections "Development of Premixed Turbulent Flame in the Flamelet Regime" in Chapter 5 and "Zimont Model of Burning Velocity in Developing Turbulent Premixed Flames" in Chapter 7.

Fourth, the dependence of U_t (or S_t) on the external pressure gradient or acceleration (Veynante and Poinsot, 1997) and mean flow velocity (Filatyev et al., 2005) is a poorly studied issue. Further research is absolutely needed in order to better understand and model these effects.

Thus, in the remainder of this section, the ratio of U_t/u' is considered to be a function of u'/S_L and L/δ_L. Such a dependence may be rewritten in different forms using the following dimensionless numbers: the Damköhler number,

$$\mathrm{Da} = \frac{\tau_t}{\tau_c}; \tag{4.11}$$

the turbulent Reynolds number,

$$\mathrm{Re}_t = \frac{u'L}{v_u} \approx \frac{u'L}{a_u}; \tag{4.12}$$

and the Karlovitz number,

$$\mathrm{Ka} = \frac{\sqrt{\mathrm{Re}_t}}{\mathrm{Da}}, \tag{4.13}$$

where

$$\tau_c = \frac{\delta_L}{S_L} = \frac{a_u}{S_L^2} \tag{4.14}$$

is the chemical timescale and

$$\tau_t = \frac{L}{u'} \tag{4.15}$$

is the turbulence timescale. It is worth stressing that the Karlovitz number (Ka) has its own physical meaning (see the section "Variations in Flamelet Structure due to Turbulent Stretching" in Chapter 5) and should not be considered only as a simple combination of the Damköhler and Reynolds numbers.

Lipatnikov and Chomiak (2002a) processed five different experimental databases on turbulent flame speeds and burning velocities using the following parameterization:

$$\ln\left(\frac{U_t - mS_L}{u'}\right) = b + d \ln\left(\frac{\mathrm{Da}}{\mathrm{Re}_t^q}\right). \tag{4.16}$$

First, each database was reduced to a database associated with $S_L < u' < u'_m$ at moderate turbulence. Here, u'_m, rather than u'_l, is used as the maximum rms velocity that bounds the domain of moderate turbulence, because the former velocity is a better-defined quantity. Second, a least square fit was applied to the database by specifying the particular values of the parameters m and q. Third, each of the parameters was varied (q with a step equal to 0.1 and m was equal to either 0 or 1) and the scatter of the experimental data around the dependence given by Equation 4.16 was calculated as follows:

$$\Xi^2 = \sum_{j=1}^{N}\left[U_t - mS_L - u'\left(\frac{\mathrm{Da}}{\mathrm{Re}_t^q}\right)^d \exp(b)\right]^2, \tag{4.17}$$

where N is the number of experimental points, U_t is the measured turbulent burning velocity or flame speed, and the parameters b and d were evaluated using a least square fit, as previously mentioned. Fourth, the dependence of the scatter Ξ on the parameters m and q was analyzed to find a set of m and q that yielded the minimum scatter.

In all the cases studied, the best approximation of the database at moderate turbulence was obtained with $m = 0$, thereby implying that different parameterizations should be applied to weakly and moderately turbulent flames. Indeed, the application of Equation 4.16 with $m = 0$ to $u' \to 0$ yields $U_t \to 0$, whereas $U_t(u' = 0)$ should be equal to S_L.

The following parameterizations of the database at moderate turbulence have been obtained by Lipatnikov and Chomiak (2002a):

$$U_t \propto u' \mathrm{Ka}^{-0.31} \qquad (4.18)$$

for the database by Karpov and Severin (1978 and 1980);

$$U_t \propto u' \mathrm{Da}^{0.44} \qquad (4.19)$$

for the database by Kido et al. (1989), who also recorded the pressure rise during the expansion of a statistically spherical, premixed turbulent flame in a fan-stirred bomb;

$$U_t \propto u' \mathrm{Da}^{0.37} \mathrm{Re}_t^{0.04} \qquad (4.20)$$

for the database by Kobayashi et al. (1996, 1998), who obtained the global turbulent burning velocities from Bunsen flames at elevated pressures;

$$S_t \propto u' \mathrm{Da}^{0.2} \mathrm{Re}_t^{-0.3} \qquad (4.21)$$

for the database by Aldredge et al. (1998), who measured the speeds of the premixed turbulent flames that propagated in a Taylor–Couette flow; and

$$S_t \propto u' \mathrm{Ka}^{-0.41} \qquad (4.22)$$

for the database by Shy et al. (2000a,b), who obtained the speeds of the expanding flames.

It is also worth noting that Bradley et al. (1992) have approximated a very extensive Leeds database on the speeds of expanding, statistically spherical flames, as follows:

$$S_t \propto u' \left(\mathrm{Ka} \cdot \mathrm{Le} \right)^{-0.3}, \qquad (4.23)$$

which is very close to Equation 4.18 if the Lewis number Le = 1. Moreover, by analyzing the turbulent flame speeds measured by the different groups, Gülder (1990a) has obtained

$$S_t = S_L + 0.62 u' \mathrm{Da}^{1/4} \mathrm{Pr}^{-1/4}. \qquad (4.24)$$

When considering any of the aforementioned parameterizations, it is worth remembering that (i) the raw experimental data are scattered around their best

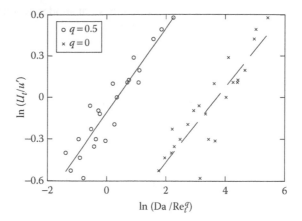

FIGURE 4.16 The approximation of the experimental data by Karpov and Severin (1978, 1980), associated with $S_L < u' < u'_m$ and Le $\approx$ 1, using Equation 4.16 with $m = 0$ and q as specified in the legends. The straight lines approximate the data (symbols) using a least square fit. (Reprinted from *Progress in Energy and Combustion Science*, 28, Lipatnikov, A.N., and Chomiak, J., Turbulent flame speed and thickness: phenomenology, evaluation, and application in multidimensional simulations, 1–74, Copyright 2002, with permission from Elsevier.)

fit (e.g., see Figure 4.16) and (ii) not only the best fit, but also other expressions can approximate the same database with almost the same accuracy. For instance, Figure 4.16 shows that the reduced ($S_L < u' < u'_m$ and Le $\approx$ 1) database by Karpov and Severin (1978, 1980) is well approximated not only by Equation 4.18 (see circles and solid line) but also by the following expression:

$$U_t \propto u'\mathrm{Da}^{0.27}, \tag{4.25}$$

(see crosses and dashed line), which is very close to Gülder's approximation given by Equation 4.24.

The aforementioned approximations further support the trends emphasized by analyzing Figures 4.12 through 4.15. First, the burning velocity is increased by u' in the range of moderate turbulence, $S_L < u' < u'_m$. Second, the dependence of U_t on u' is weaker than the linear dependence. Third, the turbulent burning velocity is increased by the laminar flame speed. Fourth, dU_t/du' is also increased by S_L.

These trends are observed for any of the databases previously analyzed. Therefore, despite the difficulties in unambiguously determining the turbulent flame speed or burning velocity, the differently defined U_t and S_t show the same trends. However, the quantitative characteristics of the behavior of U_t and S_t are different for different databases due to the aforementioned difficulties.

DEPENDENCE OF BURNING VELOCITY ON TURBULENCE LENGTH SCALE

In addition to the trends emphasized earlier, almost all of the parameterizations, with the exception of Equation 4.21, which approximates the database by Aldredge et al.

(1998), indicate an increase in U_t (or S_t) by the turbulence length scale L, with the dependence of U_t (or S_t) on L being much less pronounced than the dependence of U_t (or S_t) on S_L. The latter trend contradicts Equation 4.10 with $C = C(L/\delta_L) = (LS_L/a_u)$.

It is worth remembering, however, that for any of the aforementioned databases considered separately, the turbulence length scale was not varied and the dependencies of U_t (or S_t) on L, given by Equations 4.18 through 4.25, result from the dimensional constraint. For instance, if a best fit yields $U_t \propto u'^p S_L^q$ with $p + q \neq 1$, then a factor of $(L/a_u)^{p+q-1}$ should be inserted into the right-hand side (RHS) in order for its dimension to be correct (m/s). Target-directed research into the influence of the turbulence length scale on the flame speed and burning velocity is strongly needed. For further discussion on this issue, interested readers are referred to Sections 3.3.2 and 5.3.2 of a review paper by Lipatnikov and Chomiak (2002a).

DEPENDENCE OF TURBULENT BURNING VELOCITY ON DIFFUSIVITY, PRESSURE, AND TEMPERATURE

The parameterizations and discussion presented earlier imply that the dependence of U_t (or S_t) on L should be opposite to the dependence of U_t (or S_t) on the laminar flame thickness $\delta_L = a_u/S_L$ or the molecular heat diffusivity a_u, that is, if U_t (or S_t) is increased by L, then U_t (or S_t) is decreased if δ_L or a_u is increased. The dependence of the turbulent flame speed and burning velocity on the kinematic viscosity, ν_u, of the mixture is not addressed here, because we consider the Prandtl number to be close to unity. The influence of the differences in the molecular diffusivities of the main reactants and the heat diffusivity of the mixture on the turbulent flame speed and burning velocity will be discussed in the following section.

Unfortunately, convincing experimental data that straightforwardly show how the turbulent burning velocity or flame speed depends on the molecular heat diffusivity have not yet been reported. The matter is that (i) variations in the mixture composition typically lead to variations in the laminar flame speed much stronger than that in the heat diffusivity and (ii) U_t (or S_t) depends on S_L substantially strongly than on a_u. Therefore, an effect of the heat diffusivity on the turbulent burning velocity (or flame speed) is typically indistinguishable as compared with the stronger variations in U_t (or S_t) associated with the variations in the laminar flame speed.

Nevertheless, there is a well-documented phenomenon that reveals the dependence of the turbulent burning velocity and flame speed on the molecular heat diffusivity. The phenomenon consists of the opposite dependencies of the laminar and turbulent flame speeds on the pressure. As discussed in the section "Dependence of Laminar Flame Speed and Thickness on Mixture Composition, Pressure, and Temperature" in Chapter 2, for a typical hydrocarbon–air mixture, the laminar flame speed is decreased when the pressure is increased. On the contrary, the turbulent burning velocity and flame speed are well known to increase with pressure (e.g., cf. data 2, 3, and 4 in Figure 4.17).

Kobayashi et al. (1996, 1998) investigated the influence of pressure on the global turbulent burning velocity averaged over the mean surface of a Bunsen flame. For ethane–air and propane–air mixtures, they have reported $S_L \propto p^{-0.24}$ and $S_L \propto p^{-0.26}$, whereas $U_t \propto p^{0.23}$ and $U_t \propto p^{0.24}$, respectively. For methane–air mixtures, the

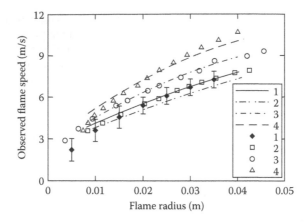

FIGURE 4.17 The observed speeds dR_f/dt of the expanding, statistically spherical, stoichiometric *iso*-octane–air turbulent flames, measured by Bradley et al. (1994b) using high-speed Schlieren cinematography at $u' = 2$ m/s and $L = 20$ mm. 1: $T_0 = 358$ K, $p = 0.1$ MPa; 2: $T_0 = 400$ K, $p = 0.1$ MPa; 3: $T_0 = 400$ K, $p = 0.5$ MPa; 4: $T_0 = 400$ K, $p = 1.0$ MPa. Curves have been computed by Lipatnikov and Chomiak (1997).

reduction effect of pressure on the laminar flame speed was more pronounced with $S_L \propto p^{-0.5}$, whereas the measured burning velocities do not indicate a substantial dependence on pressure at moderate turbulence (see Figure 4.18).

One more example is given in Figure 4.19 where two opposite dependencies of U_t on S_L are shown. When the laminar flame speed was increased by enriching a lean H_2/CO–air mixture, the turbulent burning velocity was also increased (see data at $S_L > 0.7$ m/s). On the contrary, a decrease in S_L due to an increase in the pressure resulted in increasing U_t (see solid lines).

Thus, the opposite dependencies of laminar and turbulent flame speeds on pressure are a well-documented phenomenon (see also Section 3.3.5 in a review paper

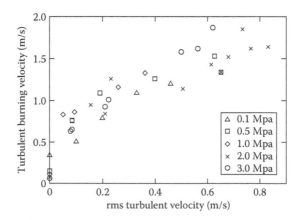

FIGURE 4.18 The turbulent burning velocities obtained by Kobayashi et al. (1996) from lean ($\Phi = 0.9$) methane–air Bunsen flames at different pressures specified in the legends.

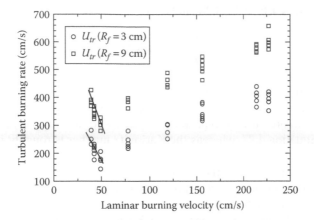

FIGURE 4.19 The turbulent burning velocities obtained by Burluka et al. (2011) from expanding, statistically spherical H_2/CO–air flames at two different mean flame radii, $R_f = 30$ mm (*circles*) and $R_f = 90$ mm (*squares*). The solid lines approximate the data obtained for a lean ($\Phi = 0.5$) mixture by increasing the pressure. Other symbols ($S_L > 0.7$ m/s) show the data obtained by increasing the equivalence ratio at $p = 0.1$ MPa. (With kind permission from Springer Science + Business Media: Flow, Turbulence and Combustion, Turbulent combustion of hydrogen–CO mixtures, 86, 2011, 735–749, Burluka A.A., El-Dein Hussin A.M.T., Sheppard, C.G.W., Liu, K., and Sanderson, V., Figure 7, Copyright Springer.)

by Lipatnikov and Chomiak [2002a]). An increase in U_t (or S_t) by S_L when varying mixture composition is another well-documented phenomenon (see Figure 4.15). These two trends may be consistent with one another if U_t (or S_t) depends on the heat diffusivity. More specifically, U_t (or S_t) should decrease when a_u increases. Indeed, let us assume that

$$U_t \propto S_L^x a_u^{-y} \tag{4.26}$$

with $x > 0$ for U_t to be increased by S_L. Furthermore, if

$$S_L \propto p^{-q} \tag{4.27}$$

with $q > 0$, then y should be greater than xq for the turbulent burning velocity to be increased with pressure, because the molecular transport coefficients are inversely proportional to p. Therefore, the well-documented increase in the turbulent burning velocity and flame speed with pressure implies that U_t (or S_t) is reduced by a_u.

It is worth noting that the effect of the pressure on the turbulent burning rate, discussed earlier, favors empirical parameterizations of the following type:

$$\frac{U_t}{u'} \propto f_1(\text{Da}) \tag{4.28}$$

(e.g., see Equations 4.19, 4.20, 4.24, and 4.25), as compared with

$$\frac{U_t}{u'} \propto f_2\left(Ka^{-1}\right) \tag{4.29}$$

(e.g., see Equations 4.18, 4.22, and 4.23). Here, f_1 and f_2 are arbitrary increasing functions. Indeed, because $Da \propto S_L^2 a_u^{-1}$ and $Ka^{-1} \propto S_L^2 a_u^{-0.5}$, the Damköhler number is increased with pressure for typical hydrocarbon–air mixtures characterized by $q \approx 0.25$ in Equation 4.27, whereas the Karlovitz number is roughly independent of the pressure for such mixtures. Therefore, Equation 4.28 shows an increase in U_t by p, in line with the experimental data discussed earlier, while Equation 4.29 does not predict a notable effect. Similarly, for methane–air mixtures, which are characterized by $q \approx 0.5$ in Equation 4.27, the Damköhler number is roughly independent of the pressure, whereas the Karlovitz number is increased by p. Accordingly, Equation 4.28 predicts a weak dependence of the turbulent burning velocity on the pressure, in line with the experimental data shown in Figure 4.18, while Equation 4.29 yields a reduced U_t under elevated pressures.

Finally, because both the laminar flame speed and the molecular heat diffusivity depend similarly on the temperature of the unburned gas, the turbulent burning velocity and flame speed should be increased with temperature if $x - y > 0$ in Equation 4.26. Unfortunately, the available experimental data on the dependence of U_t (or S_t) on T_u are not sufficient to draw solid conclusions. The circles and crosses in Figure 4.17 indicate the weak effect of T_u on the observed flame speeds, but the temperature variations were small in these measurements. Kobayashi et al. (2005) obtained the global $\rho_u U_t$, that is, the mass flux multiplied by the ratio of the area of the nozzle to the area of a mean flame surface, from slightly lean ($\Phi = 0.9$) methane–air Bunsen turbulent flames under $p = 0.1$, 0.5, or 1 MPa and $T_u = 300$ or 573 K. The data presented in the form of U_t/S_L versus u'/S_L (see Figure 6 in the cited paper) collapse to the same curve for the two temperatures but for the same p. If we assume that (i) the dependence of the turbulent burning velocity on the temperature is controlled by the dependencies of U_t on $S_L(T_u)$ and $a_u(T_u)$ and (ii) the power index q is roughly the same for $S_L \propto T_u^q$ and $a_u \propto T_u^q$, then the experimental data discussed earlier again favor Equation 4.28 as compared with Equation 4.29. Indeed,

$$\frac{U_t}{S_L} = \frac{u'}{S_L} f_1(Da) = \frac{u'}{S_L} f_1\left(\frac{LS_L^2}{u'a_u}\right) = \frac{u'}{S_L} f_1\left(\frac{S_L}{u'}\frac{LS_L}{a_u}\right) \tag{4.30}$$

depends weakly on T_u if the ratio of u'/S_L is kept constant, whereas, in the same case, the dependence of

$$\frac{U_t}{S_L} = \frac{u'}{S_L} f_2\left(Ka^{-1}\right) = \frac{u'}{S_L} f_2\left(\frac{Da}{\sqrt{Re_t}}\right) \tag{4.31}$$

on the temperature is more pronounced due to the dependence of the Reynolds number on T_u.

DEPENDENCE OF TURBULENT BURNING VELOCITY ON DIFFERENCES IN MOLECULAR TRANSPORT COEFFICIENTS

As previously discussed, the influence of the molecular heat diffusivity, a_u, on the turbulent burning velocity and flame speed is weakly pronounced under typical conditions if the pressure and the unburned gas temperature are kept constant. Nevertheless, U_t (or S_t) strongly depends on the molecular transport coefficients even in intense turbulence characterized by a turbulent diffusivity much greater than a_u.

For instance, Figure 4.20 shows the burning velocities obtained by Kido et al. (1989) from statistically spherical, premixed turbulent flames that expanded in a fan-stirred bomb. The method of measuring U_t was basically similar to the already dis cussed method by Karpov and Severin (1978). To reveal the substantial dependence of the turbulent burning velocity on the molecular transport coefficients, Kido et al. (1989) tuned the mass fractions of O_2, N_2, and He in lean ($\Phi = 0.5$) fuel/oxygen/nitrogen/helium mixtures in order to obtain the same laminar flame speed $S_L = 0.43$ m/s for four different fuels (hydrogen, methane, ethane, and propane), characterized by substantially different molecular diffusivities ($D_{C_3H_8} < D_{C_2H_6} < D_{CH_4} < D_{H_2}$). Despite the same S_L and roughly the same adiabatic combustion temperatures, density ratios, and the molecular heat diffusivities of the mixtures, the measured dependencies of U_t and dU_t/du' on u' are very different for different fuels, with the steepest (gentlest) slope dU_t/du' documented for hydrogen (propane) characterized by the largest (lowest) molecular diffusivity.

The discussed effect is commonly associated with the fact that the molecular diffusivity, D_F, of the fuel, which is the deficient reactant in a lean mixture,

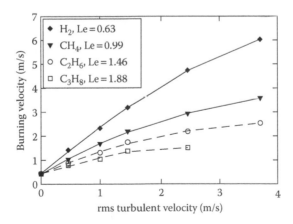

FIGURE 4.20 The burning velocities measured by Kido et al. (1989) for fuel/oxygen/nitrogen/helium mixtures with $\Phi = 0.5$ and $S_L = 0.43$ m/s. The fuels and the Lewis numbers are specified in the legends. The open (Le > 1) and filled (Le < 1) symbols show the experimental data. The dashed (Le > 1) and solid (Le < 1) curves approximate these data. (Reprinted from *Progress in Energy and Combustion Science*, 31, Lipatnikov, A.N. and Chomiak, J., Molecular transport effects on turbulent flame propagation and structure, 1–73, Copyright 2005, with permission from Elsevier.)

differs from the molecular diffusivity, D_O, of the oxidizer (O_2) and from the molecular heat conductivity, a, of the mixture. For instance, in a lean hydrogen–air mixture, $D_F \approx 0.6 \times 10^{-4}$ m²/s, whereas $D_O \approx 0.2 \times 10^{-4}$ m²/s under room conditions. On the contrary, in hydrocarbon–air mixtures under room conditions, either $D_F \approx D_O \approx a \approx 0.2 \times 10^{-4}$ m²/s for lowest hydrocarbons such as methane or $D_F < D_O \approx a \approx 0.2 \times 10^{-4}$ m²/s for higher hydrocarbons, for example, $D_{C_2H_6} \approx 0.15 \times 10^{-4}$ m²/s or $D_{C_3H_8} \approx 0.11 \times 10^{-4}$ m²/s.

When discussing premixed turbulent combustion, the difference in the molecular transport coefficients is often characterized using the Lewis number, which is defined as follows:

$$\mathrm{Le} = \frac{a}{D_d}, \tag{4.32}$$

where D_d is the molecular diffusivity of the deficient reactant. Typically, the Lewis number is evaluated in the unburned mixture; however, due to the similar dependencies of the molecular mass and heat diffusivities on the temperature, Le varies weakly in a flame. Le < 1 in lean H_2–air mixtures, Le ≈ 1 in CH_4–air mixtures, and Le > 1 in lean mixtures of most hydrocarbons and air.

Figure 4.20 clearly shows that an increase in the molecular diffusivity of the deficient reactant (or a decrease in the Lewis number) results in a significant increase in the turbulent burning velocity, as well as dU_t/du'. The physical mechanisms that control the effect of D_d on U_t are discussed in the sections "Variations in Flamelet Structure due to Turbulent Stretching," "Flame Instabilities," and "Flame Propagation along Vortex Tubes" in Chapter 5 and in the section "Models of Molecular Transport Effects in Premixed Turbulent Flames" in Chapter 7.

A basically similar, but even more pronounced, effect is shown in Figure 4.21. For instance, in lean (Φ = 0.71) and rich (Φ = 5.0) hydrogen–air mixtures, the laminar

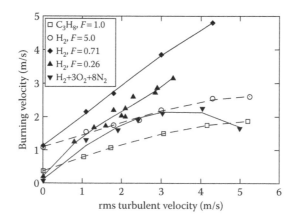

FIGURE 4.21 The burning velocity versus the rms turbulent velocity in several mixtures specified in the legends. The symbols show the data by Karpov and Severin (1978, 1980). The curves approximate the data.

flame speeds measured by Karpov and Severin (1980) are approximately the same, whereas U_t is markedly higher (cf. the open circles and filled diamonds in Figure 4.21) in the former mixture characterized by a larger D_d (0.6 and 0.2 cm²/s, respectively), with the effect being even more pronounced when discussing the slopes of the corresponding curves.

In a very lean ($\Phi = 1/6$) $H_2/O_2/N_2$ mixture characterized by a large $D_d \approx 0.6 \times 10^{-4}$ m²/s, the value of $S_L = 0.07$ m/s, reported by Karpov and Severin (1980), is lower than $S_L = 0.4$ m/s in the stoichiometric propane–air mixture by almost a factor of 6, whereas the slope dU_t/du' shows the opposite behavior and the turbulent burning velocity is higher in the former mixture at $1 < u' < 4$ m/s (cf. the open squares and filled triangles-down in Figure 4.21).

In a lean ($\Phi = 0.26$) hydrogen–air mixture, $S_L = 0.2$ m/s is lower than $S_L = 1.1$ m/s in a rich ($\Phi = 5.0$) hydrogen–air mixture by a factor of 5.5, whereas dU_t/du' is higher in the former mixture by a factor of 2.5 (cf. the open circles and filled triangles-up). Accordingly, U_t at $u' > 1$ m/s is also higher in the lean mixture. Note that the effect is well pronounced even at $u' = 4$ m/s, that is, even if the turbulent diffusivity is higher than the molecular heat diffusivity of the unburned mixture by three orders of magnitude ($L = 10$ mm in these experiments)!

Finally, the curves corresponding to a very lean ($\Phi = 1/6$) $H_2/O_2/N_2$ mixture and a rich ($\Phi = 5.0$) H_2–air mixture (see Figure 4.21) are close to one another within the interval $u' = 2$–3 m/s, despite the fact that $S_L = 0.07$ m/s in the former mixture is lower by almost a factor of 16. The effect is even more pronounced when discussing the slopes: the $U_t(u')$ curve measured in the lean mixture has a slope more than twice as high as that in the rich mixture at $1 < u' < 3$ m/s. Thus, an increase in the molecular diffusivity, D_d, of the deficient reactant by a factor of 3 (from $D_d = D_{O_2} \approx 0.2 \times 10^{-4}$ m²/s in the rich hydrogen–air mixture to $D_d = D_{H_2} \approx 0.6 \times 10^{-4}$ m²/s in the lean mixture under room conditions) affects the turbulent burning velocity more strongly than a decrease in the laminar flame speed by a factor of 16!

A number of other experimental data from Karpov's group, which show high slopes dU_t/du' in lean mixtures that contain H_2 and are characterized by a large D_d, are reported elsewhere (Betev et al., 1995; Karpov et al., 1996b, 1997). In these papers, pairs of mixtures characterized by $S_{L,1} > S_{L,2}$, but $D_{d,1} < D_{d,2}$ and $(dU_t/du')_1 < (dU_t/du')_2$ with well-pronounced differences were selected from Karpov's database and analyzed.

It is of interest to note that the effect of the Lewis number on the turbulent burning velocity, shown in Figures 4.20 and 4.21 (i.e., U_t is increased when Le is decreased), is opposite to the influence of the Lewis number on the speed, S_L, of a statistically planar laminar flame, which is increased by Le, as discussed in the section "Effect of the Lewis number on Unperturbed Laminar Flame Speed" in Chapter 2.

Finally, despite the strong effect of D_d (and, in particular, Le) on the turbulent burning velocity (shown in Figures 4.20 and 4.21), most parameterizations discussed in the section "Empirical Parameterizations of Turbulent Burning Velocity and Flame Speed," with the exception of Equation 4.23, do not involve the Lewis number. The point is that the effects associated with the difference in D_F, D_O, and a are most pronounced in the lean and rich flames, for which the deficient reactant can easily be defined. However, under room conditions, the values of u'_m for lean and rich hydrocarbon–air mixtures are low due to a low laminar flame speed (u'_m is increased by S_L, with all other things being equal, see Figure 4.13). Therefore, the ($S_L < u' < u'_m$) databases at moderate turbulence discussed in the section "Empirical Parameterizations of Turbulent Burning Velocity and Flame Speed" encompass mainly U_t and S_t obtained using near-stoichiometric hydrocarbon–air mixtures, and the effect of D_d on U_t is weakly pronounced under such conditions. Although Karpov and Severin (1978, 1980) obtained the burning velocities from a number of lean and rich hydrogen flames, Equation 4.18 and Figure 4.16 deal only with a part of their database, associated with near-stoichiometric hydrocarbon–air mixtures. The full database by Karpov and Severin (1978, 1980) will be discussed further in the section "The Concept of Leading Points" in Chapter 7, see Figures 7.33 through 7.36.

A number of other experimental data that indicate a substantial dependence of the turbulent burning velocity and flame speed on the differences in the molecular transport coefficients are reviewed in Sections 2 and 3.1 of a paper by Lipatnikov and Chomiak (2005a). Recently, Venkateswaran et al. (2011) and Daniele et al. (2011) documented such effects in open Bunsen lean H_2/CO–air flames under room conditions and in confined lean syngas–air flames under elevated pressures and temperatures, respectively (the results obtained by the latter group from a similar confined flame are shown in Figure 4.4). In both cases, the effects are well pronounced even in very intense turbulence characterized by u'/S_L as large as 100!

MEAN STRUCTURE OF A PREMIXED TURBULENT FLAME

As already discussed in the section "Mean Flame Brush Thickness," the mean thickness of a premixed turbulent flame increases as the flame develops. The burning velocity may also increase during the turbulent flame development (e.g., see Figures 4.5b and 4.17). Two more peculiarities of premixed turbulent combustion are, in fact, the self-similarity and universality of the mean structure of a developing premixed flame. What do these peculiarities mean?

The self-similarity of the mean structure of a developing premixed turbulent flame can be highlighted by replotting the experimental profiles of $\bar{c}(r,t)$, shown in Figure 4.5a, using the following normalized spatial distance:

$$\xi = \frac{r - r_f(\vartheta)}{\delta_t(\vartheta)}, \tag{4.33}$$

where r is the distance counted along the normal to the mean flame brush, that is, r is the radial distance for the expanding, statistically spherical flames addressed in Figure 4.5, $r_f(\vartheta)$ and $\delta_t(\vartheta)$ are the mean flame radius and the mean flame brush thickness, respectively, and ϑ designates either the time counted from ignition for the expanding flames (e.g., for the flames addressed in Figure 4.5) or the distance from the flame holder for the statistically stationary flames. When processing the experimental data plotted in Figure 4.5, $\delta_t(\vartheta)$ was evaluated using the maximum

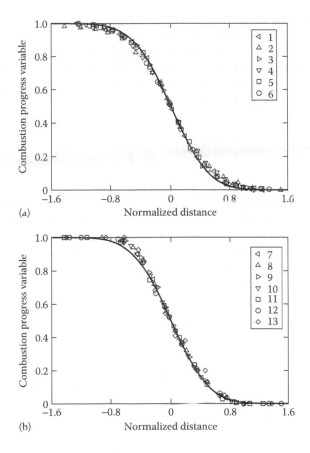

FIGURE 4.22 (a,b) The Reynolds-averaged combustion progress variable profiles normalized with the thickness δ_t calculated using Equation 4.1. The symbols show the experimental data. The curves have been calculated using Equation 4.34. 1, 2, 3, 4, 5, and 6: the data obtained by Renou et al. (2002) from an expanding, statistically spherical, stoichiometric propane–air, turbulent flame at, respectively, $t = 3, 4, 5, 6, 7$, and 8 ms after spark ignition. 7, 8, 9, and 10: the data obtained by Gouldin and Miles (1995) from a stationary, V-shaped, lean ($\Phi = 0.69$) C_2H_6–air flame at, respectively, $x = 30, 40, 50$, and 60 mm from the burner exit. 11, 12, and 13: the data obtained by Namazian et al. (1986) from a stationary, V-shaped, lean ($\Phi = 0.8$) C_2H_6–air flame at, respectively, $x = 20, 40$, and 60 mm from the burner exit.

gradient method (see Equation 4.1), and the mean flame radius, r_f, was associated with the following reference surface $\bar{c}(r = 0.5)$. The results shown with symbols 1–6 in Figure 4.22a indicate that the profile of $\bar{c}(\xi, \vartheta) = \bar{c}(\xi)$ depends solely on ξ, that is, the profiles of $\bar{c}(\xi, \vartheta)$ measured at different instants after spark ignition (or at different distances from the flame holder) collapse to the same universal curve $\bar{c}(\xi)$.

The self-similarity of the mean structure of a premixed turbulent flame, that is, the collapse of the profiles of $\bar{c}(\xi)$ measured at different distances from the flame holder, to a single curve is also indicated by data 7–10 or 11–13 shown in Figure 4.22b. More examples may be found in a review paper by Lipatnikov (2007b), where the self-similarity of the mean structure of a premixed turbulent flame is demonstrated by processing 57 profiles obtained from 16 different flames by 7 research groups.

It is also of interest to note that the self-similar profiles of $\bar{c}(\xi)$ are approximately the same for the three different flames addressed in Figure 4.22. In other words, the experimental data are scattered around a solid curve calculated using the complementary error function parameterization:

$$\bar{c} = \frac{1}{2}\mathrm{erfc}\left(\xi\sqrt{\pi}\right) = \frac{1}{\sqrt{\pi}} \int_{\xi\sqrt{\pi}}^{\infty} \exp\left(-\zeta^2\right)d\zeta. \tag{4.34}$$

The equality of the self-similar profiles of $\bar{c}(\xi)$, obtained from different flames, shows the universality of the mean flame structure. In the aforementioned review paper by Lipatnikov (2007b), this feature is indicated by processing 74 experimental profiles obtained from 33 different flames by 11 research groups. It is worth mentioning that the universal self-similar profile of $\bar{c}(\xi)$ is sometimes parameterized as follows (Shepherd, 1996):

$$\bar{c} = \frac{1}{1 + \exp\left(4\xi\right)}. \tag{4.35}$$

The curves calculated using Equations 4.34 and 4.35 are very close to one another, but Equation 4.34 can be obtained theoretically, as will be discussed in the section "Models of a Self-Similarly Developing Premixed Turbulent Flame" in Chapter 7.

SUMMARY

A premixed turbulent flame is a developing flame, while a laminar premixed flame is a fully developed flame.

The mean thickness of a premixed turbulent flame is much greater than the thickness of a premixed laminar flame, with all other things being equal.

Due to the development and significant thickness of a premixed turbulent flame, the definition of the turbulent burning velocity and flame speed is a complicated issue in a general case. From the theoretical viewpoint, the turbulent burning velocity is well defined by Equation 4.4 in a statistically planar, 1D flame. The fact that the turbulent flame speeds and burning velocities measured by different research groups, using different methods, show qualitatively similar dependencies of S_t or U_t on u', S_L, p, Le, etc., further justifies highlighting the turbulent burning velocity as a physically meaningful and valuable quantity.

The burning velocity is increased by the rms velocity at moderate turbulence ($S_L < u' < u'_m$), reaching a maximum value at a certain rms velocity u'_m, which depends on the mixture composition, pressure, and temperature, followed by a reduction in U_t and combustion extinction by strong turbulence ($u'_m < u'$). The data on the burning velocity and flame speed at weak or strong turbulence are not sufficient to draw solid conclusions.

At moderate turbulence, (i) an increase in both the burning velocity and the slope dU_t/du' by the laminar flame speed and (ii) the opposite effects of the pressure on the laminar and turbulent burning velocities are well documented.

The opposite pressure effects and empirical parameterizations of most extensive experimental databases imply that the turbulent burning velocity is decreased when the heat diffusivity of the mixture is increased. Therefore, U_t should be increased by the turbulence length scale, for dimensional reasons. Further research into the influence of L and a_u on U_t is necessary. Definitely, these two quantities affect the turbulent burning velocity substantially weakly than u' and S_L do.

The turbulent burning velocity is increased when the Lewis number is decreased and/or the ratio of the molecular diffusivity of the deficient reactant to the molecular diffusivity of the excess reactant is increased.

The turbulent burning velocity may also depend on the density ratio, external pressure gradient or acceleration (Veynante and Poinsot, 1997), mean flow velocity (Filatyev et al., 2005), etc., but these effects need more studies.

The mean structure of a premixed turbulent flame is self-similar, that is, the normal profiles of the mean combustion progress variable (or the mean temperature or the mean mass fraction of a main reactant or the mean density, etc.) measured at different stages of the flame development collapse to the same curve if the increasing mean flame brush thickness is used to normalize the spatial distance. The self-similar profile is universal, that is, roughly the same self-similar profiles have been obtained from a number of substantially different flames.

5 Physical Mechanisms and Regimes of Premixed Turbulent Combustion

In Chapter 4, we emphasized the key global features of premixed turbulent flames, which are well documented in numerous experiments. Our goal in this chapter is to discuss in detail the physical mechanisms that could control the general appearance of premixed turbulent flames and, in particular, the trends highlighted in Chapter 4. It is worth stressing that in order for turbulence to accelerate flame propagation, the turbulence should increase not only the mass burning rate within a turbulent flame brush but also the speed of the self-propagation of the leading edge of the flame brush, with the two processes in balance with each other. The focus of this chapter is on the former effect, whereas the self-propagation of the leading edge is discussed in the section "Concept of Leading Points" in Chapter 7.

FLAME WRINKLING AND FLAMELET REGIME

PHYSICAL REASONING: CLASSICAL FLAMELET REGIME

As discussed in the section "Laminar Flame Speed and Thickness" in Chapter 2, laminar premixed flames are thin under typical conditions. Therefore, we may imagine a regime of premixed turbulent combustion such that the length scale of the smallest turbulent eddies is substantially larger than the thickness of the flame in a laminar flow. Under such conditions, it is tempting (i) to solely reduce the effect of turbulence on premixed combustion to wrinkling of the instantaneous flame surface (see Figure 5.1) and (ii) to disregard perturbations in the inner structure of the instantaneous flame front by turbulent eddies in the first approximation. Such a simplification appears to be valid at least in the asymptotic case of an infinitely thin laminar flame, that is, $\delta_L/\eta \to 0$.

Moreover, as already noted in the section "Mean Flame Brush Thickness" in Chapter 4, the foregoing scenario offers an opportunity to apply Taylor's theory of turbulent diffusion, discussed in the section "Taylor's Theory" in Chapter 3, to modeling the growth of the mean flame brush thickness, thereby explaining the experimental data shown in Figure 4.6. To do so, we have only (i) to associate the instantaneous flame front (see the thin wrinkled curve in Figure 5.1) and the mean flame brush thickness, bounded by a dashed straight line in the same figure, with thin, laminar (see the thin wrinkled curve in Figure 3.9) and thick, turbulent mixing layers, respectively and (ii) to assume that the self-propagation of the flame front may be neglected in Equations 3.107 through 3.111 if $S_L \ll u'$.

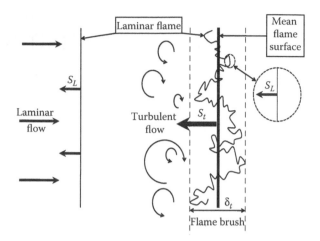

FIGURE 5.1 A sketch of laminar (*left*) and turbulent (*right*) premixed flames.

If the effect of turbulence on combustion is solely reduced to wrinkling of the inherently laminar instantaneous flame front, then the burning rate in a turbulent flow is controlled by (i) the self-propagation of the front at speed S_L with respect to unburned gas and (ii) an increase in the front surface area by turbulent eddies. If the burning rate per unit area of the instantaneous flame front is equal to $\rho_u S_L$, then the total burning rate is $\rho_u S_L A$, where A is the front area. If the instantaneous, inherently laminar flame front wrinkled by turbulent eddies is replaced by a planar mean flame surface (see the bold straight line in Figure 5.1), then the burning rate $\rho_u U_t$ per unit area of the mean flame surface should be higher than $\rho_u S_L$ in order for $\rho_u S_L A = \rho_u U_t A_t$, where $A_t < A$ is the mean flame surface area. Thus,

$$U_t = S_L \frac{A}{A_t} > S_L, \tag{5.1}$$

because turbulent eddies wrinkle the instantaneous flame front and increase its area. This mechanism, highlighted by Damköhler (1940) and Shchelkin (1943), explains a higher turbulent burning velocity as compared with the laminar flame speed.

The regime of premixed turbulent combustion discussed earlier (i.e., combustion is localized to thin, inherently laminar, flame fronts, with the influence of turbulence on the burning rate controlled by wrinkling of the fronts by turbulent eddies) is commonly called flamelet regime and thin, inherently laminar, flame fronts that separate the unburned and burned mixtures in a turbulent flow are commonly called flamelets. It is worth stressing that the term "flamelet" implies that the aforementioned fronts retain the structure of a laminar flame (unperturbed or perturbed) in a turbulent flow. The regime discussed earlier can be called an unperturbed flamelet regime, to be more precise, because the effects of turbulent eddies on the inner structure of flamelets have not yet been addressed. Nevertheless, the term "unperturbed" is commonly skipped in the literature.

How do variations in the rms velocity affect the turbulent burning rate in the unperturbed flamelet regime? To answer this question, let us first consider the asymptotic case of $S_L/u' \to 0$ in order to draw an analogy between the effect of turbulence on flamelets and the growth of the area of a material surface in a turbulent flow.

As discussed in the section "Theory of Homogeneous Isotropic Turbulence" in Chapter 3, the latter growth rate is controlled by the highest stretch rates produced by the smallest turbulent eddies. In the Kolmogorov turbulence, the highest stretch rate scales as $\tau_\eta^{-1} = \bar{\varepsilon}^{1/3}\eta^{-2/3}$ (see Equations 3.59 and 3.72). An increase in the rms velocity results in an increase in the dissipation rate and a decrease in the Kolmogorov length scale, that is, broadening of the spectral range of eddies that affect the surface. Both effects result in an increase in the highest stretch rate, with the former effect having greater importance, because $\bar{\varepsilon}^{1/3} \propto u'L^{-1/3}$ depends on u' more strongly than $\eta^{-2/3} \propto u'^{1/2}v_u^{-1/2}L^{-1/6}$. As a result, the highest stretch rate is proportional to $u'^{3/2}$, and the flamelet surface area is increased by the rms turbulent velocity.

It is also worth noting that flamelets may resist turbulent stretching, for example, by destroying the smallest-scale turbulent eddies due to heat release and thermal expansion, as will be discussed in the section "Local Quenching of Premixed Combustion in Turbulent Flows." In such a case, the highest turbulent stretch rate capable of wrinkling flamelets should scale as $\bar{\varepsilon}^{1/3}\Lambda^{-2/3}$, where the length scale $\Lambda > \eta$ is associated with the smallest eddies that survive in the vicinity of the flamelets. This length scale seems to be controlled by the physical–chemical properties of the burning mixture more than by the turbulence characteristics. Consequently, if $\eta < \Lambda$, the influence of the rms turbulent velocity on the burning rate is controlled by the influence of u' on $\bar{\varepsilon}$ in the case discussed, while a decrease in η when increasing u' does not affect the burning rate.

The considered physical mechanism explains an increase in U_t by u', which is well documented in many experiments, as discussed in Chapter 4. Nevertheless, the aforementioned explanation is oversimplified and leads to contradictions with the experimental data. First, the analogy drawn earlier between a material surface and a flamelet in a turbulent flow implies that the flamelet surface area and, hence, the turbulent burning velocity increase exponentially with time. Second, if the flamelet surface area is controlled by turbulence characteristics, as occurs for a material surface, then Equation 5.1 implies a linear dependence of U_t on S_L, whereas the experimental data discussed in the section "Empirical Parameterizations of Turbulent Burning Velocity and Flame Speed" in Chapter 4 indicate a weaker dependence. Third, because the highest stretch rate produced by the Kolmogorov turbulence is proportional to $L^{-1/2}$ and $\bar{\varepsilon}^{1/3} \propto u'L^{-1/3}$, the flamelet surface area and, hence, the turbulent burning velocity should decrease when the turbulence length scale increases, whereas the parameterizations of various experimental data, discussed in the section "Empirical Parameterizations of Turbulent Burning Velocity and Flame Speed" in Chapter 4 imply the opposite behavior.

The aforementioned problems indicate that the analogy between a material surface and a flamelet in a turbulent flow is an oversimplification even in intense turbulence characterized by $u' \gg S_L$. Even if the self-propagation of flamelets affects the flamelet surface area weakly during the early stages of wrinkling of the surface by strong turbulence, the wrinkles appearing on the surface after some time will be so small that even the slow self-propagation of a flamelet can substantially reduce its surface area. The mechanism of such a reduction in a flamelet surface area is illustrated in Figure 5.2, where the dashed and solid curves show a flamelet surface at two subsequent instances, $t = t_{k-1}$ and $t = t_k$, respectively.

If the flamelet surface is strongly wrinkled, then different segments of the surface can be very close to one another, as shown in Figure 5.2 for $t = t_1$. Due to burning, these segments collide at a certain instant $t > t_1$ so that a small-scale pocket of unburned mixture is formed behind the flamelet at $t = t_2$. Due to the self-propagation of the flamelet that bounds the pocket, the unburned mixture in the pocket is consumed, the pocket volume is decreased, and the flamelet surface area is also decreased (cf. the dashed and solid curves for $t = t_3$). Finally, the pocket is completely consumed at $t = t_4$ and the flamelet surface area at this instant is less than at $t = t_1$.

Thus, the self-propagation of flamelets reduces their surface area. As the flamelet surface is increasingly wrinkled by turbulent eddies, the probability of the scenario sketched in Figure 5.2 is increased and, hence, the role played by flamelet surface reduction due to burning is also increased. Therefore, the exponential increase in the flamelet surface area due to turbulent stretching is limited by the self-propagation of flamelets. Accordingly, an equilibrium between the two processes is likely to be reached after an initial transient stage. The flamelet surface area in the equilibrium state is controlled not only by turbulent stretching but also by flamelet self-propagation. Therefore, the dependence of the burning velocity on the laminar flame speed should be weaker than the linear dependence, because the area A in Equation 5.1 is decreased when S_L is increased due to the reduction effect of the self-propagation of flamelets on their surface area.

It is also worth noting that the collisions of laminar flamelets not only reduce the turbulent burning velocity by decreasing the flamelet surface area, but may also contribute to increasing U_t by increasing the local burning rate due to mutual heating of the colliding preheat zones (e.g., see a recent DNS study by Poludnenko and Oran [2011]).

If $S_L \ll u'$, the self-propagation of flamelets may play a substantial role only at small scales. Indeed, the time required for two flamelet elements separated by a distance l to collide due to flamelet self-propagation is of the order of l/S_L, whereas the

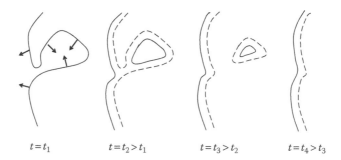

$$t = t_1 \qquad\qquad t = t_2 > t_1 \qquad\qquad t = t_3 > t_2 \qquad\qquad t = t_4 > t_3$$

FIGURE 5.2   The influence of self-propagation of a flamelet on its surface area.

lifetime of an eddy of the same scale l is of the order of $\bar{\varepsilon}^{-1/3}l^{2/3}$. If $l = l_G \equiv L(S_L/u')^3$, the two timescales are of the same order. The length scale l_G is called Gibson scale (Peters, 1986). If $l \gg l_G$, then the distance $S_L \bar{\varepsilon}^{-1/3}l^{2/3} \propto l(l_G/l)^{1/3}$ passed by a flamelet during the lifetime of a turbulent eddy of a length scale l is much less than the eddy size. Therefore, the self-propagation of flamelets may affect the topology of the flamelet surface only at small scales if $S_L \ll u'$, whereas the influence of this physical mechanism on the large-scale topology and, in particular, on the mean flame brush thickness appears to be negligible as $S_L/u' \to 0$. The aforementioned reasoning does not mean that the self-propagation of flamelets plays a negligible role if the Gibson length scale is less than the length scale of the smallest wrinkles on a flamelet surface, that is, $l_G < \eta$ or $l_G < \Lambda$. Even if this physical mechanism does play a minor role during an early stage of premixed turbulent flame development under such conditions, the exponential increase in a flamelet surface area due to turbulent stretching rapidly makes the surface so tangled that its different elements will collide due to their self-propagation for any finite Gibson length scale.

The foregoing discussion has provided no clue to clarify the dependence of U_t on L. We may note, however, that if the dependence of U_t on u' and S_L is well predicted, then dimensional reasoning may provide an insight into the dependence of the turbulent burning velocity on the length scale L, as discussed in the section "Dependence of Burning Velocity on Turbulence Length Scale" in Chapter 4.

Thus, the physical mechanism considered earlier (i.e., wrinkling of the unperturbed laminar flamelets due to turbulent stretching, with the flamelet surface area being limited by the flamelet self-propagation) can explain (i) the existence of a statistically stationary turbulent burning velocity, (ii) an increase in U_t by u', and (iii) an increase in U_t by S_L, with the dependence of the turbulent burning velocity on the laminar flame speed being weaker than the linear dependence.

Does this physical mechanism dominate in strong turbulence? The experimental data shown in Figures 4.12 and 4.13 clearly indicate that this is not so. The mechanism is not able to explain a decrease in the turbulent burning velocity by the rms velocity if $u'_m < u'$. Therefore, other physical mechanisms play a more important role, at least at $u'_m < u'$.

Because the dependence of the rms velocity u'_m on the key mixture and turbulence characteristics has not yet been quantified, the empirical criterion of $u'_m < u'$ is of no use in the literature, and the range of dominance of the physical mechanisms considered previously is commonly estimated invoking the following simple reasoning.

Because flamelets are assumed to be thin, inherently laminar, zones that separate the unburned and burned mixtures in a turbulent flow, the flamelet approach may be put into question if turbulent eddies are so small that they are able to penetrate into these zones and thicken them. The penetration of turbulent eddies into these zones will be possible only if the eddy size is smaller than the zone thickness. Therefore, the Kolmogorov length scale is commonly compared with the laminar flame thickness to estimate the boundary of the flamelet regime.

When making such estimates, numerical coefficients are typically disregarded. Accordingly, several basically similar but quantitatively different criteria may be found in the literature. For instance, the flamelet regime is associated with

$$\left(\frac{\delta_L}{\eta}\right)^2 = \text{Ka} < 1 \tag{5.2}$$

if the Karlovitz number is determined by Equation 4.13, the laminar flame thickness is evaluated using Equation 2.43, and the right-hand side (RHS) of Equation 3.64 with $v_u = a_u$ is invoked to estimate the Kolmogorov scale. Equation 5.2 clarifies the physical meaning of Ka; this number relates the scales of the smallest-scale eddies in the Kolmogorov turbulence with the counterpart scales of a laminar flame (see also Equation 5.112). Accordingly, the Karlovitz number appears to be useful for a discussion on the underlying physics of premixed turbulent combustion only if the Kolmogorov scaling holds in the flow considered! Otherwise, Ka is only a combination of the Damköhler and Reynolds numbers, but it does not have a clear basic value.

Strictly speaking, it seems to be more consistent to compare the Kolmogorov scale with the preheat zone thickness Δ_L, which is substantially larger than δ_L. If we invoke Equation 2.44 to estimate the former thickness, then the criterion changes as

$$\sigma^{1.2}\text{Ka} < 1. \tag{5.3}$$

Furthermore, alternative definitions of the Karlovitz number may be found in the literature (e.g., Abdel-Gayed et al., 1984),

$$\text{Ka}' = 0.157\left(\frac{u'}{S_L}\right)^2 \text{Re}_t^{-1/2}. \tag{5.4}$$

Equations 4.11, 4.12, 4.13, and 5.4 result in

$$\text{Ka}' = 0.157\,\text{Ka}. \tag{5.5}$$

Accordingly, Equation (5.2) read

$$\text{Ka}' < 0.157. \tag{5.6}$$

If Equation 5.2 is substituted with Equation 5.3, we obtain

$$\text{Ka}' < \frac{0.157}{\sigma^{1.2}}, \tag{5.7}$$

with the difference between the RHSs of Equation 5.2 and Equation 5.7 being as large as 50. Therefore, when analyzing the data on the Karlovitz number reported in the literature, particular attention should be paid to the method used to evaluate Ka.

In the literature, the criterion given by Equation 5.2 is used most often. It is worth stressing, however, that Equation 5.3 appears to be more consistent with the underlying physics. Furthermore, Figures 2.2 and 2.6 indicate that the real thickness of a laminar flame may be substantially larger than the thickness evaluated using the maximum gradient method. For instance, in Figure 2.6, the thickness of the recombination layer characterized by a slow temperature increase and a slow decrease in the mass fractions of radicals is larger by a factor of 10 than Δ_L calculated using Equation 2.41. If this effect is taken into account, the critical Karlovitz number should be substantially less than Ka given by Equation 5.3, that is, the flamelet regime is associated with Ka $\ll$ 1 in such a case. However, the use of the Kolmogorov length scale in the aforementioned estimates may also be put into question. For instance, Kobayashi et al. (2002) argued that the length scale of the smallest eddies in a turbulent flow is 10 times larger than η. If this is so, the underestimation of the length scale of the smallest eddies may compensate for the underestimation of the laminar flame thickness in Equation 5.2 or even in Equation 5.3.

Thus, an evaluation of the critical Karlovitz number associated with the boundary of the flamelet regime is an issue. The various criteria discussed earlier, for example, Equations 5.2 through 5.7, should be considered as scaling and training examples, rather than as well-elaborated quantitative criteria. In other words, the boundary of the flamelet regime is associated with the following criterion:

$$Ka = Ka_{cr}, \qquad (5.8)$$

but the critical number Ka_{cr} is not known and may not only be of unity order but also much less, as argued earlier, or even much larger, as will be discussed later in the text. It is worth stressing again that the use of the Karlovitz number is basically justified only for the Kolmogorov turbulence!

PHYSICAL REASONING: THIN-REACTION-ZONE REGIME

Let us assume that small-scale turbulent eddies can penetrate into flamelets and enhance mixing therein. How do such local phenomena affect the global characteristics of premixed turbulent flames and, in particular, the turbulent burning velocity?

At first sight, such effects could be associated with an increase in the local burning rate within flamelets due to an increase in the diffusivity therein (Damköhler, 1940; Summerfield et al., 1955), for example, substitution of the molecular diffusivity D_u in Equation 2.37 with the sum of $D_u + D_{t,\delta}$, where $D_{t,\delta} \propto \overline{\varepsilon}^{1/3} \Delta_L^{4/3}$ is the diffusivity due to turbulent eddies that are smaller than the laminar flame thickness Δ_L. However, within the framework of the activation-energy asymptotic (AEA) theory of laminar premixed combustion, thickening of the flamelet preheat zone and intensification of mixing within the thickened preheat zone by small-scale turbulent eddies do not affect the burning rate unless the reaction zone is affected by the eddies. Indeed, as discussed in the section "AEA Theory of Unperturbed Laminar Premixed Flame" in Chapter 2, the physical mechanism of laminar premixed combustion consists of the heat release in chemical reactions localized to a thin reaction zone and the heat transport from this zone to a much thicker preheat zone. Under stationary conditions, the laminar flame speed is controlled by the heat flux from the reaction zone to the preheat zone. Within the framework of such a two-zone concept, variations in the transport properties in the preheat zone, for example, enhanced transport caused by the small-scale turbulent eddies, change the thickness and structure of the laminar flame, but do not change its burning rate, which is controlled by the heat release rate and the heat diffusivity within the reaction zone (see Equations 2.34 and 2.35).

Thus, within the framework of the AEA theory, in order to affect the local burning rate by enhancing scalar transport, turbulent eddies should penetrate the reaction zone. Such a phenomenon might be possible only if the thickness of the reaction zone is larger than the length scale of the smallest eddies. If Equation 2.46 is invoked to estimate this thickness, then the aforementioned phenomenon may play a role if

$$Ka > Ka_{cr}Ze^2. \qquad (5.9)$$

Even if the critical number Ka_{cr} associated with the boundary of the flamelet regime is low, the RHS of Equation 5.9 is substantially larger than unity due to the large Zel'dovich number associated with a typical hydrocarbon–air flame.

Furthermore, due to the high temperature in the reaction zone, the kinematic viscosity is much larger therein as compared with the unburned mixture. Therefore, the smallest turbulent eddies may be dissipated by viscous forces before the eddies reach the reaction zone. The simplest way to address the viscous dissipation of turbulent eddies in the preheat zone is by replacing the Kolmogorov scales that characterize the unburned mixture with the Kolmogorov scales that characterize the high-temperature products. If $\nu_u \propto T^{1.6}$, then $\eta \propto T^{1.2}$ and Equation 5.9 reads

$$Ka > Ka_{cr}Ze^2\sigma^{2.4}, \qquad (5.10)$$

where the Karlovitz number is evaluated on the basis of the viscosity of the unburned mixture. Certainly, such an estimate is oversimplified, because the influence of a flamelet on small-scale turbulent eddies is a highly complex phenomenon, which depends on factors such as chemical reactions, molecular heat and mass transfer, local pressure perturbations, transient effects, flamelet curvature, etc., is not reduced solely to an increase in the molecular viscosity.

In any case, if at least $Ka < Ka_{cr}Ze^2$, turbulent eddies increase the burning rate by wrinkling a thin, inherently laminar, surface (either a flamelet if $Ka < Ka_{cr}$ or a reaction zone if $Ka > Ka_{cr}$), thereby increasing its area. Even if the local temperature (or density or mass fraction, etc.) profiles are substantially different in the cases of $Ka < Ka_{cr}$ and $Ka > Ka_{cr}$ (like the profiles in the laminar flame in the former case and much thicker in the latter case), the governing physical mechanism of an increase in U_t by u' appears to be basically the same in the two cases (within the framework of the AEA theory). A recent DNS study by Poludnenko and Oran (2011) supports the hypothesis that an increase in U_t by u' in intense turbulence is mainly controlled by wrinkling of a thin reaction zone.

It is worth noting that the distinction of the thin-reaction-zone regime of premixed turbulent combustion is basically justified only in the case of a well-pronounced scale separation between the preheat and the reaction zones, for example, in the case of a single reaction characterized by an infinitely high activation temperature, that is, within the framework of the AEA theory of laminar premixed flames. If we consider a high but finite activation temperature (see Figure 2.5) or take into account complex combustion chemistry (see Figure 2.6), then the thickness of a reaction zone is less than, but of the order of, the thickness of a preheat zone. Nevertheless, the experimental data discussed in the section "Experimental Data" (see Figures 5.15 and 5.16) do indicate that the heat release is localized to thin reaction zones in highly turbulent flames. Probably, due to an increase in the kinematic viscosity with increasing temperature and thermal expansion, small-scale turbulent eddies that enter a preheat zone are rapidly dissipated therein (see the section "Local Quenching of Premixed Combustion in Turbulent Flows") and cannot enter the reaction zone.

SIMILARITIES AND DIFFERENCES BETWEEN FLAMELET AND THIN-REACTION-ZONE REGIMES

The aforementioned physical reasoning implies that thin, inherently laminar, reacting layers exist within a turbulent flame brush and turbulence increases the burning rate by wrinkling these layers. They are either flamelets or reaction zones. In the former case (flamelet regime), associated with weaker turbulence, the layers have the structure of a laminar premixed flame and separate a cold, fresh mixture from the combustion products in the equilibrium state. In the latter case (thin-reaction-zone regime), associated with a stronger turbulence, the layers have the structure of the reaction zone of a laminar flame and separate the combustion products from a nonreacting, but preheated mixture, the composition of which varies in space due to diffusion from/to the reaction zone.

The difference between the two regimes is illustrated in Figure 5.3. In the flamelet regime, the reacting layer, which is thinner than the smallest turbulent eddies, contains both preheat and reaction zones and retains the structure of a laminar flame. In the thin-reaction-zone regime, the reaction layer contains only a reaction zone, which still retains its laminar structure, whereas the preheat zones are perturbed and thickened by small-scale turbulent eddies, with temperature fluctuations observed therein.

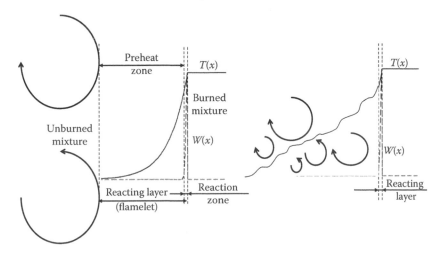

FIGURE 5.3 The structure of a reacting layer in flamelet (*left*) and thin-reaction-zone (*right*) regimes.

Furthermore, because a cold, fresh mixture and the hot combustion products are separated by a thin layer in the flamelet regime, a local instantaneous temperature record within a flame brush looks like a random telegraph signal (see Figure 5.4a), that is, the transition from low to high temperatures, associated with the unburned mixture and adiabatic equilibrium combustion products, respectively, occurs very fast in comparison with the mean time intervals during which the instantaneous temperature is equal to either T_u or T_b. On the contrary, the latter time intervals are comparable with the mean transition duration in the thin-reaction-zone regime (see Figure 5.4b), with temperature fluctuations being observed during the transition. Despite the substantial differences in local instantaneous temperature records, instantaneous records of the local heat release rate $W(t)$ look similar, that is, they contain narrow pulses, in both regimes.

It is worth noting that we use the term "thin-reaction-zone regime" following Peters (1999), who drew the attention of the combustion community to this regime (such a regime of premixed turbulent combustion was mentioned in the literature well before Peters, for example, see a reply by Chen et al. [1986] to a comment by Sirignano on p. 1490 in the cited paper). However, this term does not seem to be the best one, because the classical flamelet regime may also be called the thin-reaction-zone regime. To clearly distinguish (i) the classical flamelet regime associated with Ka < Ka_{cr} (see Equation 5.8) and (ii) the regime highlighted by Peters (1999) and associated with either Ka_{cr} < Ka < $Ka_{cr}Ze^2$ or Ka_{cr} < Ka < $Ka_{cr}Ze^2\sigma^{2.4}$ (see Equations 5.9 and 5.10, respectively), the two regimes might be called the laminar flamelet and the thickened preheat zone subregimes of a thin-reaction-zone regime.

Figure 5.4a shows the following very important peculiarity of the flamelet regime of premixed turbulent combustion. Because the time intervals associated with $T_u < T(t) < T_b$ are much shorter than the time intervals characterized by either $T(t) = T_u$ or $T(t) = T_b$, the probability, P_f, of finding the intermediate (between the

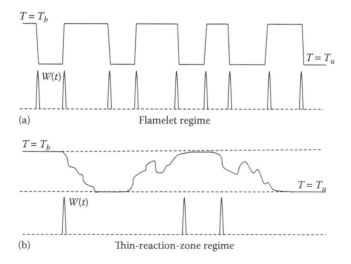

(a)   Flamelet regime

(b)   Thin-reaction-zone regime

FIGURE 5.4 Typical local instantaneous temperature $T(t)$ and heat release rate $W(t)$ records in flamelet (a) and thin-reaction-zone (b) regimes of premixed turbulent combustion.

fresh mixture and the equilibrium adiabatic combustion products) states of the reacting mixture is much less than unity in the flamelet regime of premixed turbulent
combustion. However, this is not so in the thin-reaction-zone regime (see Figure 5.4b).

The low probability, P_f, offers an opportunity to develop a very attractive and
simple approach to modeling premixed turbulent combustion in the flamelet regime.
Although models of premixed turbulent flames will be discussed in the subsequent
chapters, it is worth considering this approach here, because an experimental assessment of the flamelet concept of premixed turbulent combustion is often performed
by testing asymptotic equations valid at $P_f \to 0$.

BRAY–MOSS–LIBBY APPROACH

In the contemporary English-language literature, the approach discussed in this section is commonly called the BML model in honor of Bray, Moss, and Libby, who
contributed substantially to its development (Bray and Moss, 1977; Libby and Bray,
1977, 1980, 1981; Bray, 1980; Libby, 1985; Bray et al., 1985). It is worth recogniz-
ing, however, that Prudnikov (1960) reported many key equations of this approach
in different notations more than 15 years before Bray and Moss (1977) delivered
their seminal paper. Although the cited pioneering paper by Prudnikov was only
published in Russian, his subsequent work (Prudnikov, 1967), which was aimed, in
particular, at the development of the same approach, was translated into English well
before the appearance of the BML model. In the following text, we will use the BML
notation, because it is standard in the contemporary combustion literature.

In a general case, the probability density function (PDF) of finding the values of
q_1, q_2, etc., of the quantities Q_1, Q_2, etc., in point $\mathbf{x}$ at instant t may be decomposed
as follows:

$$P(t,\mathbf{x},q_1,q_2,\ldots) = \alpha(t,\mathbf{x}) P_u(t,\mathbf{x},q_1,q_2,\ldots) + \beta(t,\mathbf{x}) P_b(t,\mathbf{x},q_1,q_2,\ldots)$$

$$+ \gamma(t,\mathbf{x}) P_f(t,\mathbf{x},q_1,q_2,\ldots), \tag{5.11}$$

where $\alpha(t,\mathbf{x})$, $\beta(t,\mathbf{x})$, and $\gamma(t,\mathbf{x})$, are the probabilities of finding a fresh mixture, the adiabatic combustion products in the equilibrium state, and the intermediate reactants,
respectively, in point $\mathbf{x}$ at instant t. These three states of the mixture are designated
with the subscripts u, b, and f, respectively. Since any PDF satisfies the normalizing
constraint given by Equation 3.20, we have

$$\iiint P(t,\mathbf{x},q_1,q_2,\ldots)\,dq_1 dq_2\ldots = \iiint P_u(t,\mathbf{x},q_1,q_2,\ldots)\,dq_1 dq_2\ldots$$

$$= \iiint P_b(t,\mathbf{x},q_1,q_2,\ldots)\,dq_1 dq_2\ldots = \iiint P_f(t,\mathbf{x},q_1,q_2,\ldots)\,dq_1 dq_2\ldots = 1 \tag{5.12}$$

therefore,

$$\alpha(t,\mathbf{x}) + \beta(t,\mathbf{x}) + \gamma(t,\mathbf{x}) = 1. \tag{5.13}$$

In a general case, Equation 5.11 is of little value, because both the left-hand side (LHS) and the RHS involve unknown PDFs $P(t,\mathbf{x},q_1,q_2,...)$ and $P_f(t,\mathbf{x},q_1,q_2,...)$. However, Equation 5.11 may be very useful if the probability $\gamma(t,\mathbf{x})$ is much less than unity everywhere within the flame brush. In such a case, one may write

$$P(t,\mathbf{x},q_1,q_2,...) \approx \alpha(t,\mathbf{x})P_u(t,\mathbf{x},q_1,q_2,...) + \beta(t,\mathbf{x})P_b(t,\mathbf{x},q_1,q_2,...) \qquad (5.14)$$

and

$$\overline{q_l}(t,\mathbf{x}) = \iiint q_l P(t,\mathbf{x},q_1,q_2,...)dq_1dq_2... = \alpha(t,\mathbf{x})\overline{q}_{lu}(t,\mathbf{x}) + \beta(t,\mathbf{x})\overline{q}_{lb}(t,\mathbf{x}) + O(\gamma),$$

$$(5.15)$$

where

$$\overline{q}_{lu}(t,\mathbf{x}) = \iiint q_l P_u(t,\mathbf{x},q_1,q_2,...)dq_1dq_2... = \overline{(1-c)q_l} + O(\gamma) \qquad (5.16)$$

and

$$\overline{q}_{lb}(t,\mathbf{x}) = \iiint q_l P_b(t,\mathbf{x},q_1,q_2,...)dq_1dq_2... = \overline{cq_l} + q_l O(\gamma) \qquad (5.17)$$

are quantities conditioned on the unburned and burned (i.e., adiabatic combustion products in the equilibrium state) mixtures, respectively. It is worth remembering that conditioned quantities differ substantially from the counterpart mean quantities. For instance, due to the random motion of a wrinkled flamelet surface, conditional averaging does not commute with differentiation (Libby, 1975). It is also worth stressing that the BML approach discussed here addresses the case of a low but finite probability γ or, in other words, the case of thin, but not infinitely thin, flamelets. When writing Equations 5.15 through 5.17, we have taken into account that

$$\iiint P(t,\mathbf{x},q_1,q_2,...,q_{k-1},q_k,q_{k+1},...)dq_k = P(t,\mathbf{x},q_1,q_2,...,q_{k-1},q_{k+1},...) \quad (5.18)$$

due to the normalizing constraint given by Equation 3.20.

If we consider a PDF for the combustion progress variable c, the use of Equation 5.14 instead of Equation 5.11 means substitution of a real PDF $P(t,\mathbf{x},c)$, shown in Figure 5.5a, with the so-called bimodal PDF that contains two Dirac delta functions at $c = 0$ (unburned mixture) and $c = 1$ (adiabatic combustion products in the equilibrium state) (see Figure 5.5b). The two Dirac delta functions result from the

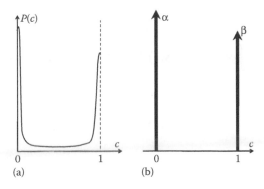

FIGURE 5.5 A combustion progress variable PDF in the flamelet regime (a), and the bimodal approximation of this PDF within the framework of the BML approach (b).

lack of fluctuations in the combustion progress variable in unburned and burned mixtures. Accordingly,

$$P_u(c) = \delta(c),$$
$$P_b(c) = \delta(1-c), \tag{5.19}$$

$$P(t,\mathbf{x},c) \approx \alpha(t,\mathbf{x})\delta(c) + \beta(t,\mathbf{x})\delta(1-c), \tag{5.20}$$

and

$$\bar{c}(t,\mathbf{x}) = \int_0^1 cP(t,\mathbf{x},c)\,dc \approx \alpha(t,\mathbf{x})\cdot 0 + \beta(t,\mathbf{x})\cdot 1 = \beta(t,\mathbf{x}), \tag{5.21}$$

that is, the Reynolds-averaged combustion progress variable is simply equal to the probability of finding burned mixture within the framework of the BML approach. Here and in the following text, asymptotically ($\gamma \to 0$) small terms O(γ) are skipped for the sake of brevity.

Similarly,

$$1 - \bar{c}(t,\mathbf{x}) = \int_0^1 (1-c)P(t,\mathbf{x},c)\,dc \approx \alpha(t,\mathbf{x})\cdot 1 + \beta(t,\mathbf{x})\cdot 0 = \alpha(t,\mathbf{x}), \tag{5.22}$$

$$\bar{\rho}(t,\mathbf{x})\cdot\tilde{c}(t,\mathbf{x}) = \int_0^1 \rho cP(t,\mathbf{x},c)\,dc \approx \alpha(t,\mathbf{x})\cdot\rho_u\cdot 0 + \beta(t,\mathbf{x})\cdot\rho_b\cdot 1 = \rho_b\beta(t,\mathbf{x}) = \rho_b\bar{c},$$

$$\tag{5.23}$$

and

$$\bar{\rho}(t,\mathbf{x})\left[1-\tilde{c}(t,\mathbf{x})\right] = \int_0^1 \rho(1-c)P(t,\mathbf{x},c)dc \approx \alpha(t,\mathbf{x})\cdot\rho_u\cdot1+\beta(t,\mathbf{x})\cdot\rho_b\cdot0 = \rho_u\alpha(t,\mathbf{x}).$$

(5.24)

Combining Equations 5.21 through 5.24, one can easily obtain

$$\bar{\rho} = \frac{\rho_u}{1+(\sigma-1)\tilde{c}},$$

(5.25)

and

$$\bar{c} = \frac{\sigma\tilde{c}}{1+(\sigma-1)\tilde{c}}$$

(5.26)

or

$$\tilde{c} = \frac{\bar{c}}{\bar{c}+\sigma(1-\bar{c})}.$$

(5.27)

It is worth remembering that the Reynolds- and Favre-averaged values of the combustion progress variable differ substantially from each other. For instance, if the density ratio $\sigma = 7.5$, then $\tilde{c}(\bar{c} = 0.5) = 0.12$, while $\tilde{c}(\bar{c} = 0.5) = 0.17$ if $\sigma = 5$. Therefore, in a typical premixed turbulent flame, an isosurface associated with equal probabilities of finding unburned and burned mixtures (i.e., $\bar{c} = 0.5$) is characterized by a low value of the Favre-averaged combustion progress variable $\tilde{c} = 0.1-0.2$.

Finally,

$$\overline{\rho'^2} = \int_0^1 (\rho-\bar{\rho})^2 P(c)dc \approx \alpha(\rho_u-\bar{\rho})^2 + \beta(\rho_b-\bar{\rho})^2 \approx \bar{\rho}(1-\tilde{c})\rho_u\left[\frac{(\sigma-1)\tilde{c}}{1+(\sigma-1)\tilde{c}}\right]^2$$

$$+ \bar{\rho}\tilde{c}\rho_b\left[\frac{(\sigma-1)(1-\tilde{c})}{1+(\sigma-1)\tilde{c}}\right]^2 \approx \frac{\bar{\rho}^3(\sigma-1)^2}{\rho_u}\tilde{c}(1-\hat{c})\left[\tilde{c}+\frac{1-\tilde{c}}{\sigma}\right] = \bar{\rho}^2(\sigma-1)^2\frac{\tilde{c}(1-\hat{c})}{\sigma},$$

(5.28)

$$\overline{\rho c''^2} = \int_0^1 \rho(c-\tilde{c})^2 P(c)dc \approx \alpha\rho_u(0-\tilde{c})^2 + \beta\rho_b(1-\hat{c})^2$$

$$\approx \bar{\rho}(1-\tilde{c})\tilde{c}^2 + \bar{\rho}\tilde{c}(1-\hat{c})^2 = \bar{\rho}\tilde{c}(1-\hat{c})$$

(5.29)

using Equations 5.23 and 5.24. Here and in the following text, the dependencies of PDFs and their moments on time and spatial coordinates are not specified for the sake of brevity.

The obtained equalities are widely used to assess the validity of the BML approach, for example, by evaluating the so-called segregation factor defined as follows:

$$g = \frac{\overline{\rho c''^2}}{\overline{\rho}\tilde{c}(1-\tilde{c})}. \tag{5.30}$$

If the segregation factor obtained from a premixed turbulent flame is close to unity (see Equation 5.29), then the BML approach is commonly claimed to hold in this particular flame. If the segregation factor is substantially less than unity, then the probability γ of finding the intermediate states of the reacting mixture is significant, the BML approach oversimplifies the problem, and combustion does not occur in the flamelet regime.

Let us consider a joint PDF $P(c,\mathbf{u})$ for the combustion progress variable and the flow velocity vector. In this case, Equation 5.14 reads

$$P(c,\mathbf{u}) \approx \alpha P_u(\mathbf{u})\delta(c) + \beta P_b(\mathbf{u})\delta(1-c). \tag{5.31}$$

Using Equation 5.31, one can easily obtain

$$\bar{\mathbf{u}} = \iiint \mathbf{u}P(c,\mathbf{u})dcd\mathbf{u} \approx \alpha\bar{\mathbf{u}}_u + \beta\bar{\mathbf{u}}_b \approx (1-\bar{c})\bar{\mathbf{u}}_u + \overline{c\mathbf{u}}_b, \tag{5.32}$$

$$\overline{\rho}\tilde{\mathbf{u}} = \iiint \rho\mathbf{u}P(c,\mathbf{u})dcd\mathbf{u} \approx \alpha\rho_u\bar{\mathbf{u}}_u + \beta\rho_b\bar{\mathbf{u}}_b \approx \overline{\rho}(1-\tilde{c})\mathbf{u}_u + \overline{\rho}\tilde{c}\tilde{\mathbf{u}}_b, \tag{5.33}$$

that is,

$$\tilde{\mathbf{u}} = (1-\tilde{c})\bar{\mathbf{u}}_u + \tilde{c}\tilde{\mathbf{u}}_b. \tag{5.34}$$

Furthermore,

$$\overline{\rho\mathbf{u}''c''} = \iiint \rho(\mathbf{u}-\tilde{\mathbf{u}})(c-\tilde{c})P(c,\mathbf{u})dcd\mathbf{u} \approx \alpha\rho_u(\bar{\mathbf{u}}_u-\tilde{\mathbf{u}})(0-\tilde{c}) + \beta\rho_b(\bar{\mathbf{u}}_b-\tilde{\mathbf{u}})(1-\tilde{c})$$

$$\approx -\overline{\rho}(1-\tilde{c})(\bar{\mathbf{u}}_u-\tilde{\mathbf{u}})\tilde{c} + \overline{\rho}\tilde{c}(\bar{\mathbf{u}}_b-\tilde{\mathbf{u}})(1-\tilde{c}) = \overline{\rho}\tilde{c}(1-\tilde{c})(\bar{\mathbf{u}}_b-\bar{\mathbf{u}}_u) \tag{5.35}$$

using Equations 5.23 and 5.24. Here,

$$(1-\bar{c})\bar{\mathbf{u}}_u \equiv (1-\bar{c})\iiint \mathbf{u}P_u(\mathbf{u})d\mathbf{u} \approx \iiint (1-c)\mathbf{u}P(\mathbf{u})d\mathbf{u} = \overline{(1-c)\mathbf{u}} \quad (5.36)$$

and

$$\overline{c}\bar{\mathbf{u}}_b \equiv \bar{c}\iiint \mathbf{u}P_b(\mathbf{u})d\mathbf{u} \approx \iiint c\mathbf{u}P(\mathbf{u})d\mathbf{u} = \overline{c\mathbf{u}} \quad (5.37)$$

are the flow velocity vectors conditioned on the unburned and burned mixtures, respectively.

Contrary to the combustion progress variable, the flow velocity fluctuates in both unburned and burned mixtures, and the conditioned velocity PDFs $P_u(\mathbf{u})$ and $P_b(\mathbf{u})$ differ from the Dirac delta function (see Figure 5.6). For instance, the Gaussian function appears to be a reasonable approximation of the velocity PDF conditioned on the unburned mixture.

The conditioned velocity PDFs are characterized not only by the conditioned velocities $\bar{\mathbf{u}}_u$ and $\bar{\mathbf{u}}_b$ but also by higher moments, for example,

$$\left(\overline{\mathbf{u}'^2}\right)_u = \iiint (\mathbf{u}-\bar{\mathbf{u}}_u)^2 P_u(\mathbf{u})d\mathbf{u} \quad (5.38)$$

and

$$\left(\overline{\mathbf{u}'^2}\right)_b = \iiint (\mathbf{u}-\bar{\mathbf{u}}_b)^2 P_b(\mathbf{u})d\mathbf{u} . \quad (5.39)$$

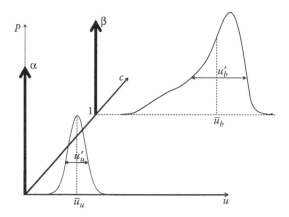

FIGURE 5.6   Velocity PDFs conditioned on unburned and burned mixtures.

Note that Equations 5.32 and 5.34 read

$$\overline{\mathbf{u}''} = \overline{\mathbf{u} - \tilde{\mathbf{u}}} = \overline{\mathbf{u}} - \tilde{\mathbf{u}} = \overline{\left(\overline{c} - \tilde{c}\right)\left(\overline{\mathbf{u}}_b - \overline{\mathbf{u}}_u\right)} = \frac{(\sigma-1)\overline{\rho}}{\rho_u}\tilde{c}\left(1-\tilde{c}\right)\left(\overline{\mathbf{u}}_b - \overline{\mathbf{u}}_u\right) = (\sigma-1)\frac{\overline{\rho\mathbf{u}''c''}}{\rho_u}$$

(5.40)

using Equations 5.26 and 5.35. Therefore, the Reynolds-averaged value of density-weighted velocity fluctuations does not vanish in a variable-density case.

The joint PDF determined by Equation 5.31 may be applied to find the Reynolds stresses and higher moments, for example,

$$\overline{\rho u_i'' u_j''} = \overline{\rho}\left(1-\tilde{c}\right)\left(\overline{u_i' u_j'}\right)_u + \overline{\rho}\tilde{c}\left(\overline{u_i' u_j'}\right)_b + \overline{\rho}\tilde{c}\left(1-\tilde{c}\right)\left(\overline{u}_{ib} - \overline{u}_{iu}\right)\left(\overline{u}_{jb} - \overline{u}_{ju}\right) \quad (5.41)$$

$$\overline{\rho u_i'' u_j'' c''} = \overline{\rho}\tilde{c}\left(1-\tilde{c}\right)\left[\left(\overline{u_i' u_j'}\right)_b - \left(\overline{u_i' u_j'}\right)_u\right] + \left(1-2\tilde{c}\right)\frac{\overline{\rho u_i'' c''}\cdot\overline{\rho u_j'' c''}}{\overline{\rho}\tilde{c}\left(1-\tilde{c}\right)}, \quad (5.42)$$

$$\overline{\rho u_i'' u_j'' u_k''} = \overline{\rho}\left(1-\tilde{c}\right)\left(\overline{u_i' u_j' u_k'}\right)_u + \overline{\rho}\tilde{c}\left(\overline{u_i' u_j' u_k'}\right)_b + \left(1-2\tilde{c}\right)\frac{\overline{\rho u_i'' c''}\cdot\overline{\rho u_j'' c''}\cdot\overline{\rho u_k'' c''}}{\left[\overline{\rho}\tilde{c}\left(1-\tilde{c}\right)\right]^2}$$

$$+\left[\left(\overline{u_i' u_j'}\right)_b - \left(\overline{u_i' u_j'}\right)_u\right]\overline{\rho u_k'' c''} + \left[\left(\overline{u_i' u_k'}\right)_b - \left(\overline{u_i' u_k'}\right)_u\right]\overline{\rho u_j'' c''}$$

$$+\left[\left(\overline{u_k' u_j'}\right)_b - \left(\overline{u_k' u_j'}\right)_u\right]\overline{\rho u_i'' c''}, \quad (5.43)$$

and so on.

It is worth stressing that the foregoing analysis dealt with asymptotically ($\gamma \to 0$) bounded quantities such as c, ρ, and u. However, there are also quantities that vary strongly within thin, laminar flamelets and tend to infinity as $\gamma \to 0$. These are the rate W of product formation in Equation 2.14, divergence $\nabla\cdot\mathbf{u}$ of the flow velocity vector, pressure gradient, etc. For such quantities, the neglect of the last term on the RHS of Equation 5.11 would be an error, because the low value of the probability γ can be overwhelmed by a high value of the considered quantity within flamelets. For instance, an attempt to evaluate the mean rate $\overline{W}$ of product formation using Equation 5.14 yields an obviously wrong result:

$$\overline{W} = \int_0^1 W(c)P(c)dc \approx \alpha\cdot 0 + \beta\cdot 0 = 0. \quad (5.44)$$

Therefore, the last term on the RHS of Equation 5.11 should be retained when averaging such asymptotically ($\gamma \to 0$) infinite quantities, for example,

$$\bar{W} = \int_0^1 W(c)P(c)dc = \gamma \int_0^1 W(c)P_f(c) \equiv \gamma(\bar{W})_f, \tag{5.45}$$

$$\overline{\nabla \mathbf{u}} = \int_0^1 \nabla \mathbf{u} P(c)dc = \gamma \int_0^1 \nabla \mathbf{u} P_f(c) \equiv \gamma(\overline{\nabla \mathbf{u}})_f, \tag{5.46}$$

$$\overline{\nabla c} = \int_0^1 \nabla c P(c)dc = \gamma \int_0^1 \nabla c P_f(c) \equiv \gamma(\overline{\nabla c})_f, \tag{5.47}$$

$$\overline{\nabla p} = \int_0^1 \nabla p P(c)dc = (1-\bar{c})(\overline{\nabla p})_u + \bar{c}(\overline{\nabla p})_b + \gamma(\overline{\nabla p})_f, \tag{5.48}$$

etc.

Experimental Data

There are experimental data that support the flamelet concept of premixed turbulent combustion for some flames, but there are experimental data that put this concept into question for other flames. In this section, we restrict ourselves to a few typical examples. For a more detailed discussion of the issue, interested readers are referred to a review paper by Driscoll (2008) and to Section 3.1 of a review paper by Lipatnikov and Chomiak (2010).

As discussed earlier, the flamelet regime of premixed turbulent combustion is characterized by (i) a random telegraph form of a record of an instantaneous gas characteristic at a point $\mathbf{x}$ within the turbulent flame brush (see Figure 5.4a) and (ii) the bimodal form of a PDF (see Figure 5.5a). Indeed, many research groups documented both the random telegraph signals and the bimodal PDFs in various flames (e.g., see Figures 5.7 and 5.8).

It is worth noting, however, that even if a measured PDF has a bimodal shape, the probability γ of finding the intermediate states of a reacting mixture is not negligible in many flames. For instance, the area under curve (c) in Figure 5.8 in the range of $340°C < T < 1700°C$ is comparable with the area under this curve in the ranges of $T < 340°C$ and $T > 1700°C$. Moreover, contrary to the Dirac delta function, the two peaks associated with unburned and burned mixtures are broad. Therefore, the probability γ is not small even in flame (c) characterized by a bimodal PDF.

The three curves plotted in Figure 5.8 also show that the probability γ is increased when the equivalence ratio is decreased in the three methane–air turbulent flames investigated by Yoshida (1988). Therefore, γ is increased when the laminar flame speed is decreased (or γ is increased by u'/S_L if dimensionless criteria are used).

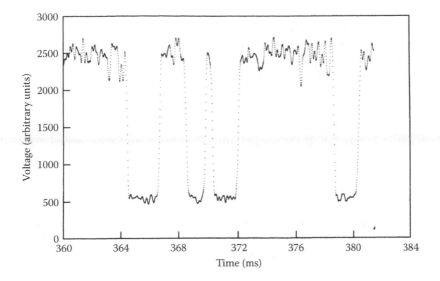

FIGURE 5.7   An instantaneous Rayleigh-scattering signal recorded from a lean ($\Phi = 0.8$) methane–air turbulent Bunsen flame. (Reprinted from Deschamps, B., Boukhalfa, A., Chauveau, C., Gökalp, I., Shepherd, I.G., and Cheng, R.K., *Proc. Combust. Inst.*, 24, 469–475, 1992. With permission.)

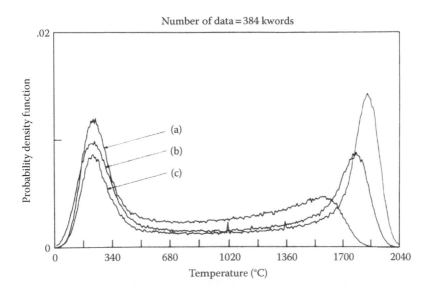

FIGURE 5.8   Temperature PDFs obtained from methane–air turbulent flames characterized by $\Phi = 0.7$ (a), 0.85 (b), and 1.0 (c). (Reprinted from Yoshida, A., *Proc. Combust. Inst.*, 22, 1471–1478, 1988. With permission.)

Such behavior of the probability γ, also indicated by other experimental and DNS data reviewed elsewhere (see Section 3.1 of a paper by Lipatnikov and Chomiak [2010]), is consistent with the discussion in the section "Physical Reasoning: Thin-Reaction-Zone Regime." Indeed, an increase in u'/S_L is associated with an increase in the Karlovitz number (see Equation 5.4). Accordingly, if the ratio of u'/S_L is increased, the critical condition given by Equation 5.8 may be reached and combustion will occur in a nonflamelet regime. Such a transition from flamelet to nonflamelet regimes of premixed turbulent burning appears to be observed in Figure 5.8. If the temperature PDF has a well-pronounced bimodal shape in the stoichiometric flame (c), characterized by lower u'/S_L and Ka, the PDF obtained from the leanest flame (a) characterized by larger u'/S_L and Ka is clearly non-bimodal.

It is worth emphasizing that obtaining a bimodal PDF from a flame is a necessary, but not a sufficient condition to claim that combustion occurs in the flamelet regime. As discussed earlier, the probability γ may be significant even if a PDF has a bimodal shape. Therefore, straightforward tests of the BML equations obtained in the section "Bray–Moss–Libby Approach" are necessary to assess the flamelet concept. Such quantitative tests commonly address either Equation 5.28 for the rms density $\left(\overline{\rho'^2}\right)^{1/2}$ or Equation 5.30 for the segregation factor.

The majority of the tests of Equations 5.28 and 5.30 have shown that these BML equations overestimate the magnitude of fluctuations in the density and combustion progress variables. For instance, Figure 5.9 indicates that the density fluctuations are well predicted by Equation 5.28 in a stoichiometric flame (triangles) characterized by lower u'/S_L and Ka but are overestimated in leaner flames characterized by larger u'/S_L and Ka. Such data again imply that premixed turbulent combustion occurs in

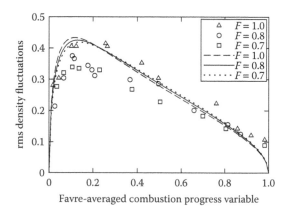

FIGURE 5.9 Normalized rms $\sqrt{\overline{\rho'^2}}/\rho_u$ of density fluctuations versus the Favre-averaged combustion progress variable. The symbols show the experimental data obtained by Driscoll and Gulati (1988) from three oblique, unconfined, statistically stationary, turbulent, methane–air flames characterized by three different equivalence ratios, specified in the legends. The curves have been computed using Equation 5.28. (Adapted from Driscoll, J.F. and Gulati, A., *Combust. Flame*, 72, 131–152, 1988. With permission.)

the flamelet regime at low Karlovitz numbers, while a transition to a nonflamelet regime happens as Ka is increased.

The segregation factor g obtained by Chen and Bilger (2002) from methane–air and propane–air Bunsen flames is substantially less than unity (see Figure 5.10), thus indicating that the BML approach overestimates the magnitude of temperature fluctuations (c was the normalized temperature in the cited work). The effect is increased by the ratio of the thermal laminar flame thickness to the Kolmogorov length scale, that is, by the Karlovitz number. Note that Ka was larger than unity in all flames investigated by Chen and Bilger (2002), thus implying that combustion should occur in a nonflamelet regime, in line with the obtained values of $g < 1$.

The validity of the flamelet concept is often assessed by measuring the instantaneous temperature gradient, evaluating the relevant length scale $l_T = (T_b - T_u)$ max$^{-1}|\nabla T|$, and comparing it with the laminar flame thickness Δ_L calculated using Equation 2.41. Such measurements have shown that the probability of $l_T > \Delta_L$ is significant (see Figures 5.11 and 5.12).

It is worth noting an important difference between the data reported in the two figures. While Figure 5.11 indicates a lower temperature gradient in the turbulent flame as compared with the laminar flame (i.e., only broadening of the instantaneous temperature profiles by turbulent eddies), Figure 5.12 shows not only broadening but also thinning of the instantaneous temperature profiles. Moreover, the latter effect (associated with stretching of laminar flamelets by turbulent eddies, as discussed in the section "Variations in Flamelet Structure Due To Turbulent Stretching"), plays a more important role, because $\bar{l}_T < \Delta_L$ in the flame addressed in Figure 5.12.

As summarized in Sections 2.1 and 3.1 of review papers by Driscoll (2008) and Lipatnikov and Chomiak (2010), respectively, there are more experimental data indicating $\bar{l}_T < \Delta_L$ and the opposite trend. Dinkelacker (2003) hypothesized that the former (latter) trend is relevant to mixtures characterized by Lewis numbers lower (larger) than unity, because stretching of laminar flames (see the section "Variations

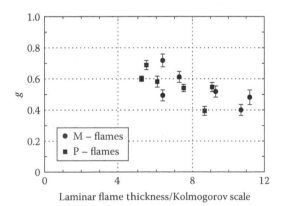

FIGURE 5.10 The average values of the segregation factor in the range of $0.4 < \bar{c} < 0.6$ versus the ratio of a laminar flame thickness (determined using the maximum temperature gradient) to the Kolmogorov length scale. (Adapted from Chen, Y.-C. and Bilger, R.W., *Combust. Flame*, 131, 400–435, 2002. With permission.)

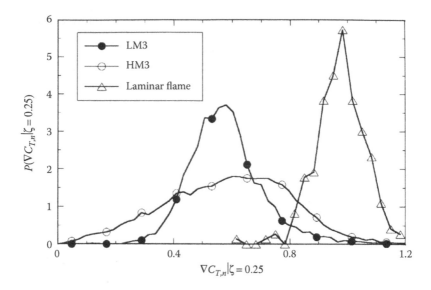

FIGURE 5.11 Probability density functions of $|\nabla c|_{c=0.25}$ obtained from high-turbulence (HM3), low-turbulence (LM3), and laminar lean ($\Phi = 0.75$) natural gas–air flames. The gradient $|\nabla c|$ is normalized using the computed thermal thickness $\Delta_L = 1.47$ mm of the laminar flame. (Reprinted from Chen, Y.-C. and Bilger, R.W., *Proc. Combust. Inst.*, 30, 801–808, 2005. With permission.)

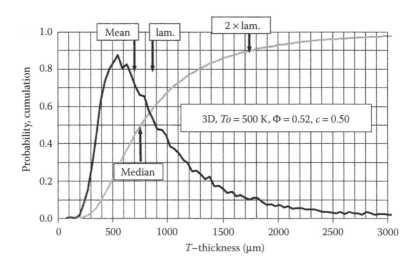

FIGURE 5.12 The probability density and cumulating functions of the instantaneous flame-front thickness obtained using 3D temperature measurements from a highly turbulent, preheated, lean ($\Phi = 0.52$) natural gas–air flame. The arrows indicate the mean and media values of l_T, as well as Δ_L and $2\Delta_L$. (Reprinted from Dinkelacker, F., Soika, A., Most, D., Hofmann, D., Leipertz, A., Polifke, W., and Döbbeling, K., *Proc. Combust. Inst.*, 27, 857–865, 1998. With permission.)

in Flamelet Structure Due to Turbulent Stretching") decreases (increases) their thickness if Le < 1 (Le > 1). Although de Goey et al. (2005) supported this hypothesis both experimentally and numerically, it does not seem to settle the problem finally. For instance, the experimental data shown in Figures 5.11 and 5.12 indicate $\bar{l}_T > \Delta_L$ and $\bar{l}_T < \Delta_L$, respectively, despite the fact that the data were obtained from lean natural gas–air flames characterized by approximately the same Le.

Recent experimental data by Dunn et al. (2010) indicate that, on the one hand, $\bar{l}_T$ is increased with increasing Re_t (see Figure 9 in the cited paper), thereby implying the increase in the thickness of instantaneous temperature profiles by very intense turbulence. On the other hand, $\bar{l}_T$ is still smaller than Δ_L in spatial regions characterized by a significant burning rate (cf. Figures 8 and 9 in the cited paper), thereby indicating the opposite trend in all the studied flames. It is worth noting, however, that (i) even if $\bar{l}_T < \Delta_L$, the thickness of highly strained laminar flames is significantly less than Δ_L and is less than the thickness $\bar{l}_T$ obtained from flame PM1-150 (cf. the dotted line in Figure 9 with lines plotted in Figure 8) and (ii) the measured two-dimensional (2D) temperature gradients reported in Figure 9 are larger than 3D temperature gradients in a turbulent flame. Moreover, Figure 10 from the cited paper shows that the instantaneous temperature gradient decreases (i.e., the thickness $\bar{l}_T$ becomes significantly larger than Δ_L) in spatial regions characterized by a low burning rate in very intense turbulence.

In summary, the influence of turbulence on the mean thickness of instantaneous temperature profiles is still an intricate issue that challenges the combustion community.

The experimental data shown in Figures 5.8 through 5.12 and numerous other similar data (e.g., see Section 3.1 of a review paper by Lipatnikov and Chomiak [2010]) may briefly be summarized as follows. At low Karlovitz numbers, premixed turbulent combustion occurs in the flamelet regime with a low probability γ of finding the intermediate states of a reacting mixture. The BML approach works well under such conditions. As Ka is increased, the instantaneous temperature profiles are disturbed and broadened by turbulent eddies, the probability γ is increased, the BML approach works worse, and the transition from the flamelet regime to the nonflamelet regime of premixed turbulent combustion appears to occur.

This summary of the experimental data agrees well with the physical arguments discussed in the sections "Physical Reasoning: Classical Flamelet Regime" and "Physical Reasoning: Thin-Reaction-Zone Regime," where another thin-reaction-zone regime, associated with larger Karlovitz numbers, was considered. Like the flamelet regime, heat release is localized to thin, inherently laminar, reacting layers in the thin-reaction-zone regime. Contrary to the flamelet regime, the instantaneous temperature profiles are perturbed and broadened by turbulent eddies in the thin-reaction-zone regime.

The latter claim is indirectly supported in Figures 5.8 through 5.12. Two more experimental data that indicate perturbations and broadening of instantaneous temperature profiles are shown in Figures 5.13 and 5.14.

It is worth stressing, however, that even if instantaneous temperature perturbations by small-scale turbulent eddies may look very strong (e.g., see curve A in Figure 5.14a), the statistical importance of these perturbations is an issue. For

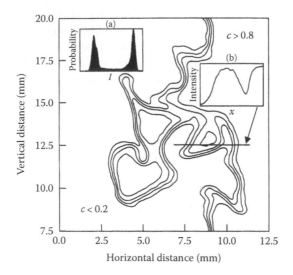

FIGURE 5.13 Instantaneous 2D contours of the temperature-based combustion progress variable ($c = 0.2, 0.4, 0.6$, and 0.8) obtained from a highly turbulent lean ($\Phi = 0.7$) methane–air flame stabilized on a low-swirl burner. (a) A PDF of scattered light intensity. (b) A scattered light intensity profile along a horizontal line. (Reprinted from Shepherd, I.G., Cheng, R.K., Plessing, T., Kortschik, C., and Peters, N., *Proc. Combust. Inst.*, 29, 1833–1840, 2002. With permission.)

instance, Shepherd et al. (2002) reported that the mean area between the c-contours 0.2 and 0.8 (or 0.3 and 0.7) of unit length was statistically independent of u'/S_L and was approximately equal to the area computed for laminar flames, despite the significant instantaneous perturbations shown in Figure 5.13b. Similarly, Kortschik et al. (2004) documented that the mean temperature profiles conditioned on the distance from the surface characterized by $T = 800$ K (see Figure 5.14b) were weakly perturbed even by intense turbulence that is characterized by u'/S_L as large as 18.7, despite the significant instantaneous perturbations shown in Figure 5.14a.

Independent of the role played by perturbations of the instantaneous c-profiles, previously discussed, the key peculiarity of the thin-reaction-zone regime is highlighted by its name. This peculiarity consists of a drastic drop in the heat release rate outside thin reaction zones even in intense turbulence. Experimental detection of such zones became possible only recently, thanks to rapid development in advanced optical methods of flame diagnostics. Today, the visualization of reaction zones is mainly performed by measuring laser-induced fluorescence (LIF) either from CH radicals or simultaneously from OH radicals and formaldehyde, CH_2O. In the latter case, the product of concentrations of OH and CH_2O is considered to be a marker of a reaction zone, because the concentration of OH drops in preheat zones, while the concentration of formaldehyde drops in combustion products. Obtaining 2D images of the local heat release rate is still at the cutting edge of contemporary experimental research into turbulent combustion, and only a few solid experimental data sets of that kind have yet been reported. Recently, these experimental data were

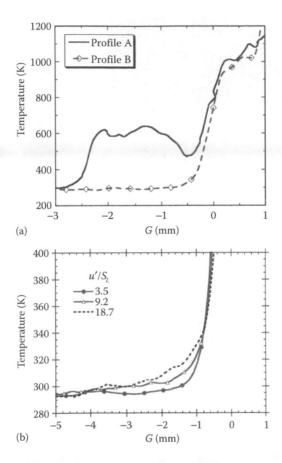

(a)

(b)

FIGURE 5.14 Instantaneous temperature profiles (a) and mean temperature profiles conditioned on the distance from the surface characterized by $T = 800$ K (b) obtained from swirl-stabilized lean ($\Phi = 0.7$) CH_4–air flames at different u'/S_L, specified in the legends. (Reprinted from Kortschik, C., Plessing, T., and Peters, N., *Combust. Flame*, 136, 43–50, 2004. With permission.)

thoroughly analyzed by Driscoll (2008), who drew a conclusion that thin reaction zones "exist for nearly all cases for which images of the reaction zone have been obtained." It is worth remembering, however, that, strictly speaking, these experimental data do not prove that thin reaction zones are inherently laminar.

As a recent example of turbulent premixed combustion in the thin-reaction-zone regime, Figure 5.15 shows instantaneous 2D images of OH radicals, formaldehyde CH_2O, and the heat release rate obtained by applying simultaneous (OH and CH_2O) planar laser-induced fluorescence (PLIF) techniques to lean premixed ethylene–air flames. The heat release images have been created by a pixel-by-pixel evaluation of the product of the OH and CH_2O PLIF signals. One can see that, although small-scale turbulent eddies significantly perturb the preheat zones and, in particular, substantially broaden the instantaneous spatial distribution of formaldehyde (see the

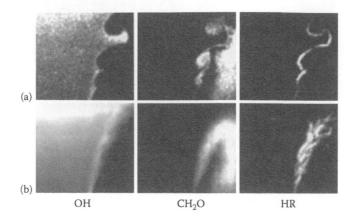

FIGURE 5.15 Instantaneous (a) and mean (b) 2D images of OH radicals (*left*), formaldehyde CH$_2$O (*center*), and heat release (*right*) obtained from lean ($\Phi = 0.45$) ethylene–air turbulent flames stabilized behind a bluff body. (Reprinted from Ayoola, B.O., Balachandran, R., Frank, J.H., Mastorakos, E., and Kaminski, C.F., *Combust. Flame*, 144, 1–16, 2006. With permission.)

central image in Figure 5.15a), which is localized to a thin zone in the counterpart laminar flame, the instantaneous heat release is confined to a thin, wrinkled zone (the right image in Figure 5.15a).

A similar phenomenon is observed in Figure 5.16. Although the instantaneous spatial distribution of CH$_2$O is substantially broadened by small-scale turbulent eddies, the concentration of CH, which is commonly considered to be a marker of

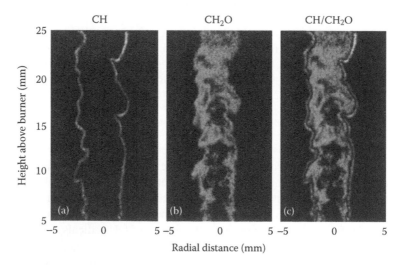

FIGURE 5.16 Instantaneous 2D images of CH (a), formaldehyde CH$_2$O (b), and both CH and CH$_2$O (c) obtained from highly turbulent, piloted jet, stoichiometric methane–air flames. (Reprinted from Li, Z.S., Li, B., Sun, Z.W., Bai, X.S., and Aldén, M., *Combust. Flame*, 157, 1087–1096, 2010. With permission.)

the heat release rate, is significant only in a thin zone, with the thickness of this zone being "independent of the turbulence intensity" (Li et al., 2010).

DEVELOPMENT OF PREMIXED TURBULENT FLAME IN THE FLAMELET REGIME

The foregoing discussion has shown that, at least at sufficiently low Karlovitz numbers, premixed turbulent combustion occurs in the flamelet regime when the unburned mixture and combustion products in the equilibrium state are separated from one another by thin, inherently laminar, self-propagating flamelets. In such a case, a high burning velocity is associated with an increase in the flamelet surface area by turbulent stretching, while a large mean flame brush thickness is due to the random advection of thin flamelets by turbulent eddies (see Figures 5.17 and 5.18).

The fact that the turbulent burning velocity and mean flame brush thickness are controlled by different turbulent phenomena (stretching and advection, respectively) is of paramount importance and manifests itself, in particular, in the substantially different rates of increase of U_t and δ_t during premixed turbulent flame development. To simplify the discussion on this issue, let us consider the asymptotic case of $\sigma \to 1$ (constant density), $\mathrm{Re}_t \to \infty$ (the Kolmogorov turbulence), $\mathrm{Ka} \to 0$ (the flamelet regime of premixed turbulent combustion), and $u'/S_L \to \infty$ (strong turbulence). These limits are consistent with one another (see Equation 5.4).

FIGURE 5.17 Superimposed contours of instantaneous flamelets. (Reprinted from Fox, M.D. and Weinberg, F.J., *Proc. R. Soc. Lond. A*, 268, 222–239, 1962. With permission.)

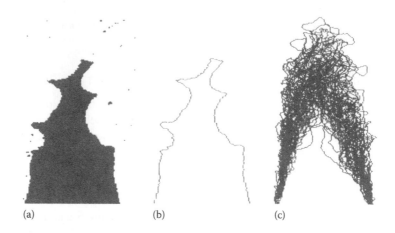

(a)   (b)   (c)

FIGURE 5.18 An instantaneous image of a methane–air turbulent Bunsen flame, obtained by recording a Mie-scattering signal (a), an instantaneous flamelet contour associated with this image (b), and superimposed contours of instantaneous flamelets (c). (Reprinted from Kobayashi, H., Tamura, T., Maruta, K., Niioka, T., and Williams, F.A., *Proc. Combust. Inst.*, 26, 389–396, 1996. With permission.)

In the limit case of strong turbulence, the self-propagation of flamelets appears to manifest itself in the consumption of the wrinkles of the flamelet surface, characterized by the shortest length scales, whereas the large-scale characteristics of the flamelet surface seem to be similar to the large-scale characteristics of a thin, inherently laminar, mixing layer advected by turbulent eddies. As discussed in the section "Taylor's Theory" in Chapter 3, the random advection of a laminar mixing layer results in turbulent diffusion so that the growth of the turbulent mixing layer (i.e., a spatial range where the probability of finding the molecular mixing layer is marked) is well described by Taylor's theory and is controlled by large-scale turbulent eddies. Therefore, it is tempting to assume that the increase in the mean flame brush thickness is also controlled by large-scale turbulent eddies in the asymptotic case considered. Consequently, the rate of increase of the thickness varies slowly in time, that is, $d\delta_t/dt$ changes notably during a long time interval of the order of τ_t (see Figures 4.6 and 4.7).

On the contrary, the physical mechanism of the development of the turbulent burning velocity is totally different. The burning velocity is high due to wrinkling of the flamelet surface by turbulent stretching, and it is the smallest-scale turbulent eddies that produce the highest instantaneous stretch rate (combustion does not affect turbulence in the constant-density case considered). Therefore, the burning velocity is mainly controlled by the small-scale turbulent eddies in the asymptotic case considered. Consequently, U_t rapidly increases during a short early stage of premixed turbulent flame development ($dU_t/dt \propto \tau_\eta^{-1}$ at $t < \tau_\eta$), followed by a saturation due to the equilibrium between an increase in the area of the flamelet surface owing to small-scale turbulent stretching and the consumption of the small-scale wrinkles of the surface owing to the self-propagation of the flamelet. During the latter, small-scale equilibrium stage, the turbulent burning velocity still increases, because large-scale eddies also contribute to an increase in the flamelet surface area. However, this increase is weakly pronounced, because, in the combustion regime discussed previously, U_t is mainly controlled by the small-scale turbulent eddies and, hence, the normalized burning velocity $u_t \equiv U_t/U_{t,\infty}$ is of unity order (but less than unity) already at $t = \tau_\eta$.

Thus, if we consider an earlier, but sufficiently long, for example, $0 < t < t_1 = O(\tau_t)$, stage of premixed turbulent flame development, then the development of $\delta_t(t)$ differs qualitatively from the development of $U_t(t)$. If the rate of increase $d\delta_t/dt \propto u'$ is roughly constant during this interval, see the section "Mean Flame Brush Thickness" in Chapter 4 and Equation 4.2, the rate of increase dU_t/dt is very high at $t < \tau_\eta$ and strongly reduces during flame development. Accordingly, the dependence of the mean flame brush thickness on the flame-development time t is roughly linear, whereas the dependence of the turbulent burning velocity on t is strongly nonlinear with the negative second derivative d^2U_t/dt^2.

The emphasized difference in the development of $\delta_t(t)$ and $U_t(t)$ at $0 < t < t_1 = O(\tau_t)$ offers the unique opportunity to assess and rank various models of premixed turbulent combustion, as discussed in the section "Modeling of Effects of Turbulence on Premixed Combustion in RANS Simulations" in Chapter 7. Even if the asymptotic case discussed here may be claimed to be unpractical, it is still within the domain of applicability for the vast majority of contemporary models of premixed turbulent combustion. Consequently, the prediction of the highlighted qualitative difference in the development of $\delta_t(t)$ and $U_t(t)$ at $0 < t < t_1 = O(\tau_t)$ is a necessary condition to consider any such model to be consistent with the underlying physics.

Although, the foregoing discussion was limited to the asymptotic case of $\sigma \to 1$, $Re_t \to \infty$, $Ka \to 0$, and $u'/S_L \to \infty$, the qualitative difference in the development of $\delta_t(t)$ and $U_t(t)$ may be pronounced in more realistic cases. The history of science gave a number of examples of asymptotic theories that yielded valuable results well outside the asymptotic limits addressed by them. For instance, the AEA theory of laminar premixed flames developed for $Ze \to \infty$ and the Kolmogorov theory of turbulence developed for $Re_t \to \infty$ are examples of such theories, which are very close to the present subject.

Unfortunately, the development of the turbulent burning velocity was outside the mainstream research into premixed turbulent combustion, and little data on the growth of U_t have yet been reported. These data discussed elsewhere (Lipatnikov, 2009b) do not contradict the foregoing discussion at least, but they are too scanty to draw solid conclusions. Definitely, the issue requires more experimental investigation. Nevertheless, it is worth giving one example here.

Figure 5.19 shows (i) the development of the normalized burning velocity (solid curve) evaluated using the following empirical parameterization (Abdel-Gayed et al., 1984):

$$u_t \equiv \frac{U_t}{U_{t,\infty}} = \left\{ 1 - \exp\left[-0.2 \left(\frac{8}{\pi} \right)^{3/8} \left(\frac{t}{\tau_t} \right)^{3/4} \right] \right\}^{1/2} \tag{5.49}$$

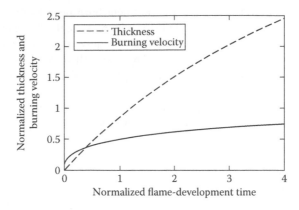

FIGURE 5.19 The development of a normalized flame brush thickness $\delta_t/(\sqrt{2\pi}u'\tau_L)$, calculated using Equation 4.2, and a normalized burning velocity evaluated using Equation 5.49.

of the experimental data obtained from statistically spherical premixed turbulent flames expanding in a Leeds fan-stirred bomb and (ii) the development of the normalized flame brush thickness $\delta_t/(\sqrt{2\pi}u'\tau_L)$, calculated using Equation 4.2, which well approximates the experimental data obtained from similar flames (see the crosses and pluses in Figure 4.6). The two curves indicate the trends emphasized earlier: a roughly linear dependence of δ_t on t and a strongly nonlinear dependence of U_t on t with the negative second derivative d^2U_t/dt^2.

SUMMARY

In many flames, premixed turbulent combustion occurs in the thin-reaction-zone regime when the heat release rate drops drastically outside thin, inherently laminar, reaction zones wrinkled by turbulent eddies. At sufficiently low Karlovitz numbers, not only the reaction zones but also the preheat zones are thin and retain the structure of a laminar flame (flamelet subregime). At a higher Ka (see Equations 5.9 and 5.10), the preheat zones may be strongly perturbed and broadened by small-scale turbulent eddies (thickened-preheat-zone subregime), but the statistical importance of such perturbations is an issue. Eventual penetration of small-scale turbulent eddies into reaction zones is another issue, which will be discussed further in the section "Large-Scale Intense Turbulence and Distributed Reaction Regime."

In the thin-reaction-zone regime, an increase in the burning velocity by the rms turbulent velocity is mainly controlled by wrinkling of the zones by turbulent eddies, while the increase in U_t with time is limited by the annihilation of reaction zones that collide due to their self-propagation. Due to the latter physical mechanism, an increase in U_t by S_L is weaker than the linear dependence.

In the considered regime, the increase in the mean flame brush thickness is likely to differ qualitatively from the increase in the turbulent burning velocity at $0 < t < t_1 = O(\tau_t)$. More specifically, the rate of increase $d\delta_t/dt \propto u'$ appears to be roughly constant during this interval, while the rate of increase dU_t/dt seems to be very high at $t < \tau_\eta$ and appears to decrease strongly during flame development.

The BML approach discussed in the section "Bray–Moss–Libby Approach" is an effective tool for modeling premixed turbulent combustion in the flamelet subregime.

VARIATIONS IN FLAMELET STRUCTURE DUE TO TURBULENT STRETCHING

In the foregoing discussion, the effect of turbulence on the burning rate was reduced to an increase in the surface area of thin, inherently laminar, flamelets or reacting zones, which were assumed to retain the structure of the unperturbed laminar flame or its reaction zone, respectively. Such a physical mechanism yields an increase in U_t by u' and S_L, but it cannot explain the existence of the maximum burning velocity at certain u'_m, a decrease in U_t by u' at strong $(u' > u'_m)$ turbulence, and the significant effect of the molecular diffusivity of the deficient reactant on the turbulent burning rate (especially, considering that U_t is decreased when Le is increased, while the unperturbed laminar flame speed is increased by Le, see the section "Dependence of Turbulent Burning Velocity on Differences in Molecular Transport Coefficients" in Chapter 4 and the section "Effect of the Lewis number on Unperturbed Laminar Flame Speed" in Chapter 2). To understand these phenomena, another physical mechanism is often highlighted, namely, perturbations of the inner structure of inherently laminar flamelets or reacting zones by turbulent stretching. Such a physical mechanism is commonly discussed by invoking the results of numerous theoretical, numerical, and experimental studies of perturbed laminar premixed flames. Definitely, such a method is only justified for the flamelet subregime of the thin-reaction-zone regime of premixed turbulent combustion. Therefore, in this section, the discussion is limited to this subregime associated with low Karlovitz numbers, whereas the influence of small-scale turbulent stretching on the structure of thin reaction zones in the thickened-preheat-zone subregime associated with larger Karlovitz numbers is still an issue.

Because theoretical studies of weakly perturbed laminar premixed flames have shown that various flame perturbations may be characterized by a single quantity called stretch rate, it is worth beginning a discussion on the influence of turbulent eddies on the local flamelet structure and burning rate with a brief introduction to stretch rate.

STRETCH RATE, STRAIN RATE, AND CURVATURE

For an arbitrary surface, the local stretch rate is defined as follows:

$$\dot{s}(\mathbf{x},t) \equiv \frac{1}{A(\mathbf{x},t)} \frac{dA}{dt}, \tag{5.50}$$

where $A(\mathbf{x},t)$ is the area of this surface in an infinitesimal volume around point $\mathbf{x}$. Therefore, the stretch rate characterizes the increase in the surface area in the logarithmic scale. As discussed in the section "Energy Cascade and Turbulent Stretching" in Chapter 3, the mean stretch rate controls the rate of increase of a material surface in isotropic homogeneous turbulence (Batchelor, 1952; Yeung et al., 1990) (see Equation 3.54).

For an arbitrary, infinitely thin, self-propagating surface in an arbitrary flow, it can be shown (e.g., see Candel and Poinsot, 1990) that the local stretch rate is equal to

$$\dot{s}(\mathbf{x},t) = n_j n_k \frac{\partial u_j}{\partial x_k} - \frac{\partial u_k}{\partial x_k} + S \frac{\partial n_k}{\partial x_k}, \qquad (5.51)$$

where S is the speed of self-propagation of the surface, the summation convention applies to the repeated subscripts j and k, and $\mathbf{n}$ is the unit vector normal to the surface, that is,

$$\mathbf{n} = -\frac{\nabla \Psi}{|\nabla \Psi|}, \qquad (5.52)$$

with the surface being determined by $\Psi(\mathbf{x},t) = 0$. In the combustion literature, $S = S_L$ is the laminar flame speed, and positive values of the function $\Psi(\mathbf{x},t)$ are commonly associated with the burned mixture; therefore, the normal vector $\mathbf{n}$ points to the unburned gas.

Equation 5.51 is often written in other forms, for example,

$$\dot{s}(\mathbf{x},t) = n_j n_k \frac{\partial u_j}{\partial x_k} + S \frac{\partial n_k}{\partial x_k}, \qquad (5.53)$$

because $\nabla \cdot \mathbf{u} = 0$ in a constant-density flow, or (Pope, 1988)

$$\dot{s}(\mathbf{x},t) = \nabla_t \cdot \mathbf{u} + S \nabla \cdot \mathbf{n}, \qquad (5.54)$$

where

$$\nabla_t \cdot \mathbf{u} = \mu_j \mu_k \frac{\partial u_j}{\partial x_k} + \eta_j \eta_k \frac{\partial u_j}{\partial x_k} = \left(n_j n_k - \delta_{jk} \right) \frac{\partial u_j}{\partial x_k} = n_j n_k \frac{\partial u_j}{\partial x_k} - \frac{\partial u_k}{\partial x_k} \qquad (5.55)$$

is the local tangential (to the surface) divergence of the flow velocity vector, and μ_i and η_i are the components of two unity vectors that are perpendicular both to each other and to the vector $\mathbf{n}$, that is, the two former vectors lie on the local tangential plane.

Alternatively, Chung and Law (1984) have derived the following expression:

$$\dot{s}(\mathbf{x},t) = \nabla_t \cdot \mathbf{v} + (\mathbf{u} + S\mathbf{n}) \cdot \mathbf{n}(\nabla_t \cdot \mathbf{n}), \qquad (5.56)$$

where $\mathbf{v} = \mathbf{u} - \mathbf{n}(\mathbf{u} \cdot \mathbf{n})$ is the tangential (to the surface) component of the flow velocity vector. Because the vector $\mathbf{n}$ is perpendicular to the local tangential plane, we have $\mathbf{n} \cdot \nabla_t q = 0$ for any scalar quantity q, for example, $q = \mathbf{u} \cdot \mathbf{n}$. Consequently, Equation 5.56 reads (Pope, 1988)

$$
\begin{aligned}
\dot{s}(\mathbf{x},t) &= \nabla_t \cdot \mathbf{v} + (\mathbf{u} + S\mathbf{n}) \cdot \mathbf{n}(\nabla_t \cdot \mathbf{n}) \\
&= \nabla_t \cdot \mathbf{u} - (\mathbf{u} \cdot \mathbf{n})(\nabla_t \cdot \mathbf{n}) + (\mathbf{u} \cdot \mathbf{n})(\nabla_t \cdot \mathbf{n}) + S\nabla_t \cdot \mathbf{n} \\
&= \nabla_t \cdot \mathbf{u} + S\nabla_t \cdot \mathbf{n}.
\end{aligned}
\tag{5.57}
$$

Because the normal vector $\mathbf{n}$ is not changed in the direction normal to the surface, $\nabla \cdot \mathbf{n} = \nabla_t \cdot \mathbf{n}$, and Equation 5.57 reduces to Equation 5.54.

Equation 5.54 is often rewritten in the following form:

$$
\dot{s}(\mathbf{x},t) = a_t + 2Sh_m,
\tag{5.58}
$$

where

$$
a_t(\mathbf{x},t) \equiv \nabla_t \cdot \mathbf{u}
\tag{5.59}
$$

is the local strain rate,

$$
2h_m \equiv \nabla \cdot \mathbf{n} = \frac{1}{R_1} + \frac{1}{R_2}
\tag{5.60}
$$

is the mean curvature, and R_1 and R_2 are the principal radii of curvature of the surface. For example, $R_1 = R$ and $R_2 \to \infty$ for a cylindrical surface and $R_1 = R_2 = R$ for a spherical surface.

Thus, stretching of a self-propagating surface results from (i) strain rates of a spatially nonuniform flow and (ii) the curvature of the surface. In the following two subsections, the physics of the influence of strain rate and surface curvature on a laminar premixed flame will be discussed by considering two simplified model problems: a planar laminar flame stabilized in a divergent flow and an expanding spherical flame. In subsequent subsections, the key results of theoretical, numerical, and experimental studies of stretched laminar premixed flames will be briefly summarized and experimental and numerical data on the influence of turbulent stretching on the local flamelet structure and burning velocity will be discussed.

STRAINED LAMINAR PREMIXED FLAMES

Twin symmetrical, stationary premixed flames stabilized in two identical counterflow streams of the same mixture are commonly considered to be the classical model problem for studying strained laminar flames following the seminal work by Klimov (1963). A sketch of the two flames is shown in Figure 5.20, where ABCD is a control volume within the preheat zone, the reaction zones are drawn in thick straight lines, the thin curved lines with arrows show streamlines, and the thick arrows indicate the

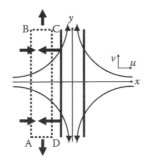

FIGURE 5.20 Two symmetrical counterflow laminar flames.

directions of the convective and conductive heat fluxes. The burned gas occupies the central region bounded by the two reaction zones.

Due to the symmetry of the considered problem with respect to the x-axis, the axial velocity u, temperature, density, and mass species of the reactants and products depend weakly on y in the vicinity of the x-axis, that is, $\partial q/\partial y \to 0$ as $y \to 0$. Consequently, the mass conservation equation

$$\frac{\partial}{\partial x}(\rho u) + \frac{\partial}{\partial y}(\rho v) = 0 \tag{5.61}$$

yields

$$\frac{\partial v}{\partial y} = -\frac{1}{\rho}\frac{\partial}{\partial x}(\rho u) \equiv \psi(x). \tag{5.62}$$

Therefore, $v = y\psi(x)$ and Equations 5.58 through 5.60 yield

$$\dot{s} = a_t = \nabla_t \cdot \mathbf{u} = \frac{\partial v}{\partial y} = \psi(x), \tag{5.63}$$

with the curvature term vanishing, because the flame surface is planar.

In the case of a constant density, it can be shown (e.g., see Dixon-Lewis, 1990) that the stretch rate is independent of the coordinate x. In the case of a variable density, the stretch rate depends on x, whereas another quantity, $y^{-1}\partial p/\partial y$, is independent of the x-coordinate (Dixon-Lewis, 1990) and characterizes the entire flow sketched in Figure 5.20. Note that $a_t = (-y^{-1}\partial p/\partial y)^{1/2}$ both in the case of a constant density and in the flow of an unburned mixture ahead of a flame, but this equality does not hold within the flame if the density changes.

To discuss the physical mechanism of variations in the mass burning rate, speed, and structure of a strained laminar flame in the simplest qualitative manner, let us assume that the stretch rate is constant, as occurs if $\rho = $ const. Then, Equation 5.62 reads

$$\rho u - \rho_b S_{L,b} = \dot{s}\zeta \tag{5.64}$$

for the left flame in Figure 5.20, that is, for the flame stabilized at $x < 0$. Here,

$$\zeta = \int\limits_{x}^{x_f} \rho dx, \tag{5.65}$$

$x_f < 0$ is the coordinate of the reaction zone, which is assumed to be infinitely thin within the framework of the AEA theory of premixed laminar combustion (see the section "AEA Theory of Unperturbed Laminar Premixed Flame" in Chapter 2), and $\rho u = \rho_b S_{L,b}$ locally in this zone in the steady case considered. Equation 5.64 shows that the magnitude of the axial velocity increases with the distance from the reaction zone of a strained flame. Such an increase in $|u|$ results in a decrease in both the combustion progress variable in the preheat zone and the flame speed as compared with the unperturbed laminar flame.

The point is that a part of the heat conducted from the reaction zone of a strained flame is convected along the flame front, because the magnitude $|v|$ of the tangential velocity increases with $|y|$. For instance, the thermal energy that enters the control volume ABCD through the side CD (see Figure 5.20) preheats not only a part of the oncoming flow that enters the reaction zone through the same side CD but also a part of the oncoming flow that is convected out of this volume through the sides AD and BC. On the contrary, in the unperturbed flame, the thermal energy conducted through the side CD preheats only the oncoming flow that enters the reaction zone through the same side CD. Accordingly, if we assume that the flame speed is not affected by the strain rate, then the temperature of the control volume ABCD in the strained flame should be lower than the temperature of the counterpart volume in the unperturbed flame, with all other things being equal, because the same flux of thermal energy from the reaction zone should preheat a larger amount of unburned mixture in the former case. The aforementioned decrease in the temperature in the preheat zone of the strained flame results in a steeper temperature gradient, that is, a decrease in the flame thickness by the strain rate.

Moreover, the tangential heat flux not only reduces the flame thickness but also decreases the flame speed, which is equal to the flow velocity at the reaction zone divided by the density ratio, that is, $S_L = \rho_b S_{L,b}/\rho_u = \rho_b u(x_f)/\rho_u$. Indeed, because the thermal energy released in the reaction zone of a strained flame not only preheats the oncoming flow, but it is also convected along the reaction zone, the normal heat flux would not be sufficient to balance the convection flux if $S_{L,b} = S^0_{L,b}$, where the superscript 0 designates the unperturbed flame. The normal conductive and convective fluxes may balance one another only if $S_{L,b} < S^0_{L,b}$ in a strained flame.

The foregoing qualitative reasoning can be supported by the following equations. In the steady, 2D case considered, the combustion progress variable balance Equation 1.79 reads

$$\frac{\partial}{\partial x}(\rho u c) + \frac{\partial}{\partial y}(\rho v c) = \frac{\partial}{\partial x}\left(\rho D \frac{\partial c}{\partial x}\right) + \frac{\partial}{\partial y}\left(\rho D \frac{\partial c}{\partial y}\right) + W. \tag{5.66}$$

Using the continuity Equation 5.61, we obtain

$$\rho u \frac{\partial c}{\partial x} = \frac{\partial}{\partial x}\left(\rho D \frac{\partial c}{\partial x}\right) + W, \tag{5.67}$$

because $\partial c/\partial y = 0$ in the vicinity of the x-axis ($y = 0$) for symmetry reasons. The sole difference between this equation and Equation 2.14, which models the unperturbed flame, consists of the dependence of u on x in the former equation, while $\rho u = \rho_b S_{L,b} = $ const in Equation 2.14. In the preheat zone, $W = 0$ and Equation 5.67 reads

$$\rho u \frac{\partial c}{\partial x} = \frac{\partial}{\partial x}\left(\rho D \frac{\partial c}{\partial x}\right). \tag{5.68}$$

Accordingly, an increase in $|u|$ with $|x|$ can only be balanced by an increase in the diffusion term, that is, by a decrease in the thickness Δ of the $c(x)$ profile, because the convection and diffusion terms are proportional to Δ^{-1} and Δ^{-2}, respectively. The decrease in Δ, required to balance the increase in $|u|$ with $|x|$, means that the slope of the $c(x)$ profile is steeper in the strained flame as compared with the unperturbed flame. Because $c \approx 1$ at the boundary of the reaction zone in both strained and unperturbed flames, a steeper c profile in the former flame means a lower value of the combustion progress variable in the preheat zone (see Figure 5.21).

Let us assume for simplicity that $\rho^2 D = \rho_u^2 D_u = $ const. Then, Equations 5.65 and 5.68 yield

$$\rho u \frac{\partial c}{\partial \zeta} = -\rho_u^2 D_u \frac{\partial^2 c}{\partial \zeta^2}. \tag{5.69}$$

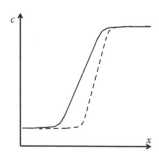

FIGURE 5.21   A sketch of c-profiles in unperturbed (*solid curve*) and strained (*dashed curve*) flames.

The solutions to this equation are

$$\frac{\partial c}{\partial \zeta} = \frac{\partial c}{\partial \zeta}\bigg|_f \exp\left(-\frac{S_L^0 \zeta}{\rho_u D_u}\right) \tag{5.70}$$

if $\rho u = \rho_u S_L^0$ (unperturbed flame) and

$$\frac{\partial c}{\partial \zeta} = \frac{\partial c}{\partial \zeta}\bigg|_f \exp\left(-\frac{\rho_b S_{L,b} \zeta}{\rho_u^2 D_u}\right) \exp\left(-\frac{\dot{s}\zeta^2}{2\rho_u^2 D_u}\right) = B \exp\left[-\frac{\left(\rho_b S_{L,b} + \dot{s}\zeta\right)^2}{2\dot{s}\rho_u^2 D_u}\right] \tag{5.71}$$

if Equation 5.64 holds (strained flame). Here, B is a constant and the subscript f designates the quantities calculated at the boundary between the preheat and the thin reaction zones, that is, at $x = x_f < 0$ or $\xi = 0$. In the remainder of this book, the symbols S_L^0 and S_L will designate the speeds of unperturbed and perturbed laminar flames, respectively, when discussing issues that are sensitive to the difference in the two speeds. If the difference in S_L^0 and S_L is not emphasized, a single symbol S_L will be used for simplicity.

Within the framework of the AEA theory, the reaction zone is not affected by straining (in the case of unity Lewis number for all reactants, while flames with Le $\neq 1$ will be discussed later), because the convection term in Equation 5.67 vanishes in the asymptotically (Ze $\to \infty$) thin reaction zone, while two other terms in this equation are not straightforwardly affected by a strain rate. In physical terms, flow nonuniformities of any finite length scale, for example, a strain rate, cannot manifest themselves at infinitely small length scales.

If the reaction zone is not affected by straining, then c_f is asymptotically equal to unity, and the integration of Equation 5.71 in the preheat zone yields

$$1 = B \int_\infty^0 \exp\left[-\frac{\left(\rho_b S_{Lb} + \dot{s}\zeta\right)^2}{2\dot{s}\rho_u^2 D_u}\right] d\zeta = -\frac{B}{\dot{s}} \int_0^\infty \exp\left[-\frac{\left(\rho_b S_{Lb} + \mu\right)^2}{2\dot{s}\rho_u^2 D_u}\right] d\mu$$

$$= -\frac{B}{\dot{s}} \int_{\rho_b S_{Lb}}^\infty \exp\left(-\frac{\eta^2}{2\dot{s}\rho_u^2 D_u}\right) d\eta = -B\rho_u \sqrt{\frac{2D_u}{\dot{s}}} \int_\chi^\infty \exp\left(-\lambda^2\right) d\lambda$$

$$= -B\rho_u \sqrt{\frac{\pi D_u}{2\dot{s}}} \operatorname{erfc}(\chi), \tag{5.72}$$

where

$$\chi \equiv \frac{\rho_b S_{L,b}}{\rho_u \sqrt{2\dot{s}D_u}} = \frac{S_L}{\sqrt{2\dot{s}D_u}} \tag{5.73}$$

and erfc is the complementary error function. Therefore,

$$\frac{\partial c}{\partial \zeta}\bigg|_f = \frac{\partial c}{\partial \zeta}\bigg|_{\zeta=0} = B\exp\left(-\frac{\rho_b^2 S_{L,b}^2}{2\dot{s}\rho_u^2 D_u}\right) = -\sqrt{\frac{2\dot{s}}{\pi D_u}}\frac{\exp\left(-\chi^2\right)}{\rho_u \mathrm{erfc}\left(\chi\right)}. \tag{5.74}$$

Similarly, the integration of Equation 5.70 yields

$$\frac{\partial c}{\partial \zeta}\bigg|_f = -\frac{S_L^0}{\rho_u D_u} \tag{5.75}$$

for the unperturbed flame. Because the infinitely thin reaction zone is not affected by straining if Le = 1, the heat fluxes from the reaction zone are equal to one another in strained and unperturbed flames, that is, the gradients on the LHSs of Equation 5.74 and 5.75 are equal to one another. Therefore,

$$\frac{S_L^0}{D_u} = \sqrt{\frac{2\dot{s}}{\pi D_u}}\frac{\exp\left(-\chi^2\right)}{\mathrm{erfc}\left(\chi\right)} \tag{5.76}$$

or

$$\frac{S_L}{S_L^0} = \sqrt{\pi}\chi\,\mathrm{erfc}\left(\chi\right)\exp\left(\chi^2\right). \tag{5.77}$$

From the mathematical analysis, it is known that

$$\frac{2}{1+\sqrt{1+\dfrac{2}{\chi^2}}} < \sqrt{\pi}\chi\,\mathrm{erfc}\left(\chi\right)\exp\left(\chi^2\right) < \frac{2}{1+\sqrt{1+\dfrac{4}{\pi\chi^2}}} < 1 \tag{5.78}$$

for $\chi > 0$. Thus, $S_L < S_L^0$ in the case considered, in line with the aforementioned physical reasoning.

In the asymptotical ($\dot{s} \to 0$ and $\chi \to \infty$) case of a weakly strained flame, Equation 5.77 tends to

$$\frac{S_L}{S_L^0} \to 1 - \frac{1}{2\chi^2} = 1 - \frac{\dot{s}D_u}{S_L^2}, \tag{5.79}$$

that is,

$$S_L \to S_L^0\left(1-\tau_c\dot{s}\right) \tag{5.80}$$

if $\tau_c \dot{s} \to 0$, where

$$\tau_c = \frac{D_u}{S_L^2}$$ (5.81)

is the chemical timescale. The results of a more rigorous theory will be reported in the section "A Brief Summary of Theory of Weakly Stretched Laminar Premixed Flames."

In the opposite asymptotical ($\tau_c \dot{s} \to \infty$ and, hence $\chi \to 0$) case of a strongly strained flame, the RHS of Equation 5.77 tends to zero. The numerical solution of Equation 5.77 shows that the flame speed (i.e., the axial flow velocity at the reaction zone multiplied by ρ_b/ρ_u) drops to zero at $\tau_c \dot{s} \approx 1$ (Zel'dovich et al., 1985). In this case, the right and left reaction zones merge together at $x = 0$, but the mass burning rate per unit flame surface area is still equal to $\rho_u S_L^0$ if the reaction zone is infinitely thin and is not affected by straining.

According to Equations 5.70 and 5.71, a decrease in S_L as compared with S_L^0 results in an increase in the slope of the c profile at $\zeta \to 0$ in the strained flame as compared with the unperturbed flame, that is, straining reduces the thickness of a laminar premixed flame, again in line with the aforementioned physical reasoning. It is worth stressing, however, that within the framework of the AEA theory, the flame thickness determined using the maximum gradient method (see Equations 2.40 and 2.41) is not affected by the strain rate if Le = 1 and the gradient of c reaches its maximum value at the boundary of the reaction zone, that is, at $\zeta = 0$. Indeed, since the heat flux from the reaction zone to the preheat zone is controlled by the burning rate, which is not affected by low strain rates if Le = 1, the gradient of c, calculated at $\zeta = 0$, does not depend on $\dot{s}$ if the strain rate is sufficiently low in order for the reaction zone to be far from the symmetry axis $x = 0$. However, at small ($\zeta \ll \rho_u S_L \dot{s}^{-1}$) but finite ζ, the gradient of c, calculated using Equation 5.71 for a strained flame, is larger than the gradient of c, calculated using Equation 5.70 for an unperturbed flame, because $\rho_b S_{L,b}/\rho_u$ in the exponent on the RHS of the former equation is less than S_L^0.

Thus, within the framework of the AEA theory and if $D_F = D_O = a$, then straining

- Reduces the thickness of a laminar premixed flame but does not affect the maximum temperature gradient reached at the boundary of the reaction zone
- Decreases the flame speed evaluated at the reaction zone
- Does not affect the reaction zone itself and, in particular, the burning rate

In a real laminar premixed flame, the strain rate not only affects the flame thickness and speed, but can also change the burning rate in the reaction zone even in the adiabatic case. The sensitivity of the burning rate to flame straining is associated with the following three effects. First, if two symmetrical reaction zones merge together at $x = 0$ in highly strained flames, a further increase in the strain rate can extinguish the combustion.

Second, if the thickness of a reaction zone is small but finite (e.g., due to a large but finite Zel'dovich number or due to complex chemistry), then straining reduces not only the thickness of the preheat zone but also the thickness, δ_r, of the reaction zone. Due to the reduction in δ_r in strained flames, the integral of W over the reaction zone and, hence, the burning velocity is decreased when the strain rate is increased.

Third, the reaction zone and, in particular, the burning rate may be sensitive to the strain rates if the molecular transport coefficients of the main reactants and heat are not equal to one another. If, in the unperturbed laminar premixed flame, molecular diffusion and heat conduction balance each other so that the local equivalence ratio and temperature are equal to $\Phi_f = \Phi^0$ and $T_f = T_b^0$, respectively, in the reaction zone, then the balance may be changed in the reaction zone of a strained flame so that $\Phi_f \neq \Phi^0$ and/or $T_f \neq T_b^0$. Indeed, since straining of a flame increases the gradients within the preheat zone, the straining also increases the molecular mass and heat fluxes toward and from the reaction zone. If the Lewis numbers are equal to unity for all the main reactants, the aforementioned balance between the fluxes still holds in a strained flame. If, however, for example, Le < 1 for a fuel in a lean mixture, then $a < D_F$, and an increase in the heat flux from the reaction zone will be lower than an increase in the flux of chemical energy (bound in the fuel) toward the zone. Accordingly, the local energy in the reaction zone will be increased by straining, hence $T_f > T_b^0$. Such an increase in the local temperature in the reaction zone of a strained flame will result in an increase in the burning rate.

The aforementioned physical reasoning may be supported by the following expressions. Because Equations 5.68 through 5.77 involve the molecular diffusivity D, they should be associated with the normalized mass fraction of the deficient reactant (e.g., fuel in a lean mixture), that is, $c = 1 - Y_F/Y_{F,u}$ in this case. If Le = 1, the same equations hold for the normalized temperature $\theta = (T - T_u)/(T_f - T_u)$, provided that c and D are substituted with θ and the heat diffusivity a, respectively. Therefore,

$$a_u \left. \frac{\partial \theta}{\partial \zeta} \right|_r = -\frac{S_L \exp\left(-\lambda^2\right)}{\sqrt{\pi} \rho_u \lambda \mathrm{erfc}\left(\lambda\right)} \equiv -\rho_u^{-1} S_L \Xi\left(\lambda\right), \qquad (5.82)$$

where

$$\lambda \equiv \frac{S_L}{\sqrt{2\dot{s}a_u}} = \frac{\chi}{\sqrt{\mathrm{Le}}}. \qquad (5.83)$$

If we invoke Equations 5.74 and 5.82 to assess qualitatively the relative changes in the heat and mass fluxes in the case of weak deviations in the Lewis number from unity, then these equations indicate that the normalized heat flux from the reaction zone to the preheat zone is less than the normalized flux of chemical energy (bounded in the fuel) toward the reaction zone if Le < 1. Indeed, because (i) the function $\Xi(\chi)$ decreases when χ increases (see Equation 5.78 written for $\Xi^{-1}(\chi)$) and (ii) $\chi < \lambda$, we have $\Xi(\chi) > \Xi(\lambda)$ if Le < 1. In the unperturbed flame, the two fluxes are equal to each other, because the gradients of c and θ in the reaction zone are inversely proportional to D and a, respectively, in this case (see Equation 5.75).

Due to a stronger (if Le < 1) flux of chemical energy into the reaction zone of a strained flame, the temperature is increased therein as compared with the adiabatic combustion temperature $T_b(\Phi^0)$. On the contrary, $T_f < T_b(\Phi^0)$ if Le > 1. Molecular transport processes play a more important role in strained flames as compared with unperturbed flames due to the higher spatial gradients in the former case.

The physical scenario discussed previously is illustrated in Figure 5.22, where the solid and dashed curves are associated with unperturbed and strained laminar premixed flames, respectively. The filled circles A and B show the temperature and the mass fraction of a deficient reactant, respectively, at a point within the reaction zone of the unperturbed flame. In the strained flame, the temperature (circle A′) is higher at a point characterized by the same mass fraction Y_B (circle B′), Therefore, the local reaction rate is higher in the strained flame as compared with the unperturbed flame. It is worth stressing, however, that, even if a moderate strain rate increases the burning rate in the case of Le < 1, a sufficiently strong strain rate can extinguish the flame. For instance, the twin flames sketched in Figure 5.20 are quenched due to incomplete reactions when the two reaction zones of a finite thickness move close to the plane of symmetry. Certainly, the extinction is promoted by an increase in Le due to a decrease in the burning rate.

The foregoing discussion dealt with temperature variations in the reaction zone of a strained flame due to nonunity Lewis number. Similarly, the local equivalence ratio in the reaction zone of a strained flame may differ from the initial value Φ^0 if the molecular diffusivities of the fuel and oxidizer differ from each other.

Thus, if $D_F > D_O$ and Le < 1 in a lean (e.g., hydrogen–air) flame, the temperature and the equivalence ratio in the reaction zone are increased by the strain rate, with both processes increasing the burning rate and impeding combustion extinction by strong strain rates. If $D_F > D_O$ and Le > 1 in a rich (e.g., hydrogen–air) flame, the temperature is reduced, while the equivalence ratio is increased in the reaction zone by the strain rate. In this case, both processes reduce the burning rate and promote flame quenching. If $D_F < D_O$ and Le < 1 in a rich (e.g., propane–air) flame, the temperature is increased, while the equivalence ratio is reduced in the reaction zone by the strain rate, with both processes increasing the burning rate and impeding combustion extinction by strong strain rates. Finally, if $D_F < D_O$ and Le > 1 in a lean (e.g., propane–air) flame, the temperature and equivalence ratio in the reaction zone are reduced by the strain rate, with both processes reducing the burning rate and promoting flame quenching.

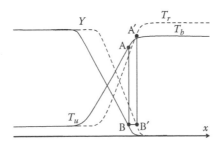

FIGURE 5.22   A sketch of the profiles of the mass fraction of a deficient reactant and temperature in unperturbed (*solid curves*) and strained (*dashed curves*) laminar premixed flames with Le < 1.

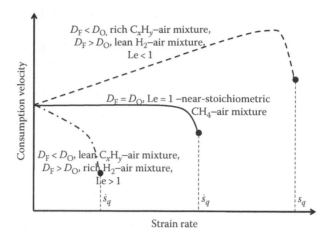

FIGURE 5.23 The dependence of the consumption speed on a strain rate in twin counterflow flames.

The foregoing discussion is summarized in Figure 5.23, which sketches the dependence of the so-called consumption speed (or consumption velocity or consumption rate or burning rate)

$$u_c \equiv \rho_u^{-1} \int_{-\infty}^{0} W dx \qquad (5.84)$$

on the strain rate for different mixtures. Here, the rate W is relevant to the combustion progress variable and the upper integration limit is set to be equal to zero, as the left flame shown in Figure 5.20 is considered.

If the diffusivity of a deficient reactant (fuel in a rich mixture or oxidizer in lean mixture) is high, the consumption velocity is increased by the strain rate (see dashed curve) until the reaction zone reaches the symmetry axis. As the strain rate is further increased, the consumption velocity is decreased due to the reduction in the thickness of the reaction zone (in the case of large but finite activation temperature) and the flame is quenched by a sufficiently high strain rate. Note that a decrease in u_c by a_t does not occur if Le < 1 and Ze $\rightarrow \infty$. In such an asymptotic case, the maximum consumption velocity is achieved when two highly strained flames merge at $x = 0$.

If the diffusivity of a deficient reactant is low, the consumption velocity is decreased when the strain rate is increased (dotted-dashed curve). Extinction occurs at a low strain rate.

In the equidiffusive ($D_F = D_O = a$) flame, the consumption velocity is weakly affected by the strain rate (solid curve) unless the reaction zone reaches the symmetry axis. As the strain rate is further increased, the consumption velocity is decreased due to the reduction in the thickness of the reaction zone (in the case of large but finite activation temperature) and the flame is quenched by a moderate strain rate. Thus, strained flames, which are characterized by the higher diffusivity of a deficient reactant, burn faster and survive at stronger strain rates, with all other things being equal.

A sketch similar to Figure 5.23 may be drawn for a flame (or displacement) speed. However, because a flame speed is evaluated with respect to the local flow, S_L is affected not only by local variations in the consumption velocity but also by the flow nonuniformity. As previously discussed in the case of Le = 1, the latter effect reduces the flame speed (see Equation 5.77) and results in $S_L < u_c = S_L^0$. If Le is sufficiently low, an increase in S_L by a_t may occur in weakly and moderately strained laminar flames, followed by the opposite trend at higher strain rates.

Finally, it is worth noting that a reduction in the consumption velocity by a strain rate (Le > 1) may result in an increase in the thickness, Δ_L, calculated using Equation 2.41. Indeed, within the framework of the AEA theory, a decrease in u_c is associated with a decrease in the heat flux from the reaction zone, that is, a decrease in the maximum temperature gradient.

Experimental, numerical, and theoretical studies of strained laminar premixed flames support the physical scenario discussed earlier. Here, we restrict ourselves to a single example and refer interested readers to review papers by Dixon-Lewis (1990), Clavin (1985), Law and Sung (2000), and Lipatnikov and Chomiak (2005a, Section 5.1). Figure 5.24 shows the spatial profiles of heat release rate calculated for various twin counterflow laminar premixed flames sketched in Figure 5.20 under the influence of

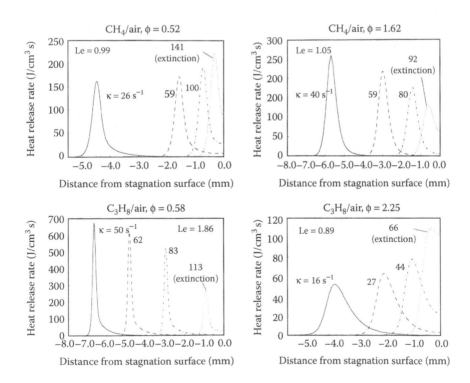

FIGURE 5.24 Heat release profiles computed for twin counterflow flames affected by various strain rates, specified in the legends. (Reprinted from *Progress in Energy and Combustion Science*, 26, Law, C.K. and Sung, C.J., Structure, aerodynamics, and geometry of premixed flames, 459–505, Copyright 2000, with permission from Elsevier.)

various strain rates. For lean methane–air ($D_F > D_O$) and rich propane–air ($D_F < D_O$) flames, an increase in the strain rate pushes the flame toward the axis of symmetry and results in an increase in the maximum heat release rate, in line with the dashed curve in Figure 5.23. When the strain rate is sufficiently high, the reaction zone approaches the axis of symmetry, and a further increase in $\dot{s}$ reduces the thickness of the reaction zone and quenches the flames. For rich methane–air and lean propane–air flames, an increase in the strain rate results in a decrease in the maximum heat release rate, followed by extinction, in line with the dotted-dashed curve in Figure 5.23.

Finally, let us estimate the magnitude of the effect of straining on laminar premixed flames. In the case of Le = 1, Equations 5.70 through 5.77 show that it is controlled by a nondimensional quantity χ, which scales as $(\tau_c \dot{s})^{-1/2}$. Therefore, the stretch rate normalized with the chemical timescale defined by Equation 5.81 is a proper quantity for characterizing the response of a laminar premixed flame to a strain rate. The response is more pronounced for mixtures characterized by a larger timescale (i.e., a lower unperturbed laminar flame speed). For instance, the circles in Figure 5.25 show that the extinction stretch rate is higher for near-stoichiometric CH_4–air mixtures characterized by a larger flame speed as compared with lean or rich mixtures. Note that the highest extinction strain rate $\dot{s}_q \approx 2$ s^{-1}, measured by Chung et al. (1996), yields $\tau_c \dot{s}_q \approx 0.25$.

Moreover, as discussed earlier, the response of a laminar premixed flame to a strain rate depends on the differences in molecular mass and heat diffusivities. For this reason, the highest extinction strain rate was measured by Chung et al. (1996) in a slightly lean methane–air mixture ($D_F > D_O$), while the maximum unperturbed laminar flame speed is associated with a slightly rich mixture (see Figure 2.8).

It is worth stressing that the effect of the differences in molecular mass and heat diffusivities on the response of a laminar premixed flame to a strain rate may be very strong. For instance, Cho et al. (2006) measured the extinction strain rates for twin counterflow flames for different fuels and equivalence ratios. The results have been reported in the form of the extinction Karlovitz number, $Ka_q = \dot{s}_q \Delta_L^0 / S_L^0$, as a function of the equivalence ratio Φ (see Figure 5.26). Here, the preheat zone thickness

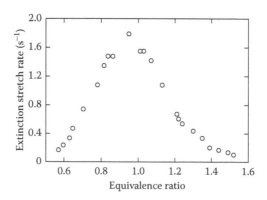

FIGURE 5.25 Extinction stretch rates obtained from twin counterflow laminar methane–air flames. (From Chung, S.H., Chung, D.H., Fu, C., and Cho, P., *Combust. Flame*, 106, 515–520, 1996.)

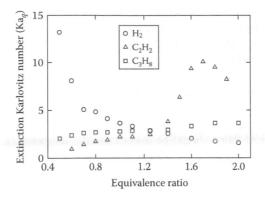

FIGURE 5.26   Extinction Karlovitz number versus the equivalence ratio obtained by Cho et al. (2006) from twin counterflow laminar flames for different fuels.

of the unperturbed flame is determined using Equation 2.41, that is, Δ_L^0/S_L^0 is larger than τ_c by several times. For flames associated with Le ≈ 1 and $D_F \approx D_O$ (C_2H_2–air mixtures), the extinction Karlovitz number is about 2 and depends weakly on the equivalence ratio. However, Ka_q may be as large as 10 and 13 for rich C_3H_8–air (Le < 1, $D_F < D_O$) and lean H_2–air (Le < 1, $D_F > D_O$) flames, respectively.

CURVED LAMINAR PREMIXED FLAMES

Akin to strain rates, wrinkling of a laminar flame surface affects the speed of the self-propagation of the flame and the local burning rate. Let us begin a discussion on these effects with the case of Le $= 1$ and consider the balance of thermal energy within small volumes ABA′B′ of the preheat zones of planar and spherical laminar flames (see Figure 5.27). The thermal energy that comes via surface AB from the reaction zone (see the bold arrows in Figure 5.27) serves to heat the gas that enters the considered volume via surface A′B′ due to convection. The rest of the energy leaves the volume via surface A′B′ and serves to heat the gas ahead of this surface. In a planar flame, the areas of the surfaces AB and A′B′ are equal to each other.

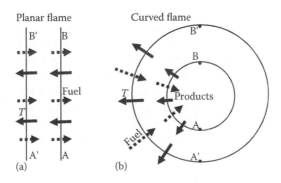

FIGURE 5.27   A sketch of planar (a) and curved (b) flames.

However, if a flame is positively curved, as sketched in Figure 5.26b, the area of the surface AB is smaller than the area of the surface A'B'. Therefore, if the speeds (with respect to the mixture) of the planar and spherical flames were the same, then, in the latter case, the thermal energy that is conducted to the control volume from the reaction zone via a smaller surface AB would not be sufficient to heat the mixture that is convected to the volume via a larger surface A'B' to the same temperature as in the planar case. Therefore, (i) the speed of the spherical flame should be lower than the unperturbed flame speed S_L^0 and (ii) the temperature at distance x from the reaction zone of the spherical flame should be lower than the temperature at the same distance from the reaction zone of the planar flame, that is, the preheat zone is thinner in the spherical case, and this effect is similar to the thinning of the preheat zone by the strain rates (see Figure 5.21).

In the case of Le = 1, the decrease in the speed of a spherical flame can also be shown by rewriting the combustion progress variable balance Equation 1.79 in the spherical coordinate framework. Using the continuity Equation 1.49, we obtain

$$\rho \frac{\partial c}{\partial t} + \rho u \frac{\partial c}{\partial r} = \frac{1}{r^2} \frac{\partial}{\partial r} \left(r^2 \rho D \frac{\partial c}{\partial r} \right) + W(c). \tag{5.85}$$

For a flame that has a stationary thickness, the density and combustion progress variable profiles should depend on a single variable $\xi = r - Vt$, where $V = dR_f/dt$ is the observed flame speed and $R_f(t)$ is the radial coordinate of the infinitely thin reaction zone. Therefore,

$$\rho(u - V) \frac{dc}{d\xi} = \frac{1}{r^2} \frac{\partial}{\partial r} \left(r^2 \rho D \frac{\partial c}{\partial r} \right) + W(c) \tag{5.86}$$

or (Zel'dovich et al., 1985)

$$\rho \left(u - V - \frac{2D}{r} \right) \frac{dc}{d\xi} = \frac{d}{d\xi} \left(\rho D \frac{dc}{d\xi} \right) + W(c). \tag{5.87}$$

Equation 5.87 is very similar to the basic Equation 2.14 of the AEA theory of the unperturbed laminar premixed flame. The sole difference between the two equations consists of the replacement of the unperturbed flame speed S_L^0 in Equation 2.14 with the term $u - V - 2D/r$ in Equation 5.87. Because within the framework of the AEA theory, the infinitely thin reaction zone is not perturbed by wrinkles of a finite length scale if Le = 1, the RHS of Equation 5.87 is equal to the RHS of Equation 2.14 within the reaction zone, and a similar equality holds for ∇c therein. Consequently, the LHSs of Equations 2.14 and 5.87 should be equal to each other at the boundary between the preheat zone and the reaction zone, that is,

$$\rho_b S_f = \rho_u S_L^0 - \frac{2 \rho_b D_b}{R_f}, \tag{5.88}$$

considering that (i) the speed of self-propagation of the reaction zone of the spherical flame is equal to $S_f = V - u(R_f)$ and (ii) in the case of an infinitely thin reaction zone, $u(R_f) = 0$, because the combustion products are quiescent for symmetry reasons.

In the case of a weak curvature, $\rho_b S_f \approx \rho_u S_L^0$, that is, the observed speed dR_f/dt of an expanding spherical flame is approximately equal to σS_L^0. Therefore, Equation 5.88 reads

$$S_L = \sigma^{-1} S_{L,b} = \sigma^{-1} S_f = S_L^0 - \sigma^{-1} \frac{2}{R_f} \frac{D_b}{D_u} \frac{D_u}{\left(S_L^0\right)^2} S_L^0 \frac{1}{\sigma} \frac{dR_f}{dt} = S_L^0 \left(1 - \frac{D_b}{\sigma^2 D_u} \tau_c \dot{s}\right), \quad (5.89)$$

because the stretch rate is equal to $2d \ln R_f/dt$ for a spherical surface (see Equation 5.50). If $\sigma^2 D_u = D_b$, then Equation 5.89 reduces to Equation 5.80, which was obtained for a weakly strained planar flame invoking the same ($\sigma^2 D_u = D_b$) simplification. The results of a more rigorous theory are reported in the section "A Brief Summary of Theory of Weakly Stretched Laminar Premixed Flames."

Thus, in the case of Le = 1, a positive (e.g., an expanding spherical flame) curvature reduces the flame speed, S_f, with the relative magnitude of the reduction effect being proportional to $\tau_c \dot{s}$. If we consider a negatively curved flame (e.g., a collapsing spherical flame with the unburned mixture occupying the center), then the flame speed is increased by the curvature, because heat enters (leaves) a control volume via a larger (smaller) surface in this case. Because an infinitely thin reaction zone is not perturbed by wrinkles of a finite length scale, the local consumption (or burning) velocity (or speed) $u_c = \rho_u^{-1} \int_0^\infty Wdr$ is not affected by the curvature if Le = 1.

The reduction effect of the flame curvature on its speed in spite of $u_c = S_L^0$ may be further supported by the following reasoning. Thermal energy released within a segment of a reaction zone that has a surface area Σ_r (see Figure 5.28) is transported by molecular conductivity and heats the unburned mixture that enters the preheat zone via a surface that has an area Σ_u. In the planar case, $\Sigma_r = \Sigma_u$, whereas $\Sigma_r < \Sigma_u$ in an expanding spherical flame. Accordingly, the speed S_d of the unburned gas

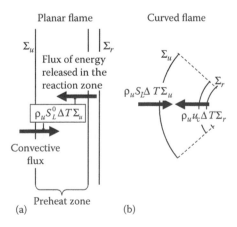

FIGURE 5.28   The effect of curvature on flame speed in planar (a) and spherical (b) flames.

with respect to the spherical flame at the surface Σ_u should be lower than $u_c = S_L^0$ in order for $\rho_u S_d \Sigma_u c_p (T_b - T_u)$ to be equal to the heat $\rho_u u_c \Sigma_r c_p (T_b - T_u)$ released in the considered segment of the reaction zone. An equality of $\rho_u S_d \Sigma_u = \rho_u u_c \Sigma_r$ yields $S_L < u_c = S_L^0$, because $\Sigma_r < \Sigma_u$ in the spherical flame sketched in Figure 5.28b.

The foregoing discussion addressed the case of equal molecular diffusivities of the main reactants and heat, that is, $D_F = D_O = a$. The effects of the differences between the molecular transport coefficients on the local burning rate and the speed of a curved flame are qualitatively similar to the phenomena already discussed for strained flames. In particular, (i) the temperature in the reaction zone of a positively (negatively) curved flame is higher (lower) than the adiabatic combustion temperature if Le < 1, while the opposite trend is observed if Le > 1 and (ii) the equivalence ratio in the reaction zone of a positively (negatively) curved flame is higher (lower) than the equivalence ratio in the unburned mixture if $D_F > D_O$, while the opposite trend is observed if $D_F < D_O$. As a result, the local consumption velocity is increased by the curvature of lean hydrogen–air ($D_F > a > D_O$) or rich hydrocarbon–air ($D_F < a < D_O$ for C_xH_y with $x > 2$) flames. The opposite effect is observed in rich hydrogen–air or lean hydrocarbon–air (again with $x > 2$) flames.

For instance, if we restrict the discussion solely to molecular transport terms and consider a segment ABA′B′ of an expanding spherical flame (see Figure 5.27b), then the flame curvature acts to decrease the thermal energy within this segment but to increase the chemical energy therein. Accordingly, if the flame speed and the gradients of the temperature and mass fractions were not affected by curvature, then the temperature (mass fraction Y_d of the deficient reactant) within this segment would be lower (higher) than the temperature (mass fraction) in the counterpart segment of the unperturbed laminar flame, because the area of the surface AB is smaller than the area of the surface A′B′ in the spherical case. Thus, heat conduction acts to increase the thermal energy in the considered segment less than in the unperturbed laminar flame, whereas molecular diffusivity acts to increase the chemical energy in the segment ABA′B′ more than in the unperturbed flame. If $D > a$ (i.e., Le < 1), the latter effect is stronger and the total enthalpy is increased within the segment.

Indeed, in the spherical coordinate framework, the enthalpy balance Equation 1.64 involves an extra (as compared with the planar case) term

$$\frac{2\mu}{r\,\text{Pr}}\left[\frac{\partial h}{\partial r} + \left(\frac{1}{\text{Le}} - 1\right)\sum_{l=1}^{N} h_l \frac{\partial Y_l}{\partial r}\right] = \frac{2\mu}{r\,\text{Pr}}\left[\frac{\partial h}{\partial r} + \left(\frac{1}{\text{Le}} - 1\right)\frac{\partial}{\partial r}\sum_{l=1}^{N} Y_l \Delta h_l^0\right] \quad (5.90)$$

on the RHS. Here, for simplicity, we assume that $\text{Le}_l = \text{Le}$ and $c_{P,l} = c_P$, for all species. Because $\sum_{l=1}^{N} Y_l \Delta h_l^0$ is larger in the unburned mixture than in the combustion products, the second term in square brackets on the RHS of Equation 5.90 is positive (i.e. a source term) in an expanding spherical flame if Le < 1 or in a collapsing spherical flame if Le > 1. In such flames, curvature increases the enthalpy and, therefore, increases the temperature and burning rate. On the contrary, curvature reduces the enthalpy, temperature, and burning rate in an expanding spherical flame if $Le_1 > 1$ or in a collapsing spherical flame if $Le_1 < 1$. Because the curvature of an expanding (collapsing) spherical flame is positive (negative), we may briefly say that the burning rate is increased (reduced) by curvature if the Lewis number is lower (larger) than unity.

From a mathematical viewpoint, an increase (decrease) in the local temperature and burning rate in a positively curved laminar spherical flame characterized by a low (large) Lewis number can easily be shown by analyzing the so-called flame ball (Zel'dovich et al., 1985) (see the section "Flame Ball" in Chapter 7 and Equation 7.194 in particular).

In summary, the effects of curvature on the flame speed and consumption velocity are qualitatively similar to the effects of the strain rate on S_L and u_c, which were discussed in the previous section. In particular, a sketch similar to Figure 5.23 may be drawn to show the effect of flame curvature on the consumption velocity.

Finally, it is worth noting that the spherical flames discussed earlier are not only curved, but also strained in the case of a variable density. The point is that heat release in a wrinkled laminar premixed flame induces an inhomogeneous flow field characterized by a nonzero strain rate. For instance, if we write the continuity Equation 1.49 in the spherical coordinate framework,

$$\frac{\partial \rho}{\partial t} + \frac{1}{r^2}\frac{\partial}{\partial r}\left(r^2 \rho u\right) = 0, \tag{5.91}$$

and assume that $\rho = \rho(r - Vt)$ for an expanding flame, then integration of Equation 5.91 yields

$$r^2 \rho u = V\int_0^r \xi^2 \frac{d\rho}{d\xi}\,d\xi = r^2 \rho V - 2V\int_0^r \rho\xi\,d\xi. \tag{5.92}$$

If an infinitely thin flame is considered and $r > R_f$, where R_f is the flame radius, then Equation 5.92 reads

$$r^2 \rho_u u = r^2 \rho_u V - V\left[\rho_b R_f^2 + \rho_u\left(r^2 - R_f^2\right)\right] = VR_f^2\left(\rho_u - \rho_b\right). \tag{5.93}$$

Because $V = S_L + u_f$, where $u_f \equiv \lim_{\Delta r \to 0} u(r = R_f + \Delta r)$, Equation 5.93 results in

$$u_f = S_L\left(\sigma - 1\right). \tag{5.94}$$

if $r \to R_f$ and $u \to u_f$ on the LHS. Subsequently, Equations 5.93 and 5.94 yield

$$r^2 u = R_f^2 u_f \tag{5.95}$$

if $r > R_f$. Equation 5.94 is not surprising. Indeed, since the combustion products are quiescent for symmetry reasons, the observed speed of the flame expansion

$$\frac{dR_f}{dt} = V = S_L(\sigma - 1) + S_L = \sigma S_L = S_{L,b} \qquad (5.96)$$

should be equal to the flame speed with respect to the quiescent products.

A comparison of Equations 5.50, 5.58, 5.60, and 5.96 shows that the strain rate

$$a_t = \frac{2}{R_f}\frac{dR_f}{dt} - \frac{2S_L}{R_f} = (\sigma - 1)\frac{2S_L}{R_f}, \qquad (5.97)$$

that is, for an expanding spherical laminar premixed flame, the strain rate contribution to the stretch rate is larger than the curvature contribution to $\dot{s}$ by a factor of $\tau = \sigma - 1$. Therefore, the effect of strain rate on a typical expanding spherical flame is significantly stronger than the effect of curvature on the flame.

Typically, the effects of stretch rates on curved laminar flames are experimentally studied by measuring the observed speed of an expanding spherical flame at different flame kernel radii. The results of such measurements are commonly consistent with Figure 5.23. For instance, Figure 5.29 shows lower (higher) speeds of highly curved, rich (lean) hydrogen–air flames as compared with weakly curved flames.

BRIEF SUMMARY OF THEORY OF WEAKLY STRETCHED LAMINAR PREMIXED FLAMES

In the two previous sections, we discussed the effects of strain rates and curvature on laminar premixed flames, based on the AEA theory developed by Zel'dovich and Frank-Kamenetskii (1938), invoking simple physical reasoning. Certainly, over more than 70 years after the development of this seminal theory, substantial progress has been made in studies of stretched premixed laminar flames and more rigorous

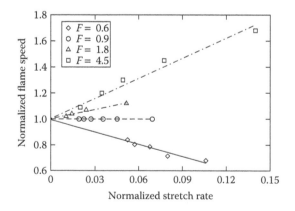

FIGURE 5.29 Normalized flame speed $S_L/S_L^0 = (1/\sigma S_L^0)(dR_f/dt)$ versus the normalized stretch rate $\mathrm{Ka} = (D_u/S_L^2)(2d \ln R_f/dt)$. The symbols show the experimental data obtained by Kwon and Faeth (2001) from expanding, spherical, laminar hydrogen–air flames characterized by various equivalence ratios, specified in the legends. The straight lines approximate the measured data.

mathematical theories based on the AEA asymptotic have been developed using multiscale perturbation methods. In this section, the key results of such theories that are of importance for modeling premixed turbulent combustion are briefly summarized. For a more detailed discussion on these results, interested readers are referred to review papers by Clavin (1985), Law and Sung (2000), and Lipatnikov and Chomiak (2005a, see Section 5.1) and to references cited therein.

First, most of the aforementioned theories address weakly stretched laminar flames, that is, the stretch rate normalized using the chemical timescale τ_c defined by Equation 5.81 is much less than unity

$$\tau_c \dot{s} \ll 1. \tag{5.98}$$

The term $\tau_c \dot{s}$ is often called the Karlovitz number, Ka, in papers that deal with laminar flames. Readers interested in a theory of perturbed laminar flames that is not limited to weak stretch rates are referred to the work by de Goey and ten Thije Boonkkamp (1999) and a recent review paper by de Goey et al. (2011). We may also note that certain perturbations of laminar flames cannot be discussed in terms of stretch rate. For instance, the flame ball (Zel'dovich et al., 1985), already mentioned earlier and discussed in detail in the section "Flame Ball" in Chapter 7, is an example of a strongly perturbed laminar flame characterized by a zero stretch rate, as it does not expand and its surface area is constant.

Second, for weakly stretched flames, the theory predicts a linear dependence of the consumption velocity and differently defined flame speeds on the stretch rates. For instance (Bechtold and Matalon, 2001),

$$\frac{u_c}{S_L^0} = 1 - \mathrm{Ma}_c \tau_c \dot{s} + \mathrm{O}(\tau_c \dot{s}), \tag{5.99}$$

$$\frac{S_d}{S_L^0} = 1 - \mathrm{Ma}_d \tau_c \dot{s} + \mathrm{O}(\tau_c \dot{s}), \tag{5.100}$$

and

$$\frac{S_h}{S_L^0} = 1 - \mathrm{Ma}_h \tau_c \dot{s} + \mathrm{O}(\tau_c \dot{s}), \tag{5.101}$$

where the so-called Markstein numbers, Ma_c, Ma_d, and Ma_h, are controlled by the mixture characteristics. If $\mathrm{Ze} \gg 1$, $\mathrm{Ze}|\mathrm{Le} - 1| = \mathrm{O}(1)$, $\rho D = \rho_u D_u$, $\rho a = \rho_u a_u$, and $\rho v = \rho_u v_u$, then (Matalon and Matkowsky, 1982; Bechtold and Matalon, 2001)

$$\mathrm{Ma}_c = \frac{\mathrm{Ze}\left(1 - \mathrm{Le}^{-1}\right)}{2} \int_0^{\sigma-1} \frac{\ln(1+x)}{x} dx, \tag{5.102}$$

$$\mathrm{Ma}_d = \mathrm{Ma}_c + \frac{\ln \sigma}{\sigma - 1}, \tag{5.103}$$

and

$$\mathrm{Ma}_h = \mathrm{Ma}_c + \frac{\sigma \ln \sigma}{\sigma - 1}. \tag{5.104}$$

Here, the displacement speed S_d characterizes the mass flux into the thin (Ze $\gg$ 1) reaction zone, that is, $\rho_u S_d = \rho_r(V - u_r) \approx \rho_b(V - u_b) = \rho_b S_{L,b}$, where the subscript r designates the reaction zone. Therefore, the displacement speed is controlled by the difference between the observed flame speed V and the mixture velocity u_r just ahead of the reaction zone. The symbol S_h designates the difference between the speed V and the extrapolation of the velocity distribution in the unburned gas to the reaction zone. For instance, for an expanding spherical flame, the extrapolated flow velocity $u_h = u(r)r^2/R_f^2$, where $r > R_f$, is higher than the flow velocity in any point within the preheat zone, that is, the speed S_h is lower than $V - u_r(r)$ at any distance r. It is worth remembering that these are the values of S_h and Ma_h that are reported in many articles following the review paper by Clavin (1985).

Equations 5.102 through 5.104 show that

$$\mathrm{Ma}_c < \mathrm{Ma}_d < \mathrm{Ma}_h; \tag{5.105}$$

therefore, if $\dot{s} > 0$, then $S_d < u_c$ in line with the previous discussion. It is worth stressing that not only the magnitudes but also the signs (if Le < 1) of differently defined Markstein numbers (Ma_c, Ma_d, and Ma_h) may be different! For instance, Varea et al. (2012) experimentally investigated expanding spherical laminar premixed flames and obtained the flame speeds with respect to the unburned $S_{L,u} = dR_f/dt - u_u$ and burned $S_{L,b} = dR_f/dt$ mixtures, with the flow velocity u_u being measured just ahead of the flame. In many cases, the reported Markstein numbers Ma_u and Ma_b relevant to $S_{L,u}$ and $S_{L,b}$, respectively, have opposite signs (e.g., see iso-octane–air mixtures with $\Phi = 0.9–1.3$ in Table 3 in the cited paper or ethanol–air mixtures with $\Phi = 0.9–1.5$ in Table 4 therein). This difference in differently defined Markstein numbers should be borne in mind when invoking such a quantity to model premixed turbulent combustion.

Equations 5.99 through 5.104 also show that the response of a laminar premixed flame to a weak (see Equation 5.98) stretch rate depends on the Lewis number. If Le = 1, the consumption velocity u_c is not affected ($\mathrm{Ma}_c = 0$) by weak stretch rates, while the flame speeds S_d and S_h are reduced ($0 < \mathrm{Ma}_d < \mathrm{Ma}_h$) by weak positive stretch rates, in line with the discussion in the previous sections and Figure 5.23, in particular. If the Lewis number is lower (larger) than unity, then the consumption velocity is increased (reduced) by a weak stretch rate, again in line with Figure 5.23. If the Lewis number is sufficiently low, not only u_c but also the speeds S_d and S_h may be increased by weak stretch rates.

Both the linear dependencies of the consumption velocity and variously defined flame speeds on weak stretch rates and the aforementioned effects of the Lewis number on these linear dependencies have been documented in numerous experimental and numerical studies (e.g., see Figure 5.29 or Section 5.1 in a review paper by Lipatnikov and Chomiak [2005a]).

STRETCHING OF LAMINAR FLAMELETS BY TURBULENT EDDIES

As discussed in the foregoing sections, the local burning rate and flame speed are substantially affected by the flame curvature and flow nonuniformities in the laminar case, with combustion being quenched by sufficiently strong strain rates. Such effects are likely to play a role in turbulent premixed flames, where the flow is highly nonuniform and the reacting layers are wrinkled (see Figures 5.30 through 5.32 as typical examples).

Due to stretching of flamelets by turbulent eddies, the local consumption velocity u_c may vary substantially as compared with S_L^0 in a turbulent flame brush, especially if the Lewis number differs significantly from unity. (In the remainder of the section "Variations in Flamelet Structure Due To Turbulent Stretching," we will talk about Lewis number effects for the sake of brevity and because the theory briefly summarized earlier addresses them but does not address the influence of the ratio of

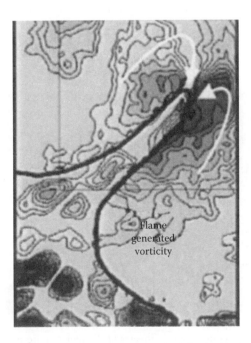

FIGURE 5.30 Contours of vorticity between 700 and –700 s^{-1} with the flame surface shown by a thick black line. (Reprinted with kind permission from Springer Science + Business Media: *Experiments in Fluids*, Measurements of turbulent premixed flame dynamics using cinema stereoscopic PIV, 44, 2008, 985–999, Steinberg, A.M., Driscoll, J.F., and Ceccio, S.L., Figure 7, Copyright Springer.)

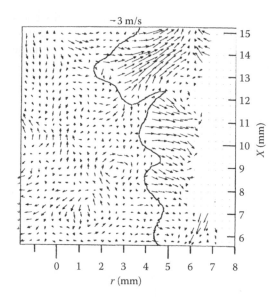

FIGURE 5.31 A combined instantaneous PIV/OH-LIF 2D image of the flow velocity (arrows) and flamelet surface (wrinkled solid line) in a lean ($\Phi = 0.7$) hydrogen–air Bunsen flame. (Reprinted from Chen, Y.-C. and Bilger, R.W., *Proc. Combust. Inst.*, 28, 521–528, 2000. With permission.)

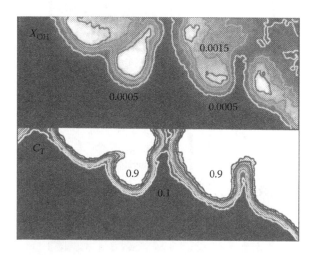

FIGURE 5.32 Instantaneous grayscale images of the temperature-based combustion progress variable, C_T, and the mole fraction, X_{OH}, of OH radicals in a lean ($\Phi = 0.325$) hydrogen–air turbulent Bunsen flame. The C_T image is overlaid with isocontours of $C_T = 0.1, 0.2, 0.3$, 0.5, 0.7, and 0.9. The X_{OH} image is overlaid with isocontours of $X_{OH} = 0.0005, 0.0010, 0.0015$, and 0.0020. (Reprinted from Chen, Y.-C. and Bilger, R., *Combust. Flame*, 138, 155–174, 2004. With permission.)

D_F/D_O on the displacement and consumption speeds of a perturbed laminar flame. Nevertheless, it is worth remembering that both Le and D_F/D_O play a role in a real flame.) Such variations in u_c have been documented in many experimental and DNS studies of premixed turbulent flames, as discussed in detail in Section 5.2 of a review paper by Lipatnikov and Chomiak (2005a). For instance, Figure 5.32 shows that the concentration of OH radicals, measured in positively curved flamelets by Chen and Bilger (2004) in a lean hydrogen–air turbulent Bunsen flame, is substantially higher (see white islands in Figure 5.32 for X_{OH}) than X_{OH} in planar flamelets and in cusps. Because the concentration of OH radicals is often considered to be an indicator of the magnitude of the heat release rate in a premixed flame, this and similar images imply an increase in the consumption velocity in positively curved flamelets, in line with the theory of perturbed laminar flames for Le < 1.

Although the theory of weakly perturbed laminar flames results in linear Equations 5.99 through 5.101, the influence of negative stretching (called compression in the following text) on the local consumption velocity does not compensate for the effects of a positive stretch rate, because (i) the mean strain rate is well known to be positive and to scale as τ_η^{-1} in turbulent flows (see Equation 3.72) and (ii) the mean curvature of flamelets should be positive (negative) at least at the leading (trailing) edge of a turbulent flame brush for geometrical reasons. Due to the former effect, the probability of finding stretched (for the sake of brevity, the term positive will be skipped in the following text) flamelets in a turbulent flow is greater than the probability of finding compressed (i.e., the local stretch rate $\dot{s}$ is negative) flamelets. Consequently, if the Lewis number is larger than unity, the mean consumption velocity $\bar{u}_c(\mathbf{x},t) \leq S_L^0$ in flamelets perturbed by turbulent eddies. If the Lewis number is substantially lower than unity, weak and moderate stretch rates increase u_c. An increase (decrease) in the mean consumption velocity by turbulence for mixtures characterized by a low (large) Le was well documented in DNS studies (e.g., see Figure 5.33).

The trends discussed earlier are briefly summarized in Figure 5.34, which is basically similar to Figure 5.23. If Le = 1, then $\bar{u}_c$ is weakly affected by low-intense turbulence but is reduced by u' in stronger turbulence due to flamelet stretching (here, the thickness of the reaction zone is assumed to be finite). If Le > 1, then $\bar{u}_c$

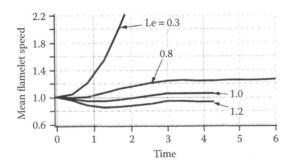

FIGURE 5.33 The development of the mean consumption velocity computed in DNS at different Lewis numbers. The quantities are normalized using S_L^0 and $\tau_t \approx L/u'$. (Reprinted from Trouvé, A. and Poinsot, T., *J. Fluid Mech.*, 278, 1–31, 1994. With permission.)

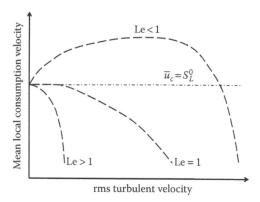

FIGURE 5.34 Typical dependencies of the mean local consumption velocity $\bar{u}_c$ on the rms turbulent velocity u' at various Lewis numbers, but the same S_L^0.

is reduced by u' due to weakening of the burning in stretched flamelets. If Le < 1, then $\bar{u}_c$ is increased by u' in moderate turbulence due to the local enhancement of burning in moderately stretched flamelets, but is reduced by u' in strong turbulence, when high stretch rates mitigate the burning in reaction zones of a finite thickness.

If the mean consumption velocity differs from the unperturbed laminar flame speed, then Equation 5.1 should be modified as

$$U_t = \bar{u}_c \frac{A_t}{A},\tag{5.106}$$

where $\bar{u}_c$ is increased when Le is decreased. Subsequently, Figure 5.34 and Equation 5.106 result in dependencies of the turbulent burning velocity on the rms velocity, sketched in Figure 5.35. The dotted-dashed curve is associated with the hypothetical

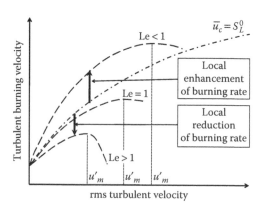

FIGURE 5.35 Typical dependencies of the turbulent burning velocity U_t on the rms turbulent velocity u' at various Lewis numbers, but the same S_L^0.

case of unperturbed flamelets, that is, $\bar{u}_c = S_L^0$ for any stretch rate. If Le $= 1$, the $U_t(u')$ curve coincides with the dotted-dashed curve in weak turbulence, followed by a reduction in dU_t/du' in more intense turbulence due to (i) the already discussed reduction in $\bar{u}_c$ by u' and (ii) the eventual local combustion quenching (this phenomenon will be discussed in the next subsection). If Le > 1, the stretch rates produced by turbulent eddies not only increase the flamelet surface area but also reduce the mean consumption velocity, with the former (latter) effect dominating in weak (stronger) turbulence. As a result, the positive slope of the $U_t(u')$ curve at low u' is reduced by Le and the maximum burning velocity is reached at lower u'_m in flames characterized by higher Le. If Le < 1, stretch rates increase both the flamelet surface area and $\bar{u}_c$ in weak and moderate turbulence, and the $U_t(u')$ curve is characterized by a steep slope. In more intense turbulence, a reduction in $\bar{u}_c$ and the eventual local combustion quenching in highly stretched flamelets may play a more important role and damp (or even reverse) the dependence of U_t on u' if $u'_m < u'$, with the velocity u'_m increased when the Lewis number is decreased.

The aforementioned physical scenario that emphasizes the effects of turbulent stretching on the local burning rate in flamelets offers an opportunity for providing a qualitative explanation of the most challenging trends in the behavior of the turbulent burning velocity, summarized in the section "Summary" in Chapter 4, that is, the influence of the Lewis number on the turbulent burning velocity even at high Reynolds numbers and a weak dependence of U_t on u' in intense turbulence. However, a qualitative explanation is a necessary but not a sufficient condition to predict these trends and to consider the concept of stretched flamelets to be well supported. There are other physical mechanisms that could be hypothesized to cause the aforementioned challenging trends. For instance, the numerical results reported by Lipatnikov and Chomiak (2000) indicate that a reduction in U_t by u' in intense turbulence, observed in expanding statistically spherical flames, may result from geometrical and memory effects. Moreover, the term "local combustion quenching" used previously is worth further discussion.

LOCAL QUENCHING OF PREMIXED COMBUSTION IN TURBULENT FLOWS

Local quenching is often associated with holes in a flamelet surface, while the occurrence of such holes may be put into question in an adiabatic, fully premixed (i.e., the field of the equivalence ratio Φ is uniform and stationary) flame. For instance, Kuznetsov and Sabel'nikov (1990) argued that a flamelet surface should be continuous at least in near-stoichiometric adiabatic premixed flames. By reviewing the recent experimental data obtained using the most advanced diagnostic techniques, Driscoll (2008) noted that holes in reacting layers had never been documented and "there is no experimental evidence that premixed flames can be shredded into thin disconnected flamelets." It is worth noting, however, that even if a flamelet surface is continuous in a highly turbulent flow, this observation does not prove negligible probability of local combustion quenching. For instance, small pockets of burned mixture may be locally quenched upstream of a continuous flamelet and such a process will reduce the area of the flamelet surface and, hence, U_t.

Simultaneous CH/CH_2O and also OH PLIF images reported recently by Li et al. (2010) indicate local combustion quenching in highly turbulent stoichiometric methane–air jet flames (see Figure 5 for flames F5 and F6 in the cited paper), but this phenomenon is substantially affected by the entrainment of ambient air to the flames.

Recently, Dunn et al. (2007, 2009, 2010) succeeded in experimentally investigating very lean ($\Phi = 0.5$) methane–air jet flames stabilized in extremely intense turbulence. Under the conditions of the measurements, the Damköhler and Karlovitz numbers evaluated using Equation 4.11 with $\tau_c = v_u/S_L^2$ and Equation 4.13, respectively, were varied from Da = 0.3 and Ka = 90 in flame PM1-50 to Da = 0.02 and Ka = 3000 in flame PM1-200. The reported PDFs $P(X_{CO}|T)$ and $P(X_{OH}|T)$ conditioned on the local temperature (see Figures 2 and 3 in Dunn et al. [2009]) and the PDF $P(X_{CO}X_{OH}|T)$ (see Figure 8 in Dunn et al. [2010]) show that (i) despite Da = 0.3 < 1 and Ka = 90 ≫ 1, the local structure of the PM1-50 flame is statistically consistent with the computed structure of counterflow laminar flames sketched in Figure 5.20 and (ii) the peak mole fractions of CO and OH and the peak product of the two concentrations are significantly but gradually reduced by the jet velocity and, hence, by the turbulent Reynolds number. These and other reported experimental data were claimed (Dunn et al., 2010) to indicate "a gradual decrease in reactedness... without any obvious local extinction" as Re_t is increased.

What physical mechanisms allow the reaction zones to survive in a turbulent flow even if the local stretch rates produced by the smallest-scale eddies are much higher than the steady stretch rate $\dot{s}_{q,s}$ that extinguishes the counterpart laminar flame? First, as discussed in the section "Physical Reasoning: Thin-Reaction-Zone Regime," small-scale turbulent eddies that produce high stretch rates may be destroyed within flamelets due to an increase in the viscosity. Moreover, due to thermal expansion, small-scale eddies become larger and their vorticity is reduced due to the conservation of the angular momentum (see the third term on the RHS of the vorticity Equation 1.55). Recent DNS data reported by Poludnenko and Oran (2010) indicate that, due to the discussed physical mechanism, small-scale turbulent eddies cannot perturb a thin reaction zone even in very intense ($u'/S_L^0 \approx 30$) small-scale ($L \approx \Delta_L$) turbulence.

Second, the stretch rates produced by turbulent eddies vary rapidly in time, while the responses of a laminar flame to steady and unsteady strain rates are substantially different (e.g., see a review paper by Law and Sung [2000]). For instance, by experimentally investigating the unsteady interaction between a laminar premixed flame and a vortex pair, Sinibaldi et al. (2003) documented perturbations in the flame speed that were much larger than perturbations that resulted from the theory of steadily stretched laminar flames (cf. the symbols with solid curves in Figure 5.36).

Transient effects are well pronounced when the mean strain rate is close to $\dot{s}_{q,s}$. For instance, numerical simulations have shown that laminar flames can survive when the strain rate oscillates about a value slightly less than $\dot{s}_{q,s}$ even if the highest instantaneous strain rate is close to $2\,\dot{s}_{q,s}$ (Lipatnikov and Chomiak, 1998) (e.g., see the solid, dotted-dashed, and dotted curves in Figure 5.37).

A flame needs finite time to respond to variations in the strain rate, and the flame cannot respond to rapid changes in $\dot{s}$ even if an instantaneous strain rate is substantially higher than $\dot{s}_{q,s}$. Mueller et al. (1996) have experimentally shown that, for a

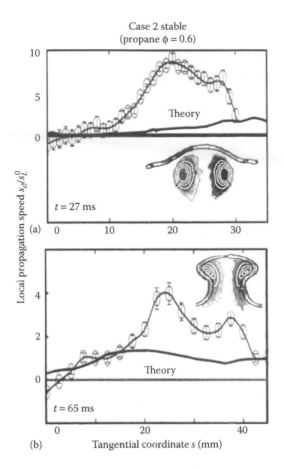

(a)

(b)

FIGURE 5.36 (a,b) Local propagation speed measured (symbols) along a lean ($\Phi = 0.6$) propane–air flame interacting with a vortex pair. The curves without symbols have been calculated substituting measured instantaneous stretch rates and the Markstein number into a linear equation similar to Equation 5.100. (Reprinted from Sinibaldi, J.O., Driscoll, J.F. Mueller, C.J., Donbar, J.M., and Carter, C.D., *Combust. Flame*, 133, 323–334, 2003. With permission.)

time lag comparable with τ_c, laminar flames are weakly affected by external stretch rates, even if the rates are 10 times greater than $\dot{s}_{q,s}$. Experimental (Samaniego and Mantel, 1999) and numerical (Mantel and Samaniego, 1999) studies of interaction between a laminar flame and a strong vortex pair did not indicate any flame quenching at strain rates much higher (by more than 10 times) than $\dot{s}_{q,s}$.

Due to the aforementioned transient effects (including the destruction of eddies within flamelets), small-scale vortexes are noneffective in stretching and thickening a laminar premixed flame, and the flame strongly resists penetration of the vortexes into it. Such phenomena were predicted numerically by Poinsot et al. (1990 and 1991) and documented experimentally by Roberts et al. (1993), who investigated the interaction of a premixed flame with two counterrotating laminar vortexes and a single

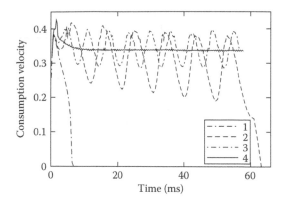

FIGURE 5.37 Oscillations in the consumption velocity versus time under the influence of the oscillating strain rate $\dot{s} = \dot{s}_{q,s}[1 + 0.92\cos(\omega t)]$, computed for a strained cylindrical flame (Adapted from Lipatnikov, A.N. and Chomiak, J., *Combust. Sci. Technol.*, 137, 277–298, 1998. With permission.)

toroidal laminar vortex, respectively. Other similar data are discussed in a review paper by Renard et al. (2000). Poinsot et al. (1990, 1991) and Roberts et al. (1993) processed their data in order to propose parameterizations $\Lambda = \delta_L(a + b\mathrm{Ka}^{-q})$ of the length scale Λ of the smallest vortex that is able to affect a laminar premixed flame. The coefficients a, b, and the power exponent q are different in the two parameterizations, for example, $q = 1/2$ (Roberts et al., 1993) or $q = 1/3$ (Poinsot et al., 1990, 1991). Although neither of the parameterizations agrees well with the experimental data obtained from premixed turbulent flames (e.g., see Figure 3 in Gülder and Smallwood [1995] or Figure 10 in Shim et al. [2011]), the idea of introducing such a length scale, which depends substantially on mixture properties, seems to be fruitful, as will be discussed further in the section "Instabilities of Laminar Flamelets in Turbulent Flows" (see Figure 5.43a in particular).

Third, the critical stretch rate depends substantially on the flamelet geometry. For instance, the values of $\dot{s}_{q,s}$ are significantly different in the two cases sketched in Figure 5.20 (two identical flames in two identical flows of unburned mixture) and in Figure 5.38 (a single flame in counterflows of unburned and burned mixtures). The latter flame is much more resistant to extinction than the former flame (Libby and Williams, 1982, 1984; Libby et al., 1983; Law and Sung, 2000). For instance, even if the strain rate is very high, the latter flame, which is modeled within the framework of the AEA theory, can survive by moving to the product side in the adiabatic case if Le = 1 (Libby and Williams, 1982).

Moreover, if the Lewis number is substantially lower than unity, the curvature of a strained laminar flame allows it to survive under the influence of high stretch rates, as documented for premixed edge flames by Liu and Ronney (1999). Simulations by Buckmaster and Short (1999) have shown that such a flame can survive at values of strain rate about three times larger than the extinction value for twin planar flames stabilized in counterflows (see Figure 5.20) of the same mixture if Le = 0.3.

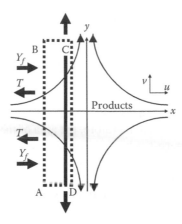

FIGURE 5.38 A strained laminar flame stabilized in counterflows of unburned (*left*) and burned (*right*) mixtures.

It is quite unclear what geometrical configuration (whether twin symmetrical planar counterflow flames sketched in Figure 5.20 or a single planar flame stabilized by the product counterflow sketched in Figure 5.38 or something else) is generally relevant in turbulent combustion. For instance, even if the probability of finding the unburned–burned (see Figure 5.38) flamelets is much greater than the probability of finding the symmetrical unburned–unburned (see Figure 5.20) flamelets inside a turbulent flame brush, the opposite relation appears to be valid at the leading edge of the flame brush due to geometrical consideration. Therefore, it is not clear which particular geometrical model of a strained laminar flame should be used to evaluate the critical stretch rate for modeling turbulent combustion.

Finally, due to the strongly uneven structure of small-scale turbulence, the discussed flamelet "quenching" may manifest itself in the inability of a flamelet to be convected into spatial regions characterized by high local stretch rates. Such phenomena may damp an increase in the flamelet surface area by turbulent stretching but do not result in local extinction of the flamelet. As far as Equation 5.106 is concerned, the latter "quenching" should affect the area ratio on the RHS, rather than the mean consumption velocity $\bar{u}_c$, which is averaged only over the burning layers.

To gain further insight into the latter interpretation of local combustion quenching, let us assume that there is a critical stretch rate $\dot{s}_q = \tau_c^{-1} f(\mathrm{Le}, D_\mathrm{F}/D_\mathrm{O}, \sigma, \mathrm{Ze}, ...)$ such that a flamelet cannot enter the spatial regions where $\dot{s} > \dot{s}_q$. Certainly, $\dot{s}_q$ should depend on transient effects, as discussed earlier. Because turbulent stretch rates scale as $\dot{s} \propto (\bar{\varepsilon}/l^2)^{1/3}$ (see Equation 3.70), an increase in u' in intense $(\tau_\eta^{-1} > \dot{s}_q)$ turbulence should narrow the spectral range of eddies that increase the area of the flamelet surface by stretching it but do not quench combustion locally. This range may be associated with $l_q < l < L$, where the lowest "surface-increasing" length scale, l_q, is determined from the following criterion:

$$\bar{\varepsilon}^{1/3} l_q^{-2/3} \propto \dot{s}_q = \tau_c^{-1} f\left(\mathrm{Le}, D_\mathrm{F}/D_\mathrm{O}, \sigma, \mathrm{Ze}, ...\right). \qquad (5.107)$$

Considering that $\bar{\varepsilon} \propto u'^3/L$, we obtain

$$\frac{l_q}{L} \propto \left(\frac{u'}{\dot{s}_q L}\right)^{3/2} = \left(\dot{s}_q \tau_t\right)^{-3/2}. \tag{5.108}$$

Accordingly, if flamelets cannot enter the spatial regions characterized by the highest stretch rates ($\tau_\eta^{-1} > \dot{s}_q$) and, therefore, the area of the flamelet surface is controlled by the highest "surface-producing" stretch rate $\dot{s} = \dot{s}_q$, then an increase in u' results in an increase in the length scale l_q, but does not affect the area, because the same stretch rate $\dot{s}_q$ is provided by larger turbulent eddies in the case of a larger u'. Accordingly, when u' is increased, the flamelet surface area reaches its maximum value if

$$\dot{s}_q \tau_\eta = O(1). \tag{5.109}$$

Consequently,

$$u'_m \propto \dot{s}_q^{2/3} \nu^{1/3} L^{1/3} \tag{5.110}$$

and a further increase in the rms velocity ($u' > u'_m$) does not result in a higher surface-producing stretch rate, because the flamelet cannot enter the spatial region characterized by $\dot{s} > \dot{s}_q$. Within the framework of the considered scenario, a decrease in U_t by $u' > u'_m$ may be caused by a reduction in the mean consumption velocity due to high (but less than $\dot{s}_q$) stretch rates.

If, due to the influence of heat release on small-scale turbulence and due to transient effects, flamelets may be unaffected by eddies that are smaller than a length scale Λ, then the aforementioned length scale l_q should be larger than Λ in order for the considered "quenching" phenomena to play a role. Accordingly, if $\eta < \Lambda$, Equations 5.109 and 5.110 should be substituted with the following estimates $\dot{s}_q \bar{\varepsilon}^{-1/3} \Lambda^{2/3} = O(1)$ and $u'_m \propto \dot{s}_q \Lambda^{2/3} L^{1/3}$, respectively.

Certainly, the aforementioned simple discussion does not address a number of important effects. First, an increase in l_q by u' would increase the scale of the smallest wrinkles on the flamelet surface. Therefore, it would reduce the annihilation of the surface due to the self-propagation of flamelets, because the latter physical mechanism plays an important role if the surface is highly wrinkled at small scales. Second, a reduction in the flamelet displacement speed by stretch rates acts in the same direction and also results in an increase in the flamelet surface. Third, due to the internal intermittency of turbulent flows, the use of the Kolmogorov scaling, that is, Equations 5.107 through 5.110, for estimating the highest local stretch rates may be disputed.

SUMMARY

The stretch rates produced by turbulent eddies not only increase the flamelet surface area but may also affect the mean local burning rate. To model the latter phenomenon, the unperturbed laminar flame speed in Equation 5.1 should be replaced with the mean consumption velocity, $\bar{u}_c$, in Equation 5.106. Based on the results of investigations of stretched laminar premixed flames, the dependencies of $\bar{u}_c$ on the rms velocity and Lewis number are sketched in Figure 5.34. To put it briefly, the mean consumption velocity may be increased when Le is decreased. Such a hypothesis explains a well-documented fact that both the turbulent burning velocity and u'_m are higher in flames characterized by a lower Lewis number, with all other things being equal.

The high stretch rates produced by turbulent eddies may locally quench combustion or damp an increase in the flamelet surface area by impeding flamelets from moving into the spatial regions characterized by locally high stretch rates. Such physical mechanisms may contribute to the reduction in the burning velocity by the rms velocity in intense turbulence.

Due to the influence of heat release on small-scale eddies and other transient effects, laminar flamelets in turbulent flows are substantially more resistant to local combustion quenching than steadily strained laminar flames. There is no evidence whatsoever to indicate that a steadily strained laminar flame is a good model for quantifying perturbed flamelets that occur in real turbulent combustion.

FLAME INSTABILITIES

Laminar premixed flames are subject to various instabilities, with two such instabilities often assumed to play a substantial role in turbulent combustion. The two phenomena, called hydrodynamic instability and diffusive-thermal instability, are briefly described in the next two subsections, followed by a discussion on their role in premixed turbulent combustion.

HYDRODYNAMIC INSTABILITY

The hydrodynamic instability of laminar premixed flames was independently predicted by Darrieus (1938) and Landau (1944), who theoretically investigated the following problem: A laminar flame is reduced to an infinitely thin surface, which propagates at a constant speed S_L^0 with respect to the unburned mixture and separates the mixture and the combustion products. The flow outside the flame is governed by the nonreacting Euler equations, that is, Equation 1.51 where the viscous stress tensor $\tau_{ij} = 0$, with the density being equal to either ρ_u or ρ_b, ahead or behind the flame, respectively. Jump conditions on the flame surface are used to close the model. A stability analysis of such a planar flame, reproduced in many textbooks (Zel'dovich et al., 1985; Landau and Lifshitz, 1987; Chomiak, 1990) and review papers (Clavin, 1985; Law and Sung, 2000), shows that the flame is unconditionally unstable to infinitesimal perturbations of any wave number κ with the growth rate $\omega \propto S_L^0 \kappa$ of the amplitude of a flame surface disturbance being linearly increased by the wave number.

The physical mechanism of hydrodynamic instability, which is often called the DL instability in honor of Darrieus and Landau, consists of the following (see Figure 5.39). Due to the flame-induced convergence (divergence) of the unburned

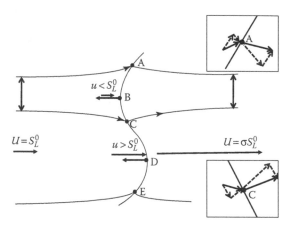

FIGURE 5.39   A sketch of a hydrodynamically unstable flame.

mixture flow upstream of concave (convex) flame fronts, the unburned flow velocity, $u_{u,f}$, at the flame surface increases (decreases) in the vicinity of point D (B), whereas the flame speed is constant. As a result, both convex (toward the unburned gas, point B) and concave (point D) bulges, characterized by $u_{u,f} < S_L^0$ and $u_{u,f} > S_L^0$, respectively, grow, that is, the amplitude of the disturbance increases and causes a further increase in the velocity perturbation. It is worth stressing that the influence of the flame on the flow is possible only in a variable-density case. In particular, streamlines that pass a flame segment AC (CE) become convergent (divergent) in the products due to an increase in the normal component of the flow velocity vector caused by the density drop in the flame, as sketched in the square insets in the right up (point A) and down (point C) corners of Figure 5.39. Such flow changes at the flame affect the pressure field, and the flame-induced pressure perturbations further perturb the upstream flow, causing the instability discussed earlier.

Because laminar flames are well known to be stable under a wide range of conditions, the theoretical prediction of the unconditional instability of the flames has been a challenge to the combustion community. The first attempt to resolve the aforementioned contradiction between theory and observations was undertaken by Markstein (1951), who suggested that flame speed is not a constant quantity but is affected by the local curvature of the flame surface (e.g., see Equations 5.88 and 5.89). This hypothesis allowed Markstein to predict the stability of the flame with respect to perturbations of short wavelengths (large wave numbers). Indeed, if Le = 1, then the flame speed is increased (reduced) in negative (positive) flame bulges (see point D (B)). If the perturbation wavelength is short, that is, the local curvature is large, such an effect of the flame curvature on its speed overwhelms the flame-induced variations in the flow velocity and the flame is stable with respect to perturbations characterized by the shortest wavelengths.

Theoretical and numerical studies reviewed elsewhere (e.g., see Section 4.1 in Lipatnikov and Chomiak [2005a]) have indicated that the instability growth rate increases linearly with wave number κ for the longest perturbations, reaches a maximum value at a certain wave number κ_m, followed by a decrease in the growth

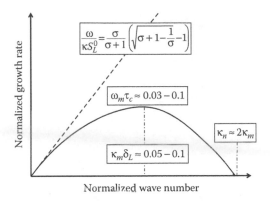

$$\frac{\omega}{\kappa S_L^0} = \frac{\sigma}{\sigma+1}\left(\sqrt{\sigma+1-\frac{1}{\sigma}}-1\right)$$

$\omega_m \tau_c \approx 0.03 - 0.1$

$\kappa_n \approx 2\kappa_m$

$\kappa_m \delta_L \approx 0.05 - 0.1$

Normalized growth rate (y-axis)

Normalized wave number (x-axis)

FIGURE 5.40 A typical dependence (*solid curve*) of the normalized growth rate $\omega\tau_c$ of the DL instability on the normalized wave number $\kappa\delta_L$. The *dashed line* shows the solution obtained by Darrieus (1938) and Landau (1944).

rate with κ for the shortest perturbations (sec Figure 5.40). The experimental data obtained by Truffaut and Searby (1999) from rich propane flames show that the normalized growth rate $\omega_m\tau_c \approx 0.03-0.1$ with $\kappa_m\delta_L \approx 0.05-0.1$. Accordingly, the so-called neutral wavelength Λ_n, which separates the ranges of stable ($\kappa > \kappa_n = 2\pi/\Lambda_n$) and unstable ($\kappa < \kappa_n$ or $\Lambda > \Lambda_n$) perturbations, is of the order of $(30-60)\delta_L$, with the laminar flame thickness being determined by Equation 2.43.

In experimental studies of laminar premixed combustion, DL instability manifests itself most often in wrinkling of the surface of an expanding spherical flame. After the pioneering experimental investigations by Karpov (1965) and Groff (1982), the formation of cellular spherical flames due to DL instability was documented in many papers. Typical images of cellular expanding spherical laminar flames are shown in the middle and right columns in Figure 5.41 (see the middle and bottom rows).

DIFFUSIVE-THERMAL INSTABILITY

The physical mechanism of diffusive-thermal instability, which manifests itself in mixtures with high molecular diffusivity of the deficient reactant (e.g., lean hydrogen–air or rich paraffin–air mixtures, with the exception of methane), consists of the local increase (decrease) in flame speed near upstream- (downstream-) pointing bulges in a wrinkled laminar flame surface due to local changes in enthalpy, mixture composition, reaction rate, and consumption velocity. If Le < 1 or the diffusivity, D_d, of the deficient reactant is higher than the diffusivity, D_e, of the excess reactant, then either the chemical energy supplied to positively curved parts of the flame surface (e.g., upstream-pointing bulge B in Figure 5.39) by molecular diffusion exceeds the heat losses due to molecular conductivity or the equivalence ratio in the bulges is shifted toward unity due to faster diffusion of the deficient reactant as compared with the excess reactant. Both processes increase the consumption velocity and flame speed locally (e.g., see Equations 5.99 through 5.104, which have been

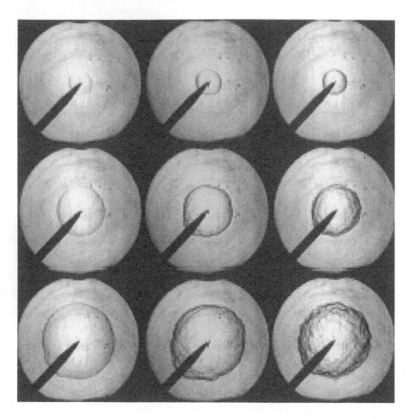

FIGURE 5.41 Schlieren images of expanding spherical laminar flames at initial pressures 0.05 (*left column*), 0.1 (*middle*), and 0.15 (*right*) MPa. (With kind permission from Springer Science + business Media: *Flow, Turbulence and Combustion*, Turbulent combustion of hydrogen–CO mixtures, 86, 2011, 735–749, Burluka, A.A., El-Dein Hussin, A.M.T., Sheppard, C.G.W., Liu, K., and Sanderson, V., Figure 1 Copyright 2011 Springer.)

derived in the case of equidiffusive reactants). The opposite phenomena occur near the downstream-pointing (negatively curved) bulges (see point D in Figure 5.39). As a result, the upstream- (downstream) pointing bulges propagate faster (slower) and the amplitude of the flame surface perturbations (bulges) grows.

On the contrary, if the Lewis number is large and/or $D_d < D_e$, then the diffusive-thermal phenomena result in decreasing (increasing) the flame speed in the vicinity of point B (D), that is, these phenomena stabilize the flame against the hydrodynamic instability discussed earlier. If the Lewis number is sufficiently large and $D_d < D_e$, a laminar flame is stable with respect to weak perturbations of the flow velocity (Clavin, 1985; Law and Sung, 2000).

INSTABILITIES OF LAMINAR FLAMELETS IN TURBULENT FLOWS

As discussed earlier, the response of a laminar flame to perturbations depends on the molecular transport properties. If Le < 1 and/or $D_d > D_e$, the hydrodynamic and diffusive-thermal instabilities enhance each other, and local perturbations in the flame surface grow, thereby increasing its area and, therefore, the global burning rate. If Le > 1 and/or $D_d < D_e$, then the diffusive-thermal effects can suppress the hydrodynamic instability, the flame surface retains a planar form, and its area remains relatively small. Similar phenomena could occur locally in turbulent flows, where laminar flamelets are perturbed by turbulent eddies.

To some degree, variations in the local consumption velocity in stretched laminar flamelets and the growth of the flamelet surface due to the diffusive-thermal instability are basically the same phenomena, because they are caused by the same physical mechanism, variations in the local enthalpy and mixture composition within stretched reaction zones. The terms "variations in $\bar{u}_c$" and "diffusive-thermal instability" designate different manifestations of this physical mechanism. The variations in the local consumption velocity in stretched laminar flamelets affect $\bar{u}_c$ on the RHS of Equation 5.106, while the hydrodynamic and diffusive-thermal instabilities can increase the area ratio in this expression.

Indeed, experimental and numerical studies reviewed elsewhere (see Sections 4.2 and 5.3 in papers by Lipatnikov and Chomiak, 2005a, 2010, respectively) have indicated that premixed turbulent flames with Le < 1 and/or $D_d > D_e$ have a higher level of surface distortion and a larger flamelet surface area than flames with Le > 1 and/or $D_d < D_e$. For instance, Figure 5.42 shows that the increase in the area of an initially laminar flame during interaction with a turbulent flow is strongly increased when Le is low ($D_d = D_e$ in the cited work).

An increase in the flame surface area with decreasing Le (or increasing D_d/D_e) is often considered to be caused by the instability of laminar flamelets and to be the manifestation of an important role played by the instability in premixed turbulent combustion. However, strictly speaking, the experimental or DNS data available today are not sufficient to draw such a conclusion. The data show only a substantial increase in the flame surface area for mixtures characterized by a faster-diffusing deficient reactant. Although the instability concept is an attractive hypothesis for explaining this effect, this hypothesis is not the only possible explanation, and therefore, alternative concepts that are able to explain the effect can be put forward

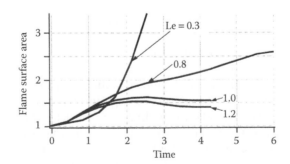

FIGURE 5.42 The development of a flamelet surface area computed in DNS at different Lewis numbers. Quantities are normalized using S_L^0 and $\tau_t = L/u'$. (Reprinted from Trouvé, A. and Poinsot, T., *J. Fluid Mech.*, 278, 1–31, 1994. With permission.)

(e.g., see Sections 4.3 and 7 in a review paper by Lipatnikov and Chomiak [2005a], as well as Section 5.3 in Lipatnikov and Chomiak [2010]).

To prove the important role played by the instabilities in premixed turbulent combustion, more sophisticated measurements and simulations are necessary. For instance, a correlation between a characteristic of a flamelet surface in a turbulent flow and the counterpart characteristic of an unstable flame in a laminar flow should be demonstrated.

On the face of it, recent experimental studies by Kobayashi et al. (2002, 2005), Soika et al. (2003), and Ichikawa et al. (2011) reached this goal. These authors investigated high-pressure premixed turbulent flames. By varying the mixture composition, turbulence characteristics, and pressure over wide ranges, they managed to substantially change (i) the length scales that characterized the smallest wrinkles on a flamelet surface in the studied flows, (ii) the Kolmogorov and Taylor turbulence length scales, and (iii) the characteristic length scale Λ_m associated with the maximum (for various wave numbers) growth rate of DL instability ($\Lambda_m \approx 0.5\Lambda_n$, see Figure 5.40).

Certain data reported by Kobayashi et al. (2002, 2005) indicate that, if $\Lambda_m > 12\eta$, then the smallest scale e_i (the so-called inner cutoff scale, see the section "Fractal Approach to Evaluating Turbulent Burning Velocity" in Chapter 7) of flamelet surface wrinkling is independent of the turbulent Reynolds number $\text{Re}_\lambda = u'\lambda/\nu_u$, which is based on the Taylor length scale λ (see the filled diamonds in Figure 5.43a). Because (i) the length scale Λ_m of DL instability does not depend on Re_λ (see the horizontal straight line), whereas (ii) the scale of the smallest turbulent eddies, which was assumed to be equal to 12η by Kobayashi et al. (2002, 2005), decreases when Re_λ is increased (see open circles), the data shown in Figure 5.43a (and other similar data reported in the cited papers) indicate that the smallest wrinkling length scale is controlled by the instability length scale if $\Lambda_m > 12\eta$. On the contrary, $e_i \approx 12\eta > \Lambda_m$ at low Re_λ. Note also that the condition of $\Lambda_m = 12\eta$ and the bending of the S_T/S_L^0 curve in Figure 5.43b are associated with roughly the same Re_λ, thereby implying a substantial influence of DL instability on the turbulent burning rate.

Soika et al. (2003) documented that the mean diameter of a flamelet curvature depended on the equivalence ratio similarly to the length scale Λ_m of DL instability, whereas turbulence length scales were independent of Φ.

OH PLIF images of Bunsen flames reported by Ichikawa et al. (2011) (cf. Figures 2d and 2e in the cited paper) show that smaller length scales of wrinkling of a flamelet surface are observed in a mixture characterized by a lower Le, with the inner cutoff scale correlating with Λ_m (see Figure 5b in the cited paper).

Even if we recognize the aforementioned experimental data to prove decisively that the length scale of the smallest wrinkling of a flamelet surface is controlled by Λ_m (provided that the latter scale is larger than the length scale of the smallest turbulent eddies), this fact does not prove that DL and thermal-diffusive instabilities play a substantial role in premixed turbulent combustion. This fact shows only that both the stabilization of a laminar premixed flame with respect to perturbations of short wavelengths and the stabilization of a laminar flamelet with respect to small-scale turbulent eddies are controlled by the same physical mechanism.

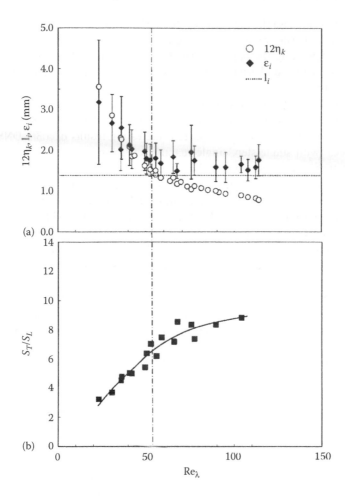

FIGURE 5.43 Length scales relevant to turbulent premixed Bunsen flames (a) and normalized turbulent burning velocity (b) versus Reynolds number $Re_\lambda = u'\lambda/\nu_u$. (Reprinted from Kobayashi, H., Seyama, K., Hagiwara, H., and Ogami, Y., *Proc. Combust. Inst.*, 30, 827–834, 2005. With permission.)

The opposite effects of pressure on laminar and turbulent flame speeds (see the section "Dependence of Turbulent Burning Velocity on Diffusivity, Pressure, and Temperature" in Chapter 4) are often considered to be a manifestation of the important role played by the instabilities in premixed turbulent combustion. Indeed, because Λ_m and Λ_n scale as the laminar flame thickness, which is reduced by pressure due to a decrease in the molecular transport properties with increasing p, the role played by the instabilities in flame propagation seems to be increased by pressure, because the spectral range of unstable perturbations is expanded when p is increased. For instance, a comparison of the left and right columns in Figure 5.41 clearly shows that the expanding spherical laminar flames become unstable under elevated pressures. Therefore, due to an increase in the neutral wave number κ_n by p, the instabilities increase the area ratio on the RHS of Equation 5.106 more at higher pressures than at lower pressures. Although this hypothesis appears to be an attractive tool for explaining the opposite effects of pressure on laminar and turbulent flame speeds, there are alternative approaches that not only explain but also predict the pressure dependence of U_t well (e.g., see Section 3.3.5 in Lipatnikov and Chomiak [2002a] or the sections "Zimont Model of Burning Velocity in Developing Turbulent Premixed Flames" and "Validation I: TFC model" in Chapter 7).

Boughanem and Trouvé (1998) reported interesting results of a 3D DNS of premixed flames (with single-step, single-reactant chemistry), embedded in decaying small-scale turbulence (see Figure 5.44). A quasi-developed phase of flame propagation, characterized by a roughly constant total burning rate, is changed (e.g., at $t \cong 2{,}5\tau_t$ in case D shown as asterisks) to an unstable phase characterized by a rapidly increasing burning rate, with this transition being promoted by an increase in the density ratio. These results were interpreted to be a manifestation of DL instability, which controlled the burning rate after the decrease in u' to an appropriate value of the order of S_L^0 (Boughanem and Trouvé, 1998). Note that the simulations have been performed for $Le = 1$ and $D_d = D_e$.

With the exception of these DNS data, evidence of an important role played by the instabilities in premixed turbulent combustion is indirect, too sparse, few, and scattered in order to draw solid conclusions, as discussed in detail in Sections 4.3 and 5.3 in review papers by Lipatnikov and Chomiak (2005a and 2010, respectively).

Furthermore, in sufficiently intense turbulence, the instabilities are likely to be suppressed due to stretching of flamelets by turbulent eddies. The stabilizing effect

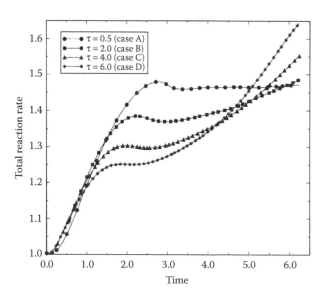

FIGURE 5.44 The effects of gas expansion on the total burning rate normalized using ρ_u and S_L^0. Time is normalized using τ_t. The density ratio is specified in the legend with $\tau = \rho_u/\rho_b - 1$. (Reprinted from Boughanem, H. and Trouvé, A., *Proc. Combust. Inst.*, 27, 971–978, 1998. With permission.)

of straining on the instabilities of laminar premixed flames has been known since a theoretical study by Sivashinsky et al. (1982). Similar phenomena are likely to occur locally inside the turbulent flames, for example, one small vortex wrinkles a flamelet and triggers the instability, but another larger vortex locally strains the flame surface and flattens it out, thereby cancelling the effects of instability.

To demonstrate this phenomenon with clarity, let us use the following very simple phenomenological model (Lipatnikov and Chomiak, 2005a). Straining flattens out the existing disturbances by providing a convection velocity toward a flame surface, with the velocity being equal to $u = \dot{s}x$, where x is the distance from the unperturbed flame position. Thus, we can write the following simple equation:

$$\frac{d\xi}{dt} = \left(\omega - \dot{s}\right)\xi \tag{5.111}$$

for the rate of increase of the amplitude ξ of a flame surface disturbance, where ω is the DL growth rate for the unstrained flame. The second term on the RHS of Equation 5.111 is associated with the flattening of the flame surface due to straining. Consequently, the criterion for damping the instability by straining is simply $\omega = \dot{s}$. In the DL solution, there are always unstable waves in the short-wave range, whatever the stretch rate is, because $\omega \propto S_L^0 \kappa$ can be very large as $\kappa \to \infty$. This is not the case for the Markstein and other, more recent, solutions, which predict the stability of short-wave perturbations. For low wave numbers, $\omega \propto S_L^0 \kappa$ is also low and even weak straining can damp the instability. For wave numbers close to the wave number κ_m associated with the highest growth rate ω_m, we have $\tau_c \omega_m = 0.03-0.1$ (see Figure 5.40) and the instability is damped if $\tau_c \tau_\eta^{-1} = 0.1-0.3$ (Lipatnikov and Chomiak, 2005a), provided that $\bar{s} = 0.28\tau_\eta^{-1}$ (Bray and Cant, 1991). This estimate yields Ka $> 0.1-0.3$ if Equations 3.65, 4.11, and 4.13 are invoked to evaluate the Karlovitz number.

Certainly, the foregoing estimate holds only if the Kolmogorov eddies are sufficiently large in order to damp the instability by flattening out the flamelet surface, that is, if the Kolmogorov wave number $\kappa_\eta \equiv 2\pi/\eta$ is less than the wave number κ_m associated with the maximum growth rate of DL instability. Although the obtained criterion, Ka $> 0.1-0.3$, implies that $\eta > \delta_L$, this inequality does not mean that $\kappa_\eta < \kappa_m$, because the wavelength $2\pi/\kappa_m$ is substantially larger than δ_L (see Figure 5.40). If $\kappa_\eta > \kappa_m$, then the maximum growth rate ω_m should be compared with the highest mean stretch rate $\bar{s}_m \propto \bar{\varepsilon}^{1/3}(2\pi/\kappa_m)^{-2/3}$ provided by eddies that are larger than the instability wavelength $2\pi/\kappa_m$. Such an estimate results in another criterion, Ka $> \mathrm{Ka}_{cr}$, for damping the instability, with the number Ka_{cr} being at least larger than 0.1 and depending substantially on the numbers $\kappa_m \delta_L$ and $\omega_m \tau_c$. Note that if the instabilities associated with the maximum growth rate are damped by the mean stretch rate provided by turbulent eddies characterized by the wave number κ_m, then the instabilities with a smaller wave number κ are damped by turbulent eddies characterized by the same wave number $\kappa < \kappa_m$, because a decrease in the mean stretch rate $\bar{s} \propto \bar{\varepsilon}^{1/3}\kappa^{2/3}$ with a decrease in κ is less pronounced than the decrease in the growth rate $\omega \propto S_L^0 \kappa$ that results from the DL solution. It is also worth remembering that if $L < 2\pi/\kappa_m$, then turbulence cannot damp the instabilities, because even the largest turbulent eddies are too small to flatten out the perturbation of a flamelet surface, characterized by the wave number κ_m.

Recently, Chaudhuri et al. (2011) also compared the growth rate of the DL instability of a wave number κ and the mean stretch rate provided by turbulent eddies of the same wave number. Note that such a comparison implies that $\kappa_\eta > \kappa_m$, while $\omega(\kappa)$ should be compared with $\bar{s} \propto \tau_\eta^{-1}$ if $\kappa_\eta < \kappa$. Under the typical conditions of $L \gg \delta_L$, a criterion obtained in the cited paper is very close to $\mathrm{Ka} = \mathrm{Ka}_{cr}$, with the critical number, Ka_{cr}, being substantially lower than unity (see the thick solid curve in Figure 5 therein). According to Chaudhuri et al. (2011), DL instability plays a minor role in premixed combustion if $\mathrm{Ka} > \mathrm{Ka}_{cr}$.

Bradley (1999) and Bradley et al. (2005) have emphasized that the stabilizing effect of straining on the instabilities manifests itself in the following phenomenon, which is well documented in experiments (e.g., cf. the top and bottom rows in Figure 5.41). When a spherical laminar flame kernel grows after spark ignition, cells appear on the flame surface only if the flame radius is sufficiently large for the stretch rate $(2/R_f)(dR_f/dt)$ to become insufficient to stabilize the flame. By processing such experimental data, Bradley et al. (2005) have estimated these critical stretch rates for different mixtures. The results calculated in the cited paper imply a notable effect of the instabilities on the burning velocity only in weak turbulence.

Furthermore, recent experimental data obtained by Al-Shahrany et al. (2006) showed that as low a value of u' as 0.6 m/s was "sufficient to reduce, and almost eradicate, the effect of these instabilities on the flame speed."

Thus, although the experimental data obtained from turbulent flames indirectly support the concept of an important role played by the DL and diffusive-thermal instabilities in weakly turbulent premixed combustion, decisive experimental evidence has not yet been reported.

Nevertheless, one manifestation of DL instability appears to be of substantial importance even in intense turbulence, at least from the theoretical viewpoint. The vast majority of models of premixed turbulent combustion are based on the consideration of a purely hypothetical case of an unconfined, unperturbed, and fully developed premixed turbulent flame, that is, a statistically planar, 1D flame that has a statistically stationary thickness and propagates at a constant speed against a statistically stationary, spatially uniform, and 1D turbulent flow of a spatially uniform unburned mixture. As will be discussed in the section "Zimont Model of Burning Velocity in Developing Turbulent Premixed Flames," in Chapter 7, the existence of such a flame is theoretically admissible in the case of a constant density. However, due to the DL instability of the entire turbulent flame brush (not the thin, inherently laminar, flamelets previously considered!), an unconfined, unperturbed, and fully developed premixed turbulent flame is unlikely to exist in the case of $\rho_u > \rho_b$ (Kuznetsov and Sabel'nikov, 1990). Indeed, if we assume that an unperturbed premixed flame (i) is fully developed after a certain finite flame-development time, which may be as long as we want, and (ii) has a certain finite mean flame brush thickness Δ_t, which may be as large as we want, then in an unconfined flow, we can always find a perturbation whose wavelength λ is much larger than Δ_t, with the ratio of λ/Δ_t being as large as we want. For such a long-scale perturbation, the premixed turbulent flame may be considered to be an infinitely thin self-propagating interface.

Thus, the problem is reduced to the classical DL problem, with the influence of the flame curvature on its speed being negligible for sufficiently large λ/Δ_l. According to the DL solution, the flame is unstable and both its speed and thickness increase as the perturbation grows. Therefore, the flame is not fully developed, contrary to the foregoing assumption (i). This contradiction proves that an unconfined, unperturbed, and fully developed premixed turbulent flame does not exist in the case of $\rho_u > \rho_b$. If the wavelength of perturbations is limited by the boundary conditions, then a flame that has a statistically stationary thickness and propagates at a constant speed against an oncoming statistically stationary, spatially uniform, 1D flow may exist; such flames were observed in some DNS studies (Nishiki et al., 2002; Treurniet et al., 2006). However, the speed and thickness of such a flame probably depend on the width of the computational domain in transverse direction.

REGIMES OF PREMIXED TURBULENT COMBUSTION

Previously, we considered the following physical mechanisms that affect the turbulent burning rate.

1. Wrinkling of laminar flamelets (or thin, laminar reaction zones in more intense turbulence) by turbulent stretching increases the area of the flamelet (or reaction zone) surface, thereby increasing the burning velocity. On the contrary, collisions and mutual annihilation of self-propagating flamelets (or reaction zones) reduce the surface area, thereby decreasing the burning velocity. On the RHS of Equation 5.106, the two mechanisms affect the area ratio, with the former mechanism being dominant. These mechanisms play an important role at least in weak and moderate turbulence.

2. Stretching of laminar flamelets (or thin, laminar reaction zones in more intense turbulence) by turbulent eddies reduces (increase) the local burning rate if the diffusivity of the deficient reactant is low (high), that is, Le > 1 and $D_d < D_e$ (Le < 1 and/or $D_d > D_e$). On the RHS of Equation 5.106, this mechanism affects the mean consumption velocity $\bar{u}_c$. Within the framework of the AEA theory, this mechanism plays a role only if Le $\neq$ 1 or $D_d \neq D_e$. For real flames characterized by a reaction zone of a finite width, the mechanism may be of importance even if Le = 1 and $D_d = D_e$. The role played by this mechanism is increased by u' in moderate turbulence.

3. Strong stretching of laminar flamelets (or thin, laminar reaction zones in more intense turbulence) by turbulent eddies may extinguish combustion locally or, more likely, counteract the first mechanism by pushing the flamelets out of the spatial regions characterized by the highest local stretch rates. On the RHS of Equation 5.106, this mechanism affects (decreases) the area ratio. The mechanism may play an important role in intense turbulence and is commonly considered to be the primary cause of a weak dependence of U_t on u' (or even a reduction in U_t by u'), followed by global combustion quenching.

4. The DL and diffusive-thermal instabilities of laminar flamelets may increase the flamelet surface area and, therefore, the turbulent burning velocity. On the RHS of Equation 5.106, this mechanism affects the area ratio. This

mechanism is likely to play a role in weakly turbulent flames (especially if Le < 1 and/or $D_d > D_e$), but it appears to be damped by turbulent stretching in more intense turbulence.

Thus, stretching of laminar flamelets (or thin, laminar reaction zones in more intense turbulence) appears to be the primary cause of the influence of turbulence on premixed combustion. On the one hand, stretching increases the burning velocity (i) by increasing the flamelet surface area and, if the diffusivity of the deficient reactant is large, (ii) by increasing the local burning rate in weakly and moderately stretched flamelets. On the other hand, stretching reduces the burning velocity (i) by damping the DL and diffusive-thermal instabilities, (ii) by restricting flamelet motion and/or by locally quenching combustion, and, if the diffusivity of the deficient reactant is low, (iii) by reducing the local burning rate.

The trends emphasized earlier are summarized in Figures 5.45 and 5.46 in two different forms. Figure 5.45 shows the typical dependencies of the burning velocity on the rms velocity and is associated with a typical set of measurements in which both u', the ratio of u'/S_L^0, and the Lewis number are varied over sufficiently wide ranges. Figure 5.46 offers another perspective, because it is not only restricted to variations in u'/S_L^0 but also allows for variations in L/δ_L. In a typical set of measurements, the integral length scale is roughly constant, while the laminar flame thickness may be changed due to variations in the equivalence ratio.

Figure 5.46 is an example of a combustion regime diagram. Such diagrams are very popular in the combustion literature and a number of them have been discussed (Bray, 1980; Williams, 1985a; Borghi, 1988; Poinsot et al., 1991; Peters, 2000; Lipatnikov and Chomiak, 2002a; Chaudhuri et al., 2011), with the diagrams proposed by Borghi (1988) and Peters (2000) being the most widely used. A typical combustion regime diagram pursues two goals: (i) to reveal the physical mechanisms that control premixed turbulent combustion under particular conditions

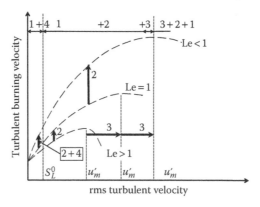

FIGURE 5.45 The influence of various physical mechanisms on the dependence of the turbulent burning velocity on the rms velocity and the Lewis number. The numbers identify the physical mechanisms enumerated at the beginning of the section "Regimes of Premixed Turbulent Combustion." The thick arrows show the changes caused by a decrease in Le, with the corresponding numbers identifying the governing physical mechanisms.

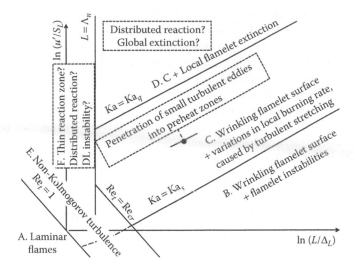

FIGURE 5.46   A regime diagram for premixed turbulent combustion.

(i.e., at particular u'/S_L^0 and L/δ_L, or Da and Ka, or Ka and Re_t, etc.) and (ii) to sketch the local structure of a premixed turbulent flame under these particular conditions. On the face of it, the two goals appear to be closely related, as a change in the governing mechanisms should somehow affect the local flame structure. However, the opposite claim is not necessarily true. For instance, as discussed in the section "Physical Reasoning: Thin-Reaction-Zone Regime," the penetration of small-scale eddies into preheat zones may change the local flame structure (thickened preheat zones), but weakly affect the burning rate, at least within the framework of the AEA theory. For these reasons, the diagram shown in Figure 5.46 and the subsequent discussion are solely aimed at revealing the governing physical mechanisms of premixed turbulent burning.

Because a typical combustion regime diagram is a tool for qualitative discussion, numerical factors of unity orders are commonly skipped therein, and the boundaries between different combustion regimes are usually associated with the equality of a criterion to unity, for example, $Re_t = 1$ or $u'/S_L^0 = 1$ or Ka = 1 or Da = 1, etc. Accordingly, the simple relations

$$Ka = \frac{Re_t^{1/2}}{Da} = \left(\frac{u'}{S_L^0}\right)^{3/2}\left(\frac{L}{\delta_L}\right)^{-1/2} = \left(\frac{u'}{S_L^0}\right)^2 Re_t^{-1/2} = \left(\frac{u'_\eta}{S_L^0}\right)^2 = \frac{\tau_c}{\tau_\eta} = \left(\frac{\delta_L}{\eta}\right)^2 = \left(\frac{\varepsilon\tau_c^3}{\delta_L^2}\right)^{1/2}$$

$$(5.112)$$

between various nondimensional numbers are invoked for a typical diagram, with the laminar flame thickness being determined by Equation 2.43 and $D_u = \nu_u$. We shall follow this common practice when discussing Figure 5.46.

If the turbulent Reynolds number is of unity order or less (see domain A in Figure 5.46), then combustion occurs in the laminar regime, the burning velocity scales as S_L^0, and the flame front is wrinkled only due to nonuniformities of the laminar flow and eventual DL and diffusive-thermal instabilities.

If Re_t is substantially larger than unity, then combustion occurs in the turbulent regime. In Figure 5.46, a weakly turbulent subregime bounded by the lines $Re_t = 1$ and $Re_t = Re_{cr}$ is shown to distinguish the ranges of the Kolmogorov turbulence ($Re_t > Re_{cr}$) and weak turbulence ($Re_t < Re_{cr}$) that do not follow the Kolmogorov scaling. The critical Reynolds number, Re_{cr}, is definitely greater than 100. For instance, a recent DNS study by Schumacher et al. (2007) of nonreacting, constant-density, homogeneous, isotropic turbulence, performed at a superfine numerical resolution, indicated deviations from the Kolmogorov scaling at the Taylor-microscale Reynolds number $Re_\lambda = u'\lambda/\nu \propto \sqrt{Re_t}$ as large as 100!

LARGE-SCALE WEAK AND MODERATE KOLMOGOROV TURBULENCE

Let us first consider premixed combustion in the Kolmogorov turbulence ($Re_t > Re_{cr}$) under conditions of $L \gg \Delta_L$, which are typical for many experiments. At low ratios of u'/S_L^0 (see domain B), the mean dissipation rate is low, the local stretch rates $\dot{s}_l \propto (\bar{\varepsilon}l^{-2})^{1/3}$ are also low, and the turbulent burning velocity is higher than S_L^0 due to an increase in the flamelet surface area by turbulent stretching and, probably, due to the DL and diffusive-thermal instabilities of flamelets. Accordingly, the dependence of U_t on u' is controlled by mechanisms 1 and 4 at weak turbulence, as sketched in Figure 5.45. An increase in U_t due to the latter (instability) mechanism is more pronounced for mixtures characterized by $Le < 1$ and $D_d > D_e$.

As turbulent stretch rates increase (either due to an increase in u' or due to a decrease in L and, hence, in η), they damp the instabilities and perturb the flamelets. Accordingly, in domain C, (i) the increase in the flamelet surface area and (ii) the perturbations in the local flamelet structure and burning rate by turbulent stretching appear to be the governing physical mechanisms of the influence of turbulence on flame propagation, while the instabilities do not seem to play a significant role. From the foregoing discussion, we may expect that the instabilities are of importance at low stretch rates $\tau_c\bar{s} < Ka_1$, while the flamelet perturbations play a significant role if $\tau_c\bar{s} > Ka_2$. In a general case, the two constants, Ka_1 and Ka_2, are not equal to each other, for example, the constant Ka_1 seems to depend more strongly on the density ratio. In Figure 5.46, for simplicity, $Ka_1 = Ka_2 = Ka_s$, where Ka_s is a constant (not Karlovitz number!) and the subscript s is associated with the terms "stable" and "stretched." Because the mean stretch rate is inversely proportional to the Kolmogorov timescale in homogeneous isotropic turbulence at high Reynolds numbers (see the sections "Energy Cascade and Turbulent Stretching" and "Kolmogorov's Theory of Turbulence" in Chapter 3), the product of $\tau_c\bar{s}$ scales as the Karlovitz number (see Equation 5.112). Consequently, the boundary between domains B and C is associated with the criterion $Ka = Ka_s$. Note that the constant Ka_s may depend on Le, D_F/D_O, the density ratio, etc. This constant appears to be markedly less than unity.

On the face of it, the critical number Ka_2 (or another critical number, e.g., Ka_q discussed later in the text) could be evaluated by simulating a strained laminar flame, for example, the flame sketched in Figure 5.20. However, due to the transient phenomena discussed in the section "Local Quenching of Premixed Combustion in Turbulent Flows," such simulations strongly overestimate the effect of turbulent stretching on a laminar flamelet. Accordingly, the numerical values of the critical numbers are poorly known and are not specified in this section.

It is also worth noting that a widely recognized criterion for assessing the role played by the instabilities in premixed turbulent combustion has not yet been elaborated. For instance, Boughanem and Trouvé (1998) proposed to use another criterion for characterizing the domain of the importance of the instabilities and this criterion is close to $u'/S_L^0 = O(1)$ under typical conditions (see Figure 2b in a review paper by Lipatnikov and Chomiak [2002a]).

Due to the lack of well-elaborated criteria, we may (especially if Le = 1 and $\bar{u}_c = S_L^0$ within the framework of the AEA theory of weakly stretched flames) imagine that neither the instabilities nor the flamelet structure perturbations substantially affect the turbulent burning velocity under certain conditions ($\text{Ka}_1 < \text{Ka} < \text{Ka}_2$). Therefore, there is a domain BC where the influence of turbulence on the burning rate is solely reduced to an increase in the flamelet surface area. Such a domain is not shown in Figure 5.46 for the sake of simplicity and because its existence and boundaries are not clear. It is also worth remembering that the equality of $\bar{u}_c = S_L^0$ is unlikely to hold in any practical turbulent flame characterized by a finite thickness of the reaction zone and by different molecular diffusivities of different species.

As the rms velocity is further increased (or L is decreased), local combustion quenching by high stretch rates (for the sake of brevity, here, the local combustion quenching means not only real extinction but also the inability of flamelets to enter the spatial regions characterized by high local stretch rates) begins to play an important role, simultaneously with an increase in the flamelet surface area and perturbations in the local burning rate. Thus, all these three physical mechanisms are of importance in domain D. Because $\bar{s} \propto \tau_\eta^{-1}$ and for the reasons discussed in the section "Strained Laminar Premixed Flames," we may expect that domain D is bounded by the line $\text{Ka} = \text{Ka}_q$ shown in Figure 5.46. The critical number Ka_q is not known, it depends on Le, D_F/D_O, σ, etc., and it appears to be substantially larger than unity for the reasons discussed in the section "Local Quenching of Premixed Combustion in Turbulent Flows."

Thus, domains B, C, and D in Figure 5.46 are bounded by straight lines associated with different constant values of the Karlovitz number. The key role played by Ka in determining the boundaries of different combustion regimes is fully consistent with the Kolmogorov theory of turbulence. Because chemical reactions that release heat are confined to thin zones if $\Delta_L \ll L$, the influence of turbulence on the zone area and local combustion rate is mainly controlled by eddies of scale $l \ll L$. Therefore, when assessing the role played by a physical mechanism of turbulent combustion under the conditions of $\text{Re}_t \gg 1$ and $\Delta_L \ll L$, the mean dissipation rate, which plays a crucial role in the Kolmogorov scaling at $l \ll L$, should be considered the most important characteristic of turbulence. In other words, the criteria discussed in this section should involve $\bar{\varepsilon} \propto u'^3/L$ (if $\Delta_L \ll L$), rather than another combination of u' and L. It is the Karlovitz number that satisfies this constraint.

The foregoing discussion explains why the domain E of weak turbulence is distinguished in Figure 5.46. In the Kolmogorov turbulence, the Karlovitz number has a clear physical meaning, as it characterizes the ratio of a Kolmogorov scale

$(u'_\eta, \eta,$ or $\tau_\eta)$ to the proper flame scale $(S_L^0, \delta_L,$ or $\tau_c,$ respectively). On the contrary, if $\mathrm{Re}_t < \mathrm{Re}_{cr}$, then the use of the Kolmogorov scaling and, in particular, the Karlovitz number, which is based on this scaling, is not justified in a general case. In other words, criteria written in the general form of $f(\mathrm{Ka}) = 0$ may be used to discuss changes in the governing physical mechanisms of premixed combustion in the Kolmogorov turbulence, while the use of the same criteria in the range of $\mathrm{Re}_t < \mathrm{Re}_{cr}$ may easily be put into question. Accordingly, the straight lines associated with different constant values of the Karlovitz number are only plotted in the range of $\mathrm{Re}_t > \mathrm{Re}_{cr}$ in Figure 5.46, whereas in the range of $\mathrm{Re}_t < \mathrm{Re}_{cr}$, these straight lines are replaced with dashed curves, with the criteria associated with these curves being unknown. To put it briefly, the same physical mechanisms appear to control the influence of turbulence on premixed combustion both at $\mathrm{Re}_t < \mathrm{Re}_{cr}$ and $\mathrm{Re}_t > \mathrm{Re}_{cr}$ (if $\Delta_L \ll L$ in the two cases), but the use of criteria that involve Ka is basically justified only in the latter domain.

COMPARISON WITH OTHER REGIME DIAGRAMS

As compared with alternative combustion regime diagrams, some widely used criteria are not shown in Figure 5.46. First, a line $u'/S_L^0 = O(1)$ is often plotted to highlight changes in the topology of a flamelet surface. If the flamelet speed with respect to the flow is high as compared with u', wrinkles in the flamelet surface, caused by turbulence, are rapidly consumed due to flamelet propagation. Therefore, an increase in the rms velocity is commonly associated with the complication of the surface topology from a slightly wrinkled surface at low u'/S_L^0 (see Figure 5.47a) to a tangled surface, which intersects a normal to the mean flame position at many points (see Figure 5.47b) at higher u'/S_L^0. Accordingly, domains above and below the line $u'/S_L^0 = O(1)$ are called the domain of "single reaction sheets" (Williams, 1985a) or "wrinkled flamelets" (Peters, 2000) and the domain of "multiple reaction sheets" (Williams, 1985a) or "corrugated flamelets" (Peters, 2000), respectively. This line, that is, the ordinate axis, is not highlighted as a boundary between two different combustion

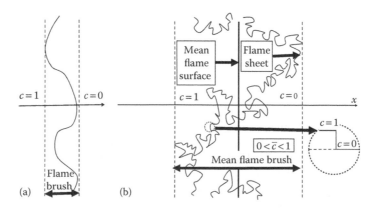

FIGURE 5.47 A sketch of a flamelet surface at (a) a low u'/S_L^0 and (b) a higher u'/S_L^0.

regimes in Figure 5.46, because the complication of a flamelet surface topology is not associated with the change in the governing physical mechanism(s) of the influence of turbulence on premixed combustion.

Second, the criterion of $\eta = \delta_L$ (or $Ka = 1$), used by Borghi (1988) and Peters (2000) to emphasize the penetration of small eddies into flamelet preheat zones, is not shown in Figure 5.46, because this phenomenon does not affect the local consumption rate within the framework of the AEA theory (see the section "Flame Wrinkling and Flamelet Regime"). Because Ka_s appears to be lower than unity, whereas Ka_q seems to be substantially larger than unity, the criterion of $\eta = \delta_L$ is associated with domain C in Figure 5.46 (see a piece of a straight line therein).

Third, a domain of global combustion extinction is drawn in certain combustion regime diagrams (Abdel-Gayed et al., 1989; Poinsot et al., 1990, 1991). However, although certain criteria for global combustion extinction have been proposed to be used either by processing particular experimental data (Abdel-Gayed and Bradley, 1985; Bradley et al., 2007) or by simulating a laminar flame–vortex interaction (Poinsot et al., 1990, 1991), the universality of these criteria may be put into question (e.g., see Section 3.1.2 in a review paper by Lipatnikov and Chomiak [2002a]). Accordingly, global combustion extinction is not addressed in Figure 5.46, and a determination of the conditions of the extinction is a subject for future studies.

Fourth, one more regime of highly turbulent premixed combustion, called "stirred reactors" by Williams (1985a), or "thickened flames" by Borghi (1988), or distributed reactions, was widely discussed by the combustion community after the pioneering work done by Damköhler (1940) and Summerfield et al. (1955). Such an eventual regime of premixed combustion in very intense turbulence is addressed in the next section.

LARGE-SCALE INTENSE TURBULENCE AND DISTRIBUTED REACTION REGIME

Let us begin our discussion on this eventual regime with the asymptotic case of $Da \to 0$. In such a case, the turbulence timescale is much shorter than the chemical timescale (see Equation 4.11), and the influence of turbulent pulsations on the mean reaction rates seems to be weak. Accordingly, the effect of turbulence on the burning velocity consist only of mixing intensification, provided that small-scale turbulent eddies are able to do so by penetrating into flamelets.

Then, Equations 1.58, 1.64 through 1.81, and 2.14 are assumed to model highly turbulent combustion in the distributed reaction regime, provided that (i) instantaneous quantities Y_k, h, T, c, ρ, W, etc., are substituted by their counterpart mean quantities $\tilde{Y}_k$, $\tilde{h}$, $\tilde{T}$, $\tilde{c}$, $\bar{\rho}$, $\overline{W}$, etc., respectively, and (ii) molecular transport coefficients are substituted by their counterpart turbulent transport coefficients. In particular, the AEA theory of laminar premixed combustion, considered in the section "AEA Theory of Unperturbed Laminar Premixed Flame" in Chapter 2, is assumed to describe a turbulent flame, provided that the laminar flame speed, S_L, and the molecular diffusivity, D, in Equation 2.14 and subsequent equations in the section "AEA Theory of Unperturbed Laminar Premixed Flame" in Chapter 2, rewritten for the mean quantities, are substituted with the turbulent burning velocity, U_t, and the turbulent diffusivity, D_t, respectively. Subsequently, applying Equation 2.38 to both laminar and turbulent flames, we finally arrive at the classical Damköhler (1940) expression

$$U_t \propto \sqrt{D_t/\tau_c} \propto S_L \sqrt{D_t/D} \propto S_L \sqrt{\mathrm{Re}_t}, \tag{5.113}$$

because $D_t \propto u'L$.

As far as the physical mechanisms of the influence of turbulence on the burning rate are concerned, the basic difference between the discussed regime of distributed reactions and the other combustion regimes previously considered consists of the following. In the former regime, the effects of turbulence on U_t are solely reduced to the intensification of mixing within thick inherently turbulent reaction zones, while the turbulent burning velocity is controlled by wrinkling, stretching, and quenching of thin, inherently laminar, reaction zones in the other regimes. Accordingly, U_t does not depend on the molecular transport properties in the regime of distributed reactions, but is sensitive to Le and D_F/D_O in the other combustion regimes, in line with the experimental data discussed in the section "Dependence of Turbulent Burning Velocity on Differences in Molecular Transport Coefficients" in Chapter 4. As far as thickening of the preheat zones by small-scale turbulent eddies is concerned, such phenomena may occur not only in the regime of distributed reactions but also in regimes D and C in Figure 5.46.

In summary, the following peculiarities of the distributed reaction regime are worth emphasizing.

- The influence of turbulence on the burning rate is solely reduced to mixing intensification.
- The scaling given by Equation 5.113 holds.
- The turbulent burning velocity depends neither on D_F/D_O nor on the Lewis number.

Moreover, because the influence of turbulent pulsations in temperature, density, and mass fractions on the mean reaction rates are assumed to be negligible, $\overline{W(T,\bar{Y}_k)} \to W(\tilde{T},\tilde{Y}_k)$ as Da $\to 0$; therefore, the maximum mean reaction rate reached in the considered turbulent flame should be close to the maximum mean reaction rate reached in the counterpart laminar flame, at least if Le $= 1$ and $D_F = D_O$ (otherwise the dependencies of $\tilde{Y}_k(\tilde{T})$ in the turbulent flame differ from the dependencies of $Y_k(T)$ in the laminar flame). Such an assumption is, in fact, invoked by the classical scaling given by Equation 5.113, where the same timescale τ_c characterizes both the laminar and turbulent premixed flames. On the contrary, in other combustion regimes discussed earlier, $\overline{W} = \gamma W_L \ll W_L$ if the probability γ of finding a thin, inherently laminar, reaction zone is much less than unity. Here, W_L is a characteristic value of the reaction rate in the reaction zone of the laminar flame. Note that $\rho_u U_t = \int_{-\infty}^{\infty} \overline{W} dx > \rho_u S_L^0 \propto W_L \delta_L$, because a turbulent flame brush is much thicker than a laminar premixed flame brush.

In the combustion regime diagram put forward by Borghi (1988), the weak influence of turbulent pulsations on the mean reaction rates is associated with Da ≤ 1.

Because, in this case, $Ka \geq \sqrt{Re_t} \gg 1$ (see Equation 4.13), the criterion of $Da \leq 1$ also implies that $\eta < \delta_L$ (see Equation 5.112), that is, the smallest turbulent eddies could penetrate into laminar flamelets. If such eddies are assumed to broaden the flamelets and transform them into thickened inherently turbulent reaction zones, then larger and larger eddies could penetrate into thicker and thicker reaction zones, thereby broadening them further and further. Due to such a penetration–broadening– penetration cascade, the reaction zones could finally reach a thickness of the order of the length scale of the largest eddies, and, therefore, turbulent eddies of all scales could intensify mixing within the thick zones.

Such a scenario is possible only if $Da \leq 1$. Indeed, if we assume that the AEA theory also describes a turbulent flame, provided that the laminar flame speed S_L and the molecular diffusivity D in Equations 2.36 through 2.43 are substituted with the turbulent burning velocity U_t and the turbulent diffusivity D_t, respectively, then Equations 2.38 and 2.43 result in $U_t \propto \sqrt{D_t/\tau_c}$ and $\delta_t \propto \sqrt{D_t \tau_c}$. Similarly, because the diffusivity associated with eddies smaller than a length scale l is $D(l) \propto (\bar{\varepsilon}l^4)^{1/3}$ within the framework of the Kolmogorov theory of turbulence (see the section "Kolmogorov's Theory of Turbulence" in Chapter 3), the contribution of turbulent eddies smaller than the length l to the burning velocity and flame thickness scales as $U(l) \propto \sqrt{(\bar{\varepsilon}l^4)^{1/3}/\tau_c}$ and $\delta(l) \propto \sqrt{(\bar{\varepsilon}l^4)^{1/3}\tau_c}$, respectively. If $Da > 1$, the latter scaling may be satisfied only if $l \propto \bar{\varepsilon}^{1/2}\tau_c^{3/2} \propto L Da^{-3/2}$ (Zimont, 1979; Ronney and Yakhot, 1992), because $\delta(l)$ should be of the order of l. If $\delta(l)$ were significantly larger than l, then turbulent eddies of larger scales would also contribute to the thickness. If $\delta(l)$ were significantly smaller than l, then turbulent eddies of the scale l would not affect the mixing within a spatial region of scale $\delta(l) < l$. Thus, if $Da > 1$, the foregoing penetration–broadening–penetration cascade mentioned previously is limited by a length scale of the order of $L Da^{-3/2}$. Significantly larger eddies cannot further broaden the reaction layers. Indeed, if the reaction layers were thicker, the heat release within them would exceed the heat flux from the layers (Zimont, 1979) (see Equation 2.33 and note that LHS and RHS scale as $l^{1/3}$ and $l^{2/3}$, respectively, if the molecular diffusivity D is substituted with the turbulent $D(l) \propto (\bar{\varepsilon}l^4)^{1/3}$).

Peters (1999, 2000) pointed out that, within the framework of the AEA theory, the laminar flame speed is controlled by the reaction rate and the molecular transport within a thin reaction zone. Accordingly, if (i) the influence of turbulence on combustion is solely reduced to mixing intensification by turbulent eddies and, subsequently, (ii) the AEA theory applies for a highly turbulent flame by assuming that $\overline{W(T,Y_k)} \to W(\tilde{T},\tilde{Y}_k)$ when $Da \to 0$, as carried out earlier, then, in order to affect U_t, the eddies should penetrate into the reaction zone, rather than the preheat zone. Accordingly, Peters (1999, 2000) put forward another criterion of $\eta = \delta_r$ (or $Ka = O(Ze^2)$, see Equation 5.9) to be the boundary of another combustion regime called the regime of "broken reaction zones" (it is worth remembering that the thickness of the reaction zone of a real laminar flame is less than but of the order of the thickness of its preheat zone, see Figures 2.5 and 2.6). Even if we assume that the penetration of the smallest eddies into the reaction zone results inevitably in reaction zones broadened by eddies of all scales due to the aforementioned penetration–broadening–penetration cascade, the regime of broken reaction zones is not equivalent to the regime of distributed reactions, because the influence of turbulent pulsations on the mean reaction rates may be substantial in the former regime. Indeed, if Ka = Ze^2, then Equation 4.13 yields Da = $\sqrt{\text{Re}_t}/\text{Ze}^2$, that is, the Damköhler number may be large, provided that the Reynolds number is very large as compared with the Zel'dovich number. In order for the two regimes to be equivalent, both the criterion of Da < 1 (Borghi, 1988) and the criterion of $\eta < \delta_r$ (Peters, 1999, 2000) should be satisfied. If $\eta < \delta_r$, but Da > 1, then the burning rate could be affected not only by mixing intensification but also by turbulent pulsations in temperature and composition.

Neither the regime of distributed reactions nor the regime of broken reaction zones are shown in Figure 5.46, because the importance (and even existence) of such regimes is an issue for a number of reasons discussed next.

First, turbulent eddies could not only thicken a reaction zone by penetrating into it but also stretch the zone and even quench it. It is not clear what physical mechanism (thickening of the reaction zone or local extinction, if any) dominates and under what conditions, but the data available today favor stretching rather than thickening of reaction zones.

Second, as discussed in the section "Local Quenching of Premixed Combustion in Turbulent Flows," laminar flames strongly resist the penetration of small-scale vortexes. Moreover, even in the case of a constant density, the lifetime of eddies of a length scale equal to δ_L may be too short for them to substantially perturb laminar flamelets. Indeed, in the Kolmogorov turbulence, the lifetime of such eddies scales $\bar{\varepsilon}^{-1/3}\delta_L^{2/3}$ and is shorter than τ_c if Ka > 1 (see Equation 5.112), whereas all eddies are larger than a flamelet thickness if Ka < 1. The experimental data by Kobayashi et al. (2002, 2005) and Ichikawa et al. (2011), discussed in the section "Instabilities of Laminar Flamelets in Turbulent Flows," see Figure 5.43a in particular, indicate that turbulent eddies do not wrinkle a flamelet surface if their length scale is smaller than the length scale Λ_m determined by mixture composition. Because $\Delta_L < \Lambda_m$, eddies smaller than the laminar flame thickness are not able to perturb flamelets either, at least under the conditions of these experiments.

To the best of the author's knowledge, statistically significant thickening of reaction zones by small-scale turbulent eddies was never documented experimentally in premixed turbulent flames characterized by a substantial local burning rate, but there is experimental evidence that reaction zones remain thin in intense turbulence (see Figures 5.15 and 5.16 and a review paper by Driscoll [2008]).

The 2D images of the local heat release rate (more precisely, the product of concentrations of OH and CH$_2$O), reported recently by Dunn et al. (2010, Figure 5, bottom row, left and middle columns), might be somehow associated with "some broadening of the reaction zones" in intense turbulence. However, this phenomenon does not seem to contribute substantially to the total burning rate in the flames studied in the cited paper, as the reaction rate in these zones is much lower than the reaction rate in thin zones shown for the same flame in the left and middle columns in the bottom row of Figure 4 in the cited paper.

Although the data by Dunn et al. (2010) show certain features commonly associated with the distributed reaction regime of premixed turbulent combustion, for example, low spatial gradients of the instantaneous temperature reported in Figure 10 in the cited paper, the strong effect of turbulence on the chemistry (in particular, a drop in the product of concentrations of OH and CO and a drop in the product of concentrations of OH and CH_2O, reported in Figures 6 and 7 therein) does not support the classical concept of the distributed reaction regime (Damköhler, 1940; Summerfield et al., 1955). The point is that, contrary to the experimental data discussed earlier, the concept assumes a weak influence of turbulence on the peak values of the mean reaction rates, as discussed at the start of this section. Therefore, the data by Dunn et al. (2010) appear to indicate a new regime of strongly weakened premixed combustion in very intense turbulence, which definitely needs further study.

All in all, more experimental data obtained from various highly turbulent premixed flames are necessary to draw far-reaching conclusions about regimes of premixed burning in very intense turbulence. Based on the available data, one may hypothesize that if (i) an integral length scale of turbulence is substantially larger than the laminar flame thickness Δ_L evaluated using Equation 2.41 and (ii) the Damköhler (Karlovitz) number is reduced (increased) by increasing u' with L being kept constant, then combustion quenching occurs well before the reaction zones become thick and inherently turbulent due to penetration of small-scale eddies. However, if $L < \Delta_L$, then local combustion quenching due to turbulent stretching appears to play a minor role, because eddies should be larger than a flamelet in order to stretch it. Accordingly, if $L < \Delta_L$, then the penetration of eddies into the reaction zones could play a role in very intense turbulence.

Indeed, recently, Aspden et al. (2008, 2010, 2011a,b,c) reported the results of a very interesting DNS study on the influence of small-scale $(L < \Delta_L)$, very intense $(u'/S_L^0$ up to 244, see flame H31 in Table 1 in Aspden et al. [2011a]) turbulence on premixed flames. Although the Karlovitz number evaluated as $Ka^2 = u'^3\Delta_L/(S_L^3 L)$ was much larger than unity in all these simulations, "the most turbulent flame does not appear to be burning in a distributed mode" (Aspden et al., 2011c). For instance, the lean $(\Phi = 0.31)$ H_2–air highly turbulent $(u'/S_L^0 = 107, L/\Delta_L = 0.5)$ flame D31 (Aspden et al., 2011c) clearly shows a thin reaction zone, a dependence of the local burning rate, temperature, and equivalence ratio on Le and D_F/D_O, including a superadiabatic temperature and an increase in the local equivalence ratio in the reaction zone (see the left Figure 5.48). Nevertheless, a few flames, for example, the richer $(\Phi = 0.40)$ hydrogen–air highly turbulent $(u'/S_L^0 = 107, L/\Delta_L = 0.5)$ flame D40 (Aspden et al., 2011c), show certain features associated with distributed burning. In particular, the temperature on the right side Figure 5.48 does not exceed the adiabatic combustion temperature, the equivalence ratio is randomly scattered around the oncoming value $\Phi = 0.40$, that is, the local structure of the flame is statistically similar to the local structure of a flame characterized by $D_F = D_O = a$, while the structure of the laminar flame (see dotted-dashed curve) is strongly different due to the significant difference in the molecular diffusivities of hydrogen and oxygen. These data imply that mixing in the reaction zone is controlled by turbulence and, thus, is sensitive neither to Le nor to D_F / D_O. However, even in such a case, the PDF of the local consumption velocity is sufficiently wide (see the dotted-dashed curve in Figure 5.49), with the probabilities of both $u_c > S_L^0$ and $u_c < S_L^0$ being substantial. While the former inequality may be attributed to the mixing intensification by turbulent eddies, the substantial probability of $u_c < S_L^0$ implies that the influence of turbulence on premixed combustion

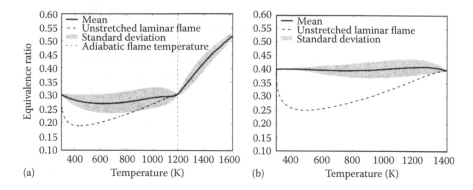

FIGURE 5.48 The conditional mean distribution of the local equivalence ratio as a function of temperature in lean hydrogen–air premixed flames characterized by $u'/S_L = 107$, $L/\Delta L = 0.5$, and either $\Phi = 0.31$ (a) or $\Phi = 0.40$ (b). (Reprinted from Aspden, A.J., Day, M.S., and Bell, J.B., *J. Fluid Mech.*, 680, 287–320, 2011. With permission.)

should not be solely reduced to mixing even at $Ka = \sqrt{u'^3 \Delta_L / (S_L^3 L)}$ as large as 410 and $Da = LS_L/(u'\Delta_L)$ as low as 0.015!

In summary, a domain associated with the regime of distributed reaction is not shown in Figure 5.46 due to the lack of experimental evidence of this regime and the lack of a well-recognized criteria for determining if the conditions under the regime are possible. In addition to criteria such as $Ka > Ka_{cr}$ or $Da < Da_{cr}$, an extra restriction such as $\Delta_L/L > 1$ appears to be necessary in order for the mixing intensification by small-scale turbulent eddies penetrating into the reaction zones to manifest itself before combustion is quenched by high stretch rates. The DNS data by Aspden et al. (2011a,b,c), obtained in the case of $L < \Delta_L$, imply that the ratio of u'/S_L should be very large in order for turbulence to be able to change the structure of the reaction zones.

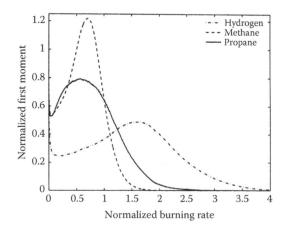

FIGURE 5.49 The normalized probability density functions of the normalized burning rate weighted by the burning rate in the laminar flame. (Reprinted from Aspden, A.J., Day, M.S., and Bell, J.B., *Proc. Combust. Inst.*, 33, 1473–1480, 2011. With permission.)

SMALL-SCALE TURBULENCE

The combustion regimes discussed earlier (domains A, B, C, and D) are widely recognized and are relevant to laboratory premixed turbulent flames characterized by a large ratio of $L/\delta_L \gg 1$. What happens as the integral length scale L is decreased and becomes comparable with the laminar flame thickness? The contemporary knowledge does not allow us to confidently answer this question.

In Figure 5.46, a boundary of small-scale premixed turbulent combustion is associated with the vertical line $L = \Lambda_n$, although alternative criteria, for example, $L = \Delta_L$, should also be borne in mind. The use of the neutral wavelength Λ_n in the former criterion is based on the experimental data by Kobayashi et al. (2002, 2005), Soika et al. (2003), and Ichikawa et al. (2011), discussed in the section "Instabilities of Laminar Flamelets in Turbulent Flows" (e.g., see Figure 5.43a). Because these data imply that turbulent eddies smaller than Λ_n cannot wrinkle a flamelet surface, one may assume that (i) flamelets resist to small-scale turbulent stretching and (ii) turbulence does not increase the area of a flamelet surface if $L < \Lambda_n$. In such an extreme case, the influence of turbulence on combustion is solely reduced to eventual intensification of mixing within flamelets, as discussed in the previous subsection. Moreover, if $L < \Lambda_n$, then turbulent eddies cannot damp the DL instability by flattening out a flamelet surface. Accordingly, the instability may also play a role in the considered case.

The aforementioned reasoning supports the use of the neutral wavelength in a criterion of small-scale premixed turbulent combustion. However, neglecting the effects of small-scale eddies on the flamelet surface area oversimplifies the problem. For instance, the DNS data by Aspden et al. (2011a,b,c), discussed in the previous subsection, show that turbulent eddies stretch and wrinkle thin, inherently laminar, reaction zones even if L is lower than not only Λ_n but also Δ_L. Moreover, even if turbulent eddies are too small to damp the DL instability if $L < \Lambda_n$, they can provide a stretch rate that is much higher than the maximum growth rate of the instability. In such a case, the eventual increase in the flamelet surface area due to the instability seems to play a minor role as compared with the increase in the area of the reaction zone due to turbulent stretching.

Therefore, it is not yet clear under what conditions turbulent eddies smaller than the neutral wavelength can wrinkle the flamelet (or reaction zone) surface. Such a criterion has to be elaborated, but it is likely to involve not only the ratio of L/Λ_n but also the normalized mean dissipation rate $\bar{\varepsilon}\tau_c^3/\delta_L^2 = Ka^2$, with smaller wrinkles being associated with a larger Ka. The parameterizations by Poinsot et al. (1990, 1991) and Roberts et al. (1993), discussed in the section "Local Quenching of Premixed Combustion in Turbulent Flows," do involve Ka, but they underestimate the smallest length scale e_i measured by Kobayashi et al. (2002, 2005) under conditions of $Ka > 1$ (see Figure 10 in a paper by Shim et al. [2011]).

Thus, in the case of $L < \Lambda_n$, the governing physical mechanisms of premixed turbulent combustion have not yet been established. Different scenarios varying from wrinkling and stretching of thin, inherently laminar, reaction zones to the lack of any effect of small-scale turbulent eddies on flamelets are admissible. Moreover, evaluation of Λ_n in the aforementioned criterion is also a challenge due to the shortage of experimental data and substantial differences in the values of the neutral wave-length, resulting from different theories of perturbed laminar premixed flames (e.g., see Figures 8 and 9 in a review paper by Lipatnikov and Chomiak, 2005a). Further research into these issues is definitely required.

Certainly, the considered case is somewhat hypothetical, because $L > \Lambda_n$ in a typical laboratory premixed turbulent flame. However, the integral length scale of turbulence is less than or of the order of the neutral wavelength in the vast major-ity of DNS studies of premixed turbulent combustion. Accordingly, care should be taken when invoking such DNS results in order to understand the governing physical mechanisms of premixed turbulent combustion in a laboratory or an engine flame. Even if the Karlovitz numbers are the same in the two (DNS and experiment) cases and $Re_t \gg 1$, the governing physical mechanisms may be different depending on the ratio of L/Λ_n. Moreover, care should be taken when interpreting the conditions of such a DNS in terms of the criteria proposed to discuss combustion regimes at $\Delta_L \ll L$. For instance, the criterion $\delta_L/\eta = O(1)$ or Ka = const is often invoked to assess whether or not the smallest-scale turbulent eddies are sufficiently small in order to enter the preheat zones of flamelets (Borghi, 1988). However, this criterion has no physical value if $\Delta_L > L$, because even large-scale turbulent eddies can enter the preheat zone in this case.

FLAME PROPAGATION ALONG VORTEX TUBES

To conclude our discussion on the physical mechanisms that may control the influ-ence of turbulence on combustion, it is worth considering one more phenomenon, which is beyond the mainstream work but could play a crucial role in premixed turbulent combustion, as hypothesized by Chomiak (1970, 1977, 1979). The phenom-enon is associated with eventual rapid flame propagation along the fine vortex tubes (vortex filaments) that exist in turbulent flows, as briefly mentioned in the section "Internal Intermittency and Vortex Filaments" in Chapter 3 (see Figure 3.8 therein).

FLAME PROPAGATION ALONG A LAMINAR VORTEX

The phenomenon of rapid flame propagation along the axis of a laminar vortex is well known, following the pioneering experiments of McCormack et al. (1972). This phenomenon was studied in a couple of papers reviewed elsewhere (Ishizuka, 2002; Kadowaki and Hasegawa, 2005; Lipatnikov and Chomiak, 2005a). Here, we will restrict ourselves to a brief summary of the key results relevant to premixed turbulent combustion.

McCormack et al. (1972) experimentally investigated flame propagation around a laminar vortex ring (see Figure 5.50a) and found that the speed, V_f, of the flame tip is linearly proportional to the vortex strength, $\Gamma = W_m d_c$, and is much higher than the laminar flame speed. Here, W_m is the maximum tangential velocity and d_c is the vortex core diameter. A similar phenomenon was documented in a number of recent measurements with vortex rings (e.g., see Figures 5.51 and 5.52) and the straight vortexes sketched in Figure 5.50b.

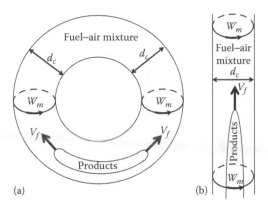

(a)   (b)

FIGURE 5.50   A sketch of flame propagation (a) around a laminar vortex ring and (b) along the axis of an elongated vortex.

Figure 5.52 also indicates that the bending and leveling-off of the $V_f(\Gamma)$ curves occurs at higher vortex strengths followed by flame quenching if the vortex ring is sufficiently strong (Asato et al., 1997; Ishizuka et al., 2002).

The magnitude, W_q, of the tangential velocity associated with quenching depends on the mixture composition and, in particular, on the differences in the molecular transport coefficients. For instance, Ishizuka et al. (2002) have found that rich $(D_d > D_e$ and Le $< 1)$ C_3H_8–air $(D_F > D_O)$ flames are more resistant to quenching than lean flames $(D_d < D_e$ and Le $> 1)$. In the former mixture, W_q is higher by a factor of 2, whereas the laminar flame speed is higher in the lean mixture. As a result, although flame speeds measured at $W_m < 3$ m/s are roughly the same in both mixtures, the maximum flame speed $V_{f,\max} = V_f(W_q)$ is much higher in the rich mixture.

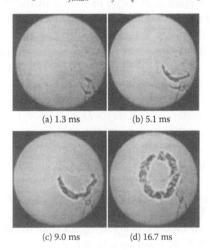

(a) 1.3 ms   (b) 5.1 ms

(c) 9.0 ms   (d) 16.7 ms

FIGURE 5.51   (a–d) High-speed shadowgraphs of the propagation of a laminar lean $(\Phi = 0.208)$ hydrogen–air flame around a vortex ring, recorded at different instants after spark ignition of the mixture inside the vortex. (Reprinted from Asato, K., Wada, H., Hiruma, T., and Takeuchi, Y., *Combust. Flame*, 110, 418–428. 1997. With permission.)

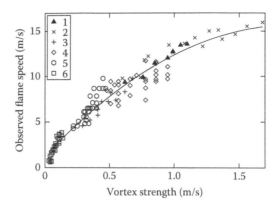

FIGURE 5.52 The speeds of flame propagation around vortex rings versus the vortex strength. 1, data obtained by McCormack et al. (1972); 2 and 3, data obtained by Asato et al. (1997) at $D_0 = 70$ and 100 mm, respectively; 4, 5, and 6, data obtained by Ishizuka et al. (1998) at $D_0 = 30, 40$, and 60 mm, respectively. D_0 is the diameter of a vortex ring. The curve approximates all data with a second-order polynomial. (Adapted from Lipatnikov, A.N. and Chomiak, J., *Prog. Energy Combust. Sci.*, 31, 1–73, 2005. With permission.)

Such behavior is associated with an increase (decrease) in the curved flame temperature if Le < 1 (Le > 1). For instance, recent numerical simulations (Nishioka and Ogura, 2005) of the propagation of a lean ($\Phi = 0.5$) hydrogen–air flame along the axis of a decaying Burgers vortex have shown that, at the tip of the flame (see Figure 5.50b), the local temperature is as high as 2100 K at $W_m = 10$ m/s, whereas the adiabatic combustion temperature is equal to 1652 K for this mixture, that is, almost 450 K lower! Such temperature variations can strongly increase the local burning rate. In particular, the displacement speed of the tip of the flame (i.e., the speed of the tip with respect to the flow in that point) is much higher than S_L^0 and is increased by the rotation velocity (Nishioka and Ogura, 2005).

Schlieren images of flames in vortex rings further support the influence of D_d on the flames. The images reported by Asato et al. (1997) and Ishizuka et al. (2002) show that the diameter, d_f, of a flame in a vortex ring (i) is less than the vortex diameter, (ii) decreases when W_m increases, and (iii) depends substantially on D_d/D_e and Le. For instance, the diameters of rich ($D_d > D_e$ and Le < 1) propane–air ($D_F > D_O$) flames are substantially smaller than those of lean propane–air flames ($D_d < D_e$ and Le > 1) and the tips of the former flames are very acute (see Figure 5.53). Therefore, the tip of a laminar flame characterized by Le < 1 can be highly curved in a vortex ring, as well as in other rotating flows (Ishizuka, 1990). For instance, Sakai and Ishizuka (1996) investigated the propagation of laminar flames in a rotating tube. The results (see Figure 5.54) show that the slope of the $V_f(W_m)$ curve measured for a rich ($\Phi = 2.0$) propane–air mixture is much higher than the dV_f/dW_m for a lean mixture ($\Phi = 0.65$), although S_L^0 is lower by a factor of 3 in the former mixture and the density ratios are almost equal in both mixtures.

The first explanation on rapid flame propagation around a laminar vortex ring was given by Chomiak (1977, 1979), who pointed out that the flame is strongly accelerated by the pressure difference between the hot products and the cold unburned

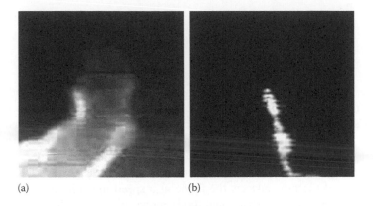

(a)   (b)

FIGURE 5.53   Schlieren images of the tips of (a) lean ($\Phi = 0.7$) and (b) rich ($\Phi = 2.2$) propane–air flames propagating around laminar vortex rings. (Adapted from Ishizuka, S., Ikeda, M., and Kameda, K., *Proc. Combust. Inst.*, 29, 1705–1712, 2002. With permission.)

mixture, which results from the reduction of the swirling motion due to the volume expansion of the products (the so-called bursting).

Indeed, the Navier–Stokes equation written for the azimuthal velocity w in the cylindrical coordinate framework reads

$$\rho \frac{w^2}{r} = \frac{\partial p}{\partial r} \tag{5.114}$$

in the simplest case of an inviscid flow with zero radial velocity. Integration of this equation from $r = 0$ to $r = \infty$ yields

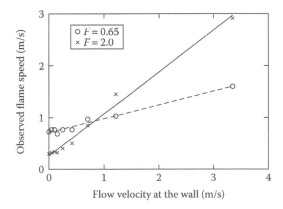

Flow velocity at the wall (m/s)

FIGURE 5.54   The observed speeds of laminar propane–air flames propagating in a rotating tube versus the azimuthal flow velocity at the wall of the tube. The symbols show the experimental data obtained by Sakai and Ishizuka (1996) with the equivalence ratio specified in the legends. The straight lines approximate the data.

$$p_\infty - p_0 = \int_0^\infty \rho \frac{w^2}{r} dr > 0, \qquad (5.115)$$

where the symbols p_0 and p_∞ designate the pressure at the vortex axis and at a far distance from the axis, respectively. Since (i) the pressure at far distances from the vortex axis should be independent of flame propagation along the vortex and (ii) the difference $p_\infty - p_0$ is larger in the unburned mixture than in the burned mixture, because $\rho_u > \rho_b$, then the pressure $p_{0,b}$ in the burned mixture on the vortex axis is higher than $p_{0,u}$. This pressure difference substantially accelerates the flow in a vortex ring, and the flow advects the flame at a speed much higher than S_L^0. It is worth remembering that, despite rapid flame propagation in the laboratory coordinate framework, the flame speed with respect to the local flow of unburned mixture is still of the order of S_L^0, at least if $D_F = D_O$ and Le = 1.

The model developed by Chomiak (1977, 1979) yields

$$V_f \propto W_m \sqrt{\sigma}. \qquad (5.116)$$

It is worth stressing that the discussed effect is possible only if the density of burned gas is lower than the density of unburned mixture, that is, $\sigma > 1$.

Over the past decade, different authors have studied the laminar flame propagation along a vortex axis, but a model that predicts all the effects observed in the measurements has not yet been elaborated. For this reason, we will not discuss such models, but refer interested readers to a recent review paper by Ishizuka (2002).

In summary, in vortexes and other rotational flows:

- A laminar flame can move at a speed controlled by the rotational velocity and much higher than the unperturbed laminar flame speed. Such rapid (in the laboratory framework) flame propagation is caused by the pressure jump ($p_{0,b} > p_{0,u}$) due to swirl and the density drop ($\rho_u > \rho_b$). The pressure jump pushes the gas from the burned to the unburned side of the flame and accelerates the flow in this direction. It is worth remembering that the opposite inequality $p_b < p_u$ holds in an unperturbed laminar flame.
- When the rotational velocity is increased, the maximum observed speed of flame propagation is limited by flame quenching and is substantially affected by the differences in the molecular transport coefficients. A flame characterized by a higher diffusivity of the deficient reactant can survive in a stronger vortex due to local enhancement of burning in the highly curved tip of the flame.

VORTEX TUBES IN TURBULENT FLOWS AND PREMIXED TURBULENT FLAMES

In nonreacting turbulent flows, vortex tubes (see Figure 3.8) have been documented in a number of experimental and DNS studies, as discussed in Section 7.3.2 in a review paper by Lipatnikov and Chomiak (2005a). In particular, the DNS study by

Jiménez and Wray (1998) has shown that (i) the mean diameter of the vortex tubes scales as the Kolmogorov length scale η, (ii) the maximum tangential velocity in the tube cross-section scales as u', and (iii) the tube length can be as large as an integral turbulence length scale L. The aforementioned scaling laws characterize averaged quantities, whereas the instant characteristics of vortex tubes are distributed in a wide domain. For instance, Jiménez and Wray (1998) have shown that the tangential velocity in vortex tubes may be as high as 2.5 u'.

Considering the existence of vortex tubes in inert turbulent flows and the well-documented rapid flame propagation of a premixed flame along the axis of a laminar vortex, it is tempting to hypothesize that rapid propagation of a flamelet along the axis of a vortex tube may (i) control the speed of the leading edge of a turbulent flame brush, (ii) increase the production of the flamelet surface area, and (iii) therefore, substantially affect the turbulent burning velocity under certain conditions. Due to the sensitivity of a laminar flame within a vortex to differences in the molecular transport coefficients, the outlined physical mechanism seems to be particularly attractive to explain the strong effect of Le and D_F/D_O on the turbulent burning velocity (see Figures 4.20 and 4.21).

It is worth stressing that vortex tubes are addressed neither by the Kolmogorov theory of turbulence nor by the Batchelor analysis of the increase in a surface area in a turbulent flow. The turbulence addressed by the Kolmogorov theory is sometimes called background turbulence. Therefore, the physical mechanisms discussed in the previous subsections are relevant to background turbulence. The most widely used statistical characteristics of isotropic homogeneous turbulence (the rms turbulent velocity, dissipation rate, energy, length scale and timescale, etc.) are almost identical to the characteristics of background turbulence, because vortex tubes and other coherent fine structures occupy a very small fraction (of the order of Re_t^{-1}, see Jiménez and Wray [1998]) of the fluid volume in a turbulent flow. For this reason, almost all models of premixed turbulent combustion deal with the Kolmogorov (or background) turbulence. Nevertheless, because of the difference in the densities of the unburned and burned mixtures, vortex tubes may play a role in premixed turbulent combustion. Rapid flamelet propagation along vortex tubes may create leading points that pull the entire flame brush. A hypothesis that the leading points control the turbulent flame speed and burning velocity was put forward by Zel'dovich and Frank-Kamenetskii (1947) and was further developed by Kuznetsov and Sabel'nikov (1990), but they did not associate the leading points with vortex tubes. The concept of leading points will be discussed further in the section "Concept of Leading Points" in Chapter 7.

Although rapid flame propagation along vortex tubes was suggested to be a governing physical mechanism of premixed turbulent combustion more than three decades ago (Chomiak, 1970), little is known about the role played by such structures in turbulent flames. Some indirect experimental evidence of vortex tubes in turbulent flames has been pointed out in a few papers (Chomiak, 1977; Abdel-Gayed et al., 1979; Yoshida et al., 1992; Buschmann et al., 1996; Kobayashi et al., 1997; Upatnieks et al., 2004). In particular, small-scale elements of a flame front, convex to the unburned mixture, have been reported to move at a speed markedly higher than u' or S_L^0 (Kobayashi et al., 1997) in weakly turbulent flames under elevated pressures.

Upatnieks et al. (2004) also documented a similar phenomenon. Such a rapid propagation of flamelets is consistent with Equation 5.116, which yields $V_f \approx 2.8\ W_m \approx 7\ u'$ if $\sigma = 8$ and $W_m = 2.5\ u'$.

Recently, Soika et al. (2003) published a 2D OH-LIPF image that looks like a vortex tube (see Figure 5.55). Nevertheless, decisive experimental data that prove an important role played by vortex tubes in premixed turbulent combustion have not yet been reported, probably because the small radii of the strongest vortex tubes impede resolving them in measurements.

It is worth noting that the small radii of vortex tubes and the considerable viscous dissipation associated with such small-scale structures do not invalidate the hypothesis, as regards the important role played by the tubes in turbulent flame propagation. First, in constant-density flows, vortex tubes survive a long time of the order of the timescale of the largest structures of the flow (Villermaux et al., 1995), because the dissipation is balanced by the energy flux from background turbulence due to stretching of the tubes. Vortex tubes most often disappear by catastrophic breaking events and due to merging with other tubes, but not by viscosity (Vincent and Meneguzzi, 1991). Since $V_f \propto u'$, a laminar flame can propagate a long way (of the order of L) in a vortex tube during the tube lifetime that scales as τ_t.

Second, in flames, viscosity increases in high-temperature products, and this process, as well as thermal expansion, might shorten the lifetime of vortex tubes. However, a recent 3D variable-density DNS study (Tanahashi et al., 2002) has indicated that vortex tubes survive in combustion products. Simulations of laminar flame propagation along the axis of an isolated Burgers vortex have shown that, despite the viscous decay of the vortex, the flame propagates a long way (at least 10 mm) before the decay begins to markedly affect the process (Nishioka and Ogura, 2005).

The foregoing discussion allows us to outline the following physical scenario (see also Section 7.4 in a review paper by Lipatnikov and Chomiak [2005a]).

First, a laminar flamelet that penetrates into a vortex tube, which exists in a turbulent unburned mixture, propagates along the tube at a high speed, V_f, of the order

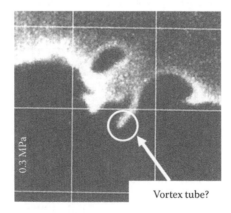

Vortex tube?

FIGURE 5.55 An OH-LIPF image of a lean ($\Phi = 0.7$) methane–air turbulent flame at $p = 0.3$ MPa. (Adapted from Soika, A., Dinkelacker, F., and Leipertz, A., *Combustion and Flame*, 132, 451–462, 2003. With permission.)

of the maximum tangential velocity, W_m, and much higher than the laminar flame speed. For any mixture, the highest local speed of the propagation, $V_{f,max}$, is associated with critical conditions (the highest $W_m = W_q$ and the extremely curved tip of the flame, under which the flame barely survives inside the tube). Accordingly, $V_{f,max}$ depends strongly on Le and D_d/D_e, in particular, $V_{f,max}$ increases with decreasing Le and increasing D_d/D_e.

Second, because the basic characteristics of vortex tubes in a turbulent flow constitute a continuous spectrum (Jiménez and Wray, 1998), the aforementioned critical conditions are reached locally in certain vortex tubes for a wide range of u', L, and mixture characteristics. If the critical conditions are reached in a tube, the laminar flame moves fast along the tube and advances furthest into the unburned mixture, provided that the tube axis is not parallel to the mean flame surface. Accordingly, the physical mechanism discussed earlier (i) controls the speed of the leading edge of a turbulent flame brush, (ii) increases the production of the flamelet surface area, and (iii) therefore, substantially affects the turbulent burning velocity.

The aforementioned hypothetical scenario strongly needs validation and further development.

CONCLUDING REMARKS

The following two physical mechanisms are widely recognized to control the influence of turbulence on the burning velocity in premixed reactants;

1. An increase in the flamelet (or reaction zone) surface area by turbulent stretching increases the turbulent burning velocity. At small scales, this process is counterbalanced by mutual annihilation of self-propagating flamelets.
2. Variations in the local consumption velocity, caused by turbulent stretching, can either increase (Le < 1, $D_d/D_e > 1$) or decrease (Le > 1, $D_d/D_e < 1$) the burning velocity.

Moreover, local flamelet quenching by strong turbulent stretching and the inability of flamelets to move into the spatial regions characterized by locally high stretch rates could reduce the burning velocity, with the effect being more pronounced for mixtures characterized by Le > 1 and/or $D_d/D_e < 1$. Because such phenomena depend substantially on transient and geometrical effects and heat release destroys small-scale eddies, laminar flamelets in turbulent flows are much more resistant to local combustion quenching than steadily strained laminar flames.

Furthermore, the flamelet surface area (and, hence, the burning velocity) in a turbulent flow may be increased due to the hydrodynamic and diffusive-thermal instabilities, with the role played by the instabilities being damped by turbulent stretching.

The conditions under which the aforementioned four physical mechanisms may play a role are summarized in Figures 5.45 and 5.46.

The mixing intensifications within the reaction zones by turbulent eddies that penetrate into the zones could also increase the burning velocity, but such phenomena were observed only by a single group in a recent DNS study on premixed combustion in very intense turbulence characterized by an integral length scale smaller than the laminar flame thickness (Aspden et al., 2011c).

Rapid propagation of flamelets along the axes of vortex tubes that exist in a turbulent flow can also increase the burning velocity, but this hypothesis has not yet been proved.

The aforementioned list of physical mechanisms of the influence of turbulence on premixed combustion may be incomplete. For instance, recent experimental data by Dunn et al. (2010) indicate a regime of strongly weakened premixed burning in intense turbulence, and the governing physical mechanisms of this regime are unclear.

Moreover, Chomiak (1977) argued that, due to the internal intermittency of turbulence, the speed of the reaction zones with respect to the mean flow may be controlled by the Kolmogorov velocity, rather than by S_L. Unfortunately, currently available experimental and DNS data are still insufficient to draw a solid conclusion about the validity of this seminal hypothesis.

Furthermore, certain 2D measurements of the flame surface density (Gülder, 2007; Gülder and Smallwood, 2007; Yuen and Gülder, 2009, 2010) indicate that Equation 5.1 significantly underestimates the burning velocity in intense turbulence. This observation may, in part, result from the application of 2D diagnostic techniques to 3D phenomena (Chen, 2009). Moreover, the flame surface density was evaluated by measuring the area of an isoscalar surface, for example, $c = 0.5$ (Yuen and Gülder, 2009, 2010), which is associated with the preheat zone, whereas the burning rate is controlled by the area of the reaction zone surface, which may be more wrinkled. For instance, a comparison of Table 3 and Figure 14 published by Tanahashi et al. (2008) with Figure 7 presented by Yuen and Gülder (2009) indicates that the lowest radii of the local curvature of the reaction zone, reported in the former paper, are significantly smaller than the radii of the local curvature of the flame surface obtained in the latter paper (note that different flames were investigated by the two research groups). Nevertheless, the data discussed by Gülder et al. may also indicate limitations of the thin-reaction-zone concept of premixed combustion in intense turbulence. Further study on this issue is definitely necessary.

All in all, although substantial progress has been made in clarifying the governing physical mechanisms of the influence of turbulence on premixed flames, a number of issues still challenge the combustion community, especially as far as burning in intense turbulence is concerned.

6 Influence of Premixed Combustion on Turbulence

When a premixed flame propagates in a turbulent flow, the turbulence affects combustion and the combustion, in turn, affects the turbulence. Chapters 4 and 5 were restricted to the phenomenology and physical mechanisms of the former effects, while the focus of this chapter is on the phenomena and physical mechanisms associated with the influence of combustion on turbulence and turbulent transport.

A premixed flame affects a turbulent flow, because the density drops by six to seven times within a typical flame. As far as an instantaneous thin flame front is concerned, local velocity gradients caused by the heat release are very high. Even if averaging over random orientation of the instantaneous flame fronts reduces the magnitude of such effects, they still play an important role when mean quantities are addressed.

In principle, any heat source (or sink) affects a turbulent gas flow by changing the density of the gas. However, the fact that strong heat release and large density drops are typically localized to thin, inherently laminar, flame fronts makes the problem of studying the influence of the heat release in flames on turbulence particularly difficult, for example, as compared with the influence of distributed heating on a turbulent flow. To place the focus of consideration on the most challenging issue, this chapter addresses the mean influence of heat release localized to thin instantaneous flame fronts on a turbulent flow. Accordingly, the Bray–Moss–Libby (BML) approach (asymptotically valid as the probability γ of finding the intermediate states of a reacting mixture tends to zero, see the section "Bray–Moss–Libby Approach" in Chapter 5) is widely used in the discussion that follows.

The problem of the influence of a premixed flame on a turbulent flow may be divided into two subproblems. First, the heat release in flames may affect and change the key turbulence characteristics, such as the rms turbulent velocity u', mean dissipation rate $\bar{\varepsilon}$, and length scale and timescale, while flame-induced variations in u' can affect the turbulent burning velocity, which depends substantially on u', as discussed in the section "Dependence of Turbulent Burning Rate on RMS Velocity and Laminar Flame Speed" in Chapter 4. Such a feedback (heat release increases the magnitude of velocity pulsations, which increase the heat release rate) can cause self-acceleration of a turbulent flame. The hydrodynamic (or Darrieus–Landau) instability discussed briefly in the section "Hydrodynamic Instability" in Chapter 5, is the most well-known example of combustion self-acceleration due to flame wrinkling caused by a flame-induced increase in the magnitude of velocity perturbations. Even if the instability is commonly associated with laminar flames, similar phenomena may locally occur in a turbulent flame brush, that is, the instantaneous flame front may perturb the oncoming flow, which wrinkles the flame front.

Second, as far as the Reynolds-averaged balance equations in a nonreacting constant-density flow are concerned, the most important manifestation of turbulence is the fact that the turbulent transfer is significantly stronger than the molecular transfer. For instance, the components of the Reynolds stress tensor $\overline{u_i'' u_j''}$ are typically much larger than the counterpart components of the viscous stress tensor $\overline{\tau}_{ij}$ in the mean Navier–Stokes Equation 3.79, or the components of the turbulent scalar flux vector $\mathbf{u}' Y'$ are much larger than the counterpart components of the flux vector $\overline{D \nabla Y}$ due to the molecular diffusion in Equation 3.102. It is worth remembering that the Reynolds number, $\text{Re}_t = u'L/v_u \gg 1$, is of the order of the ratio of turbulent viscosity, v_t, or diffusivity, D_t, to the molecular transport coefficients, v_u or D. Accordingly, the effects of combustion on turbulent transfer are of substantial importance.

The influence of turbulence on combustion and vice versa may be illustrated by considering a balance equation for the Favre-averaged combustion progress variable. Using the method discussed in the section "Favre-Averaged Balance Equations" in Chapter 3, one can easily average the combustion progress variable balance Equation 1.79 and arrive at

$$\frac{\partial}{\partial t}(\overline{\rho}\tilde{c}) + \frac{\partial}{\partial x_k}(\overline{\rho}\tilde{u}_k \tilde{c}) = -\frac{\partial}{\partial x_k}\overline{\rho u_k'' c''} + \frac{\partial}{\partial x_k}\left(\overline{\rho D \frac{\partial c}{\partial x_k}}\right) + \overline{W} \tag{6.1}$$

or

$$\frac{\partial}{\partial t}(\overline{\rho}\tilde{c}) + \frac{\partial}{\partial x_k}(\overline{\rho}\tilde{u}_k \tilde{c}) = -\frac{\partial}{\partial x_k}\overline{\rho u_k'' c''} + \overline{W} \tag{6.2}$$

by neglecting the molecular diffusion term as compared with the divergence $\nabla \cdot \overline{\rho \mathbf{u}'' c''}$ of the turbulent scalar flux vector at large Reynolds numbers. Both terms on the right-hand side (RHS) of Equation 6.2 should be closed. Evaluation of the first turbulent transport term is associated with the modeling of the influence of combustion on turbulent transfer, that is, the second subproblem raised earlier.

Evaluation of the mean reaction rate term $\overline{W}$ is the central problem of the modeling of premixed turbulent combustion, which will be discussed in Chapter 7. Such a model yields an algebraic expression (or differential equation) for determining the normalized rate $\tau_t \overline{W}/\rho_u$ as a function of the density ratio and the Reynolds, Damköhler, Karlovitz, Lewis or Markstein, and Zel'dovich numbers, with the dependence of $\overline{W}$ on the rms turbulent velocity u' (or on the ratio of u'/S_L in dimensionless terms) being pronounced particularly strong. The problem of obtaining such an expression (or differential equation) for calculating $\tau_t \overline{W}/\rho_u$ is relevant to modeling the effects of turbulence on combustion, which will be discussed in Chapter 7, while the problem of

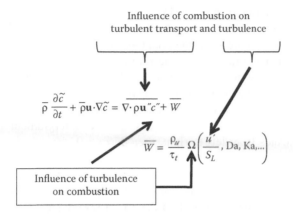

FIGURE 6.1 The influence of turbulence on combustion, and vice versa.

evaluating u' in the aforementioned expression is relevant to modeling the influence of combustion on turbulence, that is, the first subproblem raised earlier.

In summary, as far as Equation 6.2 is concerned, the influence of combustion on turbulence should be studied to (i) close the turbulent scalar flux $\overline{\rho u''c''}$ on the RHS and (ii) evaluate the turbulence characteristics (first of all, u') that are used by a model of the rate $\overline{W}$, whereas such a model itself deals with the influence of turbulence on combustion (see Figure 6.1).

This chapter is restricted to the former issue, that is, the influence of combustion on turbulence and turbulent scalar transport. Although the effects of heat release on u' (Williams et al., 1949; Karlovitz et al., 1951) drew the attention of the combustion community much earlier than the effects of combustion on the flux $\overline{\rho u''c''}$ (Clavin and Williams, 1979; Moss, 1980; Libby and Bray, 1981; Tanaka and Yanagi, 1981), we shall begin our discussion with the latter issue.

COUNTERGRADIENT SCALAR TRANSPORT IN TURBULENT PREMIXED FLAMES

STATEMENT OF THE PROBLEM

In a typical constant-density nonreacting turbulent flow, scalar transport is commonly modeled using the turbulent diffusion approximation (see Equations 3.103 and 3.104). Within the framework of such an approach, the turbulent diffusivity D_t is positive and $\overline{\mathbf{u}'Y'} \cdot \nabla \overline{Y} < 0$, that is, the turbulent flux $\overline{\mathbf{u}'Y'}$ points in the direction of decreasing mean mass fraction $\overline{Y}$ of the considered species or, in other words, turbulence transfers the species from regions characterized by a larger $\overline{Y}$ to regions characterized by a lower $\overline{Y}$ (the so-called gradient transport).

On the contrary, in premixed turbulent flames, the product $\overline{\rho u''c''} \cdot \nabla \tilde{c}$ may be positive, which is associated with the turbulent transport of combustion products from unburned to burned mixtures. The modeling of scalar transport using an expression similar to Equation 3.103 requires the invoking of negative turbulent diffusivity in such a case.

In the flamelet regime of premixed turbulent combustion, the BML expressions (see the section "Bray–Moss–Libby Approach" in Chapter 5) and, in particular, Equation 5.35 hold. Accordingly, the countergradient normal flux $\overline{\rho u''c''}$ is associated with a larger magnitude $|\overline{u}_b|$ of the normal velocity conditioned on a burned mixture as compared with the magnitude $|\overline{u}_u|$ of the normal velocity conditioned on an unburned mixture. Indeed, if a flame propagates from right to left, then $\partial \tilde{c}/\partial x > 0$, the velocity $\overline{u}_u$ is positive in the coordinate framework attached to the flame, and Equation 5.35 results in $\overline{u}_b > \overline{u}_u > 0$ in the case of the countergradient flux $\overline{\rho u''c''} > 0$. If a flame propagates from left to right, then $\partial \tilde{c}/\partial x < 0$, the velocity $\overline{u}_u$ is negative in the coordinate framework attached to the flame, and Equation 5.35 results in $\overline{u}_b < \overline{u}_u < 0$ in the case of the countergradient flux $\overline{\rho u''c''} < 0$.

Equation 5.35 with $|\overline{u}_b| > |\overline{u}_u|$ was first reported by Prudnikov (1960, 1967) in the 1960s, as discussed in detail in Section 2.2.3 of a review paper by Lipatnikov and Chomiak (2010). However, it took a further 20 years to draw the attention of the combustion community to the transport problem in premixed turbulent flames by the numerical study (Moss, 1980; Libby and Bray, 1981; Tanaka and Yanagi, 1981). Today, countergradient scalar transport in premixed turbulent flames is well documented in numerous experiments and direct numerical simulations (DNS), as reviewed in Section 3.3 of the aforementioned paper by Lipatnikov and Chomiak (2010).

SIMPLIFIED DISCUSSION OF GOVERNING PHYSICAL MECHANISMS

In order to better understand this phenomenon, let us first look at the classical problem of turbulent diffusion in a constant-density nonreacting flow from another perspective, which is typical for conducting research on premixed turbulent flames. If two mixtures are separated by an infinitely thin planar interface at $t = 0$ (e.g., $Y_1(x < 0, y, z, t = 0) = 1$, $Y_1(x > 0, y, z, t = 0) = 0$, $Y_2(x < 0, y, z, t = 0) = 0$, and $Y_2(x > 0, y, z, t = 0) = 1$ for simplicity), then molecular diffusion will increase the thickness of a molecular mixing layer (i.e., the region characterized by a finite product of local mass fractions $0 < Y_1 Y_2 < 1$), while turbulent eddies will convect the layer and wrinkle it. Initially, (i) the thickness, δ_D, of the molecular mixing layer scales as $\delta_D \propto \sqrt{Dt}$, (ii) the convection of the layer by turbulent eddies increases the thickness, δ_m, of the turbulent mixing layer (i.e., the region characterized by a finite product of the mean mass fractions, $0 < \overline{Y}_1 \overline{Y}_2 < 1$) as $\delta_m \propto u't$ (see the section "Turbulent Diffusion" in Chapter 3), and (iii) the area of the molecular mixing layer depends (increases) exponentially on t/τ_η (see the section "Energy Cascade and Turbulent Stretching" in Chapter 3). Let us neglect the thickness of the molecular mixing layer as compared with the thickness of the turbulent mixing layer, that is, consider the case of $\delta_D/\delta_m \ll 1$ for simplicity. This is admissible if, for example, $Sc \gg 1$. Then, the probability of finding the intermediate (between $Y_1 = 0$ and $Y_1 = 1$) states of the mixture may be assumed to be much less than unity everywhere within the turbulent mixing layer. In such a case, the BML expressions (see the section "Bray–Moss–Libby Approach" in Chapter 5) hold for the considered turbulent mixing layer also. In particular, Equations 3.103 and 5.35 read

$$-D_t \nabla \overline{c} = \overline{\mathbf{u}'c'} = \overline{c}\left(1 - \overline{c}\right)\left(\overline{\mathbf{u}}_b - \overline{\mathbf{u}}_u\right), \tag{6.3}$$

where $c = Y_2$ and the subscripts b and u designate mixtures 2 and 1 to draw an analogy with premixed turbulent combustion. Consequently, $\partial \bar{c}/\partial x > 0$ and $\bar{u}_b < \bar{u}_u$ in the case of turbulent mixing. If $t/\tau_\eta = O(1)$, then Equation 6.3 indicates that the magnitude of the so-called slip-velocity vector $\Delta \bar{u} \equiv \bar{u}_b - \bar{u}_u$ is of the order of $D_t/\delta_m \propto u'$ (see Equation 3.122).

To provide a physical explanation of $|\bar{u}_b| < |\bar{u}_u|$, let us consider a statistically planar, one-dimensional (1D) turbulent mixing layer (i.e., the layer bounded by dashed straight lines in Figure 6.2) in a coordinate framework selected so that the mean velocity $\bar{u}$ normal to the layer vanishes. If we draw a line $x = \text{const}$, it intersects the thin molecular mixing layer (thick solid wrinkled curve) at many points. At some intersection points, for example, point B, the local velocity u_B along the x-direction is positive and, therefore, $u_u > 0$ and $u_b > 0$ in the vicinity of point B. At other intersection points, for example, point A, the local velocity u_A is negative and, therefore, $u_u < 0$ and $u_b < 0$ in the vicinity of point A. However, statistically, the probabilities of these two events are likely to be different and depend on x. For instance, in order for a passive mixing layer to reach a line characterized by $\bar{c} > 0.5$ (or $x > 0$ if $x = 0$ is associated with the stationary mean position of the mixing layer, characterized by $\bar{c} = 0.5$), positive (like at point B) local (in the vicinity of the layer) velocities seem to be more probable. On the contrary, in order for the passive layer to reach a line characterized by $\bar{c} < 0.5$ (or $x < 0$), negative local velocities (like at point A) should prevail over positive local velocities. Accordingly, we may expect that $\bar{u}_b$ and $\bar{u}_f$ have the same sign at $x < 0$ (negative in Figure 6.2), while $\bar{u}_u$ and $\bar{u}_f$ have the same sign (positive) at $x > 0$. Here, the subscript f designates a quantity conditioned on the laminar mixing layer to draw an analogy with premixed combustion where $\bar{u}_f$ is the velocity conditioned on a flamelet. We consider $\bar{u}_f$ to be closer to $\bar{u}_b$ than to $\bar{u}_u$ at $x < 0$ (or closer to $\bar{u}_u$ at $x > 0$), because the probability of finding a burned gas is lower (larger) than the probability of finding an unburned gas at $x < 0$ ($x > 0$), and,

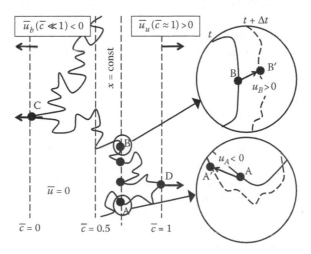

FIGURE 6.2 A simplified sketch of a molecular mixing layer (*thick solid wrinkled curve*) in a turbulent flow. Note that the surface of a real mixing layer is much more tangled, because small-scale turbulent eddies stretch the layer and increase its surface area.

hence, the events associated with crossing a line $x = \text{const}$ by the molecular mixing layer are statistically more significant for a burned mixture at $x < 0$ and for an unburned mixture at $x > 0$. For instance, the arrival of the molecular mixing layer at the leading edge of the turbulent mixing layer (see point C in Figure 6.2) does not affect $\bar{u}_u$ but controls $\bar{u}_b$, because, at the leading edge, the probability of finding $c = 0$ is much greater than the probability of finding the laminar mixing layer, while the latter probability is almost equal to the probability of finding $c = 1$. Finally, if $\bar{u}_b < 0$ at $x < 0$ and $\bar{u}_u > 0$ at $x > 0$, then $\bar{u}_u > 0$ at $x < 0$ and $\bar{u}_b < 0$ at $x > 0$, because

$$\bar{u} = \left(1 - \bar{c}\right)\bar{u}_u + \overline{cu}_b = \bar{u}_u + \bar{c}\left(\bar{u}_b - \bar{u}_u\right) = \bar{u}_b - \left(1 - \bar{c}\right)\left(\bar{u}_b - \bar{u}_u\right) \qquad (6.4)$$

and $\bar{u} = 0$ in the selected coordinate framework.

Thus, in the case of constant-density nonreacting mixing, turbulent pulsations tend to make the velocity $\bar{u}_u \geq 0$ conditioned on the "unburned" mixture larger than the velocity $\bar{u}_b \leq 0$ conditioned on the "burned" mixture, provided that the x-axis points to the burned mixture. Note that if the direction of the x-axis is reversed in Figure 6.2 and the mixture characterized by $c = 0$ is called burned mixture, then $\bar{u}_b \leq 0$ and $\bar{u}_u \geq 0$.

Such trends should be well pronounced at the leading and trailing edges of the turbulent mixing layer. Indeed, due to the well-known increase in the mean thickness δ_m of the layer, the leading (trailing) edge of the layer should statistically move to the left (right) and the velocity $\bar{u}_f$ conditioned on the molecular mixing layer should be negative (positive) at $\bar{c} \ll 1$ (or $\bar{c} \approx 1$) (see point C (or D) in Figure 6.2). Because (i) mixing does not affect the velocity field and (ii) the velocity u_b (or u_u) is always measured in the vicinity of the molecular mixing layer at the leading (trailing) edge of the turbulent mixing layer, we have $\bar{u}_b(\bar{c} \ll 1) \approx \bar{u}_f(\bar{c} \ll 1) < 0$ and $\bar{u}_u(\bar{c} \approx 1) \approx \bar{u}_f(\bar{c} \approx 1) > 0$. On the contrary, because the probability of observing the molecular mixing layer is very low at the leading and trailing edges, $\bar{u}_u(\bar{c} \ll 1) \approx \bar{u}(\bar{c} \ll 1) = 0$ and $\bar{u}_b(\bar{c} \approx 1) \approx \bar{u}(\bar{c} \approx 1) = 0$. Therefore, $\bar{u}_b < \bar{u}_u$ at both the leading and the trailing edges, in line with Equation 6.3. If we assume that the mean thickness $\delta_m \propto u't$ during an early stage of turbulent mixing (see the section "Taylor's Theory" in Chapter 3), then $\bar{u}_u - \bar{u}_b \propto u'$ at both the leading and the trailing edges.

What will change if we replace the molecular mixing layer with a flamelet in Figure 6.2? First, the flamelet self-propagates and, as a result, smoothes out the small-scale wrinkles of its surface. However, if we assume that $S_L \ll u'$ and $t = O(\tau_\eta)$, then the difference in the observed speeds of the molecular mixing layer and the flamelet will be minor, while the smoothing effect of flamelet propagation will not be pronounced, because the small-scale wrinkles do not have enough time to develop. Therefore, as far as the topology of the surfaces discussed earlier is concerned, flamelet propagation does not seem to play a substantial role under the aforementioned conditions, that is, $t = O(\tau_\eta)$. At $t \gg \tau_\eta$, the flamelet surface will be much smoother than the molecular mixing layer, that is, the thick solid wrinkled curve in Figure 6.2 sketches the flamelet surface better than the layer.

Second, due to the heat release and pressure drop in flamelets, the local velocity normal to the flamelet surface is increased from the unburned side to the burned side. Contrary to a molecular mixing layer, a flamelet affects the flow, as sketched in Figure 6.3. If, for the sake of simplicity, we neglect flamelet propagation and the influence of the heat release on the oncoming flow, then local instantaneous velocities may be assumed to be the same ahead of a molecular mixing layer (left) and flamelet (right). After averaging, the velocities $\bar{u}_u$ are positive in the two cases if we consider a point characterized by $x > 0$ and $\bar{c} > 0.5$ in Figure 6.2. Moreover, in the case of turbulent mixing, the local velocities $u_{b,f} = u_{u,f}$. In the combustion case, the instantaneous velocity, $u_{b,f}$, just behind a flamelet is not equal to the instantaneous velocity, $u_{u,f}$, just ahead of the flamelet. In particular, if a flamelet is locally normal to the x-axis, as shown in Figure 6.3, then $u_{b,f} = u_{u,f} + \tau S_L$ due to the pressure drop in the flamelet. After averaging, we obtain

$$\bar{u}_{b,f} \approx \bar{u}_{u,f} - \bar{n}_x \tau S_L, \tag{6.5}$$

where $\bar{n}_x$ is the x-component of the average unit vector $\mathbf{n}$ that is normal to flamelets and points to unburned mixture (mainly from right to left in Figure 6.3 so that $\bar{n}_x < 0$). In a constant-density flow, the heat release factor $\tau = \rho_u/\rho_b - 1 = 0$ and, therefore, $\bar{u}_{b,f} = \bar{u}_{u,f}$.

Even in the asymptotic case of $S_L/u' \to 0$, the difference in the conditioned velocities $\bar{u}_{b,f}$ and $\bar{u}_{u,f}$, modeled by Equation 6.5, may be substantial if we consider an asymptotically large heat release ($\tau \to \infty$) so that the ratio of $\tau S_L/u'$ is of unity order. Due to mixture acceleration from the unburned side to the burned side of a flamelet, the conditioned velocity $\bar{u}_b$ may exceed $\bar{u}_u$, and the BML Equation 5.35 may yield the countergradient flux.

Note that the mean velocity $\bar{u}$ should be positive in order for $\bar{u}_b > \bar{u}_u > 0$ (see Equation 6.4). Even if $S_L/u' \to 0$, the mean oncoming velocity $\bar{u}(x \to -\infty)$, which is equal to the turbulent burning velocity U_t in the coordinate framework attached to the mean flame, may be of the order of u', because U_t may be large due to a large flamelet surface area. Indeed, a slowly propagating flamelet is less effective in smoothing out small-scale wrinkles, which strongly increase the area.

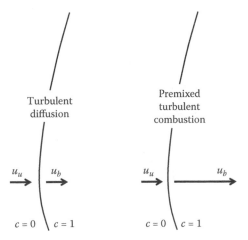

FIGURE 6.3   Local velocities normal to a molecular mixing layer (*left*) and a flamelet (*right*).

If (i) $\bar{u}_u - \bar{u}_b \propto u'$ in a turbulent mixing layer and (ii) the influence of the heat release on the conditioned velocities is modeled by Equation 6.5, then the countergradient transport seems to be pronounced if the ratio of $\tau S_L/u'$ is sufficiently large. Such a simple criterion for predicting the direction of the turbulent scalar flux is the physical basis for the well-known Bray number criterion, which will be considered in the section "Transition from Gradient to Countergradient Transport."

The foregoing discussion has clarified on a physical mechanism that can cause countergradient scalar transport in a premixed turbulent flame. The mechanism consists of local flow acceleration within thin flamelets due to the local pressure drop. Let us discuss another physical mechanism that was first highlighted by Libby and Bray (1981). It is associated with the mean pressure gradient $\nabla \bar{p}$ across the mean turbulent flame brush, rather than with the local pressure gradient $(\nabla p)_f$ across thin laminar flamelets. Note that $(\nabla p)_f$ contributes to $\nabla \bar{p}$ (see Equation 5.48). Nevertheless, the effects of $(\nabla p)_f$ on the flow will not be double-counted in the following discussion of the two physical mechanisms, because the former mechanism associated with the local pressure drop in the instantaneous flame front addresses flow perturbations in the front, whereas the latter mechanism associated with the mean pressure gradient addresses flow perturbations outside the front.

The difference between $(\nabla p)_f$ and $\nabla \bar{p}$ is sketched in Figure 6.4. The local pressure gradient (see the large circle in Figure 6.4 and the expressions within and under the circle) may be estimated using the theory of an unperturbed laminar flame (see the sections "Flow in a Laminar Premixed Flame" and "AEA Theory of Unperturbed Laminar Premixed Flame" in Chapter 2). However, similar expressions are applicable to a statistically planar, 1D, stationary turbulent flame brush, provided that the laminar flame speed, S_L, is substituted with the turbulent flame speed, S_t (see the expressions in the upper part of Figure 6.4). Accordingly, the mean pressure gradient across a statistically stationary turbulent flame brush scales as

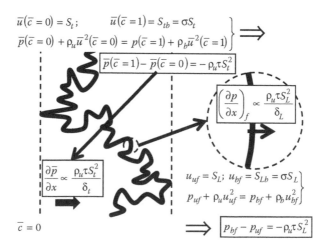

FIGURE 6.4 The difference between the local pressure gradient across a thin laminar flamelet $(\nabla p)_f$ and the mean pressure gradient $\nabla \bar{p}$ across a turbulent flame brush.

$$\frac{\partial \bar{p}}{\partial x} \propto \frac{\rho_u \tau S_t^2}{\delta_t},\tag{6.6}$$

where δ_t is the mean flame brush thickness. Note that the mean pressure gradient, $\nabla \bar{p}$, results not only from averaging the local pressure gradients, $(\nabla p)_f$, in flamelets, but also from the pressure gradients conditioned on unburned and burned mixtures (see Equation 5.48). The latter gradients are caused by the redistribution of pressure in unburned and burned mixtures due to pressure waves originating from a flamelet surface. For instance, the instantaneous pressure gradient in the unburned mixture in the left bottom part of Figure 6.4 is affected by the pressure variations in the surrounding flamelets (e.g., see the wrinkled flamelet surface in the left middle part of Figure 6.4).

To simplify the discussion on the influence of the mean pressure gradient on the mean conditioned velocities $\bar{u}_u$ and $\bar{u}_b$, let us assume that the density ratio is very large, that is, $\rho_b \ll \rho_u$. In such a case, the acceleration of the unburned gas by the pressure gradient within a turbulent flame brush may be neglected as compared with the acceleration of the burned gas, because the flow acceleration under the influence of a pressure gradient is inversely proportional to the density in a general case. Accordingly, let us assume for the sake of simplicity that the pressure gradient in a turbulent flame brush accelerates low-density combustion products but does not affect the unburned gas due to its large density. Then, the velocities $u_{u,f}$ measured in the unburned mixture are approximately equal to one another in the vicinity of points A and B in Figure 6.5. Therefore, the product velocities $u_{b,f} = u_{u,f} + \tau S_L$ are also approximately equal to one another in the vicinity of points A and B. However, the product velocity u_b measured at point B′, of which the x-coordinate is equal to the x-coordinate of point A, appears to be higher than the velocity $u_{b,f}$ at point B, because the burned gas has been accelerated by the mean pressure gradient when a product volume is convected from point B to point B′.

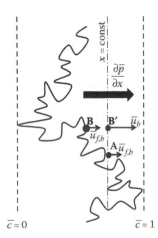

FIGURE 6.5 The effect of the mean pressure gradient on velocity $\bar{u}_b$ conditioned on low-density combustion products.

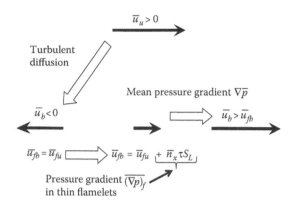

FIGURE 6.6 Three physical mechanisms that cause the difference between conditioned velocities $\bar{u}_u$ and $\bar{u}_b$.

Thus, the acceleration of low-density combustion products by the mean pressure gradient $\nabla\bar{p}$ makes the velocity $\bar{u}_b$ conditioned on the burned mixture higher than the velocity $\bar{u}_{b,f}$ conditioned on the burned side of flamelets, provided that the two velocities are measured at the same $\bar{c}$. Moreover, the acceleration of the reacting mixture by the local pressure gradient $(\nabla p)_f$ in thin flamelets makes the velocity $\bar{u}_{b,f}$ conditioned on the burned side of flamelets higher than the velocity $\bar{u}_{u,f}$ conditioned on the unburned side of flamelets (see Equation 6.5). Furthermore, turbulent pulsations (turbulent diffusion) tend to decrease $\bar{u}_b$ as compared with $\bar{u}_u$. Such a simplified physical scenario that contains three competing physical mechanisms is sketched in Figure 6.6. It is also worth noting that chemical reactions reduce the difference in $\bar{u}_u$ and $\bar{u}_b$, as will be discussed in the next section.

Physical Mechanisms and Balance Equations

Let us supplement the foregoing simplified qualitative discussion on physical mechanisms that control turbulent scalar transport in premixed flames with balance equations that are widely used to study the problem.

In the section "Favre-Averaged Balance Equations" in Chapter 3, a couple of exact but unclosed mean balance equations and, in particular, Equation 3.88 for the Favre-averaged Reynolds stresses were already derived by averaging Equations 1.49 and 1.51. A similar method may be applied to arrive at the following balance equation:

$$\frac{\partial}{\partial t}\overline{\rho u_i''c''} + \frac{\partial}{\partial x_k}\left(\tilde{u}_k\overline{\rho u_i''c''}\right) = \underbrace{-\frac{\partial}{\partial x_k}\overline{\rho u_i''u_k''c''}}_{\text{I}} \underbrace{-\overline{\rho u_i''u_k''}\frac{\partial\tilde{c}}{\partial x_k}}_{\text{II}} \underbrace{-\overline{\rho u_k''c''}\frac{\partial\tilde{u}_i}{\partial x_k}}_{\text{III}}$$

$$\underbrace{-\overline{c''}\frac{\partial\bar{p}}{\partial x_i}}_{\text{IV}} \underbrace{-\overline{c''\frac{\partial p'}{\partial x_i}}}_{\text{V}} + \underbrace{\overline{u_i''\frac{\partial}{\partial x_k}\left(\rho D\frac{\partial c}{\partial x_k}\right)}}_{\text{VI}} + \underbrace{\overline{c''\frac{\partial\tau_{ik}}{\partial x_k}}}_{\text{VII}} + \underbrace{\overline{u_i''W}}_{\text{VIII}}$$

$$(6.7)$$

by averaging Equations 1.49, 1.51, and 1.79. This equation is often used to model turbulent scalar transport in premixed flames within the framework of the Reynolds-averaged Navier–Stokes simulation (RANS) approach, but the problem of closing various terms on the RHS of Equation 6.7 has not yet been resolved, as discussed in detail in Section 4.1 of a review paper by Lipatnikov and Chomiak (2010).

The magnitudes of various terms on the RHS of Equation 6.7 were compared in a few DNS studies, reviewed in Section 4.1.1 of the aforementioned paper. The results of such DNS (e.g., see Figures 6.7 and 6.8) indicate that the pressure terms IV and V (D and E in Figure 6.7) are the main cause of countergradient scalar transport in premixed turbulent flames, in line with the foregoing qualitative discussion. Term II

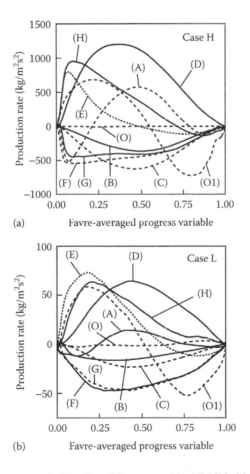

(a)

(b)

FIGURE 6.7 Various terms in Equation 6.7 computed by Nishiki (2003) in a DNS of statistically planar, 1D, stationary premixed turbulent flames characterized by low (b) and high (a) density ratios, $\rho_u/\rho_b = 2.5$ and 7.53, respectively. The positive terms cause countergradient scalar transport. Letters A–H correspond to terms I–VIII, respectively. The unsteady and convection terms on the LHS of Equation 6.7 are designated with O and O1, respectively. (Reprinted from Nishiki, S., Hasegawa, T., Borghi, R., and Himeno, R., *Combust. Theory Model.*, 10, 39–55, 2006. With permission.)

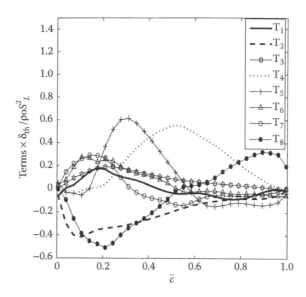

FIGURE 6.8 Various terms on the RHS of Equation 6.7. Positive terms cause countergradient scalar transport. Notations T_1–T_8 correspond to terms I–VIII, respectively. (Reprinted from Chakraborty, N. and Cant, R.S., *Combust. Flame*, 156, 1427–1444, 2009. With permission.)

(B) serves to cause gradient transport. Because the sign of term III is opposite to the sign of the flux component $\overline{\rho u_k'' c''}$, term III (C) reduces the flux magnitude. The dissipation terms VI and VII also play an important role in a typical DNS of premixed turbulent combustion, as such simulations are performed at low Reynolds numbers. As regards the reaction term VIII (H in Figure 6.7), it was positive and, therefore, served to cause countergradient transport in the majority of the DNS, while curve T_8 in Figure 6.8 indicates that the sign of this term may change from negative to positive as $\tilde{c}$ increases.

It is worth noting, however, that Equation 6.7 does not offer an opportunity to separately model the physical mechanisms discussed in the previous subsection. In other words, any of terms IV–VIII on the RHS of Equation 6.7 is affected by at least two physical mechanisms. For instance, either the pressure term IV or the pressure term V is affected both by flow acceleration due to the pressure drop in flamelets (this mechanisms results in $\overline{u}_{b,f} > \overline{u}_{u,f}$) and by flow acceleration due to the mean pressure gradient (this mechanisms results in $\overline{u}_b > \overline{u}_{b,f}$).

Indeed, within the framework of the BML approach, we can obtain

$$\underbrace{-\overline{c'' \frac{\partial \overline{p}}{\partial x_i}}}_{\text{IV}} \underbrace{-\overline{c'' \frac{\partial p'}{\partial x_i}}}_{\text{V}} + \underbrace{\overline{u_i'' \frac{\partial}{\partial x_k}\left(\rho D \frac{\partial c}{\partial x_k}\right)}}_{\text{VI}} + \underbrace{\overline{c'' \frac{\partial \tau_{ik}}{\partial x_k}}}_{\text{VII}} + \underbrace{\overline{u_i' W}}_{\text{VIII}}$$

$$= \overline{c''\left(-\frac{\partial p}{\partial x_i} + \frac{\partial \tau_{ik}}{\partial x_k}\right) + u_i''\left[\frac{\partial}{\partial x_k}\left(\rho D \frac{\partial c}{\partial x_k}\right) + W\right]}$$

$$= \overline{c\left(-\frac{\partial p}{\partial x_i} + \frac{\partial \tau_{ik}}{\partial x_k}\right) + u_i\left[\frac{\partial}{\partial x_k}\left(\rho D \frac{\partial c}{\partial x_k}\right) + W\right]}$$

$$- \tilde{c}\overline{\left(-\frac{\partial p}{\partial x_i} + \frac{\partial \tau_{ik}}{\partial x_k}\right)} - \tilde{u}_i\overline{\left[\frac{\partial}{\partial x_k}\left(\rho D \frac{\partial c}{\partial x_k}\right) + W\right]}$$

$$= \overline{c}\left(-\frac{\partial p}{\partial x_i} + \frac{\partial \tau_{ik}}{\partial x_k}\right)_b + \gamma\left(\overline{\rho c \frac{Du_i}{Dt}}\right)_f + \gamma\left(\overline{\rho u_i \frac{Dc}{Dt}}\right)_f$$

$$- \tilde{c}\overline{\left(-\frac{\partial p}{\partial x_i} + \frac{\partial \tau_{ik}}{\partial x_k}\right)} - \tilde{u}_i\gamma\left(\overline{\rho \frac{Dc}{Dt}}\right)_f$$

$$= \overline{c}\left(\overline{\rho \frac{Du_i}{Dt}}\right)_b + \gamma\left[\overline{\rho \frac{D}{Dt}(cu_i)}\right]_f - \tilde{c}\left(\overline{\rho \frac{Du_i}{Dt}}\right) - \tilde{u}_i\gamma\left(\overline{\rho \frac{Dc}{Dt}}\right)_f$$

$$= \overline{c}(1-\tilde{c})\left(\overline{\rho \frac{Du_i}{Dt}}\right)_b + \gamma\left[\overline{\rho \frac{D}{Dt}(cu_i)}\right]_f - \tilde{c}(1-\overline{c})\left(\overline{\rho \frac{Du_i}{Dt}}\right)_u$$

$$- \tilde{c}\gamma\left(\overline{\rho \frac{Du_i}{Dt}}\right)_f - \tilde{u}_i\gamma\left(\overline{\rho \frac{Dc}{Dt}}\right)_f$$

using Equations 1.51, 1.79, and 5.45 through 5.48. Here,

$$\rho \frac{Dq}{Dt} \equiv \rho \frac{\partial q}{\partial t} + \rho u_k \frac{\partial q}{\partial x_k} \tag{6.8}$$

for any quantity q and the summation convention applies for the repeated index k. Thus, we see that terms IV–VII are affected both by processes outside flamelets (see the terms with subscripts u and b) and by processes within flamelets (see terms with the subscript f).

Furthermore, term III is also affected both by processes outside flamelets and by processes within flamelets. For instance, if the density is controlled by the combustion progress variable, then the continuity Equation 1.49 reads

$$\frac{d\rho}{dc}\left(\frac{\partial c}{\partial t} + u_k \frac{\partial c}{\partial x_k}\right) + \rho \frac{\partial u_k}{\partial x_k} = 0. \tag{6.9}$$

Therefore, Equations 1.79 and 6.9 yield

$$\frac{\partial u_k}{\partial x_k} = -\frac{1}{\rho^2}\frac{d\rho}{dc}\left(\rho \frac{\partial c}{\partial t} + \rho u_k \frac{\partial c}{\partial x_k}\right) = \frac{d\rho^{-1}}{dc}\left[\frac{\partial}{\partial x_k}\left(\rho D \frac{\partial c}{\partial x_k}\right) + W\right] \tag{6.10}$$

or

$$\frac{\partial u_k}{\partial x_k} = \frac{\tau}{\rho_u}\left[\frac{\partial}{\partial x_k}\left(\rho D \frac{\partial c}{\partial x_k}\right) + W\right] \tag{6.11}$$

if $c = (T - T_u)/(T_b - T_u)$, therefore $\rho_u/\rho = T/T_u = 1 + c\tau$. By averaging Equation 6.11, we obtain (Prudnikov, 1967)

$$\nabla \cdot \bar{\mathbf{u}} = \frac{\tau}{\rho_u}\bar{W} \tag{6.12}$$

if the mean molecular diffusion term is neglected at high Reynolds numbers. Using Equations 5.40 and 6.12, term III on the RHS of Equation 6.7 reads

$$\underbrace{-\overline{\rho u''c''}\frac{\partial \tilde{u}}{\partial x}}_{\text{III}} = -\overline{\rho u''c''}\frac{\tau \bar{W}}{\rho_u} + \tau\frac{\overline{\rho u''c''}}{\rho_u}\frac{\partial}{\partial x}\overline{\rho u''c''}, \tag{6.13}$$

in the simplest case of a statistically planar, 1D turbulent flame brush. Accordingly, term III involves a reaction term, which reduces the magnitude of the turbulent scalar flux. Because $\nabla\tilde{u} > 0$ in a statistically 1D flame, the sign of term III is opposite to the sign of the flux $\overline{\rho u''c''}$ and is the same as the sign of the first (reaction) term on the RHS of Equation 6.13.

Equation 6.7 may be rewritten in another form, which allows us to better reveal the physical mechanisms discussed earlier. Using Equations 5.21 through 5.24, 5.34, and 5.35, we obtain

$$\bar{u}_{i,u} = \tilde{u}_i - \frac{\overline{\rho u_i''c''}}{\bar{\rho}(1-\tilde{c})} = \tilde{u}_i - \frac{\overline{\rho u_i''c''}}{\rho_u(1-\bar{c})} \tag{6.14}$$

and

$$\bar{u}_{i,b} = \tilde{u}_i + \frac{\overline{\rho u_i''c''}}{\bar{\rho}\tilde{c}} = \tilde{u}_i + \frac{\overline{\rho u_i''c''}}{\rho_b\bar{c}}. \tag{6.15}$$

Subsequently, the following conditioned balance equations:

$$\frac{\partial}{\partial t}(1-\bar{c}) + \frac{\partial}{\partial x_k}\left[(1-\bar{c})\bar{u}_{k,u}\right] = -\frac{\gamma}{\rho_u}\overline{\left(\rho\frac{Dc}{Dt}\right)_f}, \tag{6.16}$$

$$\frac{\partial \bar{c}}{\partial t} + \frac{\partial}{\partial x_k}(\overline{c}\overline{u}_{k,b}) = \frac{\gamma}{\rho_b}\overline{\left(\rho\frac{Dc}{Dt}\right)_f}, \tag{6.17}$$

$$\frac{\partial}{\partial t}\Big[(1-\bar{c})\bar{u}_{i,u}\Big]+\frac{\partial}{\partial x_k}\Big[(1-\bar{c})\bar{u}_{k,u}\,\bar{u}_{i,u}\Big]=-\frac{\partial}{\partial x_k}\Big[(1-\bar{c})\overline{(u_i'u_k')}_u\Big]-(1-\bar{c})\overline{\left(\frac{1}{\rho}\frac{\partial p}{\partial x_i}\right)}_u$$

$$+(1-\bar{c})\overline{\left(\frac{1}{\rho}\frac{\partial\tau_{ik}}{\partial x_k}\right)}_u+\frac{\gamma}{\rho_u}\overline{\left\{\rho\frac{D}{Dt}\Big[u_i(1-c)\Big]\right\}}_f,$$

(6.18)

and

$$\frac{\partial}{\partial t}\big(\overline{cu}_{i,b}\big)+\frac{\partial}{\partial x_k}\big(\overline{cu}_{k,b}\,\bar{u}_{i,b}\big)=-\frac{\partial}{\partial x_k}\Big[\bar{c}\,\overline{(u_i'u_k')}_b\Big]-\bar{c}\,\overline{\left(\frac{1}{\rho}\frac{\partial p}{\partial x_i}\right)}_b$$

$$+\bar{c}\,\overline{\left(\frac{1}{\rho}\frac{\partial\tau_{ik}}{\partial x_k}\right)}_b+\frac{\gamma}{\rho_b}\overline{\left[\rho\frac{D}{Dt}\big(u_ic\big)\right]}_f$$

(6.19)

may be derived by taking the time and spatial derivatives of Equations 6.14 and 6.15 and using Equations 3.78, 3.79, 6.2, and 6.7, as discussed in detail elsewhere (Lipatnikov, 2008). Note that, in conditioned moments $(\overline{u_i'u_j'})_c$, $(\overline{u_i'u_j'u_k'})_c$, etc., where subscript c refers to either unburned or burned mixture, symbol u' designates conditioned velocity fluctuations, that is, $(\overline{u_i'u_j'})_c=[\overline{(u_i-\bar{u}_{i,c})(u_j-\bar{u}_{j,c})}]_c$, $(\overline{u_i'u_j'u_k'})_c=[\overline{(u_i-\bar{u}_{i,c})(u_j-\bar{u}_{j,c})(u_k-\bar{u}_{k,c})}]_c$, etc. Combining Equations 6.16 and 6.17 with Equations 6.18 and 6.19, respectively, we obtain

$$(1-\bar{c})\frac{\partial\bar{u}_{i,u}}{\partial t}+(1-\bar{c})\bar{u}_{k,u}\frac{\partial\bar{u}_{i,u}}{\partial x_k}=-\frac{\partial}{\partial x_k}\Big[(1-\bar{c})\overline{(u_i'u_k')}_u\Big]-(1-\bar{c})\overline{\left(\frac{1}{\rho}\frac{\partial p}{\partial x_i}\right)}_u$$

$$+(1-\bar{c})\overline{\left(\frac{1}{\rho}\frac{\partial\tau_{ik}}{\partial x_k}\right)}_u+\bar{u}_{i,u}\frac{\gamma}{\rho_u}\overline{\left(\rho\frac{Dc}{Dt}\right)}_f$$

$$+\frac{\gamma}{\rho_u}\overline{\left\{\rho\frac{D}{Dt}\Big[u_i(1-c)\Big]\right\}}_f,$$

(6.20)

and

$$\bar{c}\frac{\partial\bar{u}_{i,b}}{\partial t}+\overline{cu}_{k,b}\frac{\partial\bar{u}_{i,b}}{\partial x_k}=-\frac{\partial}{\partial x_k}\Big[\bar{c}\,\overline{(u_i'u_k')}_b\Big]-\bar{c}\,\overline{\left(\frac{1}{\rho}\frac{\partial p}{\partial x_i}\right)}_b+\bar{c}\,\overline{\left(\frac{1}{\rho}\frac{\partial\tau_{ik}}{\partial x_k}\right)}_b$$

$$+\frac{\gamma}{\rho_b}\overline{\left[\rho\frac{D}{Dt}\big(u_ic\big)\right]}_f-\bar{u}_{i,b}\frac{\gamma}{\rho_b}\overline{\left(\rho\frac{Dc}{Dt}\right)}_f.$$

(6.21)

In Equations 6.20 and 6.21, terms that involve conditioned pressure gradients model the difference in $\bar{u}_u$ and $\bar{u}_{u,f}$ or in $\bar{u}_{b,f}$ and $\bar{u}_b$, while the local flow acceleration due to the pressure drop in flamelets (i.e., $\bar{u}_{b,f} > \bar{u}_{u,f}$) is associated with the flamelet terms marked with the subscript f.

To show this more clearly, let us assume that flamelets are much thinner than the Kolmogorov scale and retain the structure of the unperturbed laminar flame (such a simplification is justified at least at low Karlovitz numbers (see the sections "A Brief Summary of Theory of Weakly Stretched Laminar Premixed Flames" and "Regimes of Premixed Turbulent Combustion" in Chapter 5). Let us decompose the local instantaneous velocity in an unburned mixture in the vicinity of a flamelet into propagation and convection components:

$$\mathbf{u} = \underbrace{-S_L^0 \mathbf{n}}_{\text{propagation}} + \underbrace{\left(\mathbf{u} + S_L^0 \mathbf{n}\right)}_{\text{convection}}, \tag{6.22}$$

where the former component is normal to the flamelet. Here, the normal vector $\mathbf{n} = -\nabla c / |\nabla c|$ points to the unburned mixture. Then, the heat release solely affects the propagation component $\mathbf{u}_p$ (which is equal to $-S_L^0 \mathbf{n}$ in the unburned mixture and is increased due to the pressure drop within the flamelet), whereas the convection component $\mathbf{u}_c = \mathbf{u} + S_L^0 \mathbf{n}$ advects the flamelet as a whole but is not affected by the heat release. In the local coordinate framework that moves at a speed $-(\mathbf{u} + S_L^0 \mathbf{n})$, the flamelet looks like the classical unperturbed laminar flame (see the sections "Flow in a Laminar Premixed Flame" and "AEA Theory of Unperturbed Laminar Premixed Flame" in Chapter 2), with the normal (to the flamelet) velocities being equal to S_L^0 and τS_L^0 in the fresh mixture and equilibrium combustion products, respectively. Accordingly, in this framework,

$$\int_{-\infty}^{\infty} \rho \frac{Dc}{Dt} d\zeta = \int_{-\infty}^{\infty} \rho_u S_L^0 \frac{dc}{d\zeta} d\zeta = \rho_u S_L^0, \tag{6.23}$$

$$\int_{-\infty}^{\infty} \rho \frac{Du}{Dt} d\zeta = \int_{-\infty}^{\infty} \rho_u S_L^0 \frac{du}{d\zeta} d\zeta = \rho_u S_L^0 \left(\sigma S_L^0 - S_L^0\right) = \rho_u \tau \left(S_L^0\right)^2, \tag{6.24}$$

and

$$\int_{-\infty}^{\infty} \rho \frac{D}{Dt}(uc) d\zeta = \int_{-\infty}^{\infty} \rho_u S_L^0 \frac{d}{d\zeta}(uc) d\zeta = \rho_u S_L^0 \sigma S_L^0, \tag{6.25}$$

where the ζ-axis is locally normal to the flamelet and points toward the products. It is worth stressing again that the flamelet is assumed (i) to be convected by turbulent eddies but (ii) to retain the structure of the unperturbed laminar flame.

Returning to the laboratory coordinate framework, we have

$$\int_{-\infty}^{\infty} \rho \frac{Dc}{Dt} d\zeta = \rho_u S_L^0, \tag{6.26}$$

$$\int_{-\infty}^{\infty} \rho \frac{D\mathbf{u}}{Dt} d\zeta = \int_{-\infty}^{\infty} \rho \frac{D\mathbf{u}_p}{Dt} d\zeta = -\rho_u \tau \left(S_L^0\right)^2 \mathbf{n}, \tag{6.27}$$

and

$$\int_{-\infty}^{\infty} \rho \frac{D}{Dt} (\mathbf{u}c) d\zeta = \int_{-\infty}^{\infty} \rho \frac{D}{Dt} (\mathbf{u}_p c) d\zeta + \mathbf{u}_c \int_{-\infty}^{\infty} \rho \frac{Dc}{Dt} d\zeta$$

$$= -\rho_u \sigma \left(S_L^0\right)^2 \mathbf{n} + \mathbf{u}_c \rho_u S_L^0$$

$$= -\rho_u \sigma \left(S_L^0\right)^2 \mathbf{n} + \left(\mathbf{u} + S_L^0 \mathbf{n}\right) \rho_u S_L^0$$

$$= -\rho_u \tau \left(S_L^0\right)^2 \mathbf{n} + \mathbf{u}\rho_u S_L^0, \tag{6.28}$$

using Equation 6.27 and because the convection component $\mathbf{u}_c$ of the local velocity vector is not affected by the flamelet within the framework of the simple model considered. Furthermore, in the case of high Reynolds numbers, averaging Equation 1.79 yields

$$\overline{\gamma \left(\rho \frac{Dc}{Dt}\right)_f} = (1-\overline{c})\overline{\left(\rho \frac{Dc}{Dt}\right)_u} + \overline{c}\overline{\left(\rho \frac{Dc}{Dt}\right)_b} + \overline{\gamma \left(\rho \frac{Dc}{Dt}\right)_f} = \overline{\left(\rho \frac{Dc}{Dt}\right)} \approx \overline{W} \tag{6.29}$$

by neglecting the mean molecular transport term and considering that $Dc/Dt = 0$ outside flamelets. Based on Equations 6.26 through 6.29, we may assume that (Lipatnikov, 2008)

$$\overline{\gamma \left(\rho \frac{Du_i}{Dt}\right)_f} = -\tau S_L^0 \overline{W} \overline{n}_i \tag{6.30}$$

and

$$\overline{\gamma \left[\rho \frac{D}{Dt} (u_i c)\right]_f} = \overline{W} \left(\overline{u}_{i,u,f} - \tau S_L^0 \overline{n}_i\right). \tag{6.31}$$

Finally, substitution of Equations 6.29 through 6.31 into Equations 6.20 and 6.21 results in

$$(1-\overline{c}) \frac{\partial \overline{u}_{i,u}}{\partial t} + (1-\overline{c}) \overline{u}_{k,u} \frac{\partial \overline{u}_{i,u}}{\partial x_k} = -\frac{\partial}{\partial x_k} \left[(1-\overline{c}) \overline{\left(u_i' u_k'\right)}_u\right] - \underbrace{(1-\overline{c}) \overline{\left(\frac{1}{\rho} \frac{\partial p}{\partial x_i}\right)}_u}_{1u}$$

$$+ \underbrace{(1-\overline{c}) \overline{\left(\frac{1}{\rho} \frac{\partial \tau_{ik}}{\partial x_k}\right)}_u}_{} + \underbrace{\left(\overline{u}_{i,u} - \overline{u}_{i,u,f}\right) \frac{\overline{W}}{\rho_u}}_{3u}, \tag{6.32}$$

and

$$\bar{c}\frac{\partial \bar{u}_{i,b}}{\partial t} + \overline{cu}_{k,b}\frac{\partial \bar{u}_{i,b}}{\partial x_k} = \underbrace{-\frac{\partial}{\partial x_k}\left[\bar{c}\overline{\left(u_i'u_k'\right)}_b\right] - \bar{c}\overline{\left(\frac{1}{\rho}\frac{\partial p}{\partial x_i}\right)_b}}_{1b}$$

$$+ \underbrace{\bar{c}\overline{\left(\frac{1}{\rho}\frac{\partial \tau_{ik}}{\partial x_k}\right)_b}}_{2} - \tau S_L^0 \bar{n}_i \frac{\bar{W}}{\rho_b} + \underbrace{\left(\bar{u}_{i,u,f} - \bar{u}_{i,b}\right)\frac{\bar{W}}{\rho_b}}_{3b}. \qquad (6.33)$$

Therefore, the terms 1u and 1b in Equations 6.32 and 6.33, respectively, are associated with the acceleration of the unburned and burned mixtures by the "mean" (more precisely, conditioned) pressure gradient, that is, the physical mechanism illustrated in Figure 6.5. Accordingly, the term 1b (1a) is responsible for the difference between $\bar{u}_b$ and $\bar{u}_{b,f}$ ($\bar{u}_u$ and $\bar{u}_{u,f}$). Term 2 on the RHS of Equation 6.33 is associated with the flow acceleration due to the local pressure drop within thin flamelets, that is, the physical mechanism illustrated in Figure 6.3. Accordingly, the term 2 is responsible for the difference between $\bar{u}_{u,f}$ and $\bar{u}_{b,f}$.

The aforementioned conditioned balance equations offer an opportunity to clarify the influence of chemical reactions on the turbulent scalar flux (see the terms 3u and 3b). To do so, let us note that the available DNS data (Im et al., 2004; Chakraborty and Lipatnikov, 2011) indicate that $\bar{u}_{u,f} \approx \bar{u}_u$ in the major part of a statistically planar, 1D turbulent flame brush, with the exception of the leading edge where $\bar{u}_{u,f} < \bar{u}_u$ (see the dashed and solid curves in Figure 6.9, and also note that the curves with the circles and pluses show that $\bar{u}_{b,f} < \bar{u}_b$, in line with the foregoing discussion of product acceleration by the mean pressure gradient). Accordingly, if we consider the statistically planar, 1D case and, based on the aforementioned DNS data, assume that $\bar{u}_{u,f} \approx \bar{u}_u$, then the term 3u on the RHS of Equation 6.32 vanishes, while the sign of the term 3b on the RHS of Equation 6.33 is opposite to the sign of the slip velocity $\Delta\bar{u} = \bar{u}_b - \bar{u}_u$. Subsequently, subtraction of Equation 6.32 from Equation 6.33 yields a balance equation for the slip velocity, with the sign of the last term on the RHS of that equation being opposite to the sign of $\Delta\bar{u}$, that is,

$$\frac{\partial}{\partial t}\left(\bar{u}_b - \bar{u}_u\right) + \frac{1}{2}\frac{\partial}{\partial x}\left(\bar{u}_b^2 - \bar{u}_u^2\right) = \frac{1}{1-\bar{c}}\frac{\partial}{\partial x}\left[\left(1-\bar{c}\right)\overline{\left(u'u'\right)}_u\right] - \frac{1}{\bar{c}}\frac{\partial}{\partial x}\left[\bar{c}\overline{\left(u'u'\right)}_b\right]$$

$$-\overline{\left(\frac{1}{\rho}\frac{\partial p}{\partial x}\right)_b} + \overline{\left(\frac{1}{\rho}\frac{\partial p}{\partial x}\right)_u} + \overline{\left(\frac{1}{\rho}\frac{\partial \tau_{xk}}{\partial x_k}\right)_b} - \overline{\left(\frac{1}{\rho}\frac{\partial \tau_{xk}}{\partial x_k}\right)_u}$$

$$-\tau S_L^0 \bar{n}_x \frac{\bar{W}}{\rho_b \bar{c}} + \left(\bar{u}_u - \bar{u}_b\right)\frac{\bar{W}}{\rho_b \bar{c}}. \qquad (6.34)$$

Therefore, chemical reactions (i.e., the terms 3u and 3b) reduce the difference $(\bar{u}_b - \bar{u}_u)$ and, hence, reduce the magnitude of the turbulent scalar flux. Note that although the second last term on the RHS of Equation 6.34 is also proportional to the mean reaction rate $\bar{W}$, this term models the mixture acceleration due to the local pressure drop in the

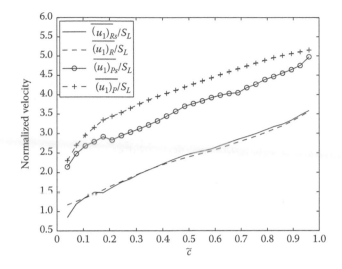

FIGURE 6.9 Normalized axial velocities conditioned on the unburned mixture (in the legends $(u_1)_R$ and _dashed curve_), unburned edges of flamelets $((u_1)_{Rs}$ and _solid curve_), burned mixture $((u_1)_P$ and _curve with crosses_), and burned edges of flamelets $((u_1)_{Ps}$ and curve with circles) versus the Favre-averaged combustion progress variable. DNS data analyzed by Chakraborty and Lipatnikov (2011).

flamelets and vanishes in the case of a constant density ($\tau = 0$), even if chemical reactions occur in the flow. As regards the reaction term VIII on the RHS of Equation 6.7, it caused countergradient scalar flux in several DNSs (e.g., see Figure 6.7), because this term models not only the reaction but also the density drop effects.

Finally, to reveal the terms that serve to cause gradient transport, let us return to Equation 6.7 and apply it to modeling constant-density nonreacting turbulent mixing in the statistically planar, 1D case. Then, Equation 6.7 reads

$$\frac{\partial}{\partial t}\overline{u'c'} = -\underbrace{\frac{\partial}{\partial x}\overline{u'^2c'}}_{\text{I}} -\underbrace{\overline{u'^2}\frac{\partial \overline{c}}{\partial x}}_{\text{II}} -\underbrace{\frac{1}{\rho}\overline{c'\frac{\partial p'}{\partial x}}}_{\text{V}}, \qquad (6.35)$$

while Equation 6.2 reduces to

$$\frac{\partial \overline{c}}{\partial t} = -\frac{\partial}{\partial x}\overline{u'c'}. \qquad (6.36)$$

Here, the molecular transfer terms VI and VII are neglected for the sake of simplicity at high Reynolds numbers due to the isotropy of small-scale turbulent motions, $\nabla \cdot \overline{\mathbf{u}} = 0$ for an incompressible flow, $\tilde{q} = \overline{q}$ if $\rho = $ const, and the coordinate framework is selected so that $\overline{u} = 0$.

Then, on the one hand, differentiating Equation 6.35 with respect to x and Equation 6.36 with respect to t and subtracting the obtained equations from one another, we arrive at

$$\frac{\partial^2 \bar{c}}{\partial t^2} = \frac{\partial^2}{\partial x^2}\overline{u'^2 c'} + \overline{u'^2}\frac{\partial^2 \bar{c}}{\partial x^2} + \frac{1}{\rho}\frac{\partial}{\partial x}\left(\overline{c'\frac{\partial p'}{\partial x}}\right) \tag{6.37}$$

if the field $\overline{u'^2}$ is spatially uniform. On the other hand, at $t \gg \tau_t$, turbulent diffusion is modeled by the following well-known telegraph equation (Goldstein, 1951; Monin and Yaglom, 1971):

$$\frac{\partial^2 \bar{c}}{\partial t^2} + \frac{\overline{u'^2}}{D_t}\frac{\partial \bar{c}}{\partial t} = \overline{u'^2}\frac{\partial^2 \bar{c}}{\partial x^2}. \tag{6.38}$$

A comparison of Equations 6.37 and 6.38 yields

$$\frac{\partial^2}{\partial x^2}\overline{u'^2 c'} + \frac{1}{\rho}\frac{\partial}{\partial x}\left(\overline{c'\frac{\partial p'}{\partial x}}\right) = -\frac{\overline{u'^2}}{D_t}\frac{\partial \bar{c}}{\partial t} = \frac{\overline{u'^2}}{D_t}\frac{\partial}{\partial x}\overline{u'c'} \tag{6.39}$$

using Equation 6.36. The integration of Equation 6.39 results in

$$\frac{\partial}{\partial x}\overline{u'^2 c'} + \frac{1}{\rho}\overline{c'\frac{\partial p'}{\partial x}} = \frac{\overline{u'^2}}{D_t}\overline{u'c'}, \tag{6.40}$$

that is, in the nonreacting, constant-density case considered, the transport and fluctuating pressure terms, I and V, respectively, are proportional to the scalar flux. Combining Equations 6.35 and 6.40, we finally obtain

$$\frac{\partial}{\partial t}\overline{u'c'} = \underbrace{-\overline{u'^2}\frac{\partial \bar{c}}{\partial x}}_{II} - \frac{\overline{u'^2}}{D_t}\overline{u'c'}. \tag{6.41}$$

This equation clearly indicates that only term II in Equation 6.35 can result in $\overline{u'c'} \neq 0$, whereas the other two terms (I and V) on the RHS just reduce the magnitude of the turbulent scalar flux. If $\bar{c}$ increases (decreases) in the direction of the x-axis, then term II on the RHS of Equation 6.41 is negative (positive) and serves to cause negative (positive) turbulent scalar flux $\overline{u'c'}$, that is, the flux generated by term II is directed from higher to lower $\bar{c}$ (gradient diffusion).

It is worth noting that Equation 6.41 may be rewritten as

$$\frac{\partial}{\partial t}\overline{u'c'} = -\left[(1-\bar{c})\left(\overline{u'^2}\right)_u + \bar{c}\left(\overline{u'^2}\right)_b\right]\frac{\partial \bar{c}}{\partial x} - \frac{\left(\overline{u'c'}\right)^2}{\bar{c}(1-\bar{c})}\frac{\partial \bar{c}}{\partial x} - \frac{\overline{u'^2}}{D_t}\overline{u'c'} \tag{6.42}$$

using Equations 5.35 and 5.41, which still hold in the considered case of a constant density. Both the first and second terms on the RHS of Equation 6.42 serve to cause gradient diffusion.

The foregoing discussion may be briefly summarized as follows. First, in the case of a spatially nonuniform concentration field, turbulent pulsations cause the well-known gradient diffusion. This physical mechanism is modeled by term II in Equation 6.7 or, in part, by the first two terms on the RHS of Equation 6.34. Second, the pressure drop within thin flamelets accelerates the local flow and results in $\bar{u}_{u,f} < \bar{u}_{b,f}$. This physical mechanism is associated with term 2 on the RHS of Equation 6.33. Third, the pressure gradient within the turbulent flame brush accelerates flow and results, in particular, in $\bar{u}_{b,f} < \bar{u}_b$. This physical mechanism is modeled by the terms 1u and 1b in Equations 6.32 and 6.33, respectively. In order for the two mechanisms associated with the pressure gradients to play a role, the density should be varied due to the heat release in the chemical reactions. Finally, the chemical reactions reduce the magnitude of the turbulent scalar flux, with the effect pronounced even in a constant-density case. This physical mechanism is associated with the terms 3u and 3b in Equations 6.32 and 6.33, respectively, or with the last term on the RHS of Equation 6.34.

It is worth noting that, using Equations 6.14 and 6.29 through 6.31, one can rewrite Equation 6.7 as

$$\frac{\partial}{\partial t}\overline{\rho u_i''c''} + \frac{\partial}{\partial x_k}\left(\tilde{u}_k \overline{\rho u_i''c''}\right) = -\frac{\partial}{\partial x_k}\overline{\rho u_i''u_k''c''} - \overline{\rho u_i''u_k''}\frac{\partial \tilde{c}}{\partial x_k} - \overline{\rho u_k''c''}\frac{\partial \tilde{u}_i}{\partial x_k}$$

$$+ \bar{\rho}\tilde{c}(1-\tilde{c})\left[\overline{\left(\frac{Du_i}{Dt}\right)_b} - \overline{\left(\frac{Du_i}{Dt}\right)_u}\right] - (1-\tilde{c})\tau S_L^0 \overline{W}\bar{n}_i$$

$$- \frac{\overline{\rho u_i''c''}}{\bar{\rho}(1-\tilde{c})}\overline{W} + \left(\bar{u}_{i,u,f} - \bar{u}_{i,u}\right)\overline{W} \tag{6.43}$$

or as

$$\frac{\partial}{\partial t}\overline{\rho u_i''c''} + \frac{\partial}{\partial x_k}\left(\tilde{u}_k \overline{\rho u_i''c''}\right) = -\frac{\partial}{\partial x_k}\overline{\rho u_i''u_k''c''} \underbrace{- \overline{\rho u_i''u_k''}\frac{\partial \tilde{c}}{\partial x_k}}_{1} \underbrace{- (1-\tilde{c})\tau S_L^0 \overline{W}\bar{n}_i}_{2}$$

$$+ \underbrace{\bar{\rho}\tilde{c}(1-\tilde{c})\left[\left(\frac{1}{\rho}\frac{\partial p}{\partial x_i}\right)_u - \left(\frac{1}{\rho}\frac{\partial p}{\partial x_i}\right)_b\right]}_{3}$$

$$\underbrace{- \overline{\rho u_k''c''}\left[\frac{\partial \tilde{u}_i}{\partial x_k} + \frac{\overline{W}\delta_{ik}}{\bar{\rho}(1-\tilde{c})}\right]}_{4} \tag{6.44}$$

if the viscous terms are neglected at large Reynolds numbers and $\bar{u}_{i,u,f}$ is assumed to be equal to $\bar{u}_{i,u}$. The latter equation looks similar to Equation 6.34, with terms 1–4 being associated with the four physical mechanisms discussed earlier. A balance equation that is basically similar to Equation 6.43, with the fourth term on the

RHS being neglected, was introduced by Masuya (1986) by developing a proposal by Kuznetsov (1979).

NUMERICAL MODELING OF TURBULENT SCALAR TRANSPORT IN PREMIXED FLAMES

Over the past three decades, the vast majority of RANS simulations of the influence of heat release and chemical reactions on turbulent scalar transport in premixed flames dealt with the second-order balance Equation 6.7. Because terms I and IV–VIII in this equation are not closed (term II is closed if the second-order balance Equation 3.88 is invoked), a number of models of these terms were proposed, as discussed in Section 4.1 of a review paper by Lipatnikov and Chomiak (2010). However, such efforts have not yet yielded a thoroughly validated, well-elaborated, widely recognized, predictive model, as summarized in Section 4.1.7 of the cited review paper. For this reason and due to the complexity of the submodels invoked to close certain terms (e.g., term V), this approach is not discussed here and interested readers are referred to Section 4.1 of the aforementioned review paper.

Recently, Weller (1993), Im et al. (2004), and Lipatnikov (2008) put forward balance equations for conditioned velocities as an alternative approach to the second-order balance Equation 6.7. In the cited papers, the conditioned balance equations were obtained in two different forms. The already discussed Equations 6.18 and 6.19 were rigorously derived by Lipatnikov (2008), and an important merit of these equations consists of an opportunity to model the influence of the processes within flamelets on the conditioned velocity (and, hence, turbulent scalar transport) by invoking the simple closure relations given by Equations 6.30 and 6.31. Moreover, as discussed in the previous section, the physical mechanisms of the influence of the heat release and chemical reactions on turbulent scalar transport are straightforwardly associated with terms 1–3 in the conditioned Equations 6.32 and 6.33, whereas such a straightforward association is difficult when using the second-order balance Equation 6.7.

Weller (1993) and Im et al. (2004) obtained conditioned balance equations in the following form:

$$\frac{\partial}{\partial t}\left[\rho_u\left(1-\bar{c}\right)\bar{u}_{i,u}\right]+\frac{\partial}{\partial x_k}\left[\rho_u\left(1-\bar{c}\right)\bar{u}_{k,u}\bar{u}_{i,u}\right]$$

$$=-\frac{\partial}{\partial x_k}\left[\rho_u\left(1-\bar{c}\right)\left(\overline{u_i'u_k'}\right)_u\right]-\frac{\partial}{\partial x_i}\left[\left(1-\bar{c}\right)\bar{p}_u\right]-\rho_u S_L^0 \bar{u}_{i,u,f}\bar{\Sigma}+\left(\overline{pn_i}\right)_{u,f}\bar{\Sigma}, \qquad (6.45)$$

$$\frac{\partial}{\partial t}\left(\rho_b \overline{cu}_{i,b}\right)+\frac{\partial}{\partial x_k}\left(\rho_b \overline{cu}_{k,b}\bar{u}_{i,b}\right)=-\frac{\partial}{\partial x_k}\left[\rho_b \bar{c}\left(\overline{u_i'u_k'}\right)_b\right]$$

$$-\frac{\partial}{\partial x_i}\left(\overline{cp}_b\right)+\rho_b S_{Lb}^0 \bar{u}_{i,b,f}\bar{\Sigma}-\left(\overline{pn_i}\right)_{b,f}\bar{\Sigma}. \qquad (6.46)$$

Here, the viscous terms are neglected for the sake of simplicity, $\bar{\Sigma}=\bar{W}/(\rho_u S_L^0)$,

$$\rho_u S_L^0 = \rho_b S_{L,b}^0, \qquad (6.47)$$

$$\rho_u S_L^0 \bar{u}_{i,u,f} + \left(\overline{pn_i}\right)_{u,f} = \rho_b S_{Lb}^0 \bar{u}_{i,b,f} + \left(\overline{pn_i}\right)_{b,f}, \tag{6.48}$$

and the correlations $\left(\overline{p\mathbf{n}}\right)_{u,f}$ or $\left(\overline{p\mathbf{n}}\right)_{b,f}$ between the pressure and the unit vector $\mathbf{n}$ locally normal to a flamelet surface are averaged along the unburned or burned edges of flamelets.

Initial tests of the conditioned balance equations were performed by simulating conditioned velocities and turbulent scalar fluxes in premixed flames stabilized in impinging jets (Lee and Huh, 2004; Lipatnikov, 2011b). Even if the results of these tests are encouraging (see Figures 40, 42, and 43 in a review paper by Lipatnikov and Chomiak [2010]) the approach definitely needs further development and tests. For instance, as far as Equations 6.18 and 6.19 are concerned, a model of the influence of combustion on the conditioned pressure gradients $(\nabla p)_u$ and $(\nabla p)_b$ has not yet been developed.

For further discussion on the conditioned balance equations, interested readers are referred to Section 4.4 of the cited review paper, while in this book, we will restrict ourselves to a much simpler approach that has been put forward very recently (Sabel'nikov and Lipatnikov, 2011a).

Let us consider an infinitely thin flamelet in a turbulent flow. If the densities of the unburned and burned mixtures are constant, then the velocity fields in the unburned and burned mixtures are incompressible and the following equality holds:

$$\nabla\left[(1-c)\mathbf{u}_u\right] = (1-c)\nabla \cdot \mathbf{u}_u - \mathbf{u}_u \cdot \nabla c = -\mathbf{u}_u \cdot \nabla c = \mathbf{u}_u \cdot \mathbf{n}|\nabla c|, \tag{6.49}$$

where the unit normal vector $\mathbf{n}$ is defined as follows:

$$\mathbf{n}|\nabla c| \equiv -\nabla c. \tag{6.50}$$

Averaging Equation 6.49 yields

$$\nabla\left[(1-\bar{c})\bar{\mathbf{u}}_u\right] = \overline{\nabla(1-c)\mathbf{u}} = \overline{\nabla(1-c)\mathbf{u}_u} = \overline{\nabla(1-c)\mathbf{u}_u} = \overline{\mathbf{u}_u \cdot \mathbf{n}|\nabla c|} \tag{6.51}$$

using Equation 5.36. Therefore,

$$(1-\bar{c})\nabla \cdot \bar{\mathbf{u}}_u = \overline{\mathbf{u}_u \cdot \mathbf{n}|\nabla c|} + \bar{\mathbf{u}}_u \cdot \nabla \bar{c}. \tag{6.52}$$

If we introduce the flamelet surface average:

$$(\bar{q})_f \equiv \frac{\overline{q|\nabla c|}}{\overline{|\nabla c|}}, \tag{6.53}$$

$$\left(q'\right)_f \equiv q - \left(\bar{q}\right)_f, \tag{6.54}$$

then, Equation 6.52 reads

$$
\begin{aligned}
\left(1-\bar{c}\right)\nabla\cdot\bar{\mathbf{u}}_u &= \left[\overline{\left(\mathbf{u}_u\cdot\mathbf{n}\right)}_f - \bar{\mathbf{u}}_u\cdot\left(\bar{\mathbf{n}}\right)_f\right]\cdot\bar{\Sigma} \\
&= \left\{\overline{\left[\left(\bar{\mathbf{u}}_u\right)_f + \left(\mathbf{u}_u'\right)_f\right]\cdot\left[\left(\bar{\mathbf{n}}\right)_f + \left(\mathbf{n}'\right)_f\right]} - \bar{\mathbf{u}}_u\cdot\left(\bar{\mathbf{n}}\right)_f\right\}\cdot\bar{\Sigma} \\
&= \left\{\left(\bar{\mathbf{u}}_u\right)_f\cdot\left(\bar{\mathbf{n}}\right)_f + \overline{\left(\mathbf{u}_u'\right)_f\cdot\left(\mathbf{n}'\right)_f} - \bar{\mathbf{u}}_u\cdot\left(\bar{\mathbf{n}}\right)_f\right\}\cdot\bar{\Sigma} \\
&= \overline{\left(\mathbf{u}_u'\right)_f\cdot\left(\mathbf{n}'\right)_f}\cdot\bar{\Sigma} - \left[\bar{\mathbf{u}}_u - \left(\bar{\mathbf{u}}_u\right)_f\right]\cdot\left(\bar{\mathbf{n}}\right)_f\bar{\Sigma} \\
&= \overline{\left(\mathbf{u}_u'\right)_f\cdot\left(\mathbf{n}'\right)_f}\cdot\bar{\Sigma} + \left[\bar{\mathbf{u}}_u - \left(\bar{\mathbf{u}}_u\right)_f\right]\cdot\nabla\bar{c}, \tag{6.55}
\end{aligned}
$$

where

$$\Sigma \equiv |\nabla c| \tag{6.56}$$

and

$$\left(\bar{\mathbf{n}}\right)_f \equiv \frac{\overline{\mathbf{n}|\nabla c|}}{|\nabla c|} = \frac{\overline{-\nabla c}}{|\nabla c|} = -\frac{\nabla\bar{c}}{\bar{\Sigma}} \tag{6.57}$$

using Equation 6.50.

The quantity $\Sigma(\mathbf{x},t)$, called the flame surface density (FSD), characterizes the area of the instantaneous flamelet surface within an infinitesimal volume around point $\mathbf{x}$ at instant t. Indeed, because $|\nabla c| = \delta(\zeta)$ for infinitely thin flamelets, where $\delta(\zeta)$ is the Dirac delta function and the distance ζ is counted along the local normal to the instantaneous flamelet surface with $\zeta = 0$ at the surface, the integration of $|\nabla c|(\mathbf{x},t)$ over the volume yields the area of the flamelet surface within that volume at instant t. The mean FSD $\bar{\Sigma} = \overline{|\nabla c|}$ plays an important role in various models of premixed turbulent combustion. For instance, if the mass rate of product formation per unit flamelet surface area is assumed to be equal to the rate $\rho_u S_L^0$ in the corresponding unperturbed laminar flame, then the mass rate $\bar{W}$ of product formation per unit volume is equal to the product of $\rho_u S_L^0$ and the area of the flamelet surface in that volume, that is, $\bar{W} = \rho_u S_L^0\bar{\Sigma}$. The FSD models of the influence of turbulence on premixed combustion will be discussed in the section "Two-Equation Models: Flame Surface Density and Scalar Dissipation Rate" in Chapter 7.

Equation 6.55 is the straightforward consequence of the incompressibility assumption. On the face of it, Equation 6.55 appears to be difficult to use because of the unclosed correlation $\overline{(\mathbf{u}_u')_f \cdot (\mathbf{n}')_f}$ and unclosed flow velocity $(\overline{\mathbf{u}}_u)_f \equiv \overline{\mathbf{u}}_{u,f}$ conditioned on the unburned edge of flamelets. However, as already discussed (see Figure 6.9), the available DNS data support a simplification of $(\overline{u}_u)_f \approx \overline{u}_u$ in the largest part of a statistically planar, 1D premixed turbulent flame, with the exception of the leading edge. Moreover, Im et al. (2004) have also reported that, for various $0 < \overline{c} < 1$, the correlation $(\mathbf{u}_u')_f \cdot (\mathbf{n}')_f$ "remains nearly constant, small, and positive." Based on these DNS data, Sabel'nikov and Lipatnikov (2011a) have proposed the following simple closure:

$$(1-\overline{c})\nabla \cdot \overline{\mathbf{u}}_u = bu'\overline{\Sigma} = \frac{bu'\overline{W}}{\rho_u S_L^0}, \qquad (6.58)$$

where b is a constant and u' is the rms turbulent velocity at the leading edge of the turbulent flame brush.

Furthermore, by determining $\nabla \cdot [(1-\overline{c})\overline{\mathbf{u}}_u]$ using the conditioned continuity Equation 6.16 supplemented with Equation 6.29, we can easily rewrite Equations 6.55 and 6.58 as

$$(\overline{\mathbf{u}}_u)_f \cdot \nabla \overline{c} = \left\{1 + \frac{\overline{(\mathbf{u}_u')_f \cdot (\mathbf{n}')_f}}{S_L^0}\right\} \frac{\overline{W}}{\rho_u} - \frac{\partial \overline{c}}{\partial t} = \left\{S_L^0 + \overline{(\mathbf{u}_u')_f \cdot (\mathbf{n}')_f}\right\}\overline{\Sigma} - \frac{\partial \overline{c}}{\partial t}, \qquad (6.59)$$

$$\overline{\mathbf{u}}_u \cdot \nabla \overline{c} = \left(1 + \frac{bu'}{S_L^0}\right)\frac{\overline{W}}{\rho_u} - \frac{\partial \overline{c}}{\partial t} = \left(S_L^0 + bu'\right)\overline{\Sigma} - \frac{\partial \overline{c}}{\partial t}. \qquad (6.60)$$

Equation 6.60 is a closure of Equation 6.59, with the latter equation being exact (in the limit case of infinitely thin unperturbed flamelets) but unclosed. Equation 6.60 offers an opportunity to evaluate the component of the conditioned velocity vector that is normal to the mean flame brush. Subsequently, the BML Equations 5.32 through 5.35 may be used to calculate the turbulent scalar flux normal to the mean flame brush. If either the transverse scalar flux is neglected as compared with the normal scalar flux or the influence of combustion on the former flux is weak (i.e., the transverse flux may be calculated using the gradient diffusion approximation), then Equation 6.60, if thoroughly validated, resolves the problem of modeling turbulent scalar transport. If the influence of combustion on the transverse scalar flux is of substantial importance, then Equation 6.60 is not sufficient to resolve the problem and more sophisticated models should be invoked; however, further discussion of such models is beyond the scope of this book.

The first test of Equation 6.60 yielded encouraging results (Sabel'nikov and Lipatnikov, 2011a). As shown in Figures 6.10 and 6.11, this simple model with the same value, $b = 1.1$, of a single constant predicts well the conditioned velocities and turbulent scalar fluxes obtained from six premixed flames stabilized in impinging jets by four research groups (see the sketch in Figure 4.8b).

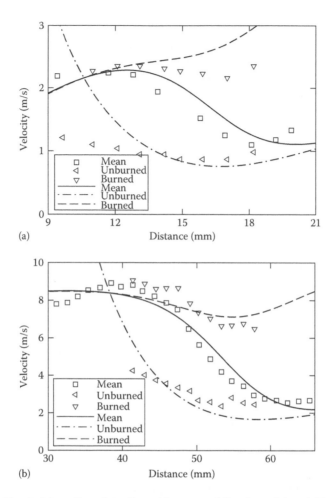

FIGURE 6.10 Axial profiles of the Reynolds-averaged $\bar{u}$ and conditioned ($\bar{u}_u$ and $\bar{u}_b$) axial velocities in case 1 studied by Cho et al. (1988) (a) and in flame s9 investigated by Cheng and Shepherd (1991) (b). The symbols show the experimental data. The curves have been computed by Sabel'nikov and Lipatnikov (2011a).

The success of the first test of the model indicates an important role played by the correlation $(\mathbf{u}'_u)_f \cdot (\mathbf{n}')_f$ in premixed turbulent flames and calls for investigating this correlation in future DNS studies. In particular, the eventual dependence of $(\mathbf{u}'_u)_f \cdot (\mathbf{n}')_f$ on $\bar{c}$, the flame-development time, and the mixture and turbulence characteristics should be addressed. Moreover, further research on the difference between $\bar{\mathbf{u}}_u$ and $(\bar{\mathbf{u}}_u)_f \equiv \bar{\mathbf{u}}_{u,f}$, especially at the leading edge of a turbulent flame brush, is necessary for improving the model.

TRANSITION FROM GRADIENT TO COUNTERGRADIENT TRANSPORT

As reviewed elsewhere (Lipatnikov and Chomiak, 2010, Section 3.3), countergradient scalar transport was observed in many premixed turbulent flames. However, in

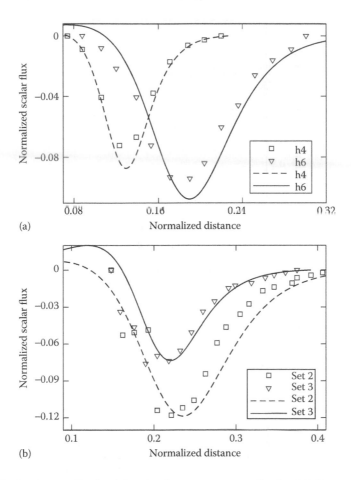

FIGURE 6.11 A normalized axial scalar flux versus a normalized axial distance in flames h4 and h6 studied by Li et al. (1994) (a) and in flames set 2 and set 3 investigated by Stevens et al. (1998) (b). The symbols show the experimental data. The curves have been computed by Sabel'nikov and Lipatnikov (2011a).

many other premixed flames, turbulent scalar transport showed the gradient behavior. For instance, Kalt et al. (2002) documented $\bar{u}_u < \bar{u}_b$ (this inequality is associated with countergradient scalar transport, see Equation 5.35) in weakly turbulent flames stabilized in impinging jets, while the opposite relation was obtained from similar flames characterized by a higher ratio of u'/S_L^0 (see Figure 6.12).

A DNS study of a laminar premixed flame embedded into decaying turbulence at $t = 0$ by Veynante et al. (1997) has shown that either countergradient or gradient transport may dominate in a turbulent flame brush under different conditions. For instance, Figure 6.13 indicates that gradient transport dominates at an early stage of flame development followed by transition to countergradient transport.

Such experimental (Figure 6.12) and DNS (Figure 6.13) data, as well as many other similar data, call for a criterion that would allow a researcher to predict the direction of the turbulent scalar flux in a particular premixed turbulent flame. The earliest and

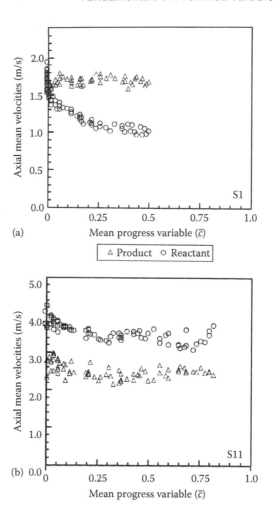

FIGURE 6.12 Axial velocities conditioned on the unburned (*circles*) and burned (*triangles*) mixtures for flames S1 ($u'/S_L^0 = 1.2$) (a) and S11 ($u'/S_L^0 = 5.4$) (b). (Reprinted from Kalt, P.A.M., Chen, Y.-C., and Bilger, R.W., *Combust. Flame*, 129, 401–415, 2002. With permission.)

still the most popular criterion of that kind is called the Bray number in honor of Bray, who first proposed it (Bray, 1995). According to the criterion, turbulent scalar transport shows countergradient (gradient) behavior if the Bray number

$$N_B \equiv \frac{\tau S_L^0}{2\alpha u'} \tag{6.61}$$

is larger (smaller) than unity. Here, α is an unknown parameter of unity order. Veynante et al. (1997) noted that their DNS data indicated an increase in the α-parameter by the ratio of L/δ_L. The Bray number criterion predicts that a transition from gradient

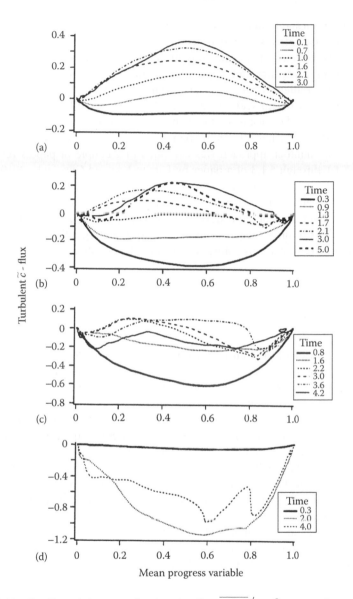

FIGURE 6.13 Profiles of the normalized scalar flux $\overline{\rho u''c''}/(\rho_u S_L^0)$ versus the mean combustion progress variable, computed at different normalized flame-development times tu_0'/L_0 specified in the legends and different ratios of $u_0'/S_L^0 = 2$ (a), 3 (b), 5 (c), and 10 (d) in the statistically planar 1D case. (Reprinted from Veynante, D., Trouvé, A., Bray, K.N.C., and Mantel, T., *J. Fluid Mech.*, 332, 263–293, 1997. With permission.)

to countergradient transport occurs if the ratio of u'/S_L^0 is decreased, for example, due to enrichment of a lean mixture. Such a trend was well documented in various experiments and DNS (e.g., see Figure 6.12), thereby supporting this criterion.

However, as already discussed in the section "A Simplified Discussion of Governing Physical Mechanisms," the Bray number, in fact, highlights two physical mechanisms, turbulent diffusion and local flow acceleration within flamelets, and compares the magnitudes of the two effects (scalar transport shows countergradient behavior if the latter effect is stronger). Thus, the Bray number approach oversimplifies the problem by ignoring other important physical mechanisms (e.g., flow acceleration by the mean pressure gradient within a turbulent flame brush and the influence of chemical reactions on turbulent scalar transport). As a result, the Bray number criterion fails in many cases. For instance, it does not allow for the dependence of the direction of the turbulent scalar flux on the flame-development time, indicated in Figure 6.13. The Bray number approach was extended in a couple of papers reviewed in Sections 5.1.1 through 5.1.3 of a paper by Lipatnikov and Chomiak (2010), but the approach still cannot predict the effects associated with flame development.

Recently, an alternative criterion was proposed (Lipatnikov, 2011c) to highlight the link between turbulent flame development and flux direction. Because the product $\overline{\rho \mathbf{u}''c''} \cdot \nabla \tilde{c}$ is positive (negative) in the case of countergradient (gradient) turbulent scalar transport, it seems natural to consider the equality

$$\int_{-\infty}^{\infty} \overline{\rho \mathbf{u}''c''} \cdot \nabla \tilde{c}\, dx = 0 \qquad (6.62)$$

as a global criterion of the transition from gradient to countergradient scalar transport. Here, integration is performed along the normal to mean flame brush. The scalar flux mainly shows the countergradient (gradient) behavior if the integral on the LHS is positive (negative). Certainly, this is a global, integral criterion, that is, even if the countergradient (gradient) transport dominates in the major part of a turbulent flame brush and the integral on the LHS of Equation 6.62 is positive (negative), the opposite, gradient (countergradient) transport, may be observed in certain parts of the flame brush. For instance, the dotted curve in Figure 6.13a indicates countergradient scalar transport in the major part of a turbulent flame brush, whereas the transport shows the gradient behavior at the leading and trailing edges of the flame brush.

To evaluate the integral on the LHS of Equation 6.62, let us consider a statistically planar, 1D premixed turbulent flame that moves from right to left. Then, taking the integral by parts and using Equation 6.2, we obtain

$$0 = \int_{-\infty}^{\infty} \overline{\rho u''c''} \frac{\partial \tilde{c}}{\partial x}\, dx = -\int_{-\infty}^{\infty} \tilde{c} \frac{\partial \overline{\rho u''c''}}{\partial x}\, dx = \int_{-\infty}^{\infty} \tilde{c}\left[\bar{\rho}\frac{\partial \tilde{c}}{\partial t} + \bar{\rho}\tilde{u}\frac{\partial \tilde{c}}{\partial x} - \overline{W}\right] dx$$

$$= \frac{1}{2}\int_{-\infty}^{\infty} \frac{\partial}{\partial t}\left(\bar{\rho}\tilde{c}^2\right) dx + \frac{1}{2}\int_{-\infty}^{\infty} \frac{\partial}{\partial x}\left(\bar{\rho}\tilde{u}\tilde{c}^2\right) dx - \int_{-\infty}^{\infty} \tilde{c}\overline{W} dx$$

$$= -\frac{1}{2}\frac{\partial}{\partial t}\int_{-\infty}^{\infty} \bar{\rho}\tilde{c}\left(1-\tilde{c}\right) dx + \frac{1}{2}\frac{\partial}{\partial t}\int_{-\infty}^{\infty} \bar{\rho}\tilde{c}\, dx + \frac{1}{2}\rho_b U_b - \int_{-\infty}^{\infty} \tilde{c}\overline{W} dx, \qquad (6.63)$$

where U_b is the mean flow velocity at $x \rightarrow \infty$, that is, in the products. The integration of Equation 6.2 yields

$$\frac{\partial}{\partial t} \int_{-\infty}^{\infty} \overline{\rho} \tilde{c} dx + \rho_b U_b = \int_{-\infty}^{\infty} \overline{W} dx, \tag{6.64}$$

and we finally obtain

$$\int_{-\infty}^{\infty} \overline{\rho u'' c''} \frac{\partial \tilde{c}}{\partial x} dx = -\frac{1}{2} \frac{\partial}{\partial t} \int_{-\infty}^{\infty} \overline{\rho} \tilde{c} (1 - \tilde{c}) dx + \int_{-\infty}^{\infty} \left(\frac{1}{2} - \tilde{c} \right) \overline{W} dx. \tag{6.65}$$

By virtue of Equations 6.62 and 6.65, the transition from gradient to countergradient scalar transport is associated with the following criterion:

$$\frac{1}{2} \frac{\partial}{\partial t} \int_{-\infty}^{\infty} \overline{\rho} \tilde{c} (1 - \tilde{c}) dx = \int_{-\infty}^{\infty} \left(\frac{1}{2} - \tilde{c} \right) \overline{W} dx. \tag{6.66}$$

The turbulent scalar flux shows the gradient (countergradient) behavior if the LHS of Equation 6.66 is greater (smaller) than the RHS.

Equation 6.66 is a straightforward consequence (i) of Equation 6.2, which is asymptotically exact at high Reynolds numbers and (ii) of the chosen global criterion given by Equation 6.62. Let us further simplify the obtained result by invoking certain model assumptions.

First, as discussed in the section "Mean Structure of a Premixed Turbulent Flame" in Chapter 4, the mean structure of a typical premixed turbulent flame is self-similar. Therefore, it is tempting to assume that $\overline{\rho}(x,t) = \overline{\rho}(\xi)$ and $\tilde{c}(x,t) = \tilde{c}(\xi)$ in Equation 6.66, with the normalized distance ξ being determined by Equation 4.33. Then, Equation 6.66 reads

$$\frac{1}{2} \frac{d\delta_t}{dt} \int_{-\infty}^{\infty} \overline{\rho} \tilde{c} (1 - \tilde{c}) d\xi = \int_{-\infty}^{\infty} \left(\frac{1}{2} - \tilde{c} \right) \overline{W} dx \tag{6.67}$$

and the time derivative of the mean flame brush thickness may be calculated by invoking Equation 4.2.

Second, considering that that the turbulent burning velocity is controlled by the integrated mean reaction rate, that is,

$$\rho_u U_t = \int_{-\infty}^{\infty} \overline{W} dx \tag{6.68}$$

in the statistically planar, 1D flame studied here, Equation 6.67 results in

$$
\frac{1}{2}\frac{d\delta_t}{dt} = \rho_u U_t \frac{\displaystyle\int_{-\infty}^{\infty}\left(\frac{1}{2}-\tilde{c}\right)\bar{W}dx}{\displaystyle\int_{-\infty}^{\infty}\bar{W}dx \cdot \int_{-\infty}^{\infty}\bar{\rho}\tilde{c}\left(1-\tilde{c}\right)d\xi}. \tag{6.69}
$$

Third, if we assume that a fully developed turbulent burning velocity is reached after a sufficiently long flame-development stage, that is, $U_t(t \to \infty) \to U_{t,\infty}$, then Equation 6.69 may be further simplified as

$$
\frac{1}{u_t}\frac{dd_t}{d\vartheta} = \frac{U_{t,\infty}}{u'}\Xi, \tag{6.70}
$$

where

$u_t \equiv U_t/U_{t,\infty}$ is the normalized burning velocity
$d_t \equiv \delta_t/L$ is the normalized flame brush thickness
$\vartheta = tu'/L$ is the normalized time

and the parameter

$$
\Xi \equiv \frac{\displaystyle\int_{-\infty}^{\infty}\left(1-2\tilde{c}\right)\bar{W}dx}{\displaystyle\int_{-\infty}^{\infty}\bar{W}dx \cdot \int_{-\infty}^{\infty}\frac{\bar{\rho}}{\rho_u}\tilde{c}\left(1-\tilde{c}\right)d\xi} \tag{6.71}
$$

is increased by the density ratio σ due to the second integral in the denominator. If the mean reaction rate depends on x and t only because of the dependence of $\bar{W}$ on $\bar{\rho}$ and $\tilde{c}$, then

$$
\bar{W}\left[\tilde{c}(x,t),\frac{\bar{\rho}(x,t)}{\rho_u}\right] = \bar{W}\left[\tilde{c}(\xi),\frac{\bar{\rho}(\xi)}{\rho_u}\right] = \Omega(\xi,\sigma) \tag{6.72}
$$

and the parameter Ξ depends solely on the density ratio.

Finally, if Equation 4.2 is invoked to evaluate the rate of increase of the thickness and Equation 7.29, discussed in the next chapter, is used, then the LHS of Equation 6.70 depends solely on the normalized time, that is,

$$
\frac{1}{u_t}\frac{dd_t}{d\vartheta} = \Gamma(\vartheta). \tag{6.73}
$$

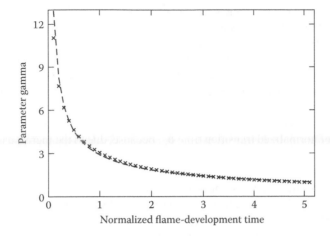

FIGURE 6.14 The function $\Gamma(\vartheta)$ calculated using Equations 4.2, 6.73, and 7.29 (*crosses*) and its approximation (*dashed curve*) given by Equation 6.74. (With kind permission from Springer Science + Business Media: *Flow, Turbulence and Combustion*, Transient behavior of turbulent scalar transport in premixed flames, 86, 2011, 609–637, Lipatnikov, A.N., Figure 7, Copyright Springer.)

The function $\Gamma(\vartheta)$ is shown as crosses in Figure 6.14. The dashed curve approximates this function using the following expression:

$$\Gamma(\vartheta) \approx 2.9\vartheta^{-0.65}. \tag{6.74}$$

The unsteady terms on the LHSs of Equation 6.66 or 6.70 clearly indicate the effect of premixed turbulent flame development on the transition from gradient to countergradient scalar transport. As a flame develops, both the burning velocity and the mean flame brush thickness increase, but these processes slow down with time (see the section "Development of Premixed Turbulent Flame in the Flamelet Regime" and, in particular, Figure 5.19 in Chapter 5. Accordingly, the LHS of Equation 6.70 decreases (initially, it is large due to a low u_t) and the transition from gradient to countergradient transport occurs at a certain instant ϑ_{tr}, provided that the parameter $\Xi > 0$.

Let us compare the criterion given by Equation 6.69 or Equations 6.70 and 6.71 with the available experimental and DNS data.

First, the transition from gradient to countergradient scalar transport during premixed turbulent flame development is observed in Figure 6.13, as well as in Figure 1b published by Swaminathan et al. (1997), in Figure 12 reported by Veynante and Poinsot (1997), and in Table 4 (see Le = 0.34 and 0.6) from the paper by Chakraborty and Cant (2009). The transition from gradient to countergradient scalar flux with the distance from a flame holder was experimentally documented by Veynante et al. (1996) and by Pfadler et al. (2007), but the fluxes measured by them were almost tangential to the mean flame brushes. A similar transition was obtained in a DNS by Domingo et al. (2005b) (see Figure 8 in the cited paper).

Second, if the laminar flame speed is increased, then both the turbulent burning velocity (see the section "Dependence of Turbulent Burning Rate on RMS Velocity and Laminar Flame Speed" in Chapter 4) and the RHSs of Equations 6.69 and 6.70 are increased. An increase in the density ratio acts in the same way, because the second integral in the denominator on the RHS of Equation 6.69 is decreased when σ is increased and the parameter Ξ is also increased by σ. Therefore, the magnitude of the LHS of Equation 6.70, that is, the function Γ defined by Equation 6.73, evaluated at the transition instant, should be increased by S_L^0 and/or σ. A higher Γ is associated with a shorter normalized transition time ϑ_{tr}, because, due to the increase in $u_t(\vartheta)$ and the decrease in $d\delta_t/dt$ with time (see Figures 4.6 and 4.7), the function Γ is decreased when ϑ is increased (see Figure 6.14 and Equation 6.74).

If, moreover, the flame-development time is weakly affected by variations in S_L^0 and/or σ, we may observe gradient ($t < t_{tr}$) transport for lean mixtures characterized by a lower σS_L^0 and, hence, by a longer t_{tr}, but countergradient ($t > t_{tr}$) transport for near-stoichiometric mixtures characterized by a higher σS_L^0, and, hence, by a shorter t_{tr}. The discussed model problem is relevant to the following experiments. Kalt et al. (1998) and Frank et al. (1999) measured turbulent scalar fluxes in various Bunsen flames at a fixed distance X from the burner exit. When the equivalence ratio was varied in the same oncoming flow (natural gas–air flames A, B, D, and E or propane–air flames A, B, and D), these authors documented $\left.|\overline{u}|\right._b < \left.|\overline{u}|\right._u$ in the leanest flames D ($\Phi = 0.6$) characterized by the lowest product σS_L^0, whereas $\left.|\overline{u}|\right._b$ was higher than $\left.|\overline{u}|\right._u$ in the flames A, B, and E. Troiani et al. (2009) documented countergradient scalar flux from near-stoichiometric open methane–air flames stabilized by a bluff body, whereas the behavior of the flux $\overline{\rho u'' c''}$ was gradient in a lean flame ($\Phi = 0.67$). Because neither the oncoming turbulence characteristics nor the flame-development time was substantially varied in these experiments, the observed transition from gradient to countergradient transport with the enrichment of lean mixtures is consistent with the trend discussed earlier, thereby qualitatively supporting the criterion given by Equation 6.69 or Equations 6.70 and 6.71.

Third, because the turbulent burning velocity is increased when the Lewis number is decreased, see the section "Dependence of Turbulent Burning Velocity on Differences in Molecular Transport Coefficients" in Chapter 4, the RHS of Equation 6.70 seems to be greater for mixtures characterized by a lower Le, that is, a decrease in the Lewis number should promote the transition (i.e., decrease in ϑ_{tr}) from gradient to countergradient scalar transport. This trend agrees with the results of a recent DNS by Chakraborty and Cant (2009) that yielded a countergradient scalar flux for Le ≤ 0.6, but a gradient flux if Le ≥ 0.8, with all other things being equal.

Fourth, if the turbulence length scale L is increased, with all other things being equal, then the ratio of $U_{t,\infty}/u'$ on the RHS of Equation 6.70 also appears to be increased (e.g., see Equations 4.18 through 4.25). Accordingly, a shorter normalized transition time, ϑ_{tr}, is associated with the transition from gradient to countergradient transport at a larger L, provided that the LHS of Equation 6.70 depends solely on ϑ in a first approximation (e.g., see Equation 6.73).

However, the dimensional transition time appears to show the opposite behavior and to be increased by L, because the time $t_{tr} = \vartheta_{tr} L/u'$ is proportional to the length scale. For instance, Equation 6.74 yields $\Gamma \propto L^{0.65}$, while various empirical parameterizations of the turbulent burning velocity result in $U_t \propto L^q$ with $q \leq 1/4$ (see Equations 4.18 through 4.25). Therefore, the ratio of $\Gamma u'/U_{t,\infty}$ is increased by L, that is, the transition discussed earlier is impeded by an increase in the turbulence length scale (the transition time should be longer to compensate for the increase in $\Gamma u'/U_{t,\infty}$ by L).

Such a trend is consistent with the DNS data obtained by Veynante et al. (1997), which indicate an increase in the α parameter in the Bray number (see Equation 6.61) by the ratio of L/δ_L (see Figure 15 in the cited paper). If the same value of the Bray number, N_B, is associated with the studied transition under various conditions, then an increase in α is equivalent to a decrease in σS_L^0, that is, the transition is impeded by an increase in L/δ_L, in line with the present analysis. Moreover, the empirical parameterization of the Bray number reported by Chen and Bilger (2000) (see Equation 4 in the cited paper) and supported experimentally by Kalt et al. (2002) also indicates that the transition is impeded by an increase in L/δ_L.

Fifth, if u' is increased, with all other things being equal, then the ratio of $U_{t,\infty}/u'$ on the RHS of Equation 6.70 seems to be decreased (see Equations 4.18 through 4.25). Accordingly, a longer normalized transition time, ϑ_{tr}, is associated with the transition from gradient to countergradient transport at a higher u', provided that the LHS of Equation 6.70 depends solely on ϑ in a first approximation.

This trend agrees with the aforementioned DNS data obtained by Veynante et al. (1997). Inspection of Figure 6.13 shows that the transition discussed earlier occurred at $\vartheta_{tr} \approx 0.7$ if $u'/S_L^0 = 2$, $1.3 < \vartheta_{tr} < 1.7$ if $u'/S_L^0 = 3$, and $1.6 < \vartheta_{tr} < 2.2$ if $u'/S_L^0 = 5$, that is, an increase in the initial (turbulence decayed in these simulations) rms turbulent velocity u'_0 resulted in an increase in the normalized transition time, in line with the criterion given by Equation 6.70.

Sixth, on the contrary, if the increase in the mean flame brush thickness is modeled by invoking Equation 4.2 and the parameter Ξ is assumed to be independent of u', then the dimensional transition time $t_{tr} = \vartheta_{tr}L/u'$ yielded by Equations 6.70 through 6.74 appears to decrease with increasing u'. For instance, Equation 6.74 yields $\Gamma \propto u'^{-0.65}$, while various empirical parameterizations of the turbulent burning velocity imply that $U_{t,\infty}/u' \propto u'^q$ with $-0.5 < q < 0$ (see Equations 4.18 through 4.25). Therefore, when the rms turbulent velocity is increased, the ratio of $\Gamma u'/U_{t,\infty}$, given by Equations 6.70 through 6.74, is decreased and the transition from gradient to countergradient scalar transport may be predicted for a fixed flame-development time.

Such a trend does not seem to agree with the aforementioned experiments by Kalt et al. (1998) and Frank et al. (1999), who increased u' in the oncoming flow by moving a grid that generated turbulence toward the burner exit. Such measurements were performed in lean ($\Phi = 0.7$) natural gas–air and propane–air flames B and C. For both fuels, the increase in u' substantially reduced the magnitude of the slip velocity while $|\overline{u}|_b < |\overline{u}|_u$ in all the cases cited. Thus, these experimental data are associated with the transition from gradient to countergradient transport with a decreasing rms turbulent velocity, contrary to the trend that results from Equations 6.70 through 6.74, as discussed earlier.

Moreover, a decrease in the dimensional transition time, t_{tr}, with an increase in u', yielded by Equations 6.70 through 6.74, seems to be inconsistent with the DNS data obtained by Veynante et al. (1997). For instance, the aforementioned Figure 6.13 and Table 1 from the cited paper show that the transition from gradient to counter-gradient transport occurs at $t_{tr}S_L^0/\delta_L \approx 3.9$ if $u_0' = 2S_L^0$ and at $4.8 < t_{tr}S_L^0/\delta_L < 6.2$ if $u_0' = 3S_L^0$, that is, the transition time is increased by u_0'.

The aforementioned inconsistency between Equations 6.70 through 6.74 and the experimental data obtained by Kalt et al. (1998) and Frank et al. (1999) from flame C, characterized by a higher u' as compared with the other flames investigated by them, may be caused, in particular, by the use of Equation 4.2 to calculate the rate of increase of thickness and by an assumption that the parameter Ξ is independent of t, u', etc. For instance, recent numerical simulations (Lipatnikov, 2011c) have shown that the influence of combustion on turbulence not only causes countergradient scalar transport but also reduces the increase in the normalized mean flame brush thickness δ_t/L as compared with Equation 4.2, with the effect being more pronounced at a lower u'/S_L^0. Accordingly, the time derivative on the LHS of Equation 6.70 is increased by u'/S_L^0. Therefore, at a larger u'/S_L^0, a longer flame-development time is required in order for the LHS of Equation 6.70 to be smaller than the RHS. Consequently, the weakening of the influence of combustion on the rate of increase of the normalized thickness $d_t = \delta_t/L$ contributes to an increase in the transition time by u'.

Furthermore, the previously discussed DNS data obtained by Veynante et al. (1997) imply that the parameter Ξ depends on the flame-development time and u'. First, Figure 6.13c indicates that turbulent scalar transport shows gradient behavior at $\vartheta = 0.8$ and 1.6; countergradient behavior at $\vartheta = 2.2$, 3.0, and 3.6; and, then, again gradient behavior at $\vartheta = 4.2$. In order for Equation 6.70 (which is asymptotically exact at high Reynolds numbers for a flame with a self-similar mean structure, provided that a finite fully developed turbulent burning velocity is reached at $t \to \infty$) to yield the latter transition from countergradient to gradient transport at $3.6 < \vartheta < 4.2$, we have to assume that Ξ decreases as the flame develops. Indeed, the LHS of Equation 6.70 monotonically decreases with time, because the increase in the mean flame brush thickness slows down. The ratio of $U_{t,\infty}/u'$ on the RHS of Equation 6.70 may depend on time due to the decay of turbulence in the discussed DNS. However, because the rms velocity decreases in decaying turbulence, while the integral length scale increases, the ratio of $U_{t,\infty}/u'$ should increase with time (see Equations 4.18 through 4.25). Therefore, the LHS of Equation 6.70 may become greater than the RHS at large ϑ only if the parameter Ξ decreases (and, maybe, become negative) as the flame develops. In recent simulations by Lipatnikov and Sabel'nikov (2011), a decrease in Ξ during flame development and a negative Ξ were observed.

Second, Figure 6.13d indicates that scalar transport always shows gradient behavior at $u_0'/S_L^0 = 10$. However, if Ξ were positive and depended neither on t nor on u', then a transition from gradient to countergradient scalar transport would be observed at any u_0'/S_L^0, as large as we want, provided that the fully developed stage of turbulent flame propagation is reached at $t \to \infty$, and, therefore, $d\delta_t/dt = 0$ and the LHS of Equation 6.70 becomes infinitesimally small as $t \to \infty$.

This claim may be supported by the following simple example (Zimont et al., 2001). In the statistically stationary, planar, 1D case, Equation 6.2 reads

$$\frac{d}{dx}\left(\overline{\rho}\tilde{u}\tilde{c}\right) = -\frac{d}{dx}\,\overline{\rho u''c''} + \overline{W}, \tag{6.75}$$

while the Favre-averaged continuity equation has the following solution:

$$\rho_u U_u \equiv \rho_u \tilde{u}\left(x \to -\infty\right) = \overline{\rho}\tilde{u} = \rho_b \tilde{u}\left(x \to \infty\right) \equiv \rho_b U_b \tag{6.76}$$

for a flame that moves from right to left. Integration of Equation 6.75 from $-\infty$ to x yields

$$\overline{\rho u''c''} = \int_{-\infty}^{x}\left(\overline{W} - \rho_u U_t\,\frac{d\tilde{c}}{d\xi}\right)d\xi. \tag{6.77}$$

Therefore, if the mean reaction rate depends on x by means of the dependence of $\overline{W}$ on $\tilde{c}$ and $\overline{\rho}$ and, hence, the parameter Ξ defined by Equation 6.71 depends solely on the density ratio, as assumed earlier, then in the fully developed flame considered, the sign of the flux $\overline{\rho u''c''}$ depends solely on the sign of the difference between $\tilde{c}_{\overline{W}}$ and $\tilde{c}_{\nabla\tilde{c}}$. Here, $\tilde{c}_{\overline{W}}$ and $\tilde{c}_{\nabla\tilde{c}}$ are the values of the Favre-averaged combustion progress variable, associated with the maxima of dependencies of $\overline{W}(\tilde{c})$ and $\nabla\tilde{c}(\tilde{c})$, respectively. If $\tilde{c}_W < \tilde{c}_{\nabla\tilde{c}}$ (cf. the dashed and solid curves in Figure 6.15), then Equation 6.77 yields $\overline{\rho u''c''} > 0$, that is, countergradient scalar transport. On the contrary, if $\tilde{c}_W > \tilde{c}_{\nabla\tilde{c}}$ (cf. the dotted-dashed and solid curves in Figure 6.15), then $\overline{\rho u''c''} < 0$ and scalar transport shows the gradient behavior. It is worth remembering that the areas under the three curves in Figure 6.15 are equal to one another by virtue of Equation 6.77 with $x = \infty$.

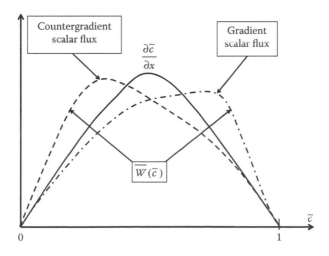

FIGURE 6.15 The relation between the maxima of the dependencies of $\overline{W}(\tilde{c})$ and $\nabla\tilde{c}(\tilde{c})$ and the direction of the turbulent scalar flux in a fully developed, statistically planar, 1D flame.

Thus, in a fully developed, statistically planar, 1D flame, the direction of the turbulent scalar flux is controlled solely by the relative positions of the maxima of dependencies of $\bar{W}(\tilde{c})$ and $\nabla\tilde{c}(\tilde{c})$ (Zimont et al., 2001). In such a flame, the flux direction may depend on u'/S_L^0 (or the Bray number) only if either $\tilde{c}_{\bar{W}}$ or $\tilde{c}_{\nabla\tilde{c}}$ depends on u'/S_L^0. Note that the experimental data discussed in the section "Mean Structure of a Premixed Turbulent Flame" in Chapter 4 imply that at least $\tilde{c}_{\nabla\tilde{c}}$ is independent of this ratio. Accordingly, in order for turbulent scalar transport to show the countergradient behavior at low u'/S_L^0 and the gradient behavior at large u'/S_L^0, we have to assume that $\tilde{c}_{\bar{W}}$ is increased by u', at least in the case of a fully developed, statistically planar, 1D flame.

The criterion given by Equations 6.70 and 6.71 is fully consistent with the aforementioned simple reasoning. Indeed, if a fully developed flame is assumed to be reached at $t \to \infty$, then $d\delta_t/dt \to 0$ and the transition to countergradient transport is inevitable if $\Xi(t \to \infty) > 0$. Because

$$\int_{-\infty}^{\infty} (1-2\tilde{c})\frac{\partial\tilde{c}}{\partial x} dx = 1 - \int_{-\infty}^{\infty} \frac{\partial\tilde{c}^2}{\partial x} dx = 0, \tag{6.78}$$

Equation 6.71 may be rewritten as

$$\Xi = \frac{\int_{-\infty}^{\infty} (1-2\tilde{c})\left(\bar{W} - \frac{\partial\tilde{c}}{\partial x}\right) dx}{\int_{-\infty}^{\infty} \bar{W} dx \cdot \int_{-\infty}^{\infty} \frac{\bar{\rho}}{\rho_u} \tilde{c}(1-\tilde{c}) d\xi}. \tag{6.79}$$

Since the denominator of Equation 6.79 is always positive, the sign of Ξ is controlled by the sign of the integral in the nominator. This integral is positive (negative) if $\tilde{c}_W < \tilde{c}_{\nabla\tilde{c}}$ ($\tilde{c}_W > \tilde{c}_{\nabla\tilde{c}}$) (see Figure 6.15) in line with the foregoing discussion.

In summary, the direction of turbulent scalar transport in a premixed flame is substantially affected not only by the ratio of $\tau S_L/u'$ but also by the flame development and shape of the $\bar{W}(\tilde{c})$ curve (more precisely, by the value of $\tilde{c}_{\bar{W}}$).

Even if the vast majority of the currently available closure relations for $\bar{W}(\tilde{c})$ (see the so-called algebraic models discussed in the section "Algebraic Models" in Chapter 7) yield $\tilde{c}_{\bar{W}}$ independent of the flame-development time and u', such dependencies are theoretically admissible. For instance, recently, Lipatnikov and Sabel'nikov (2011) combined an extended Equation 6.60 with the so-called TFC model discussed in the section "Models of a Self-Similarly Developing Premixed Turbulent Flame" in Chapter 7, in order to study a statistically planar, 1D premixed flame that propagated in frozen turbulence. The numerical results obtained indicate that $\tilde{c}_{\bar{W}}$ is increased as such a flame develops, and the $\tilde{c}_{\bar{W}}(t)$ dependence is able to cause a transition from countergradient to gradient transport. Definitely, further research on this issue is strongly required.

TURBULENCE IN PREMIXED FLAMES

FLAME-GENERATED TURBULENCE

Generally speaking, the effects of a premixed flame on turbulence can be divided into indirect and direct effects. The former effects are not specific to combustion and similar phenomena exist in nonreacting flows. For instance, first, turbulence decays due to viscous dissipation in both flames and nonreactive flows; however, if Re_t is not very large, combustion can accelerate this process by increasing the viscosity of the mixture. Second, turbulence is commonly generated due to the spatial nonuniformities of the mean flow (e.g., see the first two terms on the RHS of Equation 3.88 or term P in Equation 3.95), but heat release can enhance such turbulence generation due to an increase in the mean flow velocity and its gradients within a flame brush (Williams et al., 1949). Such indirect effects are of minor interest, and, therefore, the focus of the following discussion will be on the latter, direct effects, which are specific to flames.

There are different approaches to discussing such direct effects. On the one hand, we may consider simple physical models of them. On the other hand, the influence of heat release on various terms in the balance equation for turbulent kinetic energy could be studied to highlight the terms that vanish in a constant-density flow but play an important role in flames. Alternatively, a similar analysis could be applied to the vorticity equation as turbulence is the rotational flow. In the following text, we will use all the three approaches and we will try to reveal the links between the simple models and the behavior of the aforementioned equations in flames. Let us begin with discussing simple models.

In the contemporary combustion literature, posing the issue of the influence of combustion on turbulence is often attributed to Karlovitz, who first put forward a simple model of the direct effect of a premixed flame on a turbulent flow (Karlovitz et al., 1951). It is worth remembering, however, that an important role played by the so-called "flame-generated" turbulence in premixed combustion was earlier hypothesized by Scurlock (1948), who restricted himself to discussing an indirect effect of combustion on turbulence, already mentioned earlier, that is, the enhancement of turbulence production due to an increase in the mean velocity gradients by heat release. Moreover, a simple model (see the section "Hydrodynamic Instability" in Chapter 5) straightforwardly relevant to the discussed issue was developed by Darrieus (1938) and Landau (1944) well before Karlovitz. Nevertheless, let us consider Karlovitz's model at first.

Owing to the heat release, density drop, and pressure decrease, an instantaneous flame front is "a flow source that introduces a velocity into a gas flow" (Karlovitz et al., 1951). If perturbations in the internal structure of the front and in the local burning rate, caused by turbulent eddies, are neglected, then in the coordinate framework attached to the flame front, the normal (to the front) component of the local flow velocity jumps from S_L^0 just upstream of the front to σS_L^0 just downstream of the front. On account of random orientation of the instantaneous flame front (see the wrinkled solid curve in Figure 6.16) with respect to the normal $\mathbf{n}_t = \nabla \bar{c} / |\nabla \bar{c}|$ to a mean flame surface $\bar{c} = $ const (the thick dashed straight line is parallel to the mean

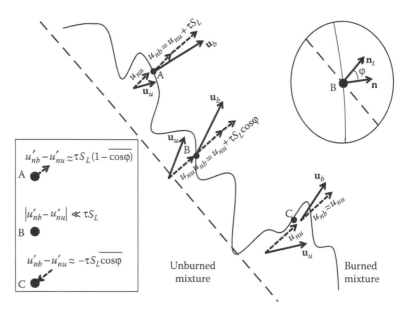

FIGURE 6.16 The generation of velocity oscillations by a wrinkled, thin laminar flame front.

flame surface in Figure 6.16), the projection δu_n of the local velocity induced by the flame front on the normal $\mathbf{n}_t$ oscillates from τS_L^0 to $-\tau S_L^0$. For instance, at point A, the flame front is almost parallel to the mean flame surface and $\delta u_n \approx \tau S_L^0$, that is, $u_{n,b} \approx u_{n,u} + \tau S_L^0$. Note that, in the laboratory coordinate framework, $u_{n,u}$ does not vanish, but is equal to the velocity at which the flame front is convected by the flow, that is, $u_{n,u} = \mathbf{u}_u \cdot \mathbf{n}_t - S_L^0$. At point B (see the circle inset in the upper right corner of Figure 6.16), $\delta u_n = \tau S_L^0 \cos\varphi = \tau S_L^0 \mathbf{n} \cdot \mathbf{n}_t$, where $\mathbf{n}$ is the unit normal to the local flame front (see Equation 6.50). At point C, the flame front is almost perpendicular to the mean flame surface and $|\delta u_n| \ll \tau S_L^0$. After averaging along the flame front, we obtain $\overline{\delta u_n} = \tau S_L^0 \overline{\cos\varphi} = \tau S_L^0 \overline{\mathbf{n}} \cdot \mathbf{n}_t$, that is, the discussed mechanism accelerates the mean flow in the $\mathbf{n}_t$-direction. Moreover, owing to random orientation of the front, the same physical mechanism also contributes to velocity fluctuations. Indeed, at point A, $\delta u_n - \overline{\delta u_n} = \tau S_L^0 - \tau S_L^0 \overline{\cos\varphi}$, that is, the velocity fluctuation of the magnitude $\tau S_L^0 (1 - \overline{\cos\varphi})$ is induced by the flame in the burned gas in the vicinity of point A (see the rectangular inset in the down left corner of Figure 6.16). At point B, where $\cos\varphi \approx \overline{\cos\varphi}$, the flame-induced fluctuation in the normal velocity u_n is weak, while the fluctuation is negative ($-\tau S_L^0 \overline{\cos\varphi}$) at point C.

Although Karlovitz et al. (1951) obtained an expression for estimating flame-generated rms velocity in burned mixture (see Equation 16 in the cited paper), we will not discuss this part of his model, because the estimate was not supported by subsequent experimental and numerical studies discussed in Section 2.2.1 of a review paper by Lipatnikov and Chomiak (2010). For our own purposes, it is sufficient to stress that the previously discussed physical mechanism, that is, local flow accelerations in randomly oriented instantaneous flame fronts due to the local pressure drop across the fronts, can generate velocity fluctuations in combustion products. However, in order for a flame to self-accelerate, turbulence should also be generated in the upstream flow of the unburned mixture. Although Karlovitz et al. (1951) aimed at studying combustion self-acceleration due to flame-generated turbulence, they did not develop a model of enhanced turbulence production in the unburned mixture within a turbulent flame brush, but they restricted themselves to assuming that turbulence produced in the products "diffuses against the flow before the flame front."

It is also worth noting that Karlovitz's model highlights the local pressure gradient in the flame front but does not allow for perturbations in the upstream pressure field induced by heat release in the curved front. In other words, the model neglects the effects associated with the mean pressure gradient in a flame brush, whereas such effects may be of importance at least as far as scalar transport in turbulent flames is concerned (see Figures 6.4 through 6.6).

Furthermore, an increase in the rms velocity yielded by Karlovitz's model in the burned mixture does not necessarily mean turbulence generation in a general case. Because turbulence is a rotational motion, measuring high rms velocity in a flow is not sufficient to claim that the flow is turbulent. The high rms velocity may result from fluctuations in a potential flow field. For instance, near the outer edge of a free turbulent shear flow (e.g., a jet or a wake), the rms velocity weakly varies, while the velocity gradients drop sharply, thereby indicating that the ambient fluid is nonturbulent even if the potential velocity fluctuations are large therein (Townsend, 1976; Kuznetsov and Sabel'nikov, 1990). This well-known phenomenon is called intermittency of shear flows. The Darrieus–Landau (DL) instability of laminar flames, briefly considered in the section "Hydrodynamic Instability" in Chapter 5 and further discussed later, is another example of potential velocity fluctuations that should not be confused with turbulent motion.

The problem of the hydrodynamic instability of a laminar flame, posed by Darrieus (1938) and Landau (1944), offers us the opportunity to attain another insight into the influence of combustion on turbulence. The statement of the problem and the physical mechanism of the instability were discussed at the start of the section "Hydrodynamic Instability" in Chapter 5. For the present purposes, it is also worth reporting the following equations:

$$\zeta = a e^{\omega t} \cos(\kappa y),$$

$$u_u = a\omega e^{\kappa x + \omega t} \cos(\kappa y),$$

$$v_u = -a\omega e^{\kappa x + \omega t} \sin(\kappa y),$$

$$u_b = \underbrace{a\omega\sigma \frac{S_L^0 \kappa + \omega}{\sigma S_L^0 \kappa - \omega} e^{-\kappa x + \omega t} \cos(\kappa y)}_{u_{b1}} - \underbrace{a\omega \frac{(\sigma+1)\omega}{\sigma S_L^0 \kappa - \omega} e^{\omega\left[t - x/\left(\sigma S_L^0\right)\right]} \cos(\kappa y)}_{u_{b2}},$$

$$v_b = \underbrace{a\omega\sigma \frac{S_L^0 \kappa + \omega}{\sigma S_L^0 \kappa - \omega} e^{-\kappa x + \omega t} \sin(\kappa y)}_{v_{b1}} - \underbrace{a\omega \frac{\omega}{\sigma S_L^0 \kappa} \frac{(\sigma+1)\omega}{\sigma S_L^0 \kappa - \omega} e^{\omega\left[t - x/\left(\sigma S_L^0\right)\right]} \sin(\kappa y)}_{v_{b2}},$$

$$p_u = -\rho_u a\omega \frac{S_L^0 \kappa + \omega}{\kappa} e^{\kappa x + \omega t} \cos(\kappa y),$$

$$p_b = -\rho_u a\omega \frac{S_L^0 \kappa + \omega}{\kappa} e^{-\kappa x + \omega t} \cos(\kappa y),$$

$$\omega = \frac{\sigma S_L^0 \kappa}{\sigma + 1} \left(\sqrt{1 + \sigma - \frac{1}{\sigma}} - 1 \right), \tag{6.80}$$

which result straightforwardly from the analysis by Darrieus (1938) and Landau (1944) in the 2D case. Here, the x-axis is normal to the mean flame and points to the burned mixture, the y-axis is tangential to the mean flame surface $x = 0$, ζ is a perturbation of the x-coordinate of the flame, a and κ are the amplitude and wave number of the perturbation with $a\kappa \ll 1$, u and v are the x- and y-components, respectively, of the flame-induced perturbations in the flow velocity vector, p_u and p_b are the flame-induced perturbations in the pressure field, and $\omega > 0$ is the perturbation growth rate. For derivation of these equations, interested readers are referred, for example, to textbooks by Zel'dovich et al. (1985), Landau and Lifshitz (1987), and Chomiak (1990).

The following points are worth emphasizing. First, the growth rate ω is positive for any wave number if the density ratio $\sigma > 1$ but vanishes in the constant-density case.

Second, the flow velocity $\mathbf{u}_b$ in a burned mixture is decomposed into potential contribution $\mathbf{u}_{b1}$ and solenoidal contribution $\mathbf{u}_{b2}$, while the velocity perturbations in an unburned gas are potential. Indeed, one can easily show that $\boldsymbol{\omega}_u = \nabla \times \mathbf{u}_u = 0$ and $\boldsymbol{\omega}_{b1} = \nabla \times \mathbf{u}_{b1} = 0$, while the z-component of the vorticity vector $\boldsymbol{\omega}_{b2} = \nabla \times \mathbf{u}_{b2}$ is as follows:

$$\boldsymbol{\omega} \cdot \mathbf{e}_z = -a\omega \frac{(\sigma + 1)\omega\kappa}{\sigma S_L^0 \kappa - \omega} \left[\left(\frac{\omega}{\sigma S_L^0 \kappa} \right)^2 - 1 \right] e^{\omega \left[t - x / \left(\sigma S_L^0 \right) \right]} \sin(\kappa y). \tag{6.81}$$

Therefore, the flame generates vorticity downstream. Such vorticity generation may straightforwardly be associated with turbulence production, but this process is restricted to burned mixtures within the framework of the DL analysis. As will be discussed later, the vorticity production by a perturbed laminar flame is controlled by the pressure gradient within the flame. The same pressure gradient also plays a crucial role in Karlovitz's model.

Third, the flame perturbs the pressure field and the pressure perturbations, in turn, induce potential velocity oscillations both upstream and downstream, with the magnitude of the downstream oscillations being larger, that is, $|\mathbf{u}_{b1}| > |\mathbf{u}_u|$ at $x \to 0$. The potential velocity oscillations are controlled not only by the pressure gradient within the flame but also by the pressure gradients in the unburned and burned mixtures. The latter gradients are not addressed by Karlovitz's model. Certainly, potential velocity oscillations in a laminar flow should not be confused with turbulence, but potential velocity perturbations induced by a flame in a turbulent flow can be transformed to turbulence by viscous forces.

Fourth, Equation 6.80 shows that the perturbation growth rate increases linearly with the wave number. On account of the finite thickness, δ_L, of a laminar flame (the DL theory deals with an infinitely thin flame), such a linear dependence holds for the longest perturbations. As shown for a typical laminar flame in Figure 5.40, the growth rate attains a maximum at $\kappa_m \delta_L \approx 0.05-0.1$, while a laminar flame is stable with respect to short perturbations ($\kappa > \kappa_n \approx 2\kappa_m$). Therefore, if, due to the interaction between the DL mechanism and the viscous forces, a rotational random motion is induced in the unburned mixture within a turbulent flame brush, such a flame-generated "turbulent" motion basically differs from the Kolmogorov turbulence, because the energy sources are mainly associated with small and large scales in the former and latter cases, respectively.

If we compare Karlovitz's model and the DL theory, they supplement each other by revealing different phenomena relevant to flame-generated turbulence. Karlovitz et al. (1951) (i) considered a randomly oriented (turbulent) flame front, (ii) restricted himself to the pressure gradient within the front, and (iii) showed the enhancement of velocity fluctuations in a burned mixture. Darrieus (1938) and Landau (1944) analyzed the influence of a flame not only on a downstream but also on an upstream velocity field, but restricted themselves to deterministic wrinkling of the flame surface. At the same time, both approaches highlight the velocity jump at the flame front due to the pressure gradient therein (e.g., the refraction of streamlines at a flame front, shown in Figure 5.39, is caused by that pressure gradient).

The BML approach discussed in the section "Bray–Moss–Libby Approach" in Chapter 5 seems to be one more simple model relevant to the problem of flame-generated turbulence. Within the framework of the approach (see Equation 5.41) the Reynolds stresses are equal to

$$\underbrace{\overline{\rho u_i'' u_j''}}_{\text{Turbulence?}} = \underbrace{\overline{\rho}\left(1-\tilde{c}\right)\left(\overline{u_i' u_j'}\right)_u + \overline{\rho}\tilde{c}\left(\overline{u_i' u_j'}\right)_b}_{\text{Turbulence?}} + \underbrace{\overline{\rho}\tilde{c}\left(1-\tilde{c}\right)\left(\overline{u}_{ib}-\overline{u}_{iu}\right)\left(\overline{u}_{jb}-\overline{u}_{ju}\right)}_{\substack{\text{Reactant–product} \\ \text{intermittency}}}. \quad (6.82)$$

This equation clearly shows that the Reynolds stresses can be increased by a flame due to the influence of the flame on the difference between the conditioned velocities $\overline{\mathbf{u}}_u$ and $\overline{\mathbf{u}}_b$. The latter effect was discussed in the section "Countergradient Scalar Transport in Turbulent Premixed Flames," and the crucial role played by the pressure gradients both in a flame front and in unburned or burned mixtures was stressed. From this perspective, Karlovitz's model, the DL theory, and the BML approach highlight the same driving force: pressure gradient.

Moreover, the BML approach offers an opportunity to pose the following important problem. Because the last term on the RHS of Equation 6.82 is controlled by reactant–product intermittency and by the influence of heat release on conditioned velocities, this term does not characterize turbulence and, hence, the Reynolds stresses on the LHS of Equation 6.82 do not seem to be proper turbulence characteristics within a premixed flame brush. For this reason, the conditioned Reynolds stresses $(\overline{u_i' u_j'})_u$ and $(\overline{u_i' u_j'})_b$ are widely used to characterize turbulence, but such a standpoint will be disputed in the section "Can We Characterize Turbulence in Premixed Flames?"

Let us now discuss the problem of the influence of premixed combustion on turbulence using another widely accepted method: an analysis of the effects of a flame on various terms in the second-moment balance equations. Note that such a method does not allow for the aforementioned problem of using the mean Reynolds stresses for characterizing turbulence in flames.

The balance Equation 3.88 for the mean Reynolds stresses $\overline{\rho u_i'' u_j''}$ in flames was derived in the section "Favre-Averaged Balance Equations" in Chapter 3. The summation of Equation 3.88 written for $\rho u_1'' u_1''$, $\rho u_2'' u_2''$, and $\rho u_3'' u_3''$ yields the following balance equation:

$$\underbrace{\frac{\partial}{\partial t}\left(\bar{\rho}\tilde{k}\right)+\frac{\partial}{\partial x_k}\left(\bar{\rho}\tilde{u}_k\tilde{k}\right)}_{\mathrm{I}}=\underbrace{-\overline{\rho u_j'' u_k''}\frac{\partial \tilde{u}_j}{\partial x_k}}_{}-\underbrace{\frac{\partial}{\partial x_k}\overline{\rho u_k'' k}}_{\mathrm{II}}+\underbrace{\overline{u_j''\frac{\partial \tau_{jk}}{\partial x_k}}}_{\mathrm{III}}-\underbrace{\overline{u_k''\frac{\partial p'}{\partial x_k}}}_{\mathrm{IV}}-\underbrace{\overline{u_k''}\frac{\partial \bar{p}}{\partial x_k}}_{\mathrm{V}} \quad (6.83)$$

for a quantity

$$\tilde{k} \equiv \frac{\overline{\rho u_k'' u_k''}}{2\bar{\rho}} \quad (6.84)$$

often associated with turbulent kinetic energy. It is worth reminding ourselves that $\tilde{k}$ defined by Equation 6.84 does not seem to be a proper characteristic of turbulence in flames due to the last term on the RHS of Equation 6.82. For this reason, in the following text, we will call $\tilde{k}$ an apparent turbulent kinetic energy (ATKE) in flames.

In the constant-density case, Equation 6.83 reads

$$\frac{\partial}{\partial t}\left(\bar{\rho}\tilde{k}\right)+\frac{\partial}{\partial x_k}\left(\bar{\rho}\tilde{u}_k\tilde{k}\right)=\underbrace{-\overline{\rho u_j'' u_k''}\frac{\partial \tilde{u}_j}{\partial x_k}}_{\mathrm{i}}-\underbrace{\frac{\partial}{\partial x_k}\left(\overline{\rho u_k'' k}+\overline{p' u_k''}-2\mu\overline{u_j'' s_{jk}}\right)}_{\mathrm{ii}}\underbrace{-\bar{\rho}\varepsilon}_{\mathrm{iii}} \quad (6.85)$$

(see Equations 3.91 through 3.98 and note that $\overline{pu_k} = \overline{p}\overline{u_k''} + \overline{p' u_k''} = \overline{pu_k''} + \overline{p' u_k''} = \overline{p' u_k''}$ if $\rho = $ const and, therefore, $\overline{u_k''} = \tilde{u}_k' = 0$). Here, we still use the Favre-averaged quantities and do not skip the constant $\bar{\rho}$ so that Equations 6.83 and 6.85 look as similar as possible. For this purpose, let us also rewrite Equation 6.83 as

$$\underbrace{\frac{\partial}{\partial t}\left(\bar{\rho}\tilde{k}\right)+\frac{\partial}{\partial x_k}\left(\bar{\rho}\tilde{u}_k\tilde{k}\right)}_{\mathrm{I}}=\underbrace{-\overline{\rho u_j'' u_k''}\frac{\partial \tilde{u}_j}{\partial x_k}}_{}-\underbrace{\frac{\partial}{\partial x_k}\left(\overline{\rho u_k'' k}+\overline{p' u_k''}\right)}_{\mathrm{II'}}+\underbrace{\overline{u_j''\frac{\partial \tau_{jk}}{\partial x_k}}}_{\mathrm{III}}+\underbrace{\overline{p'\frac{\partial u_k''}{\partial x_k}}}_{\mathrm{IV'}}-\underbrace{\overline{u_k''}\frac{\partial \bar{p}}{\partial x_k}}_{\mathrm{V}}.$$

$$(6.86)$$

The most important difference between Equation 6.85, which holds in a constant-density reacting flow, and Equation 6.86, valid in a flame, consists of terms IV′ and V on the RHS of the latter equation. In a constant-density flow, $\nabla \cdot \mathbf{u} = 0$, $\overline{\mathbf{u}''} = \tilde{\mathbf{u}}' = 0$, and the two terms vanish.

Term IV' in Equation 6.86 results from the fluctuating pressure-gradient term IV in Equation 6.83:

$$-\overline{u_k''\frac{\partial p'}{\partial x_k}} = -\overline{\frac{\partial}{\partial x_k}\left(p'u_k''\right)} + \overline{p'\frac{\partial u_k''}{\partial x_k}} = -\frac{\partial}{\partial x_k}\overline{p'u_k''} + \overline{p'\frac{\partial u_k''}{\partial x_k}} \qquad (6.87)$$

and is equal to $\overline{p'\nabla \cdot \mathbf{u}}$, because $\overline{p'\nabla \cdot \tilde{\mathbf{u}}} = \overline{p'}\nabla \cdot \tilde{\mathbf{u}} = 0$. Therefore, term IV' is controlled by the dilatation $\nabla \cdot \mathbf{u}$ in the instantaneous flame front if flows of unburned and burned mixtures are incompressible and, hence, $\nabla \cdot \mathbf{u} = 0$ outside the front. It is worth remembering, however, that $p' = p - \overline{p}$ in term IV' depends not only on the local pressure p at the front, but also on the mean pressure $\overline{p}$, which is affected by processes that occur in the unburned and burned mixtures.

Dilatation $\nabla \cdot \mathbf{u}$ in term IV' is straightforwardly associated with the pressure gradient in the thin flame front. Because this pressure gradient (i) controls the increase in the magnitude of the velocity fluctuations in the burned mixture within the framework of Karlovitz's model, (ii) controls the increase in the magnitude of the velocity fluctuations both in unburned and burned mixtures within the framework of the DL theory, and (iii) substantially affects the difference between the conditioned velocities $\overline{\mathbf{u}}_b$ and $\overline{\mathbf{u}}_b$ (see the discussion in the section "Simplified Discussion of Governing Physical Mechanisms" and Figure 6.3 in particular) and, hence, the magnitude of the last term on the RHS of Equation 6.82, we may assume that term IV' on the RHS of Equation 6.86 or the counterpart term IV on the RHS of Equation 6.83 plays a crucial role in Equation 6.86, as far as an increase in ATKE in a flame is concerned.

Indeed, DNS results (e.g., see Figure 6.17 or Section 4.1.1 in a review paper by Lipatnikov and Chomiak, 2010) indicate that the term IV plays an important role in the production of ATKE by a premixed flame. The mean pressure gradient term V was also large in such DNS studies and contributed substantially to an increase in $\tilde{k}$. For instance, term V (III in Figure 6.17) was even larger than term IV in the DNS performed by Nishiki (2003) in case H, characterized by a large density ratio $\sigma = 7.53$ (see Figure 6.17a).

It is worth remembering, however, that the sign of term V is controlled by the behavior of the turbulent scalar flux, because vectors $\mathbf{u}''$ and $\overline{\rho \mathbf{u}''c''}$ point in the same direction (see Equation 5.40). If we consider a statistically planar, 1D premixed turbulent flame, addressed in a typical DNS and, in particular, by Nishiki (2003) and Nishiki et al. (2002), then $\nabla \overline{p} < 0$ for a flame that moves from right to left and term V is positive if $\overline{u''} > 0$, that is, if the flux $\overline{\rho u''c''}$ is countergradient. On the contrary, if the flux shows the gradient behavior in another statistically planar, 1D flame, then term V serves to reduce the ATKE in the latter flame.

Such an effect may be explained using the BML Equation 6.82. In the statistically planar, 1D case considered, the last (intermittency) term on the RHS of Equation 6.82 is positive within the flame brush. Because the mean pressure gradient increases $\overline{u}_b$ as compared with $\overline{u}_u$, as discussed in the section "Simplified Discussion of Governing Physical Mechanisms," term V serves to increase the intermittency contribution $\tilde{c}(1 - \tilde{c})(\overline{u}_b - \overline{u}_u)^2$ to $\tilde{k}$ if $\overline{u}_b > \overline{u}_u$, that is, if the scalar transport is countergradient. However, in the opposite case of $\overline{u}_b < \overline{u}_u$, term V serves to decrease the difference in $(\overline{u}_b - \overline{u}_u)^2$ by preferentially increasing the former conditioned velocity.

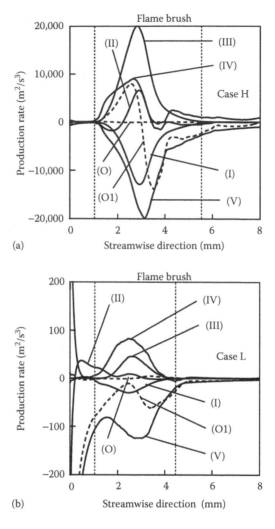

FIGURE 6.17 Various terms in the balance Equation 6.83 for $\tilde{k}$ evaluated in case H ($\sigma = 7.53$) (a) and L ($\sigma = 2.5$) (b). The letters U and C designate the unsteady and convection terms, respectively. The roman figures III and V designate the terms V and III, respectively. (Reprinted from Nishiki, S., Hasegawa, T., Borghi, R., and Himeno, R., *Proc. Combust. Inst.*, 29, 2017–2022, 2002. With permission.)

It is worth emphasizing that the influence of term V on the ATKE is not restricted to the aforementioned simple scenario (i.e., the effect of the mean pressure gradient on the difference in the conditioned velocities) and term V also serves to change the conditioned Reynolds stresses in a general case. For instance, within the framework of the DL theory, nonuniform pressure perturbations in the unburned mixture increase the velocity oscillations in the oncoming flow, and such pressure perturbations are addressed by term V (and also by term IV). Thus, both the pressure terms on the RHS of Equation 6.83 affect all the terms on the RHS of the BML

Equation 6.82 in a general case. Sometimes (see the next paragraph), terms IV and V may act in opposite directions and balance each other.

As far as the simple physical mechanisms discussed earlier are concerned, Karlovitz's model is mainly associated with term IV, more precisely, with the flame front contribution to this term. The same term appears to control the production of the ATKE within the framework of the DL theory. For instance, by averaging the fields $u(x,y,t)$, $v(x,y,t)$, $p(x,y,t)$, etc., given by Equation 6.80, in the y-direction, one can show that term V is counterbalanced by the contribution to term IV from unburned and burned mixtures in the DL case (see Figure 2 in Lipatnikov and Chomiak, 2010). Therefore, the production of the ATKE is controlled by the flame front contribution to term IV in this case.

Thus, as far as the foregoing direct effects of combustion on turbulence are concerned, the pressure terms on the RHSs of Equations 6.83 and 6.86 appear to play a crucial role. The indirect effects of combustion on turbulence are mainly associated with terms I and III on the RHSs of Equation 6.83 or 6.86. The dissipation term III plays an important role and serves to reduce the ATKE in a typical DNS run at a moderate turbulent Reynolds number (e.g., see Figure 6.17). Indeed, an increase in the molecular viscosity due to heat release can increase the magnitude of term III. However, if the Reynolds number is sufficiently high and the Kolmogorov theory of turbulence is assumed to be valid within a flame brush, then an increase in viscosity increases the scale of the smallest turbulent eddies (i.e., the Kolmogorov length scale) but does not seem to straightforwardly affect the mean dissipation rate, which is controlled by the rate of energy withdrawal from the mean flow by large-scale turbulent eddies.

Combustion can also increase term I by increasing the mean flow velocity and its gradient. Scurlock (1948) and Williams et al. (1949) highlighted this indirect mechanism of flame-generated turbulence. The mechanism appears to play a substantial role in confined flames stabilized by a recirculation zone (e.g., behind a bluff body) in a high-velocity flow. In such a case, the mean flame surface is almost parallel to the mean flow direction and the transverse gradient $-\partial \bar{u}/\partial y$ of the mean axial velocity is large and is increased by the heat release. Accordingly, term I serves to increase the ATKE in the considered flow.

However, in other flows, term I may serve to decrease the ATKE due to the following two mechanisms. First, in a statistically planar, 1D case, term I reduces to $-\overline{\rho u''^2}\, \partial \tilde{u}/\partial x < 0$ in a flame and serves to decrease the ATKE (e.g., see Figure 6.17), while such a dilatation term vanishes in a constant-density flow.

Second, due to the intermittency term on the RHS of Equation 6.82, quantities $\overline{\rho u_i'' u_j''}$ and $\partial \tilde{u}_i/\partial x_j$ (with $i \neq j$) may have the same sign in a turbulent premixed flame. This phenomenon is basically similar to countergradient scalar transport in flames and is sometimes called negative turbulent viscosity. It is worth remembering that the product of $\overline{\rho u_i'' u_j''}(\partial \tilde{u}_i/\partial x_j)$ is negative (if $i \neq j$) in a typical constant-density turbulent flow characterized by $v_t > 0$ (see Equations 3.89 and 3.90). If $\overline{\rho u_i'' u_j''}(\partial \tilde{u}_i/\partial x_j) > 0$ in a flame, then term I serves to reduce the ATKE.

Finally, the transport term II in Equation 6.83 is commonly considered to redistribute the ATKE within a flame brush but does not contribute to the net ATKE in the flame. Strictly speaking, the integration of term II from the leading edge to the trailing edge of a flame brush yields $(\overline{\rho u'' k})_0 - (\overline{\rho u'' k})_1$ in the statistically planar, 1D case, where the subscripts 0 and 1 designate the leading and trailing edges, respectively. However, both terms vanish in isotropic incompressible turbulence commonly addressed in the theory. Within a flame brush, term II is not negligible (e.g., see Figure 6.17) and the influence of combustion on it may be of importance (e.g. see Section 4.1.2 in a review paper by Lipatnikov and Chomiak [2010]).

Thus, the foregoing discussion on the influence of combustion on various terms in the balance Equations 6.83 and 6.86 for the ATKE $\tilde{k}$ shows that the problem is intricate. In addition to the generation of ATKE due to the local pressure drop in the instantaneous flame front (see Karlovitz's model and the DL theory), combustion can reduce $\tilde{k}$ due to the mean dilatation (term I), negative turbulent viscosity (term I), and, if Re_t is moderate, due to an increase in the molecular viscosity by the heat release (term III). Moreover, the mean pressure gradient induced by a flame (term V) may also reduce ATKE if turbulent scalar transport shows the gradient behavior. It is worth noting that, because countergradient scalar transport and negative turbulent viscosity are associated with the same effect (a preferential increase in $\left|\overline{\mathbf{u}}_b\right|$ by the heat release) within the framework of the BML approach (see Equations 5.35 and 5.41), the two phenomena should be observed under similar conditions, that is, if the scalar flux $\overline{\rho u_i'' c''}$ shows countergradient behavior, then the sign of the Reynolds stress $\overline{\rho u_j'' u_i''}$ ($i \neq j$) is associated with the negative viscosity and vice versa. Therefore, term V and the part of term I calculated for $i \neq j$ are likely to act in opposite directions; if term V serves to increase ATKE, then the discussed part of term I serves to decrease $\tilde{k}$ and vice versa.

Finally, the summation of Equation 6.82 written for $\overline{\rho u_1'' u_1''}$, $\overline{\rho u_2'' u_2''}$, and $\overline{\rho u_3'' u_3''}$ yields

$$\tilde{k} = \left(1 - \tilde{c}\right)\overline{k}_u + \tilde{c}\overline{k}_b + \frac{1}{2}\tilde{c}\left(1 - \tilde{c}\right)\underbrace{\sum_{j=1}^{3}\left(\overline{u}_{j,b} - \overline{u}_{j,u}\right)^2}_{\substack{\text{reactant–product} \\ \text{intermittency}}}, \tag{6.88}$$

where $\overline{k}_u \equiv \left(\overline{u_j' u_j'}\right)_u / 2$, $\overline{k}_b \equiv \left(\overline{u_j' u_j'}\right)_b / 2$, and the summation convention applies to the repeated index j. Because the last (intermittency) term on the RHS of Equation 6.88 cannot be negative but does not seem to characterize turbulence, the real energy of turbulent motion appears to be lower than $\tilde{k}$ in a flame brush. Accordingly, even if term IV or IV' on the RHS of Equation 6.83 or 6.86, respectively, overwhelms other terms and causes an increase in $\tilde{k}$ in a flame brush under certain conditions, such an effect does not necessitate the generation of turbulence by the flame.

Because the heat release (i) affects all terms on the RHSs of Equations 6.83 and 6.86, (ii) increases not only the source terms but also the sink terms, and (iii) may even change the sign of terms I and V, particular features of the influence of premixed combustion on a turbulent flow may be substantially different for different types of flames, depending on what terms dominate under particular conditions and how these terms are affected by the flame. Accordingly, the diverse effects of combustion on mean and conditioned components of the Reynolds stress tensor were documented in different flames, as discussed in detail in Section 3.2 of a review paper by Lipatnikov and Chomiak (2010). Based on that discussion on the available experimental and DNS data, the following features are worth emphasizing:

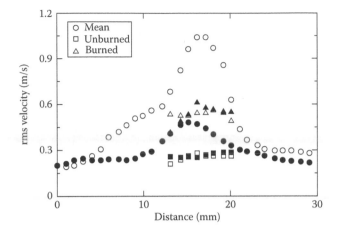

FIGURE 6.18 Profiles of unconditioned and conditioned rms velocities normal (open symbols) and tangential (filled symbols) to a mean flame brush. Experimental data have been obtained by Cheng and Shepherd (1986) from a lean ($\Phi = 0.7$), V-shaped ethane–air turbulent flame along a normal to the mean flame brush.

First, combustion substantially affects mean second moments. In particular, $\overline{\rho u''^2}$ (here, the component u of the flow velocity vector is normal to the mean flame brush) and $\tilde{k}$ often reach peak values within a flame brush (e.g., see the open circles in Figure 6.18).

Second, such effects are typically controlled by the intermittency term on the RHS of Equation 6.82 or 6.88, while the influence of combustion on conditioned Reynolds stresses is significantly less pronounced (e.g., cf. the circles and squares or triangles in Figure 6.18).

Therefore, third, variations (and peak values) in the mean second moments within a turbulent flame brush are commonly more pronounced under conditions associated with a relatively high slip velocity $\Delta \mathbf{u} \equiv \overline{\mathbf{u}}_b - \overline{\mathbf{u}}_u$; that is, (i) in the velocity-vector component that is closer to the normal to the flame brush, (ii) in mixtures characterized by a higher density ratio and a higher laminar flame speed, and (iii) in weakly turbulent flows. Moreover, the mean Reynolds stress tensor becomes anisotropic within a turbulent flame brush, with $\overline{\rho u''^2} > \overline{\rho v''^2}$ in particular (e.g., cf. the open and filled circles in Figure 6.18).

Fourth, an increase in $(\overline{u'^2})_b$ or $\overline{k}_b$ was documented in many flames (e.g., see the triangles in Figure 6.18), whereas experimental evidence of the influence of combustion on $(\overline{u'^2})_u$ or $\overline{k}_u$ is much less cogent. The hypothesis that a premixed flame may self-accelerate by increasing turbulence in the unburned mixture (Williams et al., 1949; Karlovitz et al., 1951) does not seem to be supported by the available experimental and DNS data, as discussed in detail in Section 3.2 of a paper by Lipatnikov and Chomiak (2010).

Because turbulence is the rotational motion, further insight into the problem of flame-generated turbulence may be attained by analyzing the balance equation for vorticity, $\boldsymbol{\omega} \equiv \nabla \times \mathbf{u}$. As discussed in the section "Balance Equations" in Chapter 1, the balance equation reads

$$\frac{\partial \boldsymbol{\omega}}{\partial t} + (\mathbf{u} \cdot \nabla) \boldsymbol{\omega} = \underbrace{(\boldsymbol{\omega} \cdot \nabla) \mathbf{u}}_{1} + \underbrace{\nu \nabla^2 \boldsymbol{\omega}}_{2}. \tag{6.89}$$

in a constant-density flow and

$$\frac{\partial \boldsymbol{\omega}}{\partial t} + (\mathbf{u} \cdot \nabla) \boldsymbol{\omega} = \underbrace{(\boldsymbol{\omega} \cdot \nabla) \mathbf{u}}_{1} + \underbrace{\nu \nabla^2 \boldsymbol{\omega}}_{2} - \underbrace{\boldsymbol{\omega} (\nabla \cdot \mathbf{u})}_{3} + \underbrace{\frac{1}{\rho^2} \nabla \rho \times \nabla p}_{4} \tag{6.90}$$

in a general case.

Term 1 on the RHSs of Equations 6.89 and 6.90 is associated with vorticity generation or damping due to the stretching or compressing of vortexes by the flow. In turbulent flows, the former process dominates, and generally, this term is a production term. Term 2 is a viscous term. In a turbulent flame brush, this term may play an important role by controlling the viscous diffusion of vorticity generated by instantaneous flame fronts to the unburned mixture. In other words, term 2 is associated with the eventual enhancement of vorticity by a flame in the upstream flow.

Equation 6.90 involves two more terms, 3 and 4, which vanish in the case of $\nabla \cdot \mathbf{u} = 0$ and $\rho = \text{const}$. The terms 3 and 4 are associated with (i) a decrease in vorticity when vortex sizes are increased due to the expansion of the hot gas and (ii) vorticity generation by pressure forces, respectively. Depending on the relative magnitudes of the two terms, a flame may either dampen vorticity due to the conservation of the angular momentum during the expansion of the hot gas (term 3) or produce vorticity due to baroclinic torque (term 4).

The counteraction of the two physical mechanisms may be shown by rewriting Equation 6.90 as

$$\frac{\partial}{\partial t} \left(\frac{\boldsymbol{\omega}}{\rho} \right) + \mathbf{u} \cdot \nabla \left(\frac{\boldsymbol{\omega}}{\rho} \right) = \frac{1}{\rho} \left[\underbrace{(\boldsymbol{\omega} \cdot \nabla) \mathbf{u}}_{1} + \underbrace{\nu \nabla^2 \boldsymbol{\omega}}_{2} - \underbrace{\nabla \left(\frac{1}{\rho} \right) \times \nabla p}_{4'} \right] \tag{6.91}$$

using the continuity Equation 1.49 and moving term 3 from the RHS to the LHS. In the hypothetical case of a 2D inviscid flow, Equation 6.91 reads

$$\frac{\partial}{\partial t} \left(\frac{\boldsymbol{\omega}}{\rho} \right) + \mathbf{u} \cdot \nabla \left(\frac{\boldsymbol{\omega}}{\rho} \right) = -\frac{1}{\rho} \nabla \left(\frac{1}{\rho} \right) \times \nabla p. \tag{6.92}$$

If baroclinic torque is of minor importance in a 2D flame, then the ratio of ω/ρ remains constant along a flow line and vorticity decreases due to the density drop in the products. Even if baroclinic torque plays a substantial role and markedly increases the ratio of ω/ρ, vorticity may still decrease in such a flame. In order for a 2D flame to generate vorticity, baroclinic torque should be sufficiently strong in order to overwhelm the density drop and to increase not only ω/ρ but also ω. Certainly, if the oncoming flow is irrotational, as in the DL case, then a wrinkled 2D laminar flame always generates vorticity due to baroclinic torque.

Thus, in line with the previous discussion on Karlovitz's model, the DL theory, and various terms in the ATKE Equation 6.83 or 6.86, the vorticity Equations 6.90 through 6.92 also highlight the crucial role played by the pressure gradient in turbulence generation by a flame (certainly, variations in term 1, induced by heat release, can also serve to generate vorticity in a general 3D case). Moreover, the vorticity Equations 6.90 through 6.92 show that this pressure gradient should not point normal to density isolines and, hence, should differ from the pressure gradient in the counterpart unperturbed laminar flame. If the local pressure gradient within flamelets in a turbulent flow were equal to the pressure gradient in the counterpart unperturbed laminar flame, then the vectors ∇p and $\nabla \rho$ would be locally parallel to each other, and term 4 on the RHS of Equation 6.90, as well as term 4′ on the RHS of Equation 6.91, would vanish. Therefore, a misalignment of pressure and density isolines is of vital importance for vorticity generation by a flame. Such an effect may occur only if the local pressure gradient is controlled not only by the local flamelet structure but also by the pressure field induced by other elements of the flamelet. The importance of the misalignment of ∇p and $\nabla \rho$ is not addressed by Karlovitz's model. Moreover, if the flamelet contribution to term IV on the RHS of Equation 6.83 is modeled by associating the pressure gradient in the flamelets with the pressure gradient in the counterpart unperturbed laminar flame, that is, $(\nabla p)_f \propto \mathbf{n}\rho_u\tau S_L^2/\delta_L$, then such a simple model does not allow for vorticity generation by the flamelets, because the misalignment of ∇p and $\nabla \rho$ is disregarded.

Further insight into the problem of vorticity generation by a flamelet may be attained by considering a hypothetical case of a premixed flame stabilized in a 2D laminar flow (Uberoi et al., 1958). Let us rewrite the Euler equation, that is, Equation 1.51 without the viscous term, as

$$\frac{\partial \mathbf{u}}{\partial t} - \mathbf{u} \times \boldsymbol{\omega} = -\frac{1}{\rho}\nabla p - \frac{1}{2}\nabla(\mathbf{u}\cdot\mathbf{u}). \tag{6.93}$$

In the steady 2D case considered, the tangential (to the flame) component of Equation 6.93 reads

$$\rho u_n \omega_z = -\frac{\partial}{\partial \eta}\left(p + \rho u_n^2\right) + u_n \frac{\partial}{\partial \eta}\left(\rho u_n\right) - \frac{\rho}{2}\frac{\partial u_t^2}{\partial \eta} \tag{6.94}$$

outside the flame, that is, if $\rho = \text{const}$. Here, u_n and u_t are, respectively, the normal and tangential components of the velocity vector and η is the distance along the flame. Because quantities u_t, $\rho u_n = \rho_u S_L$, and $p + \rho u_n^2$ are continuous on a stationary, infinitely thin flame surface (see the section "Flow in a Laminar Premixed Flame" in Chapter 2), we can easily obtain (Uberoi et al., 1958)

$$\rho_n S_L \left(\omega_b - \omega_z \right) = \left(\sigma - 1 \right) \rho_u S_L \frac{\partial S_L}{\partial \eta} - \frac{\rho_u - \rho_b}{2} \frac{\partial u_t^2}{\partial \eta} \tag{6.95}$$

by subtracting Equation 6.94 written at two points immediately upstream and downstream of the flame. If the mass flux $\rho_u S_L$ varies weakly along the flame surface, that is, if the perturbations of the flame structure are weak, then the flame-induced vorticity is controlled by the tangential variations in the tangential flow velocity. It is worth remembering that such tangential velocity variations control the rate of strain (see term $\nabla_t \cdot \mathbf{v}$ on the RHS of Equation 5.56).

Owing to the counteraction of dilatation and baroclinic torque (see terms 3 and 4, respectively,) in Equation 6.90, a wrinkled, inherently laminar flame front may either generate or dampen vorticity depending on the initial and boundary conditions. Studies of the influence of premixed combustion on vorticity, reviewed in Section 5.5 of a paper by Lipatnikov and Chomiak (2010), have not yet shown that the net effect of a flame on vorticity in a turbulent flow results in vorticity generation. On the contrary, the attenuation of turbulent eddies in the products (e.g., see the left top and right bottom regions in Figure 5.30) was observed more often than the strengthening of vorticity. As yet, the latter effect has been documented only in localized spatial regions (e.g., see vorticity generation behind a flame cusp in the right top region in Figure 5.30).

Furthermore, vorticity generation behind a flame front does not necessitate an increase in the burning rate due to flame-generated turbulence. As pointed out by Driscoll (2008), flame-generated vorticity may be "a stabilizing factor that tends to decrease the amplitude of the wrinkles" of the instantaneous flame front. For instance, the white arrows in Figure 5.30, which show the direction of the rotational flow induced by the heat release in the vicinity of a flame cusp, indicate that the flow flattens the cusp out. In such a case, the flame-generated vorticity increases neither the flame front area nor the turbulent burning velocity.

We may also note that the stabilizing effect of flame-generated vorticity on a weakly perturbed laminar flame was, in fact, predicted by Darrieus (1938) and Landau (1944). Indeed, Equation 6.80 shows that the potential ($\mathbf{u}_{b1} = \{u_{b1}, v_{b1}\}$) and rotational ($\mathbf{u}_{b2} = \{u_{b2}, v_{b2}\}$) components of the flow velocity vector have opposite signs at the mean flame surface ($x = 0$), with the magnitude of the potential component being larger. Accordingly, the rotational flow reduces the destabilizing effect of the potential flow in the DL case.

Thus, due to a number of competing effects, which manifest themselves differently under different conditions, the classical problem of the influence of a flame on a turbulent flow, posed by Scurlock (1948), Williams et al. (1949), and Karlovitz et al. (1951) in the middle of the last century, remains among the most important challenges to the combustion community. Numerical research on this problem has dealt mainly with the second-moment balance Equation 3.88 or 6.83 or 6.86. Although various submodels have been proposed to close various terms on the RHSs of these equations (see Section 4.1 in a review paper by Lipatnikov and Chomiak, 2010), a well-recognized predictive model has not yet been elaborated. For this reason, a discussion on the aforementioned submodels is beyond the scope of this book and interested readers are referred to the review paper and references cited therein.

Finally, it is worth emphasizing the following point. The earlier discussion shows the important role played by such a small-scale effect as the influence of the heat release, density drop, dilatation, and pressure gradient, all localized at thin flame fronts, on small-scale turbulent eddies. If this small-scale effect contributes substantially to flame-induced velocity fluctuations, then the physical mechanism of the generation of these fluctuations is very different from the physics of constant-density turbulence. It is worth remembering that, according to the theory of Kolmogorov turbulence (see the section "Kolmogorov's Theory of Turbulence" in Chapter 3), large-scale eddies withdraw energy from the mean flow; then, this energy is transported to smaller and smaller eddies due to the vortex stretching mechanism. Finally, the energy is converted into thermal energy by the smallest eddies of the Kolmogorov scale. Thus, for constant-density turbulence, the energy source is associated with the low-frequency, large-scale part of the velocity spectrum, while the source of the energy of the flame-induced velocity fluctuations may be distributed over a smaller scale range of the spectrum. Accordingly, flame-induced velocity fluctuations may differ substantially from the Kolmogorov turbulence.

This issue was theoretically studied by Kuznetsov, as reviewed elsewhere (see Kuznetsov and Sabel'nikov, 1990, Section 6.5). In particular, Kuznetsov obtained a scaling for "the conditionally averaged structure function" $D_{uu}^{(b)}(l) \equiv [u(x) - u(x+l)]_b^2 \propto \ln^2 l$ (see Equation 6.34 in the cited book), which differed substantially from the counterpart Kolmogorov scaling $D_{uu}(l) \propto (\overline{\varepsilon}l)^{2/3}$. The author is not aware of any target-directed experimental or DNS research on the scaling of the spectral characteristics of flame-induced velocity fluctuations. A few experimental data on the spectra of turbulence in flames, reviewed in Section 3.2.4 of a paper by Lipatnikov and Chomiak (2010), are not sufficient to lead to a solid conclusion.

It is worth stressing once again that because the focus of premixed turbulent combustion theory is placed on predicting the burning rate, the effects of combustion on turbulent eddies *upstream* of the instantaneous flame fronts appear to be of greater importance than the effects of combustion on velocity fluctuations in the burned gas, because it is the former eddies that stretch the fronts and increase their surface area. The author is not aware of any study that casts solid doubts on the applicability of the Kolmogorov scaling to the modeling of the former eddies. Note that flow perturbations induced by the DL instability of a flame front are potential in the unburned mixture. They can only contribute to turbulence after a long transformation time and, because they are advected by the mean flow at a large distance during this time, they do not seem to substantially affect the Kolmogorov scaling in the unburned gas upstream of the flame brush.

CAN WE CHARACTERIZE TURBULENCE IN PREMIXED FLAMES?

As discussed earlier, the Reynolds stresses $\overline{\rho u_i'' u_j''}$ are affected not only by turbulence but also by the unburned–burned intermittency (see Equation 6.82), whereas, in order to model the effects of turbulence on premixed combustion, we have to know the rms velocity u_t' that solely characterizes the turbulence. Therefore, the simple evaluation

$$u'_t = \sqrt{\frac{\overline{\rho u''_k u''_k}}{3\bar{\rho}}} \qquad (6.96)$$

which is widely used to characterize turbulence in nonreacting flows, does not seem to be consistent with the underlying physics of flames if the last (intermittency) term on the RHS of Equation 6.82 plays a substantial role. In many flames, this term contributes significantly to $\overline{\rho u''_i u''_j}$ (e.g., see Figure 6.18), and Equation 6.96 may substantially overestimate the true turbulence velocity.

A commonly accepted way of resolving the problem consists of using conditioned Reynolds stresses to characterize the true turbulence, for example,

$$u'_t = \sqrt{\frac{\left(\overline{u''_k u''_k}\right)_u}{3}}. \qquad (6.97)$$

Although the hypothesis that the conditioned Reynolds stresses characterize the true turbulence appears to be obvious at first sight, it was recently put into question (Lipatnikov, 2009a, 2011a) by studying certain very simple constant-density problems. Here, the density is assumed to be constant in order for the flow to be unaffected by "combustion," thereby allowing the true "turbulence" characteristics in the "flame" to be controlled by the oncoming "turbulence" and to be easily determined.

For instance, let us consider a planar, infinitely thin, self-propagating interface between reactants and products, which oscillates from $x = -U/\omega$ to $x = U/\omega$ in a 1D flow, $u(t) = S_L + U \sin(\omega t)$. If we take a point x within the "flame brush" $(-U/\omega \leq x \leq U/\omega)$, then products are observed therein from instant $t_f = \arccos(-\omega x/U)/\omega$ to instant $T - t_f$ during a single period $T = 2\pi/\omega$ of the oscillations. Accordingly, conditioned quantities are calculated as

$$\bar{q}_u = \frac{1}{T - 2t_f} \int_{t_f}^{T - t_f} q \, dt;$$

$$\bar{q}_b = \frac{1}{2t_f} \left(\int_0^{t_f} q \, dt + \int_{T - t_f}^{T} q \, dt \right) \qquad (6.98)$$

for any q. In particular,

$$\left(\overline{u'^2}\right)_u = \frac{U^2}{2} \left[1 - \frac{\sin\left(2 \arccos\left(\omega x/U\right)\right)}{2 \arccos\left(\omega x/U\right)} \right];$$

$$\left(\overline{u'^2}\right)_b = \frac{U^2}{2} \left[1 + \frac{\sin\left(2 \arccos\left(\omega x/U\right)\right)}{2\pi - 2 \arccos\left(\omega x/U\right)} \right], \qquad (6.99)$$

that is, the conditioned rms velocities $(\overline{u'^2})_u^{1/2}$ and $(\overline{u'^2})_b^{1/2}$, evaluated at the same point, differ significantly from each other and from $u_t' = U/\sqrt{2}$. For instance, the ratio of $(\overline{u'^2})_u/(\overline{u'^2})_b$ may be as low as 0.3 at one point and as large as 3 at another point (Lipatnikov, 2009a). This simple example clearly shows that the conditioned Reynolds stresses $(\overline{u_i'u_j'})_u$ and $(\overline{u_i'u_j'})_b$ are affected by the intermittency of the unburned and burned mixtures.

An infinitely thin "flame" stabilized in a 1D shear flow $u(y) = U - (S_L - U)\cos(\kappa y)$ with $U > S_L$ (see Figure 6.19) is another simple problem that puts Equation 6.97 into question.

To draw an analogy with premixed turbulent combustion, let us (i) average all quantities in the y-direction and (ii) consider the periodic spatial variations in $u(y) - U$ and in the interface position $\xi_f(y) - \overline{\xi}$ to be a perturbation that increases the mean flame speed and thickness in the x-direction in the same way as turbulent eddies increase the surface area, burning velocity, and mean thickness of a premixed flame. Here,

$$\overline{\xi} = \frac{\kappa}{2\pi} \int_{y}^{y+2\pi/\kappa} \xi_f(\eta)d\eta . \qquad (6.100)$$

Then, the mean value $\overline{q}(x)$ of a quantity q at a point x is evaluated by averaging q along the y-direction, that is, along the interval AF in Figure 6.19, whereas the conditioned quantities $\overline{q}_u(x)$ and $\overline{q}_b(x)$ are obtained by averaging q along parts of

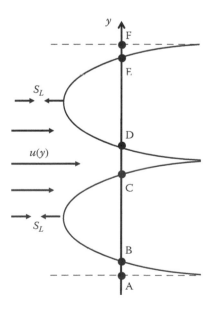

FIGURE 6.19 A sketch of a laminar flame stabilized in a 1D shear flow.

this line, associated with the unburned (AB, CD, and EF) and burned (BC and DE) mixtures.

Some results of such calculations (Lipatnikov, 2009a) are reported in Figure 6.20. Curve 1 in Figure 6.20b shows the Reynolds stress $\overline{u'^2} = \overline{[u(y) - U]^2}$, which characterizes the true "turbulence," because the flame does not affect the flow in the constant-density case considered. A comparison of $(\overline{u'^2})_u = \overline{[u(y) - U]_u^2}$ (curve 2) and $(\overline{u'^2})_b = \overline{[u(y) - U]_b^2}$ (curve 3) with curve 1 again shows that the conditioned Reynolds stresses are affected by the reactant–product intermittency.

Lipatnikov (2011a) studied a more general case of a statistically planar, 1D, constant-density premixed "flame" that propagates in statistically stationary, homogeneous, isotropic turbulence, by numerically integrating conditioned Equations 6.16 through 6.19 supplemented with the following balance equations:

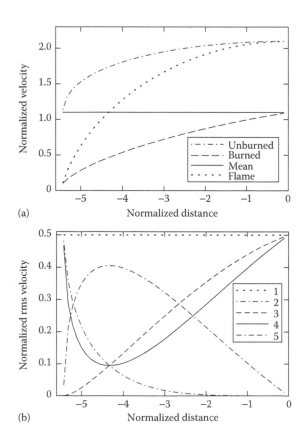

(a)

(b)

FIGURE 6.20 Velocities averaged as specified in the legends (a) and the magnitudes of velocity fluctuations (b), each normalized with $U - S_L$, versus the normalized distance $\kappa x/2\pi$. $U/S_L = 11$. 1, $\overline{u'^2}$; 2, $(\overline{u'^2})_u$; 3, $(\overline{u'^2})_b$; 4, $(1 - \overline{c})(\overline{u'^2})_u + \overline{c}(\overline{u'^2})_b$; 5, $\overline{c}(1 - \overline{c})(\overline{u}_b - \overline{u}_u)^2$. (Reprinted from Lipatnikov, A.N., *Combust. Flame*, 156, 1242–1247, 2009. With permission.)

$$\frac{\partial}{\partial t}\left[(1-\overline{c})\left(\overline{u_i'u_j'}\right)_u\right]+\frac{\partial}{\partial x_k}\left[(1-\overline{c})\overline{u}_{k,u}\left(\overline{u_i'u_j'}\right)_u\right]$$

$$=-\frac{\partial}{\partial x_k}\left[(1-\overline{c})\left(\overline{u_i'u_j'u_k'}\right)_u\right]-(1-\overline{c})\left[\left(\overline{u_i'u_k'}\right)_u\frac{\partial\overline{u}_{j,u}}{\partial x_k}+\left(\overline{u_j'u_k'}\right)_u\frac{\partial\overline{u}_{i,u}}{\partial x_k}\right]$$

$$-(1-\overline{c})\left[\left(\overline{\frac{u_j'}{\rho}\frac{\partial p}{\partial x_i}}\right)_u+\left(\overline{\frac{u_i'}{\rho}\frac{\partial p}{\partial x_j}}\right)_u-\left(\overline{\frac{u_j'}{\rho}\frac{\partial\tau_{ik}}{\partial x_k}}\right)_u-\left(\overline{\frac{u_i'}{\rho}\frac{\partial\tau_{jk}}{\partial x_k}}\right)_u\right]$$

$$+\gamma\overline{\left\{\frac{D}{Dt}\left[u_iu_j\left(1-c\right)\right]\right\}}_f-\overline{u}_{j,u}\gamma\overline{\left\{\frac{D}{Dt}\left[u_i\left(1-c\right)\right]\right\}}_f$$

$$-\overline{u}_{i,u}\gamma\overline{\left\{\frac{D}{Dt}\left[u_j\left(1-c\right)\right]\right\}}_f-\overline{u}_{i,u}\overline{u}_{j,u}\gamma\overline{\left(\frac{Dc}{Dt}\right)}_f \qquad (6.101)$$

and

$$\frac{\partial}{\partial t}\left[\overline{c}\left(\overline{u_i'u_j'}\right)_b\right]+\frac{\partial}{\partial x_k}\left[\overline{c}\overline{u}_{k,b}\left(\overline{u_i'u_j'}\right)_b\right]$$

$$=-\frac{\partial}{\partial x_k}\left[\overline{c}\left(\overline{u_i'u_j'u_k'}\right)_b\right]-\overline{c}\left[\left(\overline{u_i'u_k'}\right)_b\frac{\partial\overline{u}_{j,b}}{\partial x_k}+\left(\overline{u_j'u_k'}\right)_b\frac{\partial\overline{u}_{i,b}}{\partial x_k}\right]$$

$$+\left[\left(\overline{\frac{u_j'}{\rho}\frac{\partial p}{\partial x_i}}\right)_b+\left(\overline{\frac{u_i'}{\rho}\frac{\partial p}{\partial x_j}}\right)_b-\left(\overline{\frac{u_j'}{\rho}\frac{\partial\tau_{ik}}{\partial x_k}}\right)_b-\left(\overline{\frac{u_i'}{\rho}\frac{\partial\tau_{jk}}{\partial x_k}}\right)_b\right]$$

$$+\gamma\left\{\overline{\left[\frac{D}{Dt}\left(u_iu_jc\right)\right]}_f-\overline{u}_{j,b}\overline{\left[\frac{D}{Dt}\left(u_ic\right)\right]}_f-\overline{u}_{i,b}\overline{\left[\frac{D}{Dt}\left(u_jc\right)\right]}_f+\overline{u}_{i,b}\overline{u}_{j,b}\overline{\left(\frac{Dc}{Dt}\right)}_f\right\} \qquad (6.102)$$

for conditioned Reynolds stresses $(\overline{u_i'u_j'})_b$ and $(\overline{u_i'u_j'})_u$. Equations 6.16 through 6.19 and Equations 6.101 and 6.102 were derived using the same method (Lipatnikov, 2008). The aforementioned assumption of a constant density was again invoked to obtain the reference value of u'_t, which is controlled by oncoming turbulence and is not affected by a constant-density "flame." The results of the simulations (see Figure 6.21) also indicate that conditioned Reynolds stresses in a flame brush are not equal to the counterpart Reynolds stresses in the oncoming flow. Consequently, within a flame brush, conditioned Reynolds stresses do not characterize turbulence in a consistent manner.

The matter is that even if the addressed constant-density flame does not affect a flow, the method of conditional averaging divides the flow field into two regions: the flow in the products and the flow in the unburned mixture. Because the instantaneous flame surface has to be closer to (farther from) the leading edge of the mean flame brush in order for us to observe the burned (unburned) mixture at a point x, the flow velocity in the vicinity of this point should be statistically lower (higher) than the

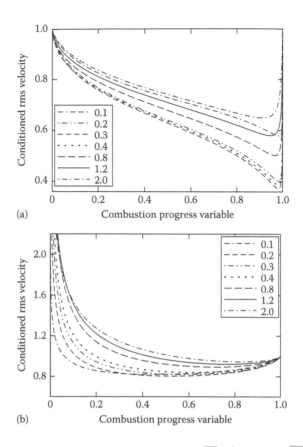

FIGURE 6.21 Normalized conditioned rms velocities $\overline{(u'^2)}_u/u_t'^2$ (a) and $\overline{(u'^2)}_b/u_t'^2$ (b) computed at various normalized flame-development times t/τ_t specified in the legends. (Reprinted from Lipatnikov, A.N., *Proc. Combust. Inst.*, 33, 1489–1496, 2010. With permission.)

turbulent flame speed S_t. Thus, the conditional averaging in the burned (unburned) mixture is associated with averaging over low-speed (high-speed) regions of the entire flow field in the aforementioned constant-density case. For instance, if we consider point A (or B) in Figure 6.22, then the flow velocity conditioned on the burned (unburned) mixture at $x = x_A$ (or $x = x_B$) should be statistically lower (higher) than S_t, because the burned (unburned) mixture is convected to this point by strong velocity fluctuations directed to the leading (trailing) edge of the flame.

Moreover, due to the random motion of an interface between two fluids (e.g., unburned and burned mixtures, see the thick wrinkled curve in Figure 6.22), taking a conditional mean does not commute with differentiation. For instance, by studying the intermittency in constant-density, free, shear flows, Libby (1975) has shown that "conditioned velocity components are not divergence-free," while $\nabla \cdot \mathbf{u} = 0$ everywhere in the flow. Consequently,

$$\nabla \cdot \overline{\mathbf{u}}_c \neq \left(\overline{\nabla \cdot \mathbf{u}}\right)_c = 0, \tag{6.103}$$

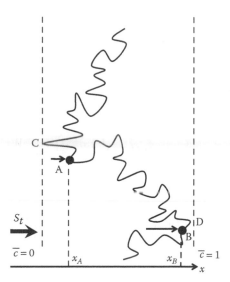

FIGURE 6.22 Conditioned velocities at different points within a flame brush.

where the subscript c designates a conditioned velocity, for example, velocity conditioned either on turbulent fluid or on ambient irrotational fluid in the case studied by Libby (1975) or velocity conditioned either on unburned or burned mixture in a premixed turbulent flame.

To stress that the behavior of conditioned moments may substantially differ from the behavior of the counterpart conventional moments, let us further discuss a statistically planar, 1D, constant-density, premixed "flame" that propagates in a statistically stationary, homogeneous, isotropic Kolmogorov turbulence.

If $\rho = $ const, then the instantaneous flame front does not perturb the velocity field; therefore,

$$\gamma\overline{\left[\frac{D}{Dt}(\rho u_i c)\right]}_f = \gamma\overline{u}_{i,f}\left(\overline{\rho\frac{Dc}{Dt}}\right) = \overline{u}_{i,f}\overline{W},$$

$$\gamma\overline{\left[\frac{D}{Dt}(\rho u_i u_j c)\right]}_f = \gamma\overline{u}_{i,f}\overline{u}_{j,f}\left(\overline{\rho\frac{Dc}{Dt}}\right) = \overline{u}_{i,f}\overline{u}_{j,f}\overline{W}, \tag{6.104}$$

using Equation 6.29 and provided that the front is much thinner than the Kolmogorov length scale. Here, the subscript f designates the velocity conditioned on the instantaneous flame surface. Accordingly, in the simple 1D case considered, Equations 6.16 through 6.19 and Equations 6.101 and 6.102 read

$$\frac{\partial}{\partial t}(1-\overline{c}) + \frac{\partial}{\partial x}\left[(1-\overline{c})\overline{u}_u\right] = -\frac{\overline{W}}{\rho}, \tag{6.105}$$

$$\frac{\partial \bar{c}}{\partial t} + \frac{\partial}{\partial x}\left(\overline{cu_b}\right) = \frac{\overline{W}}{\rho},$$

(6.106)

$$\frac{\partial}{\partial t}\left[(1-\bar{c})\bar{u}_u\right] + \frac{\partial}{\partial x}\left[(1-\bar{c})\bar{u}_u^2\right] = -\frac{\partial}{\partial x}\left[(1-\bar{c})\left(\overline{u'^2}\right)_u\right] - (1-\bar{c})\overline{\left(\frac{1}{\rho}\frac{\partial p}{\partial x}\right)_u}$$

$$+ (1-\bar{c})\overline{\left(\frac{1}{\rho}\frac{\partial \tau_{1k}}{\partial x_k}\right)_u} - \bar{u}_f\frac{\overline{W}}{\rho},$$

(6.107)

$$\frac{\partial}{\partial t}\left(\overline{cu_b}\right) + \frac{\partial}{\partial x}\left(\overline{cu_b^2}\right) = -\frac{\partial}{\partial x}\left[\bar{c}\left(\overline{u'^2}\right)_b\right] - \bar{c}\overline{\left(\frac{1}{\rho}\frac{\partial p}{\partial x}\right)_b} + \bar{c}\overline{\left(\frac{1}{\rho}\frac{\partial \tau_{1k}}{\partial x_k}\right)_b} + \bar{u}_f\frac{\overline{W}}{\rho},$$

(6.108)

$$\frac{\partial}{\partial t}\left[(1-\bar{c})\left(\overline{u'^2}\right)_u\right] + \frac{\partial}{\partial x}\left[(1-\bar{c})\bar{u}_u\left(\overline{u'^2}\right)_u\right] = -\frac{\partial}{\partial x}\left[(1-\bar{c})\left(\overline{u'^3}\right)_u\right] - 2(1-\bar{c})\left(\overline{u'^2}\right)_u\frac{\partial \bar{u}_u}{\partial x}$$

$$-2(1-\bar{c})\overline{\left(\frac{u'}{\rho}\frac{\partial p}{\partial x}\right)_u} + 2(1-\bar{c})\overline{\left(\frac{u'}{\rho}\frac{\partial \tau_{1k}}{\partial x_k}\right)_u}$$

$$-\left(\bar{u}_f - \bar{u}_u\right)^2\frac{\overline{W}}{\rho},$$

(6.109)

and

$$\frac{\partial}{\partial t}\left[\bar{c}\left(\overline{u'^2}\right)_b\right] + \frac{\partial}{\partial x}\left[\overline{cu_b}\left(\overline{u'^2}\right)_b\right] = -\frac{\partial}{\partial x}\left[\bar{c}\left(\overline{u'^3}\right)_b\right] - 2\bar{c}\left(\overline{u'^2}\right)_b\frac{\partial \bar{u}_b}{\partial x}$$

$$-2\bar{c}\overline{\left(\frac{u'}{\rho}\frac{\partial p}{\partial x}\right)_b} + 2\bar{c}\overline{\left(\frac{u'}{\rho}\frac{\partial \tau_{1k}}{\partial x_k}\right)_b} + \left(\bar{u}_f - \bar{u}_b\right)^2\frac{\overline{W}}{\rho}.$$

(6.110)

Multiplying Equation 6.107 with $2\bar{u}_u$, summing up the obtained equation with Equation 6.109, and using Equation 6.105, one can easily arrive at

$$\frac{\partial}{\partial t}\left[(1-\bar{c})\left(\overline{u^2}\right)_u\right] = -\frac{\partial}{\partial x}\left[(1-\bar{c})\left(\overline{u^3}\right)_u\right] - 2(1-\bar{c})\overline{\left(\frac{u}{\rho}\frac{\partial p}{\partial x}\right)_u}$$

$$+ 2(1-\bar{c})\overline{\left(\frac{u}{\rho}\frac{\partial \tau_{1k}}{\partial x_k}\right)_u} - \bar{u}_f^2\frac{\overline{W}}{\rho}.$$

(6.111)

When deriving Equation 6.111, we have taken into account that

$$2\bar{u}_u \frac{\partial}{\partial t}\left[(1-\bar{c})\bar{u}_u\right] + 2\bar{u}_u \frac{\partial}{\partial x}\left[(1-\bar{c})\bar{u}_u^2\right]$$

$$= 2(1-\bar{c})\bar{u}_u \frac{\partial \bar{u}_u}{\partial t} + 2(1-\bar{c})\bar{u}_u^2 \frac{\partial \bar{u}_u}{\partial x} - 2\bar{u}_u^2 \frac{\bar{W}}{\rho}$$

$$= (1-\bar{c})\frac{\partial \bar{u}_u^2}{\partial t} + (1-\bar{c})\bar{u}_u \frac{\partial \bar{u}_u^2}{\partial x} - 2\bar{u}_u^2 \frac{\bar{W}}{\rho}$$

$$= \frac{d}{\partial t}\left[(1-\bar{c})\bar{u}_u^2\right] + \frac{\partial}{\partial x}\left[(1-\bar{c})\bar{u}_u^3\right] - \bar{u}_u^2 \frac{\bar{W}}{\rho} \qquad (6.112)$$

and

$$\frac{\partial}{\partial x}\left[(1-\bar{c})\bar{u}_u^3\right] + 2\bar{u}_u \frac{\partial}{\partial x}\left[(1-\bar{c})\left(\overline{u'^2}\right)_u\right]$$

$$+ \frac{\partial}{\partial x}\left[(1-\bar{c})\bar{u}_u\left(\overline{u'^2}\right)_u\right] + \frac{\partial}{\partial x}\left[(1-\bar{c})\left(\overline{u'^3}\right)_u\right] + 2(1-\bar{c})\left(\overline{u'^2}\right)_u \frac{\partial \bar{u}_u}{\partial x}$$

$$= \frac{\partial}{\partial x}\left\{(1-\bar{c})\left[\bar{u}_u^3 + 2\bar{u}_u\left(\overline{u'^2}\right)_u + \bar{u}_u\left(\overline{u'^2}\right)_u + \left(\overline{u'^3}\right)_u\right]\right\} = \frac{\partial}{\partial x}\left[(1-\bar{c})\left(\overline{u^3}\right)_u\right]. \quad (6.113)$$

Let us select the coordinate framework so that $\bar{u} = 0$. If the closure relations developed for conventional moments held for the counterpart conditioned moments, Equation 6.111 would read

$$\frac{\partial}{\partial t}\left[(1-\bar{c})\left(\overline{u^2}\right)_u\right] = 2(1-\bar{c})\left(\frac{u}{\rho}\frac{\partial \tau_{1k}}{\partial x_k}\right)_u - \overline{u_f^2}\frac{\bar{W}}{\rho}, \qquad (6.114)$$

because $\overline{u^3} = \overline{u'^3} = 0$ and $\overline{\mathbf{u}\cdot\nabla p} = \overline{\mathbf{u}'\cdot\nabla p} = 0$ in a homogeneous isotropic turbulent flow (Jones, 1994). However, Equation 6.114 does not hold in a developing premixed turbulent flame in a general case. Indeed, due to the growth of a mean flame brush thickness (see the section "Mean Flame Brush Thickness" in Chapter 4), one can find such a point x_1 and such an instant t that $\bar{c}(x_1,0) = 1$, while $\bar{c}(x_1,t) < 1$ (at least during an early stage of premixed turbulent flame development characterized by $\delta_t \propto u't$ and $U_t \approx S_L \ll u'$). Consequently, the LHS of Equation 6.114 should be positive somewhere at the trailing edge of the flame, whereas the RHS of the considered equation is negative (because a constant-density flame does not affect the flow, we can always study such a fast burning mixture that the negative reaction term dominates on the RHS).

Thus, an assumption that closure relations valid for conventional moments may be applied to the counterpart conditioned moments has straightforwardly led us to a wrong Equation 6.114. Consequently, the aforementioned assumption is wrong, that is, the behavior of conditioned moments differs substantially from the behavior of the counterpart conventional moments. In particular, the first term on the RHS of Equation 6.111 should play an important role at least at the trailing edge of the flame.

Equation 6.111 also shows that chemical reactions (see the last term) reduce $\overline{(u^2)}_u$. In the balance equation for $\overline{(u^2)}_b$:

$$\frac{\partial}{\partial t}\left[\overline{c}\overline{\left(u^2\right)}_b\right] = -\frac{\partial}{\partial x}\left[\overline{c}\overline{\left(u^3\right)}_b\right] - 2\overline{c}\overline{\left(\frac{u}{\rho}\frac{\partial p}{\partial x}\right)}_b + 2\overline{c}\overline{\left(\frac{u}{\rho}\frac{\partial \tau_{1k}}{\partial x_k}\right)}_b + \overline{u}_f^2\frac{\overline{W}}{\rho},\qquad (6.115)$$

which results from Equations 6.106, 6.108, and 6.110, the reaction term is positive and increases the rms velocity conditioned on burned mixture. Summing up Equations 6.111 and 6.115 and using the BML Equations 5.15, 5.21, and 5.22, we can easily obtain the balance Equation 3.82 for $\overline{u^2}$, with the latter equation involving no reaction term, contrary to the counterpart conditioned Equations 6.111 and 6.115. Therefore, in the constant-density case considered, chemical reactions do not affect $\overline{u^2}$, but reduce $\overline{(u^2)}_b$ and increase $\overline{(u^2)}_u$, thereby making the conditioned second moments different from the conventional moment.

The difference between conditioned and conventional moments can easily be shown by considering the leading edge of a hypothetical, fully developed, statistically stationary, planar, 1D constant-density "flame" brush formed due to the random convection of an infinitely thin instantaneous flame front by turbulent eddies (see Figure 6.22). If a point C is the leading edge of the flame brush, then the x-component, u_C, of the axial flow velocity at point C should be zero (in the coordinate framework attached to the flame), otherwise the considered element of the flame front would reach a point $x < x_C$ at a previous or subsequent instant if $u_C > 0$ or $u_C < 0$, respectively. However, the instantaneous flame front should not be observed at $x < x_C$, because the flame element C has already reached the leading edge of the flame brush. Therefore, the velocity evaluated at the instantaneous flame front is always equal to zero at $x = x_C$. Because the burned mixture may be observed at point C only simultaneously with the front, we have $(\overline{u^n})_b \to (\overline{u^n})_f \to 0$ when $\overline{c} \to 0$ for any power exponent n. Consequently, $\overline{u}_b(\overline{c} \to 0) \to 0$ and $(\overline{u'^2})_b(\overline{c} \to 0) \to 0$, whereas $\overline{u}_u(\overline{c} \to 0) \to S_t$ and $(\overline{u'^2})_u(\overline{c} \to 0) \to u'^2$, because quantities conditioned on the unburned mixture tend to the counterpart mean quantities at the leading edge of the flame brush (events associated with finding the flame front at the leading edge are statistically insignificant as far as mean quantities or quantities conditioned on the unburned mixture are concerned). Certainly, the foregoing sketch is oversimplified and is not applicable to a developing flame characterized by a growing flame brush thickness. Nevertheless, this simple model does seem to provide a useful insight into the behavior of $(\overline{u^n})_f$ and $(\overline{u^n})_b$ at the leading edge of a premixed turbulent flame brush. In particular, the simple model explains the inequality of $0 < \overline{u}_{fu} < \overline{u}_u$

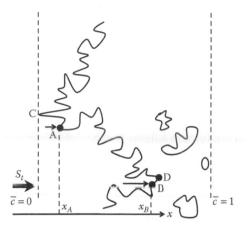

FIGURE 6.23 Pockets at the trailing edge of a turbulent flame brush.

obtained in a DNS by Im et al. (2004) and Chakraborty and Lipatnikov (2011) (cf. the solid and dashed curves in Figure 6.9). It is worth noting that, in a general case, u_f may not vanish at the leading or trailing edge of a fully developed statistically stationary premixed turbulent flame. For example, at $\bar{c} \to 1$, the consumption of the unburned mixture may occur in small pockets that are convected by turbulent eddies with sufficiently high velocities and disappear at the trailing edge (see Figure 6.23). Accordingly, the observed speed of flamelets may be substantial at $\bar{c} \to 1$.

All in all, the aforementioned simple models, numerical simulations, and asymptotically ($\gamma \to 0$) exact balance Equations 6.105 through 6.111 and 6.115 prove that the behavior of conditioned moments in a premixed turbulent flame differs substantially from the behavior of the counterpart conventional moments. In particular, taking conditional averages does not commute with differentiation, and conditioned Reynolds stresses do not characterize turbulence in a consistent manner.

Further studies (especially DNS) are necessary to develop a well-justified method for characterizing turbulence in premixed flames. Until such a method is elaborated in weakly turbulent flames characterized by $u' < \tau S_L$ and by a relatively large magnitude of the last (intermittency) term on the RHS of Equation 6.82, the mean heat release rate $\bar{W}(u'_t,\dots)$ and, especially, the turbulent burning velocity $U_t(u'_t,\dots)$ could be evaluated using the value of u' (i.e., $u'_t = u'$) calculated at the leading edge of the turbulent flame brush. As discussed in the section "Concept of Leading Points" in Chapter 7, the leading edge appears to strongly affect the processes within the turbulent flame brush and control the turbulent burning velocity. We may also note that numerous experimental data on the turbulent burning velocity and flame speed have already been reasonably well parameterized (see the section "Empirical Parameterizations for Turbulent Burning Velocity and Flame Speed" in Chapter 4) using the rms velocity u' evaluated in an upstream fresh mixture. As regards intense turbulence characterized by $u' \gg \tau S_L$, the last (intermittency) term on the RHS of Equation 6.82 seems to be substantially lower than the LHS and Equation 6.96 could be invoked to characterize turbulence in this case.

SUMMARY

A premixed flame can substantially affect a turbulent flow. In particular, the second moments $\overline{\rho u_i'' c''}$ and $\overline{\rho u_i'' u_j''}$ vary strongly in a turbulent flame brush (e.g., even the direction of the turbulent scalar flux $\overline{\rho u'' c''}$ may be reversed by the heat release), while the influence of combustion on conditioned Reynolds stresses $(\overline{u_i' u_j'})_b$ and, especially, $(\overline{u_i' u_j'})_u$ is much less pronounced. Such a difference in the behavior of the mean and conditioned Reynolds stresses is associated with the intermittency of the unburned and burned mixtures in a flame brush.

A number of physical mechanisms affect the behavior of the second moments in a flame, but pressure perturbations caused by the heat release appear to play a crucial role in both countergradient scalar transport and generation of velocity fluctuations in a premixed turbulent flame. Accordingly, in a typical open flame, the sum of the terms that involve pressure gradients is the major source term in the balance Equations 6.7 and 3.88 for $\overline{\rho u_i'' c''}$ and $\overline{\rho u_i'' u_j''}$, respectively. However, in a high-speed, confined flame, turbulence production may be controlled by shear stresses (see term I in Equation 6.83), which are increased due to the heat release. On account of the counteraction of several physical mechanisms, variations in the second moments within a flame brush are strongly sensitive to boundary and initial conditions.

Thin flamelets can strongly affect vorticity in a turbulent flame, with the attenuation of the incident eddies appearing to be more probable than the local generation of vorticity. The variations in vorticity in the vicinity of thin flamelets are mainly controlled by (i) a decrease in vorticity due to the conservation of the angular momentum in expanding hot gas and (ii) an increase in vorticity due to baroclinic torque, with the misalignment of the local pressure and density gradients being of crucial importance for vorticity production. Moreover, the influence of combustion on the stretching of vortices by the flow (see term 1 in Equation 6.90) may also be of importance for vorticity production/attenuation in a flame brush.

The direction of turbulent scalar transport is significantly affected not only by the ratio of $\tau S_L / u'$ but also by premixed turbulent flame development and by the shape of the $\overline{W}(\bar{c})$ curve.

The following basic issues still pose a challenge to the combustion community:

- Can a premixed turbulent flame self-accelerate by generating more intense turbulence?
- Does flame-generated turbulence (if any) follow the Kolmogorov scaling?
- What quantities should be used to characterize turbulence in a flame brush?

Moreover, predicting the influence of a premixed flame on velocity fluctuations is also an unresolved basic issue, while substantial progress has been made in modeling countergradient scalar transport by studying the behavior of conditioned velocities in the flamelet regime of premixed turbulent combustion. As a whole, the influence of combustion on turbulence is an intricate and basic problem that strongly needs further research.

7 Modeling of Premixed Burning in Turbulent Flows

When a premixed flame propagates in a turbulent flow, the heat release in the flame affects the turbulence, which, in turn, affects the burning rate. The former effects were discussed in Chapter 6, while the physical mechanisms of the latter effects were considered in Chapter 5. The focus of this chapter is on modeling the effects of turbulence on a premixed flame.

There are several approaches to modeling these effects. Starting from the pioneering work of Damköhler (1940) and Shchelkin (1943), the focus of theoretical studies of premixed turbulent combustion was on obtaining an analytical expression for the burning velocity, U_t, of an unperturbed, fully developed flame. Such efforts were particularly popular over the second half of the past century and a number of analytical expressions for U_t were obtained, as discussed in many textbooks on combustion (Williams, 1985a; Chomiak, 1990; Warnatz et al., 1996). Despite 70 years of research on turbulent burning velocity, the problem has not yet been solved and new theoretical expressions for U_t are still published in leading journals (e.g., see the papers by Peters [1999], Bychkov [2003], Aldredge [2006], Kolla et al. [2009], and Chaudhuri et al. [2011]). Such models typically suffer from two limitations: (i) turbulence is assumed to be known, that is, the effects of combustion on turbulence, discussed in Chapter 6, are not addressed with a few exceptions (Kuznetsov and Sabel'nikov, 1990; Bychkov, 2003; Kolla et al., 2009; Chaudhuri et al., 2011) and (ii) the development of premixed turbulent flames, highlighted in Chapter 4, is not addressed with a single exception (Zimont, 1979).

In this chapter, we will not follow common practice and will not discuss a number of turbulent burning velocity models. Interested readers are referred to the aforementioned books and to Appendix B in a review paper by Lipatnikov and Chomiak (2002a). Here, we restrict ourselves to considering (i) a fractal approach, which seemed to be a very promising and simple tool for evaluating the turbulent burning velocity but has not yet justified hopes and (ii) the Zimont model, which (a) addresses developing flames, contrary to the vast majority of alternative models, and (b) agrees qualitatively with numerous experiments. These two approaches are discussed in the sections "Fractal Approach to Evaluating Turbulent Burning Velocity" and "Zimont Model of Burning Velocity in Developing Turbulent Premixed Flames," respectively.

Other approaches to modeling premixed turbulent combustion are numerical tools that have been developed over the past few decades, thanks to the rapid growth of available computer resources. These numerical approaches may be briefly summarized as follows:

The so-called RANS (Reynolds-averaged Navier–Stokes) simulations of premixed turbulent flames deal with the Favre-averaged balance equations, some of which were derived in the section "Favre-Averaged Balance Equations" in Chapter 3. There were two different directions in the approach development: (i) numerical methods and computer hardware and (ii) modeling of unclosed terms in the balance equations, for example, the two terms on the RHS of Equation 6.2. This book deals only with the latter approach. Over more than 30 years of development, the RANS approach has yielded some models that capture the basic physics of premixed turbulent combustion. For these reasons, a substantial part of this chapter is devoted to the RANS models of the influence of turbulence on combustion.

It is worth stressing that the models discussed in the sections "Fractal Approach to Evaluating Turbulent Burning Velocity," "Zimont Model of Burning Velocity in Developing Turbulent Premixed Flames," and "Modeling of Effects of Turbulence on Premixed Combustion in RANS Simulations" have been developed for flames that are characterized by a single global reaction, Le $= 1$, and $D_O = D_F$. Models that address the Lewis number and preferential diffusion effects, which play an important role in premixed turbulent combustion (see the section "Dependence of Turbulent Burning Velocity on Differences in Molecular Transport Coefficients" in Chapter 4), are considered in the section "Models of Molecular Transport Effects in Premixed Turbulent Flames." Approaches aimed at using detailed combustion chemistry in numerical simulations of premixed turbulent flames are briefly addressed in the section "Chemistry in RANS Simulations of Premixed Turbulent Combustion. Flamelet Library."

While the RANS approach dominated in numerical simulations of turbulent combustion in the 1980s and 1990s, another approach called large-eddy simulation (LES) is becoming the leading numerical tool for combustion modeling in this century. LES is based on resolving large-scale turbulent eddies and modeling the influence of small-scale eddies on premixed flames. The approach is very attractive for practical applications, especially for modeling phenomena that involve the unsteady dynamics of large-scale vortexes (e.g., thermoacoustic instabilities in gas–turbine combustors or in the afterburners of aerojet engines and cyclic variability in spark ignition engines). However, the combustion models used in LES are commonly based on models developed earlier for RANS applications and LES has not yet contributed much to the basic understanding of the physics of premixed turbulent combustion. For these reasons, the discussion of the LES approach will be very brief in the section "Large Eddy Simulation of Premixed Turbulent Combustion."

Direct numerical simulation (DNS) resolves not only large-scale eddies, as LES does, but also both smallest turbulent eddies and flamelets by integrating the instantaneous balance equations using a very fine numerical grid. This approach requires no closure of unknown terms in the equations and, therefore, is often considered to be a tool for obtaining "exact" (with the exception of numerical errors) solutions to the basic balance equations. In other words, a DNS study may be considered to be a numerical experiment. Accordingly, this book is restricted solely to using the

DNS results in addition to the experimental data for analyzing the physical mechanisms of premixed turbulent combustion and assessing various models, but neither DNS nor the experimental techniques are discussed. Readers interested in the DNS techniques are referred to review papers by Poinsot et al. (1995), Poinsot (1996), and Chen (2011). Although DNS has substantially contributed to our understanding of turbulent combustion, the currently available DNS data are still not fully equivalent to the experimental data. For instance, recent DNS results obtained by Day et al. (2012) by simulating a low-swirl flame differ from the experimental data obtained from a similar flame, probably, due to differences in the initial conditions for the measurements and simulations (Day et al., 2012). Moreover, it is worth remembering that the vast majority (with a few very recent exceptions) of the available DNS data on premixed turbulent combustion have been computed in the case of $L/\Delta_L = O(1)$ (see domain F in Figure 5.46).

Numerical solution to a balance equation for a joint probability density function (PDF) $P(\mathbf{u},c)$ and stochastic simulations with synthetic velocity fields are two more approaches to combustion modeling. These approaches are not considered in this book, and interested readers are referred to books by Kuznetsov and Sabel'nikov (1990) and Poinsot and Veynante (2005), as well as to review papers by O'Brien (1980), Pope (1985, 1994), Dopazo (1994), Givi (1994), Ashurst (1994), Kerstein (2002), and Haworth (2010).

FRACTAL APPROACH TO EVALUATING TURBULENT BURNING VELOCITY

Mandelbrot (1975, 1983) developed the theory of fractals to characterize naturally occurring geometries that display a wide range of self-similar shapes and forms. The theory deals with fractal objects, the peculiarity of which consists of the power-law dependence of the measured object size on the resolution scale.

For instance, let us consider a surface (e.g., the surface of a cloud) within a cube L^3 and measure the surface area using the box-counting algorithm. If we divide the cube into subcubes of side l and count the number $N(l)$ subcubes that contain the surface, then $A(l) = l^2 N(l)$ provides a measure of the surface area. For an ideal fractal surface, see the straight dashed line in Figure 7.1,

$$N \propto \left(\frac{l}{L}\right)^{-D} \tag{7.1}$$

and

$$A(l) \propto L^2 \left(\frac{l}{L}\right)^{2-D}, \tag{7.2}$$

where $2 < D < 3$ is a constant called the fractal dimension. Natural fractal objects (see the solid curve in Figure 7.1) such as clouds, seacoasts, snowflakes, show the

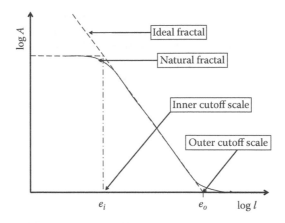

FIGURE 7.1   Ideal and natural fractal objects.

fractal behavior in a bounded domain $e_i < l < e_o$, where e_i and e_o are called the inner and outer cutoff scales, respectively.

Mandelbrot (1975) has hypothesized that isoscalar surfaces in nonreacting, isotropic, homogeneous turbulence are fractal objects with $e_i \propto \eta$ and $e_o \propto L$. He evaluated D = 8/3 for Kolmogorov turbulence and D = 5/2 for Burgers turbulence. Procaccia (1984) and Hentschel and Procaccia (1984) obtained D = 2.37–2.41 at the high Reynolds number limit. A variety of turbulent flows has since been observed to show fractal behavior. Sreenivasan and Meneveau (1986) investigated turbulent boundary layers, axisymmetric jets, plane wakes, and mixing layers and showed the fractal nature of such constant-property surfaces as isovelocity and isoconcentration surfaces, as well as the so-called viscous superlayer, that is, a thin interface that separates the ambient irrotational fluid and the turbulent fluid in an intermittent free shear flow. The fractal behavior was observed in the range of scales $\eta < l < L$. For the viscous superlayer, the measured values of D varied from 2.32 to 2.40. Sreenivasan and Prasad (1988) showed the fractal nature of scalar interfaces for nonreacting jets, with D varying from 2.0 to 2.37 with the downstream distance. Sreenivasan (1991) reviewed the studies of the fractal behavior of scalar interfaces in nonreacting, isotropic, homogeneous turbulence.

Based on Mandelbrot's papers, Gouldin (1987) has hypothesized that the instantaneous flame front is a fractal surface in a turbulent flow. Subsequently, Gouldin has proposed a method to evaluate the turbulent burning velocity using the following simple expression:

$$U_t = S_L \left(\frac{e_i}{e_o} \right)^{2-\mathrm{D}}, \qquad (7.3)$$

which results (provided that a flame front is a fractal surface) from Equations 5.106 and 7.2 in the simplest case of $\bar{u}_c = S_L$.

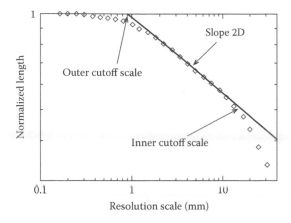

FIGURE 7.2 Measured length of a wrinkled 2D image of a flame front versus resolution scale. (Adapted from Foucher, F. and Mounaïm-Rousselle, C., *Combust. Flame*, 143, 323–332, 2005. With permission.)

Following this hypothesis, many experimental investigations of the behavior of the instantaneous flame surface were performed using the fractal methods. Such measurements are usually based on laser tomographic technique, which yields two-dimensional (2D) images. Accordingly, the fractal analysis is commonly applied to a rough curve on an image. Then, a unity is added to the obtained fractal dimension, D_{2D}, that is, $D_{3D} = D_{2D} + 1$. Mandelbrot (1975) substantiated such a correction method for isotropic surfaces, but the flame front may be an anisotropic surface (Shepherd et al., 1992) because of its self-propagation.

Although numerous measurements have confirmed that the instantaneous flame fronts show the fractal behavior in some bounded range of scales in turbulent flows (e.g., see Figure 7.2), attempts to evaluate the turbulent burning velocity using the simple Equation 7.3 faced problems, because this model involves three unknown quantities, the fractal dimension D and the inner e_i and outer e_o cutoff scales, and evaluation of any of these three quantities is an issue.

First, although e_o is often associated with an integral length scale of turbulence, this hypothesis definitely needs further study, considering the increase in the mean flame brush thickness discussed in the section "Mean Flame Brush Thickness" in Chapter 4. Recent data by Cohé et al. (2007) indicate variations in the ratio of e_o to L from 1.5 to 4.67, thereby putting a simple estimate $e_o = L$ into question.

Second, Figure 7.3 shows a significant scatter in the magnitudes of the ratio of $e_i/\delta_L = e_i a_u/S_L$, obtained from various flames (note that the data are plotted in the logarithmic scale). Moreover, Figure 7.3 indicates that the inner cutoff scale decreases when the Karlovitz number increases and Ka ≪ 1 (see the triangles and filled circles), but a dependence of e_i on Ka is weakly pronounced at the Karlovitz numbers of unity order (crosses). Moreover, e_i/δ_L was increased when Cohé et al. (2007) reduced Ka by decreasing the equivalence ratio in lean methane–air mixtures (filled circles), but e_i/δ_L was almost constant when Ka was increased by increasing pressure (open circles).

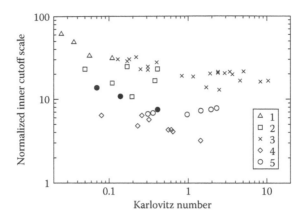

FIGURE 7.3 Normalized inner cutoff scale $e_i/\delta_L = e_i a_u/S_L$ versus Karlovitz number. 1: Shepherd et al. (1990); 2: Smallwood et al. (1995); 3: Gülder et al. (2000); 4: Das and Evans (1997); 5: Cohé et al. (2007).

Scaling e_i/δ_L as a decreasing function of the Karlovitz number was proposed by Gülder and Smallwood (1995) by processing earlier experimental data on the inner cutoff scale and by pointing out that such scaling is consistent with the results of numerical and experimental studies of the interaction between a laminar premixed flame and a vortex pair or a laminar toroidal vortex performed by Poinsot et al. (1991) and Roberts et al. (1993), respectively. Other experimental data (Kobayashi et al., 2002, 2005; Soika et al., 2003; Ichikawa et al., 2011) imply that the inner cutoff scale may be controlled by the neutral wavelength Λ_n of the hydrodynamic instability (see the section "Flame Instabilities" in Chapter 5 and, in particular, Figure 5.43a). The latter hypothesis agrees with a model by Bychkov (2003) and the simulations by Akkerman et al. (2007).

Thus, not only the magnitude but also the scaling of the inner cutoff scale is still not well recognized. Nevertheless, even a significant error in evaluating e_i may result in an acceptable error in the turbulent burning velocity calculated using Equation 7.3, because the difference D-2 is typically small (less than 1/3) in flames.

Third, the scatter in the data on the fractal dimension is a much more challenging problem for applications of the fractal approach to evaluating turbulent burning velocity. Figure 7.4 shows that the fractal dimensions obtained from various flames differ substantially from one another. Some data, for example, data 1 and 2, imply an increase in D by u'/S_L, whereas more recent data, data 7, by Gülder et al. (2000) indicate weak variations in D within a wide range of u'/S_L (see open circles). Gülder et al. (2000) argued that both the increase in D by u'/S_L and the larger fractal dimensions, reported in earlier papers by Mantzaras et al. (1989), North and Santavicca (1990), and Yoshida et al. (1994), resulted from flaws in the image analysis method used in those papers.

The recent data by Cohé et al. (2007) indicate variations in D with u'/S_L, with the sign of the correlation between these two quantities depending on the way it is used to change u'/S_L. The measured fractal dimension was increased both when u'/S_L was

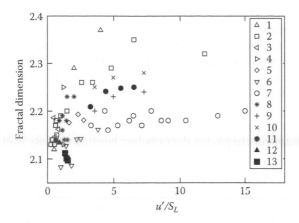

FIGURE 7.4 Fractal dimension data obtained from different flames. 1: Mantzaras et al. (1989); 2: North and Santavicca (1990); 3: Shepherd et al. (1990); 4: Goix and Shepherd (1993); 5: Üngüt et al. (1993); 6: Das and Evans (1997); 7 and 8: Gülder et al. (2000), nozzle diameters are equal to 11.2 and 22.4 mm, respectively; 9 and 10: Foucher and Mounaïm-Rousselle (2005), two different planes in an SI engine; 11, 12, and 13: Cohé et al. (2007).

increased by increasing the pressure from 0.1 to 0.9 MPa (see data 11 in Figure 7.4) and when u'/S_L was decreased by increasing the equivalence ratio from 0.6 to 0.8 (12) or the mole fraction of hydrogen (13) in a lean ($\Phi = 0.6$) H_2/CH_4–air mixture. Note that data 7 and 8 by Gülder et al. (2000) and 11–13 by Cohé et al. (2007) do not contradict one another, because, in the former study, the ratio of u'/S_L was mainly varied by changing the root mean square (rms) turbulent velocity, whereas the latter authors kept u' constant but changed the laminar flame speed by varying the pressure, equivalence ratio, and hydrogen mole fraction.

It is also worth noting that the increase in D by u'/S_L, hypothesized by North and Santavicca (1990) and still invoked by some models (see the paper by Fureby [2005] as a recent example), does not seem to be consistent with the Kolmogorov theory of turbulence, at least if $e_o < L$ and $e_i \ll L$. Indeed, in such a case, the fractal dimension is associated with the inertial range of the turbulence spectrum, but the characteristics of eddies from the inertial range depend on the mean dissipation rate $\bar{\varepsilon}$, rather than on u'. Therefore, if D is increased by u', the effect should be parameterized using another nondimensional quantity $\bar{\varepsilon}\delta_L/S_L^3$, rather than u'/S_L. If, moreover, the flame front is considered to be infinitely thin, then the aforementioned ratio should not include δ_L. In such a case, the problem may be resolved by invoking a quasi-fractal approach, that is, by assuming that the fractal dimension D in Equation 7.2 may weakly depend on a length scale l used in evaluating the flame surface area $A(l)$. Accordingly, $D = D(\bar{\varepsilon}l/S_L^3)$ for dimensional reasoning (Lipatnikov and Chomiak, 1999) and "the apparent u'/S_L dependence of D is a measure of the curvature" of the dependence of D on l (Niemeyer and Kerstein, 1997). The experimental and numerical data processed by Lipatnikov and Chomiak (1999) were consistent with the quasi-fractal approach. However, the quasi-fractal approach and other fractal models of premixed turbulent combustion are challenged by the recent data by Cohé et al. (2007), which indicate a decrease in D when increasing u'/S_L under certain conditions (see symbols 12 and 13 in Figure 7.4).

There are neither models nor parameterizations that could satisfactorily approximate the data shown in Figure 7.4. While the difference between D = 2.18 and D = 2.33 might appear to be small at first glance, this difference is significant, because (i) Equation 7.3 involves D-2 = 0.18 and 0.33, respectively, and (ii) the ratio of the outer and inner cutoff scales is large at high Re_t. Therefore, due to substantial scatter in the data on the fractal dimension of premixed turbulent flames, Equation 7.3 is unlikely to predict the turbulent burning velocity well under a wide range of conditions.

Moreover, even in a single flame, D is not a constant number but develops, that is, it increases either with time in expanding flames (Kwon et al., 1992; Erard et al., 1996) or with distance from a flame holder in statistically stationary flames (Murayama and Takeno, 1988; Goix et al., 1989; Takeno et al., 1990; Wu et al., 1991; Chen and Mansour, 1999, 2003) (e.g., see Figure 7.5). Considering the developing nature of premixed turbulent flames (see the section "Mean Flame Brush Thickness" in Chapter 4), the development of the fractal dimension does not seem to be a surprise, but there is no model of premixed turbulent combustion that addresses the development of D.

Furthermore, by measuring the global burning velocities, inner cutoff scales, and fractal dimensions of turbulent Bunsen flames, Gülder et al. (2000) and Cintosun et al. (2007) showed that the RHS of Equation 7.3 substantially underestimated U_t under the experimental conditions.

In summary, although the instantaneous flame front in turbulent flows does show the fractal behavior, Equation 7.3 is not a predictive tool for evaluating the turbulent

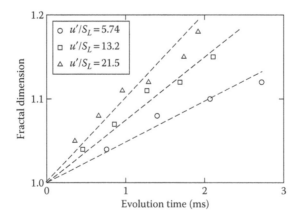

FIGURE 7.5 Development of 2D fractal dimension obtained from three different Bunsen flames. The evolution time characterizes the distance from the burner exit divided by the mean axial flow velocity. (Adapted from *Experimental Thermal and Fluid Science*, 27, Chen, Y.-C. and Mansour, M.S., Geometric interpretation of fractal parameters measured in turbulent premixed Bunsen flames, 409–416, Copyright 2003, with permission from Elsevier.)

burning velocity due to difficulties in quantifying the fractal dimension, as well as the inner and outer cutoff scales relevant to premixed turbulent flames. Nevertheless, the fractal approach is sometimes used in applied simulations and qualitative discussions.

ZIMONT MODEL OF BURNING VELOCITY IN DEVELOPING TURBULENT PREMIXED FLAMES

Because the author considers the development of premixed flames to be the key peculiarity of turbulent combustion, this section is restricted to discussing the sole model for evaluating the turbulent burning velocity, which straightforwardly allows for flame development. Prudnikov (1960, 1967) laid the foundations of this model, and Zimont (1979) reported the final result for U_t. The following discussion on the model is mainly based on the section "Flame Wrinkling and Flamelet Regime" in Chapter 5 and Appendix D of a review paper by Lipatnikov and Chomiak (2002a), where Zimont's arguments were critically analyzed and further developed.

The key idea of the Zimont model is the different treatment of the effects of small-scale and large-scale turbulent eddies on the burning velocity. Why are the small-scale and large-scale effects basically different? The point is that a premixed turbulent flame is a developing flame, as discussed in the section "Mean Flame Brush Thickness" in Chapter 4 and in the section "Development of Premixed Turbulent Flame in the Flamelet Regime" in Chapter 5. Therefore, the large-scale characteristics (e.g., mean thickness) of such a developing flame should depend on the flame-development time. However, the interaction between small-scale turbulent eddies and flamelets is characterized by a much shorter timescale ($\tau_l \propto l^{2/3} \varepsilon^{-1/3} \ll \tau_t = L/u'$ within the framework of the Kolmogorov theory if $l \ll L$). Accordingly, a statistical equilibrium between the chemical reactions confined to a thin reaction zone and small-scale turbulent eddies may be rapidly reached so that the statistical characteristics of small-scale phenomena do not depend straightforwardly on the flame-development time, contrary to the statistical characteristics of large-scale processes. In other words, when discussing large-scale phenomena, the flame-development time is considered to be an important parameter, but this is not so for small-scale phenomena.

Originally, Zimont (1979) assumed that (i) small-scale eddies enter into the preheat zone (in the case of $\eta < \delta_L$, addressed by Zimont) and enhance mixing, thereby increasing the burning rate and flamelet thickness, while (ii) large-scale eddies increase the burning rate by wrinkling the thickened flamelets. However, the former scenario (in particular, an assumption that the enhancement of the mixing in the preheat zone increases the burning rate) is inconsistent with the activation energy asymptotics (AEA) theory of laminar premixed flames, as discussed in the sections "Physical Reasoning: Thin-Reaction-Zone Regime" and "Regimes of Premixed Turbulent Combustion" in Chapter 5. It is worth remembering that the burning rate is controlled by the diffusivity in a very thin reaction zone within the framework of the AEA theory. Therefore, the small-scale part of the original

Zimont model was revised and generalized by Lipatnikov and Chomiak (2002a) as follows:

Let us assume that

- The influence (e.g., wrinkling and straining of thin reaction zones and enhancement of small-scale mixing) of turbulent eddies on flamelets is solely controlled by the mean dissipation rate $\bar{\varepsilon}$ and the eddy length scale l, provided that $\eta \ll l \leq \lambda_Z \ll L$.
- In this range of scales, all the statistical properties of highly turbulent premixed combustion are universally and unambiguously controlled by l, $\bar{\varepsilon}$, and the chemical timescale τ_c, which is defined by Equation 4.14 in the case of Le $= 1$ and $D_F = D_O$ addressed by the Zimont model (accordingly, $\bar{u}_c = S_L = S_L^0$ in this section).

The former assumption results straightforwardly from the Kolmogorov theory of turbulence. It is worth stressing that this assumption is not restricted to any particular governing physical mechanism of the effects of small-scale turbulent eddies on the burning rate. The latter assumption extends the second similarity hypothesis by Kolmogorov (1941a) to premixed turbulent combustion, that is, the second controlling parameter τ_c is added to a single controlling parameter $\bar{\varepsilon}$ highlighted by Kolmogorov for a nonreacting flow.

Then, from dimensional reasoning, we straightforwardly obtain the following estimates:

$$\lambda_Z \propto \left(\bar{\varepsilon}\tau_c^3\right)^{1/2} \propto \delta_L \mathrm{Ka} \propto \eta \mathrm{Ka}^{3/2} \propto L\mathrm{Da}^{-3/2} \qquad (7.4)$$

of the Zimont length scale λ_Z and

$$u_Z \propto \left(\bar{\varepsilon}\tau_c\right)^{1/2} \propto u_\eta' \mathrm{Ka}^{1/2} \propto u'\mathrm{Da}^{-1/2} \propto S_L \mathrm{Ka} \qquad (7.5)$$

of the contribution of small-scale ($l \leq \lambda_Z$) eddies to the burning velocity. Note that Equation 7.5 yields a decrease in the small-scale burning velocity u_z when S_L increases. This trend may, in part, be associated with the smoothing effect of flamelet self-propagation on the wrinkling of the flamelet surface by turbulent eddies (see the section "Physical Reasoning: Classical Flamelet Regime" in Chapter 5). Moreover, considering the significant decrease in the length scale λ_Z when S_L increases, u_Z decreases, because the spectral range $l \leq \lambda_Z$ of eddies that contribute to u_Z is reduced by S_L. However, because this reduction is accompanied by the extension of the spectral range $l > \lambda_Z$ of larger-scale eddies that also contribute to the burning velocity, the model yields an increase in U_t by S_L, as we will see later in the text.

Because the small-scale burning velocity u_Z cannot be lower than the laminar flame speed (in the case of Le $= 1$, $D_F = D_O$, and negligible quenching probability), the discussed approach is valid only in sufficiently strong turbulence characterized by a large Karlovitz number. In particular, the Kolmogorov length scale η should be lower than the laminar flame thickness δ_L (see Equation 5.112). Moreover, the discussed model addresses flames characterized by $Re_t \gg 1$ and $Da \gg 1$. Therefore, $\lambda_Z \ll L$ and $u_Z \ll u'$.

The earlier substantiation of Equations 7.4 and 7.5 is more general as compared with the model by Zimont (1979), but the penalty is less clarity of the governing physical mechanisms. If Zimont (1979) hypothesized a governing mechanism (enhancement of mixing in preheat zones), the present, more general analysis does not do so and may accept various physical mechanisms. The inability to clarify the underlying physical mechanism is an issue that needs further studies and shows that our understanding of the physics of premixed turbulent combustion is still limited.

When discussing the large-scale, that is $l > \lambda_Z$, phenomena, we will follow Zimont (1979) and assume that the influence of large-scale turbulent eddies on the burning rate is reduced to wrinkling of thickened flamelets. Here, the term "thickened flamelet" means a small-scale reacting structure of the length scale λ_Z, which is in the statistical equilibrium state and contains a thin, wrinkled, and stretched, inherently laminar, reaction zone (the aforementioned use of the chemical timescale τ_c as a controlling parameter implies that the reaction zones retain the laminar structure).

If the effects of eddies larger than λ_Z on the burning rate are solely reduced to an increase in the area of thickened flamelets, then the turbulent burning velocity can be evaluated as

$$U_t = u_Z \frac{\overline{A_t}}{A_0}, \tag{7.6}$$

where A_t is the area of the thickened flamelet surface and A_0 is the area of the projection of this surface onto a plane normal to the direction of the flame propagation. Zimont (1979) evaluated the area ratio on the RHS of Equation 7.6 as

$$\frac{\overline{A_t}}{A_0} = \frac{\Lambda_f}{\lambda_f} \tag{7.7}$$

by referring to the theory of random surfaces. Here, Λ_f is the dispersion of the random oscillations of the thickened flamelets, which is associated with the mean flame brush thickness δ_t, and λ_f is "the microscale of the length of the random front surface" (Zimont, 1979).

Because Zimont (1979) addressed such a regime of premixed turbulent combustion that the mean flame brush thickness increases according to the turbulent diffusion law (see the section "Taylor's Theory" in Chapter 3 and the section "Mean Flame Brush Thickness" in Chapter 4) with D_t being the fully developed diffusivity reached at $t \gg \tau_t$, he assumed that

$$\Lambda_f \propto \delta_t \propto \sqrt{2D_t t} = \sqrt{2u'Lt} \tag{7.8}$$

(see Equations 3.118 and 3.120).

As Λ_f depends on the flame-development time, the length scale λ_f may also depend on that time. For instance, if we assume that the area ratio on the LHS of Equation 7.7 is controlled by turbulent eddies with a length scale of the order of $\lambda_Z \ll L$ (considering that an increase in the surface area in turbulent flows is controlled by high stretch rates produced by small-scale eddies), then the timescale of these eddies is much shorter than τ_t, and the statistical characteristics of these eddies are unlikely to depend on time if $t > \tau_t$. Accordingly, if the LHS of Equation 7.7 is independent of the flame-development time, but the scale Λ_f depends on it, then the scale λ_f should also depend on t. Therefore, from dimensional reasoning, we have

$$\lambda_f = \lambda_Z \Psi_1\left(u', L, t, u_Z, \lambda_Z\right) = \lambda_Z \Psi_2\left(\frac{u't}{\lambda_Z}, \frac{u'}{u_Z}, \frac{L}{\lambda_Z}\right) = \lambda_Z \Psi_3\left(\frac{u't}{\lambda_Z}, \mathrm{Da}\right), \qquad (7.9)$$

where Ψ_k are unknown functions and the fact that the ratios of both λ_Z/L and u_Z/u' are controlled by the Damköhler number by virtue of Equations 7.4 and 7.5, respectively, is taken into account. Note that the length scale λ_f depends on the large-scale turbulence characteristics u' and L, rather than on the mean dissipation rate $\bar{\varepsilon}$, by virtue of Equation 7.7. Indeed, even if the area ratio on the LHS of this equation is solely controlled by small-scale ($l \propto \lambda_Z \ll L$) eddies, the statistical characteristics of which are controlled by $\bar{\varepsilon} \propto u'^3/L$, the length scale λ_f should depend on both u' and L, because the length scale Λ_f depends on u' and L.

The key characteristic of the Zimont approach is a hypothesis of an intermediate steady propagation (ISP) flame, that is, a flame that is characterized by an increasing mean flame brush thickness but a fully developed burning velocity. Such a regime of premixed turbulent combustion was highlighted by Prudnikov (1960, 1967) based on an analysis of numerous early Russian experimental data. The dominant contribution of small-scale eddies to the increase in the area of a material surface in a turbulent flow is the physical foundation of this hypothesis. Indeed, since small-scale eddies are characterized by a short timescale, an equilibrium between the increase in the flamelet surface area due to turbulent stretching and the area consumption due to flamelet self-propagation will be rapidly reached. After reaching this statistical equilibrium, the flamelet surface area will continue to increase slowly due to large-scale wrinkling of the flamelet surface, but this process may be neglected in a first approximation, because the rate of increase dU_t/dt is much higher at $t = O(\lambda_Z/u_Z) = O(\tau_t \mathrm{Da}^{-1}) \ll \tau_t$ than at $t = O(\tau_t)$ (see the section "Development of Premixed Turbulent Flame in the Flamelet Regime" in Chapter 5).

If the turbulent burning velocity and, hence, the area ratio on the RHS of Equation 7.6 do not depend on the flame-development time, then Equations 7.4 and 7.7 through 7.9 yield

$$\lambda_f = \lambda_Z \left(\frac{u't}{\lambda_Z}\right)^{1/2} \Psi_4(\mathrm{Da}), \qquad (7.10)$$

and

$$\frac{\overline{A_t}}{A_0} = \left(\frac{2u'Lt}{u'\lambda_z t}\right)^{1/2} \Psi_4^{-1}(\text{Da}) = \left(\frac{L}{\lambda_z}\right)^{1/2} \Psi_5^{-1}(\text{Da}) = \text{Da}^{3/4}\Psi_5^{-1}(\text{Da}). \quad (7.11)$$

Finally, Zimont (1979) has assumed that the function $\Psi_5(\text{Da})$ tends to a finite number as $\text{Da} \to \infty$. Then, Equations 7.4 through 7.6 and Equation 7.11 result in

$$\frac{A_t}{A_0} \propto \text{Da}^{3/4} \propto \left(\frac{L}{\lambda_z}\right)^{1/2} \propto \left(\frac{u'}{u_z}\right)^{3/2} \quad (7.12)$$

and

$$U_t \propto u_z \text{Da}^{3/4} \propto u' \text{Da}^{1/4} \propto u' \left(\frac{\tau_t}{\tau_c}\right)^{1/4} \propto u'^{3/4} L^{1/4} S_L^{1/2} a_u^{-1/4}. \quad (7.13)$$

A decrease in the area ratio on the LHS of Equation 7.12 when u_z increases may be associated with the reduction of the surface area of thickened flamelets due to the self-propagation of the thickened flamelets (see the section "Physical Reasoning: Classical Flamelet Regime" in Chapter 5 and Figure 5.2 in particular).

It is also worth noting that Equations 7.4 and 7.10 yield

$$\frac{\lambda_f}{L} \propto \text{Da}^{-3/2} \left(\frac{t}{\tau_t}\text{Da}^{3/2}\right)^{1/2} = \left(\frac{t}{\tau_t}\text{Da}^{-3/2}\right)^{1/2}. \quad (7.14)$$

Because the length scale λ_f is unlikely to be larger than an integral length scale of turbulence, the foregoing analysis may be valid if $t < \tau_t \text{Da}^{3/2}$. All the conditions of the applicability of the aforementioned model may be summarized as

$$\text{Re}_t \gg 1,$$

$$\text{Da} \gg 1,$$

$$\text{Ka} > 1, \quad (7.15)$$

$$\tau_t < t \ll \tau_t \text{Da}^{3/2}.$$

Interested readers are referred to recent papers by Zimont (2000, 2006) for further discussion on the aforementioned approach.

Although the discussed model appears to be logical, solid, and consistent with the underlying physics, there are certain points that definitely need further study. First, if the steadiness of the turbulent burning velocity, despite the increase in the mean flame brush thickness, is claimed (Zimont, 2006) to be a consequence of the dominant role played by small-scale eddies in the increase in the surface area of thickened flamelets in a turbulent flow, then Equation 7.11 does not seem to be fully consistent with this claim. Indeed, if $\lambda_z \ll L$, that is, if thickened flamelets are associated with the inertial range of the Kolmogorov turbulence, then the area ratio on the LHS of Equation 7.11 should depend on the length scale λ_z, speed u_z of the thickened flamelets, and mean dissipation rate $\bar{\varepsilon}$. However, the sole dimensionless ratio $(\bar{\varepsilon}\lambda_z)^{1/3}/u_z$, which can be combined from these three dimensional quantities, does not depend on the Damköhler number by virtue of Equations 7.4 and 7.5, contrary to Equation 7.11. If the area ratio depends on both u' and L, rather than solely on $\bar{\varepsilon} \propto u'^3/L$, then this does not seem to be consistent with the dominant role played by small-scale eddies in the increase in the surface area of thickened flamelets. Accordingly, it is not clear why the area ratio should be independent of time.

Second, the joint use of Equations 7.9 through 7.11 and an assumption that $\Psi_5(\mathrm{Da} \to \infty) \to$ const may also be disputed. Indeed, using Equation 7.4, which yields $\lambda_z/L \propto \mathrm{Da}^{-3/2}$, one may rewrite Equations 7.9 through 7.11 as

$$\lambda_f = \lambda_z \Psi_1\left(u', L, t, u_z, \lambda_>\right) = L\Psi_6\left(\frac{u't}{L}, \mathrm{Da}\right), \tag{7.16}$$

$$\lambda_f = L\left(\frac{u't}{L}\right)^{1/2} \Psi_7(\mathrm{Da}), \tag{7.17}$$

and

$$\frac{\overline{A_t}}{A_0} = \Psi_8^{-1}(\mathrm{Da}). \tag{7.18}$$

A comparison of Equations 7.11 and 7.18 shows that $\Psi_8 \propto \mathrm{Da}^{-3/4}$ if $\Psi_5(\mathrm{Da} \to \infty) \to$ const. However, it is not clear (i) why the asymptotic behavior of these two unknown functions should be so different and (ii) why Ψ_5, rather than Ψ_8, was assumed to be independent of $\mathrm{Da} \gg 1$. Zimont (1979) argued that the scale λ_f "cannot depend explicitly" on L, because large-scale eddies convect vortices of scale $\lambda_f \ll L$ as a whole. Therefore, the function Ψ_4 on the RHS of Equation 7.10 and the function $\Psi_5 = \Psi_4/\sqrt{2}$ (see Equation 7.11) "cannot depend explicitly" on L, that is they are constant at $\mathrm{Da} \gg 1$. However, this assumption does not seem to be fully consistent with the Kolmogorov theory of turbulence. It is worth remembering that the statistical characteristics of eddies of scale $l \ll L$ depend explicitly on $\bar{\varepsilon} \propto u'^3/L$ within the framework of this theory.

Third, Equation 7.12 is consistent with the well-documented fractal behavior of the instantaneous flame surface in a turbulent flow, provided that $D = 2.5$, $e_o \propto L$, and $e_i \propto \lambda_z \propto \delta_L$ Ka (see Equation 7.4). Even if such a large fractal dimension as 2.5 has never been documented in premixed turbulent combustion (see Figure 7.4), the value of $D = 2.5$ may be acceptable at $t \gg \tau_t$, considering the well-known increase in the fractal dimension in developing flames (see Figure 7.5). However, the scaling of $e_i \propto \lambda_z \propto \delta_L$ Ka is qualitatively contradicted by the experimental data shown in Figure 7.3.

It is worth stressing that the foregoing issues were emphasized solely to show that our understanding of the physics of premixed turbulent combustion is still limited and much has to be done. We did not try to somehow debunk the Zimont model, which seems to be the best model available for simply evaluating the turbulent burning velocity. It is also worth noting that other researchers have obtained Equation 7.13 based on totally different reasoning (Gülder, 1990b; Ashurst et al., 1994). Nevertheless, the Zimont model is the sole model that directly allows for the developing nature of premixed turbulent flames.

Note that the domain of validity of the Zimont model (given by Equation 7.15) is very restrictive, and it is very difficult to find experimental data that will satisfy all these criteria. Nevertheless, the history of science knows a lot of valuable asymptotic theories that predict well the experimental data obtained under conditions that are well beyond the domain of validity of the theory. Equation 7.13 seems to belong to such a group of theoretical results, because this equation agrees well with numerous experimental data obtained under conditions that do not satisfy Equation 7.15.

Indeed, Equation 7.13 predicts

- An increase in the burning velocity by the rms turbulent velocity, $U_t \propto u'^{3/4}$ (see Figure 4.12)
- Bending of $U_t(u')$ curves, $dU_t/du' \propto u'^{-1/4}$ (see Figure 4.12)
- An increase in the turbulent burning velocity by the laminar flame speed, $U_t \propto S_L^{1/2}$ (see Figure 4.15), as well as an increase in $dU_t/du' \propto S_L^{1/2}$
- The effects of differences in the molecular transport coefficients on U_t, provided that a proper submodel for the chemical timescale τ_c is invoked (see the section "Concept of Leading Points")
- The opposite effects of pressure on S_L and U_t (see Figures 4.17 through 4.19)

The pressure effects (i) are of particular importance for industrial applications such as reciprocating engines and gas turbine combustors and (ii) challenge many other models of premixed turbulent combustion. If the turbulence characteristics are kept constant, then pressure variations may affect the turbulent burning velocity calculated using Equation 7.13 via the heat diffusivity $a_u \propto p^{-1}$ and the laminar flame speed. For typical hydrocarbon–air mixtures, $S_L^0 \propto p^{-q}$ with $q \approx 0.5$ for methane and $q \approx 0.25$ for heavier paraffins. Accordingly, Equation 7.13 predicts a weak effect of pressure on the turbulent burning velocities of methane–air mixtures (see Figure 4.18), and a slow increase in U_t by p for heavier paraffins (see Figure 4.17). These predictions agree qualitatively with the available experimental data, as discussed in detail in Section 3.3.5 of a review paper by Lipatnikov and Chomiak (2002a; see Table 1 therein).

The model discussed earlier addresses the so-called ISP flames (Zimont, 2000), which are characterized by a fully developed burning velocity but an increasing mean flame brush thickness $\delta_t \propto \sqrt{D_t t}$. What occurs at the earlier and later stages of flame development? Contrary to turbulent mixing, which is characterized by unbounded growth (in unbounded space) of the mixing layer, an increase in the mean flame brush thickness may be limited by the self-propagation of the flamelets. Zimont (2000) has hypothesized that, in order for such a stabilization mechanism to play an important role, the rate of increase of the thickness due to turbulent diffusion should be of the order of the flamelet speed. If the latter speed is associated with S_L, then from the equality

$$\frac{d\delta_t}{dt} \propto S_L \tag{7.19}$$

we obtain

$$\sqrt{\frac{u'L}{t}} \propto S_L, \tag{7.20}$$

that is, a fully developed mean flame brush thickness is reached during a time interval of the order of $\tau_t(u'/S_L)^2 \propto \tau_t \mathrm{KaRe}_t^{1/2} \propto \tau_t \mathrm{Ka}^2\mathrm{Da} \gg \tau_t$ (see Equation 5.112) and is of the order of

$$\delta_t(t \to \infty) \propto L\mathrm{KaDa}^{1/2} \propto L\frac{u'}{S_L}. \tag{7.21}$$

Therefore, the last inequality in Equation 7.15, which summarizes the domain of validity of the Zimont model, should be rewritten as

$$\tau_t < t \ll \tau_t \mathrm{min}\left\{\mathrm{Da}^{3/2}, \mathrm{Ka}^2\mathrm{Da}\right\}. \tag{7.22}$$

It is worth stressing that the foregoing estimates given by Equations 7.19 through 7.21 are justified only in the case of a constant density. In the case of $\rho_b < \rho_u$, an unconfined, unperturbed, fully developed premixed turbulent flame cannot exist due to the Darrieus–Landau (DL) instability of the entire turbulent flame brush, as discussed in the section "Instabilities of Laminar Flamelets in Turbulent Flows" in Chapter 5.

It is worth noting that if the flame thickness is assumed to be stabilized due to the self-propagation of thickened flamelets with velocity u_Z, rather than flamelets with speed S_L, then,

$$\frac{d\delta_t}{dt} \propto u_Z \propto u'\mathrm{Da}^{-1/2}, \tag{7.23}$$

$$\sqrt{\frac{u'L}{t}} \propto u'\mathrm{Da}^{-1/2}, \tag{7.24}$$

and, at $t = O(\tau_t\mathrm{Da})$,

$$\delta_t \propto L\mathrm{Da}^{1/2}. \tag{7.25}$$

However, Equation 7.25 seems to be wrong, because it results in $\delta_t \to 0$ as $S_L \to 0$.

As regards an earlier stage of premixed turbulent flame development, the Zimont model may be extended as follows (Lipatnikov and Chomiak, 1997, 2002a). Equations 7.6 and 7.7 yield

$$\frac{u_Z}{\lambda_f} \propto \frac{U_t}{\Lambda_f}. \tag{7.26}$$

Moreover, by virtue of Equation 7.14, $\lambda_f \propto L\text{Da}^{-3/4} \ll L$ at $t \propto \tau_t$. Therefore, the LHS of Equation 7.26 is controlled by eddies much less than L, while the RHS is affected by eddies of various scales, including L. It is worth stressing that, because our goal is to allow for the contribution of large-scale eddies to U_t at $t \leq \tau_t$, the turbulent burning velocity may depend on the characteristics of the large-scale eddies and on time during an early stage of flame development.

Because the LHS of Equation 7.26 is controlled by eddies that are much smaller than L at $t \propto \tau_t$ and such eddies are characterized by a short (in comparison with τ_t) timescale, Zimont has hypothesized that, for the ratio of u_Z/λ_f, an intermediate equilibrium state would be reached after a short (as compared with τ_t) initial stage of flame development. Therefore, an estimate of

$$\frac{u_Z}{\lambda_f} \propto \frac{U_{t,\infty}}{\sqrt{u'Lt}}, \tag{7.27}$$

which results from the Zimont model at $t \gg \tau_t$ (see Equations 7.6 through 7.8), holds even at $t \leq \tau_t$. Here, $U_{t,\infty}$ is an intermediate, fully developed turbulent burning velocity reached at $t \to \infty$, provided that the hydrodynamic instability of the entire flame brush is not taken into account.

If Equation 7.27 is considered to be valid at $t \leq \tau_t$ and Equation 4.2 is invoked to model the increase in the mean flame brush thickness according to the turbulent diffusion law, that is,

$$\Lambda_f^2 \propto \delta_t^2 \propto u'^2 \tau_L t \left\{ 1 - \frac{\tau_L}{t} \left[1 - \exp\left(-\frac{t}{\tau_L} \right) \right] \right\} \tag{7.28}$$

then Equations 7.5 through 7.7 yield

$$U_t = u_Z \frac{\Lambda_f}{\lambda_f} = U_{t,\infty} \frac{\Lambda_f}{\sqrt{u'Lt}} = U_{t,\infty} \left\{ 1 - \frac{\tau_L}{t} \left[1 - \exp\left(-\frac{t}{\tau_L} \right) \right] \right\}^{1/2}, \tag{7.29}$$

where $U_{t,\infty}$ is determined by Equation 7.13 and the Lagrangian timescale of turbulence is of the order of $\tau_t = L/u'$. The factor of $2\sqrt{\pi}$ from Equation 4.2 is absent in Equations 7.28 and 7.29, because the ratio of $\Lambda_f(t < \tau_t)/\Lambda_f(t \gg \tau_t)$ does not involve $2\sqrt{\pi}$.

In the limit case of $t \ll \tau_t$, Equations 4.2 and 7.29 read

$$\delta_t \approx \sqrt{2\pi} u' t \tag{7.30}$$

and

$$U_t \approx U_{t,\infty} \left(\frac{t}{2\tau_L} \right)^{1/2}. \tag{7.31}$$

In the opposite limit case of $t \gg \tau_t$, we have

$$\delta_t = \sqrt{4\pi u' L t} \tag{7.32}$$

and

$$U_t = U_{t,\infty}. \tag{7.33}$$

Figure 7.6 shows the development of the normalized turbulent burning velocity $U_t/U_{t,\infty}$ and mean flame brush thickness δ_t/L, calculated using Equations 4.2 and 7.29.

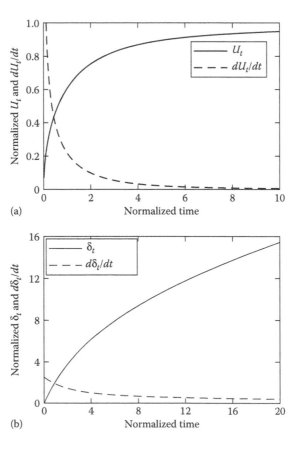

(a)

(b)

FIGURE 7.6   The normalized turbulent burning velocity $U_t/U_{t,\infty}$ (a) and mean flame brush thickness δ_t/L (b) versus normalized time t/τ_L.

Initially, the burning velocity increases rapidly and the normalized time derivative $(\tau_L/U_{t,\infty})dU_t/dt$ is high (see dashed curve), whereas the mean flame brush thickness depends linearly on t. Then, $(\tau_L/U_{t,\infty})dU_t/dt$ decreases dramatically with time, while a decrease in $d\delta_t/dt$ is moderate. As a result, the ratio of $U_t/U_{t,\infty}$ reaches 0.5 and 0.75 at $t/\tau_L = 0.6$ and 2, respectively, whereas δ_t at $t/\tau_L = 2$ is much less than δ_t at $t/\tau_L = 20$ and the increase in the mean flame brush thickness is well pronounced even at $t/\tau_L > 20$. In other words, this model yields a qualitatively different development of the turbulent burning velocity and mean flame brush thickness. Such an effect was highlighted and discussed in the section "Development of Premixed Turbulent Flame in the Flamelet Regime" in Chapter 5. It is worth remembering that this qualitative difference stems from the fact that the turbulent burning velocity is mainly controlled by small-scale turbulent eddies, which are characterized by short timescales, while the increase in the mean flame brush thickness is controlled by large-scale eddies, which are characterized by a large timescale.

MODELING OF EFFECTS OF TURBULENCE ON PREMIXED COMBUSTION IN RANS SIMULATIONS

BASIC BALANCE EQUATIONS

A typical RANS simulation of premixed turbulent combustion deals with the balance equations for the Favre-averaged density (continuity)

$$\frac{\partial \bar{\rho}}{\partial t} + \frac{\partial}{\partial x_k}\left(\bar{\rho}\tilde{u}_k\right) = 0, \tag{7.34}$$

velocity (Navier–Stokes)

$$\frac{\partial}{\partial t}\left(\bar{\rho}\tilde{u}_i\right) + \frac{\partial}{\partial x_k}\left(\bar{\rho}\tilde{u}_k\tilde{u}_i\right) = -\frac{\partial}{\partial x_k}\overline{\rho u_k'' u_i''} + \frac{\partial \overline{\tau_{ik}}}{\partial x_k} - \frac{\partial \bar{p}}{\partial x_k}, \tag{7.35}$$

and the combustion progress variable

$$\frac{\partial}{\partial t}\left(\bar{\rho}\tilde{c}\right) + \frac{\partial}{\partial x_k}\left(\bar{\rho}\tilde{u}_k\tilde{c}\right) = -\frac{\partial}{\partial x_k}\overline{\rho u_k'' c''} + \frac{\partial}{\partial x_k}\overline{\left(\rho D \frac{\partial c}{\partial x_k}\right)} + \bar{W} \approx -\frac{\partial}{\partial x_k}\overline{\rho u_k'' c''} + \bar{W}, \tag{7.36}$$

supplemented with the Bray–Moss–Libby (BML) approach discussed in the section "Bray–Moss–Libby Approach" in Chapter 5. Accordingly, the BML state Equation 5.25 is invoked and the five (in a statistically 3D case or three in a statistically 1D case) equations presented earlier involve five unknown quantities $\tilde{c}$, $\bar{p}$, and $\tilde{\mathbf{u}} = \{\tilde{u}, \tilde{v}, \tilde{w}\}$. If these quantities are computed, then the mean temperature and the mass fractions of reactants are calculated using the bimodal PDF given by Equation 5.31 with the probabilities α and β being evaluated using Equations 5.22 and 5.21, respectively. Equations 7.34 and 7.35 have been derived and discussed in the section "Favre-Averaged Balance Equations" in Chapter 3. Equation 7.36 can be obtained using the same method. The molecular transport term in Equation 7.36 is commonly neglected at high Reynolds numbers.

Even such a strongly simplified problem is challenging for two reasons. First, due to the influence of the heat release on turbulence, closures of the second moments $\overline{\rho u_i'' u_j''}$ and $\overline{\rho u_i'' c''}$, developed for constant-density nonreacting flows, may yield wrong (even qualitatively wrong!) results in premixed turbulent flames. This problem associated with the influence of combustion on turbulence was discussed in Chapter 6, while the remainder of this section will be aimed at resolving another problem associated with the influence of turbulence on combustion and stated in the next subsection.

KEY PROBLEM

In Equations 7.34 through 7.36, the effect of turbulence on flame propagation manifests itself in the mean reaction rate. The modeling of a single term $\overline{W}$ has posed a challenge to the combustion community for decades, due to the highly nonlinear dependence of the reaction rate on temperature.

To further illustrate the problem, let us compare $\overline{\exp(-\Theta/T)}$ and $\exp(-\Theta/\overline{T})$ in a point within a turbulent flame brush where the probabilities of finding unburned and burned mixtures are equal to 0.5. If $\Theta = 20{,}000$ K, $T_u = 300$ K, and $T_b = 2200$ K, then $\overline{T} = 0.5 \times 300 + 0.5 \times 2200 = 1250$ K, $\Theta/\overline{T} = 16$, and $\exp(-\Theta/\overline{T}) = 1.1 \times 10^{-7}$, while $\overline{\exp(-\Theta/T)} \approx \exp(-\Theta/T_b) \approx 0.5\exp(-9.1) = 5.6 \cdot 10^{-5}$, that is, $\overline{\exp(-\Theta/T)}$ is larger than $\exp(-\Theta/\overline{T})$ by a factor of 500!

If we try to estimate the effects of turbulent fluctuations on $\overline{\exp(-\Theta/T)}$ by expanding the exponent into the Taylor series with respect to $T'/\overline{T}$, followed by averaging, then

$$\overline{\exp\left(-\frac{\Theta}{T}\right)} = \overline{\exp\left\{-\frac{\Theta}{\overline{T}}\left[1 - \frac{T'}{\overline{T}} + \left(\frac{T'}{\overline{T}}\right)^2 - \left(\frac{T'}{\overline{T}}\right)^3 + \left(\frac{T'}{\overline{T}}\right)^4 + \ldots\right]\right\}}$$

$$= \exp\left(-\frac{\Theta}{\overline{T}}\right)\left[1 + \frac{3}{2}\frac{\overline{T'^2}}{\overline{T}^2} - \frac{13}{6}\frac{\overline{T'^3}}{\overline{T}^3} + \frac{73}{24}\frac{\overline{T'^4}}{\overline{T}^4} + \ldots\right]. \tag{7.37}$$

One can easily show that, in the considered point ($\overline{c} = 0.5$) within the mean flame brush, $\overline{T'^{(2n+1)}} = 0$ and $\overline{T'^{2n}}/\overline{T}^{2n} = [(T_b - T_u)/2\overline{T}]^{2n} = 0.76^{2n}$, where n is a positive integer number. Therefore, the use of the first-order terms in the Taylor series does not allow us to substantially improve the evaluation of $\overline{\exp(-\Theta/T)}$. For instance, the fourth-order term on the RHS of Equation 7.37 is of the order of (but larger than) the second-order term on the RHS.

Thus, the mean reaction rate cannot be calculated using the mean temperature because the instantaneous reaction rate depends very strongly on the instantaneous temperature and the magnitude of the temperature pulsations in a premixed turbulent flame is large, for example, as large as $(T_b - T_u)/2$, due to random motion of a thin instantaneous flame front. For these reasons, the standard perturbation method is useless, and a more advanced physical modeling is required to close the mean rate of product formation. Several approaches to resolve the problem are discussed in the sections "Algebraic Models," "Presumed PDF Models," "Two-Equation Models: Flame Surface Density and Scalar Dissipation Rate," and "Models of a Self-Similarly Developing Premixed Turbulent Flame."

However, before doing so, it is worth highlighting another, commonly disregarded, issue associated with modeling of the mean rate of product formation in a turbulent flow.

MEAN REACTION RATE AND TURBULENT DIFFUSION

In a laminar flame, the transport and source terms on the RHS of Equation 2.14 model two totally different physical mechanisms, molecular transport and chemical reactions, respectively. On the contrary, the turbulent transport and source terms on the RHS of Equation 7.36 are associated with basically the same physical mechanism, that is, random convection of flamelets by turbulence, even if there are two different manifestations of this mechanism: (i) an increase in the flame brush thickness, controlled by large-scale eddies (turbulent diffusion), and (ii) an increase in the flamelet surface area due to wrinkling of the surface by turbulence.

Let us consider a simple case of an interface that self-propagates at a speed S_L in constant-density Kolmogorov turbulence so that the "flame brush" formed due to random motion of the interface is statistically stationary, planar, 1D, and normal to the x-axis. The speed, S_t, of the leading edge of the flame brush is mainly (if $u' \gg S_L$) controlled by moving the interface far into the unburned mixture by large-scale turbulent eddies. This mechanism is associated with turbulent diffusion because the same physical mechanism controls the growth rate of a turbulent mixing layer. As soon as the interface arrives at a point $\mathbf{x}_1$ associated with $\bar{c}(x = x_1) \ll 1$, the mean reaction rate $\bar{W}(x_1)$ becomes positive. Therefore, in the considered constant-density case, turbulent diffusion affects not only the propagation of the leading edge of the flame brush but also the mean flame surface density $\bar{\Sigma} = \overline{|\nabla c|}$ and the mean rate $\bar{W} = \rho_u S_L \overline{|\nabla c|}$ at the leading edge. Thus, even if the mean rate $\bar{W}$ is mainly controlled by the balance between an increase in $\bar{\Sigma}$ due to wrinkling of the interface by small-scale turbulent eddies and a decrease in $\bar{\Sigma}$ due to collisions of different segments of the highly wrinkled interface, the mean rate $\bar{W}$ is also affected straightforwardly by turbulent diffusion at least at $\bar{c} \ll 1$.

Such effects are obvious in the following simple case (Sabel'nikov, private communication, 2011). Let us consider a planar laminar flame that oscillates in a 1D oncoming flow $u = S_L + s(t)$, where $s(t)$ is a random function such that (i) $\overline{s(t)} = 0$, (ii) $\overline{s(t)s(t)} = u'^2$ and $\int_0^\infty \overline{s(t)s(t+\vartheta)}d\vartheta = u'^2 \tau_s$ do not depend on time and (iii) the length scale $L = u'\tau_s$ of the velocity oscillations is much larger than the laminar flame thickness δ_L. In this simple case, (i) Equation 7.36 may be applied to model the spatial profile of $\tilde{c}(x)$; (ii) the BML equations hold, because $L \gg \delta_L$; (iii) the flame front retains its planar shape with $\bar{c}(x) = \int_{-\infty}^x \gamma d\zeta$; (iv) the probability γ of finding the flame front is equal to $d\bar{c}/dx$; and, therefore, (v) the mean rate $\bar{W} = \gamma \rho_u S_L$ is controlled by the gradient of $\bar{c}$, rather than by $\bar{c}$, whereas the instantaneous rate W is controlled by c in the laminar flame. Thus, even if the flame surface area is not affected by the considered large-scale oscillations of a laminar flame, they make the averaged dependence of $\bar{W}(\bar{c})$ very different from the dependence of $W(c)$ in the laminar flame before averaging.

The direct influence of large-scale turbulent eddies (turbulent diffusion) on the mean rate of product formation seems to be a very important peculiarity of premixed turbulent combustion as compared with laminar premixed flames, and this peculiarity should be borne in mind when modeling turbulent burning.

ALGEBRAIC MODELS

In this section, we will briefly consider models that yield an algebraic closure for the mean reaction rate $\bar{W}$. The first model of that kind, the famous eddy-break-up (EBU) model, was developed by Spalding (1971, 1976) and Mason and Spalding (1973), who assumed that the burning rate in a turbulent premixed flame is controlled by turbulent mixing. Accordingly,

$$\bar{W} = \frac{\bar{\rho}}{\tau_t} \psi\left(\bar{c}, \overline{c'^2}, \bar{\rho}\right), \tag{7.38}$$

where different functions $\psi(\bar{c}, \overline{c'^2}, \bar{\rho})$ were used for different applications, for example, Mason and Spalding (1973) set $\psi \propto (\overline{c'^2})^{1/2}$, and $\tau_t = L/u' \propto \bar{k}/\bar{\varepsilon} \propto v_t/\bar{k}$ is a turbulent timescale ($\bar{k}$ is the mean turbulent kinetic energy, $\bar{\varepsilon}$ is its dissipation rate, and $v_t \propto \bar{k}^2/\bar{\varepsilon}$ is the turbulent viscosity). Among the various submodels for the function $\psi(\bar{c}, \overline{c'^2}, \bar{\rho})$, the submodel developed by Magnussen and Hjertager (1976) became most popular in multidimensional numerical simulations and Equation 7.38 supplemented with it dominated in industrial applications in the last two decades of the past century.

Equation 7.38 was further substantiated by Bray (1979), who showed the proportionality of the mean reaction rate $\bar{W}$ to the mean scalar dissipation:

$$\bar{\chi}_c = 2D \overline{\frac{\partial c''}{\partial x_k} \frac{\partial c''}{\partial x_k}} \tag{7.39}$$

within the framework of the BML approach. Here, the summation convention applies to the repeated index k and D is the molecular diffusivity. To show that $\bar{W} \propto \bar{\chi}_c$, let us first obtain a balance equation for the variance $\overline{\rho c''^2}$. Multiplying Equation 1.79 with c and using the continuity Equation 1.49, we obtain

$$\frac{1}{2} \frac{\partial \overline{\rho c^2}}{\partial t} + \frac{1}{2} \frac{\partial}{\partial x_k} \left(\rho u_k c^2\right) = c \frac{\partial}{\partial x_k} \left(\rho D \frac{\partial c}{\partial x_k}\right) + cW \tag{7.40}$$

or

$$\frac{1}{2}\frac{\partial}{\partial t}\left(\bar{\rho}\tilde{c}^2\right)+\frac{1}{2}\frac{\partial}{\partial t}\overline{\rho c''^2}+\frac{1}{2}\frac{\partial}{\partial x_k}\left(\bar{\rho}\tilde{u}_k\tilde{c}^2\right)+\frac{\partial}{\partial x_k}\left(\tilde{c}\overline{\rho u_k''c''}\right)+\frac{1}{2}\frac{\partial}{\partial x_k}\left(\tilde{u}_k\overline{\rho c''^2}\right)$$

$$+\frac{1}{2}\frac{\partial}{\partial x_k}\overline{\rho u_k''c''^2}=\overline{\tilde{c}\frac{\partial}{\partial x_k}\left(\rho D\frac{\partial c}{\partial x_k}\right)}+\overline{c''\frac{\partial}{\partial x_k}\left(\rho D\frac{\partial c}{\partial x_k}\right)}+\tilde{c}\overline{W}+\overline{c''W} \qquad (7.41)$$

after averaging. Subtraction of Equation 7.36 multiplied by $\tilde{c}$ from Equation 7.41 yields the balance equation

$$\frac{\partial}{\partial t}\overline{\rho c''^2}+\frac{\partial}{\partial x_k}\left(\tilde{u}_k\overline{\rho c''^2}\right)=-2\overline{\rho u_k''c''}\frac{\partial\tilde{c}}{\partial x_k}-\frac{\partial}{\partial x_k}\overline{\rho u_k''c''^2}+2\overline{c''\frac{\partial}{\partial x_k}\left(\rho D\frac{\partial c}{\partial x_k}\right)}+2\overline{c''W}$$

$$(7.42)$$

for the variance $\overline{\rho c''^2}$.

Second, within the framework of the BML approach (see the section "Bray–Moss–Libby Approach" in Chapter 5),

$$\overline{\rho u_k''c''^2}=\rho_u\left(\bar{u}_{k,u}-\tilde{u}_k\right)\tilde{c}^2\left(1-\overline{c}\right)+\rho_b\left(\bar{u}_{k,b}-\tilde{u}_k\right)\left(1-\tilde{c}\right)^2\overline{c}$$

$$=\bar{\rho}\tilde{c}\left(1-\tilde{c}\right)\left[\left(\bar{u}_{k,u}-\tilde{u}_k\right)\tilde{c}+\left(\bar{u}_{k,b}-\tilde{u}_k\right)\left(1-\tilde{c}\right)\right]$$

$$=\bar{\rho}\tilde{c}\left(1-\tilde{c}\right)\left[\bar{u}_{k,u}\tilde{c}+\bar{u}_{k,b}\left(1-\tilde{c}\right)-\tilde{u}_k\right]$$

$$=\bar{\rho}\tilde{c}\left(1-\tilde{c}\right)\left[\bar{u}_{k,u}\tilde{c}+\bar{u}_{k,b}\left(1-\tilde{c}\right)-\bar{u}_{k,u}\left(1-\tilde{c}\right)-\bar{u}_{k,b}\tilde{c}\right]$$

$$=\bar{\rho}\tilde{c}\left(1-\tilde{c}\right)\left(\bar{u}_{k,b}-\bar{u}_{k,u}\right)\left(1-2\tilde{c}\right)$$

$$=\overline{\rho u_k''c''}\left(1-2\tilde{c}\right) \qquad (7.43)$$

and

$$\overline{c''W}=\left(c_m-\tilde{c}\right)\overline{W}, \qquad (7.44)$$

where

$$c_m=\frac{\gamma\int_0^1 cWP_f dc}{\overline{W}}=\frac{\int_0^1 cWP_f dc}{\int_0^1 WP_f dc}, \qquad (7.45)$$

with the probability γ and PDF P_f being determined by Equation 5.11.

Third, substitution of Equations 5.29, 5.35, 7.43, and 7.44 into Equation 7.42 yields

$$
(1-2\tilde{c})\frac{\partial}{\partial t}(\bar{\rho}\tilde{c})+(1-2\tilde{c})\frac{\partial}{\partial x_k}(\bar{\rho}\tilde{u}_k\tilde{c})
$$

$$
=-(1-2\tilde{c})\frac{\partial}{\partial x_k}\overline{\rho u_k''c''}+2\overline{c''\frac{\partial}{\partial x_k}\left(\rho D\frac{\partial c}{\partial x_k}\right)}+2(c_m-\tilde{c})\bar{W} \qquad (7.46)
$$

using the Favre-averaged continuity Equation 7.34. Finally, subtraction of Equation 7.36 multiplied by $(1-2\tilde{c})$ from Equation 7.46 results in

$$
(2c_m-1)\bar{W}=-2\overline{c''\frac{\partial}{\partial x_k}\left(\rho D\frac{\partial c}{\partial x_k}\right)}+(1-2\tilde{c})\frac{\partial}{\partial x_k}\overline{\left(\rho D\frac{\partial c}{\partial x_k}\right)}
$$

$$
=2\overline{\rho D\frac{\partial c''}{\partial x_k}\frac{\partial c}{\partial x_k}}-2\frac{\partial}{\partial x_k}\overline{\left(\rho c''D\frac{\partial c}{\partial x_k}\right)}+(1-2\tilde{c})\frac{\partial}{\partial x_k}\overline{\left(\rho D\frac{\partial c}{\partial x_k}\right)}
$$

$$
=\bar{\rho}\tilde{\chi}_c+2\overline{\rho D\frac{\partial c''}{\partial x_k}\frac{\partial \tilde{c}}{\partial x_k}}-2\frac{\partial}{\partial x_k}\overline{\left(\rho c''D\frac{\partial c}{\partial x_k}\right)}+(1-2\tilde{c})\frac{\partial}{\partial x_k}\overline{\left(\rho D\frac{\partial c}{\partial x_k}\right)}.
$$

$$ (7.47) $$

At sufficiently high Reynolds and Damköhler numbers, instantaneous variations in c are localized at the thin zones (e.g., flamelets) characterized by large (as compared with $\nabla\bar{c}$) instantaneous spatial gradients ∇c. In such a case, the first term on the RHS of Equation 7.47 dominates (it scales as δ_L^{-2}, while other terms scale as $\delta_L^{-1}\delta_t^{-1}$) and, therefore,

$$
\bar{W}\approx\frac{\bar{\chi}_c}{(2c_m-1)}. \qquad (7.48)
$$

If the following closure

$$
\frac{\overline{\rho\chi_c}}{\overline{\rho c''^2}}\propto\frac{\bar{\varepsilon}}{k}, \qquad (7.49)
$$

which is widely used when modeling nonreacting turbulent mixing layers, is invoked, the EBU Equation 7.38 is recovered from Equations 7.48 and 7.49.

In addition to these closure relations for the mean reaction rate, there are a number of alternative algebraic models, some of which are listed in Appendix B of a review paper by Lipatnikov and Chomiak (2002a). All the algebraic models that the author is aware of suffer from two important drawbacks. First, no algebraic model has yet been thoroughly validated against a wide set of representative experimental data obtained from substantially different flames under well-defined simple conditions. As far as most of the algebraic models are concerned, many of them are yet to be applied to a few laboratory flames. As regards a few algebraic models that are widely used in industrial applications, for example, the model by Magnussen and Hjertager (1976), the model constants are commonly tuned and substantially different for different flames. Moreover, multidimensional numerical simulations of an engine do not seem to be a solid test of a combustion model, because (i) the experimental data obtained from an engine are typically limited and (ii) the computed results are sensitive to many other (not combustion) models that are required to perform such a simulation. To thoroughly validate a premixed turbulent combustion model, simple laboratory flames should be computed first of all, but a few such quantitative tests have been carried out. For instance, a recent review paper by Veynante and Vervisch (2002) discusses algebraic closure relations for the mean reaction rate but does not provide even a single figure that quantitatively compares the computed and the measured results obtained from a simple premixed turbulent flame.

Second, the algebraic closure relations for $\overline{W}$ that are available today cannot predict the peculiarities of premixed turbulent flame development, highlighted in the section "Mean Structure of a Premixed Turbulent Flame" in Chapter 4 and in the section "Development of Premixed Turbulent Flame in the Flamelet Regime" in Chapter 5. Indeed, all currently used algebraic closure relations may be subsumed as follows (Lipatnikov, 2009b):

$$\overline{W} = \frac{\rho_u}{\tau_f} \Psi(\overline{\rho}, \tilde{c}), \qquad (7.50)$$

with different functions Ψ being invoked by different models. A flame timescale, τ_f, introduced for dimensional reasons, may depend on τ_t, Da, Ka, Le, etc., but neither τ_f nor Ψ depends on the flame-development time. Then, in a statistically planar, 1D case, integration of Equation 7.50 from $-\infty$ to ∞ yields

$$\rho_u U_t \equiv \int\limits_{-\infty}^{\infty} \overline{W} dx = \int\limits_{-\infty}^{\infty} \frac{\Psi(\overline{\rho}, \tilde{c})}{\tau_f} dx = \frac{\delta_t}{\tau_f} \int\limits_{-\infty}^{\infty} \Psi_1(\xi) d\xi \propto \frac{\delta_t(t)}{\tau_f} \qquad (7.51)$$

using the normalized distance ξ determined by Equation 4.33.

Equation 7.51 shows that the development of the turbulent burning velocity is solely controlled by the development of the mean flame brush thickness, contrary to the significant qualitative difference in the development of $U_t(t)$ and $\delta_t(t)$, highlighted in the section "Development of Premixed Turbulent Flame in the Flamelet Regime" in Chapter 5. To predict this peculiarity of premixed turbulent combustion, the timescale τ_f should directly depend on the flame-development time, but such a submodel for τ_f has not yet been proposed for use.

Presumed PDF Models

If the PDF $P(c,\mathbf{x},t)$ is known, then the mean reaction rate can easily be evaluated as

$$\bar{W}(\mathbf{x},t) = \int_0^1 W(c)P(\mathbf{x},t,c)dc. \tag{7.52}$$

Therefore, to solve the problem of modeling the influence of turbulence on premixed flames, one may try to guess the general shape of $P(c)$, invoking some mathematical or physical reasoning. In fact, one of the first attempts to presume $P(c)$ was undertaken by Bray, Moss, and Libby (see Equation 5.11 in the section "Bray–Moss–Libby Approach" in Chapter 5), but the BML approach does not allow us to evaluate $\bar{W}$ using Equation 7.52, because the shape of the function P_f on the RHS of Equation 5.11 is not presumed.

Another way of presuming the PDF $P(c)$ is to associate it with a beta function:

$$\tilde{P}(\mathbf{x},t,c) \equiv \frac{\rho(c)}{\bar{\rho}(\mathbf{x},t)}P(\mathbf{x},t,c) = \frac{\rho(c)}{\bar{\rho}(\mathbf{x},t)}P\left[\tilde{c}(\mathbf{x},t), \frac{\overline{\rho c''^2}}{\bar{\rho}}(\mathbf{x},t), c\right]$$

$$= \frac{\Gamma(a+b)}{\Gamma(a)\Gamma(b)}c^{a-1}(1-c)^{b-1}, \tag{7.53}$$

where

$$\Gamma(a) = \int_0^\infty z^{a-1}e^{-z}dz \tag{7.54}$$

is the gamma function, and the parameters a and b can be evaluated as

$$a(\mathbf{x},t) = \tilde{c}\left[\frac{\bar{\rho}\tilde{c}(1-\tilde{c})}{\overline{\rho c''^2}} - 1\right],$$

$$b(\mathbf{x},t) = (1-\tilde{c})\left[\frac{\bar{\rho}\tilde{c}(1-\tilde{c})}{\overline{\rho c''^2}} - 1\right] \tag{7.55}$$

using the fields $\tilde{c}(\mathbf{x},t)$, $\overline{\rho c''^2}(\mathbf{x},t)$, and $\bar{\rho}(\mathbf{x},t)$, which were obtained by solving Equations 7.36, 7.42, and 5.25, respectively. Note that the term $\overline{c''W}$ on the RHS of Equation 7.42 can easily be closed:

$$\overline{c''W}(\mathbf{x},t) = \int_0^1 cW(c)P(\mathbf{x},t,c)dc - \tilde{c}\int_0^1 W(c)P(\mathbf{x},t,c)dc \tag{7.56}$$

if $P(c)$ is presumed. As discussed in the following chapter, a basically similar model, that is, a presumed beta function $P(f, \tilde{f}, \rho f''^2)$ for a mixture fraction f, is widely used for numerical simulations of turbulent diffusion flames after the pioneering work of Janicka and Kollmann (1979). For premixed flames, the approach was first applied by Peters (1982) and was used by other researchers from time to time (e.g., see the papers by Bradley et al. [1994a], Bray et al. [2006], Jin et al. [2008], Vreman et al. [2009], and Kolla and Swaminathan [2010a,b]).

On the face of it, such an approach appears to be very attractive due to its simplicity and because, depending on the particular values of a and b, the shape of a beta function varies significantly, from a bimodal-like PDF if $a \ll 1$ and $b \ll 1$ to a Gaussian-like PDF that has a single peak and is distributed in a wide range of c (see Figure 8.11). Therefore, Equations 7.52 through 7.56 seem to be applicable to modeling various regimes of premixed turbulent combustion, from flamelet regime to distributed reaction regime (if the latter regime exists).

However, if we consider the flamelet regime of premixed turbulent combustion more carefully, then the beta-function approach may result in substantial errors in evaluating the mean reaction rate $\overline{W}$. Indeed, if the probability $\gamma \ll 1$, then the beta-function parameters $a \ll 1$ and $b \ll 1$, because the first term in square brackets on the RHS of Equation 7.55 is very close to unity (see Equation 5.29). Accordingly, even a small error in computing $\tilde{c}(\mathbf{x}, t)$ and/or $\rho c''^2(\mathbf{x}, t)$ by solving Equations 7.36 and 7.42, respectively, can result in substantial errors in calculating the small parameters a and b and, hence, substantial errors in evaluating $\overline{W}$. As regards the former (small) errors, they are inevitable, not only for numerical reasons but also and mainly because of the limitations of the models invoked to close the RHSs of Equations 7.36 and 7.42.

In other words, in the flamelet regime, a small quantity γ cannot be determined accurately by subtracting one finite quantity from another if the two finite quantities are calculated even with small errors. Therefore, the mean reaction rate, which is equal to the probability $\gamma \ll 1$ multiplied by the reaction rate integrated over the flame front, cannot be determined accurately too.

Thus, the beta-function approach does not seem to be a tool for simulating premixed turbulent combustion in the flamelet regime. A possible way of resolving the problem consists of replacing a balance equation for $\rho c''^2$ with a balance equation for the segregation factor $g = \rho c''^2 / \overline{\rho} \tilde{c}(1 - \tilde{c})$. Recently, Bray et al. (2011) considered such a balance equation.

Libby and Williams (2000) proposed another presumed PDF for use in modeling partially premixed turbulent combustion, and it was invoked in a couple of papers, as discussed in the section "Use of Presumed PDFs for Modeling Both Premixed and Diffusion Burning Modes" in Chapter 9. For a fully premixed flame, the PDF is bimodal:

$$P(\mathbf{x}, t, c) = \alpha(\mathbf{x}, t)\delta(c - c_1) + \left[1 - \alpha(\mathbf{x}, t)\right]\delta(c - c_2), \qquad (7.57)$$

but contrary to the BML PDF given by Equation 5.11, the Dirac delta functions are localized at $0 \leq c_1 < c_2 \leq 1$ with $0 < c_1$ or/and $c_2 < 1$. Accordingly, $W(c_1) \neq 0$ or/and

$W(c_2) \neq 0$ and the term γP_f (see the last term on the RHS of Equation 5.11) is not required in order for $\overline{W} \neq 0$.

As compared with the beta-function approach, the bimodal PDF given by Equation 7.57 is much less flexible but yields simpler closure relations:

$$\overline{W}(\mathbf{x},t) = \alpha(\mathbf{x},t)W(c_1) + [1-\alpha(\mathbf{x},t)]W(c_2), \tag{7.58}$$

$$\overline{c''W}(\mathbf{x},t) = \alpha(\mathbf{x},t)(c_1-\tilde{c})W(c_1) + [1-\alpha(\mathbf{x},t)](c_2-\tilde{c})W(c_2) \tag{7.59}$$

of the reaction terms in Equations 7.36 and 7.42. However, because the PDF given by Equation 7.57 involves three unknown parameters, α, c_1, and c_2, the knowledge of the two first moments $\tilde{c}$ and $\overline{\rho c''^2}$ is not sufficient to determine the PDF. Either an extra balance equation or an extra closure relation is necessary to solve the problem, as will be discussed in the section "Use of Presumed PDFs for Modeling Both Premixed and Diffusion Burning Modes" in Chapter 9.

Like the beta-function approach, the bimodal PDF given by Equation 7.57 may result in large errors when modeling premixed turbulent flames characterized by a low probability $\gamma \ll 1$ of finding intermediate states of a reacting mixture. Indeed, using Equation 7.57, one can easily obtain

$$\overline{c} = \alpha c_1 + (1-\alpha)c_2, \tag{7.60}$$

$$\overline{c}(1-\overline{c}) = [\alpha c_1 + (1-\alpha)c_2][1-\alpha c_1 - (1-\alpha)c_2], \tag{7.61}$$

and

$$\begin{aligned}
\overline{c'^2} &= \alpha(c_1-\overline{c})^2 + (1-\alpha)(c_2-\overline{c})^2 \\
&= \alpha(1-\alpha)^2(c_1-c_2)^2 + (1-\alpha)\alpha^2(c_1-c_2)^2 \\
&= \alpha(1-\alpha)(c_1-c_2)^2
\end{aligned} \tag{7.62}$$

in the simplest case of a constant density. Since $\overline{c'^2} \to \overline{c}(1-\overline{c})$ as $\gamma \to 0$, we have

$$\alpha \to \frac{c_2(1-c_2)}{c_2(1-c_2)-c_1(1-c_1)}. \tag{7.63}$$

in this limit. Equation 7.63 is consistent with the constraint of $0 \leq \alpha \leq 1$ (otherwise, one of the two terms on the RHS of Equation 7.57 is negative, but a PDF cannot be negative by definition): (i) if either $c_1 \to 0$ or $c_2 \to 1$, provided that $c_1 < c_2$, or (ii) if $c_1 \to 0$ and $c_2 \to 1$. In the former case, $\alpha \to 1$ or $\alpha \to 0$, and Equation 7.60

yields either $\bar{c} \to 0$ or $\bar{c} \to 1$, respectively. Both results are obviously wrong within a flame brush. In the latter case, Equation 7.58 yields $\bar{W} \to 0$. Thus, the discussed model cannot be used to model premixed turbulent combustion in the flamelet regime.

If the probability γ is sufficiently large, then an important merit of the two presumed PDFs given by Equations 7.53 and 7.57 is the opportunity to allow for a detailed chemical scheme of combustion, for example, by simulating the relevant complex-chemistry laminar premixed flame for parameterizing $W(c)$, followed by substituting $W(c)$ into the integral on the RHS of Equation 7.52. It is worth noting, however, that such an approach indirectly implies that the mean rate on the LHS of Equation 7.52 is weakly sensitive to variations in the shape of the presumed PDF, provided that these variations change neither $\tilde{c}$ nor $\overline{\rho c''^2}$. Such a weak sensitivity of $\bar{W}$ to the shape of $P(c)$ has not yet been shown. On the contrary, considering the highly nonlinear behavior of $W(c)$, the mean rate is likely to depend substantially not only on the first, $\tilde{c}$, and second, $\overline{\rho c''^2}/\bar{\rho}$, moments of the PDF but also on higher moments. As a result, the discussed sensitivity may be significant (e.g., see Figure 9.15). In fact, a model that determines a presumed PDF $P(c)$ based solely on its first and second moments implies indirectly that the knowledge of the two moments is sufficient for evaluating the mean reaction rate. Such an assumption is associated with retaining only the first two terms in square brackets on the RHS of Equation 7.37 and may result in large errors in averaging $W(c)$, as discussed in the section "A Key Problem."

TWO-EQUATION MODELS: FLAME SURFACE DENSITY AND SCALAR DISSIPATION RATE

The models discussed in this section invoke an extra balance equation to evaluate the mean reaction rate $\bar{W}$. In principle, a balance equation for the instantaneous reaction rate W may be derived from Equations 1.49, 1.51, and 1.79 and subsequently averaged; however, two alternative approaches are more popular in the contemporary literature. One approach invokes the so-called mean flame surface density (FSD) $\bar{\Sigma}$ to evaluate $\bar{W}$, while another approach exploits the proportionality of the mean reaction rate to the mean scalar dissipation rate (see Equation 7.48), shown by Bray (1979) within the framework of the BML approach discussed in the section "Bray–Moss–Libby Approach" in Chapter 5.

The FSD models, which are sometimes called the coherent flame models (CFM), assume that the influence of turbulent eddies on the local burning rate within thin instantaneous flame fronts may be decoupled from the influence of turbulent eddies on the topology of the front. In other words, (i) the mean mass rate of consumption of fresh reactants per unit area of the instantaneous flame front is assumed to be independent of the probability of finding the front and (ii) this probability is characterized in terms of the mean flamelet surface density $\bar{\Sigma}(\mathbf{x}, t)$, with the instantaneous $\Sigma(\mathbf{x}, t) = \delta A(\mathbf{x}, t)/\delta V(\mathbf{x}, t)$ being equal to the area $\delta A(\mathbf{x}, t)$ of the front surface within an infinitesimal volume $\delta V(\mathbf{x}, t)$, divided by this volume. The FSD is measured in meters (m^{-1}). Mathematically, the FSD may be defined as $\Sigma \equiv |\nabla c|$ for an infinitely thin flame front (in this case, $\Sigma(\mathbf{x},t)$ is the Dirac delta function) or by the following equation (Pope, 1988):

$$\Sigma\left(c^*,\mathbf{x},t\right)=\overline{\left|\nabla c\right|\delta\left(c-c^*\right)} \qquad (7.64)$$

for a front of a finite thickness. Here, δ is the Dirac delta function and c^* is a reference value of the combustion progress variable. The eventual dependence of Σ on c^* is commonly ignored by FSD models. Such a simplification seems to be justified at least if the instantaneous flame front, which involves both preheat and reaction zones, is sufficiently thin and retains the laminar structure, for example, at the BML limit, that is, if the probability $\gamma \ll 1$ in Equation 5.11.

If the mean mass rate of consumption of fresh reactants per unit area of the instantaneous flame front is equal to $\rho_u \bar{u}_c(\mathbf{x},t)$, then the mean mass rate of consumption of fresh mixture within a volume $\delta V(\mathbf{x},t)$ is equal to $\rho_u \bar{u}_c(\mathbf{x},t)\delta A(\mathbf{x},t)$, but the same rate is also equal to $\overline{W}(\mathbf{x},t)\delta V(\mathbf{x},t)$. Therefore,

$$\overline{W}\left(\mathbf{x},t\right)=\rho_u \bar{u}_c\left(\mathbf{x},t\right)\frac{\delta A\left(\mathbf{x},t\right)}{\delta V\left(\mathbf{x},t\right)}=\rho_u \bar{u}_c\left(\mathbf{x},t\right)\overline{\Sigma}\left(\mathbf{x},t\right), \qquad (7.65)$$

that is, the mean reaction rate is proportional to the mean FSD. Obviously, Equation 7.65 is valid only if the turbulent pulsations in the instantaneous consumption velocity, u_c, and in the FSD are statistically independent of one another, that is, if $\overline{u_c'\Sigma'} = 0$. Such a simplification is justified at least if $\bar{u}_c = S_L^0$, that is, if the effects of turbulent stretching on the local burning rate within thin flamelets are negligible, as often assumed when using FSD models.

Equation 7.65 shows that the problem of closing the mean reaction rate may be reduced to evaluation of the mean FSD, provided that a submodel for the mean consumption velocity, for example, $\bar{u}_c = S_L^0$, is invoked. The mean FSD may be modeled either by invoking an algebraic closure relation (e.g., see a paper by Bray [1990]) or by solving an extra balance equation for $\overline{\Sigma}(\mathbf{x},t)$. All closure relations for $\overline{W}$ that have been obtained within the framework of the former approach are algebraic models subsumed by Equation 7.38, that is, they cannot yield substantially different time-dependencies of $U_t(t)/U_{t,\infty}$ and $\delta_t(t)/L$, as discussed in the section "Algebraic Models." Therefore, only the latter approach will be addressed briefly in the following text. For a more detailed discussion on the FSD models, interested readers are referred to a book by Poinsot and Veynante (2005) and to a review paper by Veynante and Vervisch (2002).

In principle, the exact balance equations for $\Sigma(\mathbf{x},t)$ and $\overline{\Sigma}(\mathbf{x},t)$ can be derived rigorously (Pope, 1988), but the latter equation involves a number of unclosed terms, which are difficult to model. The FSD models used in practical applications deal with the following phenomenological equation:

$$\frac{\partial\overline{\Sigma}}{\partial t}+\frac{\partial}{\partial x_k}\left(\tilde{u}_k\overline{\Sigma}\right)=\frac{\partial}{\partial x_k}\left(\frac{v_t}{Sc_t}\frac{\partial\overline{\Sigma}}{\partial x_k}\right)+Q-\Xi, \qquad (7.66)$$

where

$Q \propto \bar{\Sigma}$ and $\Xi \propto \bar{\Sigma}^2$ are the source and sink terms, respectively
Sc_t is a turbulent Schmidt number
v_t is the turbulent viscosity (see Equation 3.90).

Different models for the source and sink terms may be found in the literature (e.g., see Table 5 in a paper by Veynante and Vervisch [2002] or Table 2 in a paper by Lipatnikov [2009b]), but all the models are generalized as

$$\frac{\partial \bar{\Sigma}}{\partial t} + \frac{\partial}{\partial x_k}(\tilde{u}_k \bar{\Sigma}) = \frac{\partial}{\partial x_k}\left(\frac{v_t}{Sc_t}\frac{\partial \bar{\Sigma}}{\partial x_k}\right) + \Omega\left(\frac{u'}{S_L^0}, Re_t\right)\frac{\bar{\Sigma}}{\tau_\eta} - \Psi\left(\frac{u'}{S_L^0}, \frac{\bar{\rho}}{\rho_u}, \tilde{c}\right)S_L^0\bar{\Sigma}^2, \quad (7.67)$$

with particular expressions for the functions Ω and Ψ being reported in Table 2 in the aforementioned paper by Lipatnikov (2009b).

The physical meaning of most of the terms in Equation 7.67 is clear. The two terms on the LHS are associated with the transient effects and convection, while the last two terms on the RHS model the flamelet surface area production due to turbulent stretching and the flamelet surface area consumption due to collisions of the flamelets, in line with the discussion on the primary physical mechanisms of the influence of turbulence on combustion in the section "Physical Reasoning: Classical Flamelet Regime" in Chapter 5. The production term is multiplied by a function Ω to allow for differences between the increase in the area of a material surface in a constant-density turbulent flow and the increase in the area of a self-propagating interface that separates gases with different densities. In the most popular FSD models (Boudier et al., 1992; Duclos et al., 1993), the function $\Omega \propto (\tau_\eta/\tau_t)\Gamma$, where the so-called efficiency function $\Gamma(u'/S_L^0, L/\delta_L)$, introduced by Meneveau and Poinsot (1991), parameterizes the results of numerical simulations of the interaction between a laminar flame and a vortex pair, reported by Poinsot et al. (1990, 1991).

The first term on the RHS of Equation 7.67 looks like a turbulent diffusion term, while the physical meaning of turbulent diffusion of the mean FSD is not clear.

Like the algebraic models, the FSD models suffer from the same two drawbacks: (i) scanty validation against experimental data obtained from well-defined simple cases and (ii) inability to predict a substantial difference between the development of the turbulent burning velocity and the development of the mean flame brush thickness, highlighted in the section "Development of Premixed Turbulent Flame in the Flamelet Regime" in Chapter 5.

First, as yet, no FSD model has been thoroughly validated against a wide set of representative experimental data obtained from substantially different flames under well-defined simple conditions. For instance, a review paper by Veynante and Vervisch (2002) discusses a number of FSD models in detail but does not provide even one figure that quantitatively compares the computed and the measured results obtained from a simple premixed turbulent flame. The author is aware of only a few tests of various FSD models (Duclos et al., 1993; Choi and Huh, 1998; Lee et al., 1998; Prasad and Gore, 1999) that were performed using the experimental data obtained in well-defined simple cases, with the models tested by Duclos et al. (1993)

and Prasad and Gore (1999) differing from the models tested by Choi and Huh (1998) and Lee et al. (1998).

Second, to study the development of the mean flame brush thickness and turbulent burning velocity invoking an FSD model, let us multiply Equation 7.67 by $\rho_u S_L^0$. Then, using Equation 7.65 with $\bar{u}_c = S_L^0$, we obtain the following equation for the mean reaction rate:

$$\frac{\partial \bar{W}}{\partial t} + \frac{\partial}{\partial x_k}\left(\tilde{u}_k \bar{W}\right) = \frac{\partial}{\partial x_k}\left(\frac{\nu_t}{\mathrm{Sc}_t}\frac{\partial \bar{W}}{\partial x_k}\right) + \Omega\left(\frac{u'}{S_L^0},\mathrm{Re}_t\right)\frac{\bar{W}}{\tau_\eta} - \Psi\left(\frac{u'}{S_L^0},\frac{\bar{\rho}}{\rho_u},\tilde{c}\right)\frac{\bar{W}^2}{\rho_u}. \tag{7.68}$$

Integration of this equation from $x = -\infty$ to $x = \infty$ yields

$$\frac{dU_t}{dt} = \frac{\Omega}{\tau_\eta}U_t - \int_{-\infty}^{\infty}\Psi\frac{\bar{W}^2}{\rho_u^2}dx \tag{7.69}$$

in the statistically planar, 1D case, because $\bar{W}(-\infty) = \bar{W}(\infty) = 0$. Here, for simplicity, we consider the case of statistically stationary, homogeneous, and isotropic Kolmogorov turbulence not affected by combustion (e.g., in the case of a constant-density "flame"), because such a hypothetical case is fully consistent with assumptions invoked to arrive at a typical FSD model. Accordingly, neither the function Ω nor the timescale τ_η depends on x.

If we introduce the length scale

$$\Lambda_\Sigma \equiv U_t^2\left(\int_{-\infty}^{\infty}\Psi\frac{\bar{W}^2}{\rho_u^2}dx\right)^{-1} = \left(\int_{-\infty}^{\infty}\frac{\bar{W}}{\rho_u}dx\right)^2\left(\int_{-\infty}^{\infty}\Psi\frac{\bar{W}^2}{\rho_u^2}dx\right)^{-1}, \tag{7.70}$$

then Equation 7.69 reads

$$\frac{dU_t}{dt} = \frac{\Omega}{\tau_\eta}U_t - \frac{U_t^2}{\Lambda_\Sigma}. \tag{7.71}$$

If we further assume that there is a limit $U_{t,\infty} = U_t(t \to \infty)$, then

$$\Lambda_{\Sigma,\infty} \equiv \Lambda_\Sigma\left(t \to \infty\right) = \frac{\tau_\eta U_{t,\infty}}{\Omega} \tag{7.72}$$

and Equation 7.71 reads

$$\frac{1}{u_t}\frac{du_t}{d\theta} = 1 - \frac{u_t}{\lambda_\Sigma}, \tag{7.73}$$

where $u_t \equiv U_t/U_{t,\infty}$, $\lambda_\Sigma \equiv \Lambda_\Sigma/\Lambda_{\Sigma,\infty}$, and $\theta \equiv \Omega t/\tau_\eta$ are the normalized burning velocity, length scale, and time, respectively. Note that the aforementioned assumption about the existence of a fully developed premixed turbulent flame at $t \to \infty$ appears to be justified for a constant-density "flame," which is not subject to the DL instability.

Equation 7.73 shows that the turbulent burning velocity increases only if $u_t < \lambda_\Sigma$. Since the length scale Λ_Σ, defined by Equation 7.70, appears to be substantially affected by the large-scale turbulent eddies, as well as the mean flame brush thickness, the trend yielded by Equation 7.73 does not seem to be consistent with the physical arguments discussed in the section "Development of Premixed Turbulent Flame in the Flamelet Regime" in Chapter 5. For example, if we (i) consider the self-similar regime of premixed turbulent flame development, that is, $\bar{\rho}(x,t) = \bar{\rho}(\xi)$ and $\tilde{c}(x,t) = \tilde{c}(\xi)$ with the normalized distance ξ being defined by Equation 4.33, and (ii) assume that $\bar{\Sigma}(\mathbf{x},t) = \mathrm{E}(\bar{\rho},\tilde{c})/L_\Sigma$ with the length scale L_Σ being independent of x (but may depend on the flame-development time, u'/S_L^0, Re_t, etc.), then the two integrals on the RHS of Equation 7.70 scale as $(S_L^0 \delta_t/L_\Sigma)^2$ and $\delta_t(S_L^0/L_\Sigma)^2$, respectively, using Equation 7.65. Therefore, $\Lambda_\Sigma \propto \delta_t$, and Equation 7.73 reads

$$\frac{1}{u_t}\frac{du_t}{d\theta} = 1 - \frac{u_t}{d_t}, \tag{7.74}$$

where $d_t \equiv \delta_t/\delta_{t,\infty}$ is the normalized mean flame brush thickness. Equation 7.74 shows that the turbulent burning velocity increases only if the mean flame brush thickness increases faster, but such a trend is totally inconsistent with the physical arguments discussed in the section "Development of Premixed Turbulent Flame in the Flamelet Regime" in Chapter 5.

Even if the foregoing qualitative analysis invoked a few assumptions, numerical simulations by Lipatnikov (2009b) that did not use these simplifications have confirmed that the currently available FSD models cannot predict the key peculiarities of the premixed turbulent flame development highlighted in the section "Development of Premixed Turbulent Flame in the Flamelet Regime" in Chapter 5, that is, the substantial difference between $u_t(\theta)$ and $d_t(\theta)$, with u_t being significantly larger than d_t in a typical flame.

It is also worth noting that Equation 7.67 does not seem to be applicable to modeling the early stage of the interaction between an initially ($t = 0$) laminar premixed flame and turbulence. If the laminar flame is assumed to be infinitely thin, then on the flame surface, $\bar{\Sigma} = \Sigma = |\nabla c|$ is the Dirac delta function. In such a case, the last, nonlinear term on the RHS of Equation 7.67, which involves $\bar{\Sigma}^2 = \Sigma^2 = |\nabla c|^2$, is mathematically meaningless, because the Dirac delta function may not be multiplied by itself. If a laminar flame of a finite, but small thickness is considered, then $\bar{\Sigma}^2 = \Sigma^2 \propto \Delta_L^{-2}$ at $t \to 0$. Therefore, the nonlinear term may dominate on the RHS of Equation 7.67 and, after integration from $x = -\infty$ to ∞, may result in a decrease in U_t with time (see Equation 7.69). Such an unphysical result is associated with the following self-inconsistency of the FSD models. On the one hand, the nonlinear term discussed previously is associated with the consumption of the flamelet surface area due to collisions of the flamelets, but this term is largest at the laminar flame. On the other hand, any term associated with the flamelet collisions should vanish during an early stage of the interaction between a laminar flame and a turbulent flow, because the flame surface is weakly wrinkled and the collision probability is negligible.

A balance equation for the mean reaction rate may also be obtained from a balance equation for the mean scalar dissipation rate using the linear relation between $\overline{W}$ and $\overline{\chi}_c$, given by Equation 7.48. Such an approach was put forward by Borghi (1990), developed by Mantel and Borghi (1994) and Mura and Borghi (2003), and further developed in recent papers that are cited in the following text. A balance equation for $\overline{\chi}_c$ may be derived straightforwardly from Equations 1.49, 1.51, and 1.79, but such an equation contains unclosed terms that are difficult to model even in the constant-density, nonreacting case (turbulent mixing). Heat release in a flame further complicates the problem by substantially affecting the unclosed terms, as discussed in detail by Swaminathan and Bray (2005), Swaminathan and Grout (2006), Chakraborty and Swaminathan (2007a,b), Chakraborty et al. (2008), Hartung et al. (2008), Mura et al. (2008, 2009), Mura and Champion (2009), and Kolla et al. (2009). As a result, models of premixed turbulent combustion that involve a balance equation for the mean scalar dissipation rate appear to be more cumbersome than competitive approaches, for example, FSD models. For instance, $\overline{\chi}_c$ equation models yield Equation 7.68 with several extra terms on the RHS (see Equation 37 and Table 3 in a paper by Lipatnikov [2009b]).

Such models strongly need thorough validation against a wide set of representative experimental data obtained from substantially different flames under well-defined and substantially different conditions. The author is aware of the sole quantitative test of such a model against the experimental data. Kolla et al. (2009, 2010) compared the fully developed turbulent burning velocity yielded by their model in the statistically planar, 1D case with a little amount of the experimental data on turbulent flame speeds. However, due to the development and substantial thickness of a typical premixed turbulent flame, a quantitative comparison of the theoretical value of $U_{t,\infty} = U_t(t \to \infty)$ with a measured S_t does not seem to be a solid validation (see the section "Determination of Turbulent Flame Speed and Burning Velocity" in Chapter 4 and Section 5.2 in a recent review paper by Lipatnikov and Chomiak [2010], as well as two more papers by Lipatnikov and Chomiak [2004a, 2007]).

The discussed models have also been tested against the DNS data; however, despite the substantial differences between the two competitive $\overline{\chi}_c$ equations closed by Mura et al. (2008, 2009) and Swaminathan and coworkers (Swaminathan and Bray, 2005; Swaminathan and Grout, 2006; Chakraborty and Swaminathan, 2007a,b; Chakraborty et al., 2008; Kolla et al. 2009), both the equations were claimed to be supported by the DNS data when using two different DNS databases for testing two differently closed $\overline{\chi}_c$ equations. In such a situation, it is very difficult to draw solid conclusions about the validity of the two models.

The capability of the $\overline{\chi}_c$ equation models to predict the key peculiarities of premixed turbulent flame development, highlighted in the section "Development of Premixed Turbulent Flame in the Flamelet Regime" in Chapter 5), has not yet been shown. On the contrary, a numerical study of statistically planar and 1D, developing premixed turbulent flames by Lipatnikov (2009b) has indicated that the $\overline{\chi}_c$ equation models yield qualitatively similar time-dependencies of $U_t(t)/U_{t,\infty}$ and $\delta_t(t)/\delta_{t,\infty}$.

Further discussion on the $\overline{\chi}_c$ equation models is beyond the scope of this book. Interested readers are referred to the previously cited papers.

Because the author considers flame development to be a very important feature of premixed turbulent combustion and for other reasons previously discussed, the currently available two-equation models do not seem to be a well-elaborated predictive tool. The author is not aware of any clear advantage of a two-equation model as compared with the models discussed in the next subsection.

MODELS OF SELF-SIMILARLY DEVELOPING PREMIXED TURBULENT FLAME

General Features

Let us consider the following balance equation for the Favre-averaged combustion progress variable:

$$\frac{\partial}{\partial t}\left(\overline{\rho}\tilde{c}\right)+\frac{\partial}{\partial x_k}\left(\overline{\rho}\tilde{u}_k\tilde{c}\right) = \frac{\partial}{\partial x_k}\left(\overline{\rho}D\frac{\partial\tilde{c}}{\partial x_k}\right)+\rho_u U_t\left|\nabla\tilde{c}\right|, \tag{7.75}$$

where D is the diffusivity, U_t is the turbulent burning velocity, and

$$\left|\nabla\tilde{c}\right| \equiv \sqrt{\sum_{k=1}^{3}\left(\frac{\partial\tilde{c}}{\partial x_k}\right)^2}. \tag{7.76}$$

A balance equation, which was basically similar to Equation 7.75 but was written for the Reynolds-averaged temperature $\overline{T}$, was first proposed to be used by Prudnikov (1960 and 1967) and later by Zimont (1977) and Zimont and Mesheryakov (1988). For the Favre-averaged combustion progress variable, Equation 7.75 was put forward by Zimont and Lipatnikov (1992, 1993, 1995) (see also papers by Karpov et al. [1994, 1996a]) and independently by Weller (1993) (see also Weller et al. [1994]). Accordingly, in the combustion literature, Equation 7.75 is known as the Zimont model by referring to papers by Zimont et al. (1998) and Zimont (2000), or a model by Zimont and Lipatnikov (1995), or the turbulent flame closure (TFC) model, or a model by Karpov et al. (1996a), or the Weller (1993) model. From the author's standpoint, Equation 7.75 should be called the Prudnikov equation by referring to pioneering papers by Prudnikov (1960, 1967).

For a statistically planar, 1D, premixed turbulent flame that propagates from left to right, Equation 7.75 has the following solution:

$$\overline{c} = \frac{1}{2}\mathrm{erfc}\left(\xi\sqrt{\pi}\right) = \frac{1}{\sqrt{\pi}}\int_{\xi\sqrt{\pi}}^{\infty} e^{-\zeta^2}\,d\zeta, \tag{7.77}$$

$$\xi = \frac{x - x_f(t)}{\delta_t(t)}, \tag{7.78}$$

$$x_f(t) = x_f(t = 0) + \int_0^t U_t d\theta, \tag{7.79}$$

and

$$\delta_t^2 = 4\pi \int_0^t D_t d\theta, \tag{7.80}$$

provided that the BML state Equations 5.25 through 5.27 hold. It is worth stressing that (i) Equation 7.79 holds in the coordinate framework attached to the unburned gas, that is, $\bar{u}(x \to \infty) \to 0$, and (ii) the foregoing solution is written for the Reynolds-averaged combustion progress variable, while Equation 7.75 involves the Favre-averaged one.

Indeed, substitution of Equation 7.77 into the Favre-averaged continuity Equation 7.34 yields

$$\frac{d}{d\xi}(\bar{\rho}\tilde{u}) = \left(U_t + \xi \frac{d\delta_t}{dt}\right) \frac{d\bar{\rho}}{d\xi} \tag{7.81}$$

or

$$-\bar{\rho}\tilde{u} = U_t(\rho_u - \bar{\rho}) + \frac{d\delta_t}{dt} \int_\xi^\infty \zeta \frac{d\bar{\rho}}{d\zeta} d\zeta \tag{7.82}$$

after integration from ξ to ∞ in the aforementioned coordinate framework. Because

$$-\frac{1}{\rho_u - \rho_b} \frac{d\bar{\rho}}{d\xi} = \frac{\bar{\rho}^2}{\rho_u \rho_b} \frac{d\tilde{c}}{d\xi} = \frac{d\bar{c}}{d\xi} = -e^{-\pi\xi^2} \tag{7.83}$$

by virtue of Equations 5.25 through 5.27 and Equation 7.77, the integral on the RHS of Equation 7.82 can easily be taken and

$$-\bar{\rho}\tilde{u} = U_t(\rho_u - \bar{\rho}) + \frac{d\delta_t}{dt} \frac{\rho_u - \rho_b}{2\pi} e^{-\pi\xi^2}. \tag{7.84}$$

Finally, substitution of Equation 7.77 into Equation 7.75 yields

$$-\overline{\rho}U_t\frac{d\tilde{c}}{d\xi} - \overline{\rho}\xi\frac{d\delta_t}{dt}\frac{d\tilde{c}}{d\xi} - U_t\left(\rho_u - \overline{\rho}\right)\frac{d\tilde{c}}{d\xi} - \frac{d\delta_t}{dt}\frac{\rho_u - \rho_b}{2\pi}e^{-\pi\xi^2}\frac{d\tilde{c}}{d\xi}$$

$$= \frac{D_t}{\delta_t}\frac{d}{d\xi}\left(\overline{\rho}\frac{d\tilde{c}}{d\xi}\right) - \rho_u U_t\frac{d\tilde{c}}{d\xi}, \tag{7.85}$$

that is,

$$-\overline{\rho}\xi\frac{d\delta_t}{dt}\frac{d\tilde{c}}{d\xi} - \frac{d\delta_t}{dt}\frac{\rho_u - \rho_b}{2\pi}e^{-\pi\xi^2}\frac{d\tilde{c}}{d\xi} = \frac{D_t}{\delta_t}\frac{d}{d\xi}\left(\overline{\rho}\frac{d\tilde{c}}{d\xi}\right). \tag{7.86}$$

Using Equations 7.80 and 7.83, the LHS and RHS of Equation 7.86 read

$$-\overline{\rho}\xi\frac{d\delta_t}{dt}\frac{d\tilde{c}}{d\xi} - \frac{d\delta_t}{dt}\frac{\rho_u - \rho_b}{2\pi}e^{-\pi\xi^2}\frac{d\tilde{c}}{d\xi} = \frac{\rho_u\rho_b}{\overline{\rho}}\frac{d\delta_t}{dt}\left(\xi e^{-\pi\xi^2} + \frac{\rho_u - \rho_b}{2\pi\overline{\rho}}e^{-2\pi\xi^2}\right) \tag{7.87}$$

and

$$\frac{D_t}{\delta_t}\frac{d}{d\xi}\left(\overline{\rho}\frac{d\tilde{c}}{d\xi}\right) = \frac{1}{2\pi}\frac{d\delta_t}{dt}\frac{d}{d\xi}\left(\frac{\rho_u\rho_b}{\overline{\rho}}\frac{d\tilde{c}}{d\xi}\right) = \frac{d\delta_t}{dt}\frac{\rho_u\rho_b}{2\pi\overline{\rho}}\left(2\pi\xi e^{-\pi\xi^2} + \frac{\rho_u - \rho_b}{\overline{\rho}}e^{-2\pi\xi^2}\right), \tag{7.88}$$

respectively. Therefore, Equation 7.86 holds and Equations 7.77 through 7.80 are an analytical solution to Equation 7.75 supplemented with Equations 5.25 through 5.27.

The solution given by Equations 7.77 through 7.80 indicates the following features of Equation 7.75:

First, the model predicts the self-similarity of the mean flame structure and, in particular, the complementary error function profile of the mean combustion progress variable in line with numerous experimental data discussed in the section "Mean Structure of a Premixed Turbulent Flame" in Chapter 4 (see Figure 4.22).

Second, the increase in the mean flame brush thickness yielded by the model is controlled by the turbulent diffusivity. For instance, if the diffusivity is approximated as

$$D_t = D_{t,\infty}\left[1 - \exp\left(-\frac{t}{\tau_L}\right)\right] \tag{7.89}$$

invoking the Taylor theory (see the section "Taylor's Theory" and Equation 3.122 in Chapter 3), then Equation 7.80 predicts

$$\delta_t^2 = 4\pi D_{t,\infty}t\left\{1 - \frac{\tau_L}{t}\left[1 - \exp\left(-\frac{t}{\tau_L}\right)\right]\right\} \tag{7.90}$$

in line with numerous experimental data discussed in the section "Mean Flame Brush Thickness" in Chapter 4. Here, $D_{t,\infty} = u'^2 \tau_L$ is the fully developed turbulent diffusivity.

Third, because Equations 7.79 and 7.80 are independent of each other, Equation 7.75 can yield substantially different time-dependencies of $U_t(t)/U_{t,\infty}$ and $\delta_t(t)/L$. For instance, if Equation 7.29 is invoked to close U_t, then the difference between $U_t(t)$ and $\delta_t(t)$, resulting from such a model, is shown in Figure 7.6.

Fourth, because Equation 7.75 yields a flame with an increasing mean flame brush thickness (see Equation 7.90), a consistent submodel for the turbulent burning velocity in the last term on the RHS of Equation 7.75 should address a developing flame. In particular, the original (Equation 7.13) and extended (Equation 7.29) Zimont models satisfy this requirement.

The features of Equation 7.75 emphasized earlier make this approach well tailored for simulating the developing premixed turbulent flames with the self-similar mean structure. For this reason, in the following text, we shall call this approach the self-similarly developing flame (SSDF) model.

Derivation

The foregoing advantages of the SSDF model were sometimes claimed to be devalued by the lack of a derivation of Equation 7.75. To address such a concern, let us derive the 1D Equation 7.75 from the following quite general balance equation:

$$\frac{\partial}{\partial t}(\bar{\rho}\tilde{c}) + \frac{\partial}{\partial x}(\bar{\rho}\tilde{u}\tilde{c}) = \underbrace{\chi D_{t,\infty}\frac{\partial}{\partial x}\left(\bar{\rho}\frac{\partial\tilde{c}}{\partial x}\right)}_{\text{I}} + \underbrace{\rho_u V \psi \frac{\partial f}{\partial x}}_{\text{II}} + \underbrace{\frac{\rho_u \omega \Omega}{\tau_f}}_{\text{III}}, \tag{7.91}$$

which subsumes many currently available models of premixed turbulent combustion, as discussed elsewhere (Lipatnikov and Chomiak, 2005b; Lipatnikov, 2007b).

Terms I, II, and III on the RHS of Equation 7.91 are associated with turbulent diffusion, the effects of heat release on the turbulent scalar transport (the so-called pressure-driven or countergradient transport, see the section "Countergradient Scalar Transport in Turbulent Premixed Flames" in Chapter 6), and the mean reaction rate, respectively. Accordingly, the sum of terms I and II is a generalized closure of the turbulent transport term on the RHS of Equation 7.36:

$$-\overline{\rho u'' c''} = \bar{\rho}\chi D_{t,\infty}\frac{\partial\tilde{c}}{\partial x} + \rho_u V \psi f, \tag{7.92}$$

and term III is a closure of the reaction term on the RHS of Equation 7.36, that is,

$$\overline{W} = \frac{\rho_u \omega \Omega}{\tau_f}. \tag{7.93}$$

Here, $\chi(t)$, $\psi(t)$, and $\omega(t)$ are arbitrary time-dependent, positive, bounded functions such that $\chi(t \to \infty) \to 1$, $\psi(t \to \infty) \to 1$, and $\omega(t \to \infty) \to 1$. These functions are associated with the development of turbulent diffusivity (see Equation 7.89), development of pressure-driven transport, and development of the mean reaction rate, respectively. The scales for the diffusivity $D_{t,\infty}$, slip velocity V, and mean reaction rate τ_f, measured in meters squared per second (m^2/s), meters per second (m/s), and second (s^{-1}), respectively, are introduced for dimensional reasons. Two arbitrary bounded functions, $f(\tilde{c},\bar{\rho})$ and $\Omega(\tilde{c},\bar{\rho}) \geq 0$, satisfy the conditions $f(0,\rho_u) = f(1,\rho_b) = \Omega(0,\rho_u) = \Omega(1,\rho_b) = 0$, because both the turbulent scalar flux and the reaction rate vanish in an unburned or burned mixture. The function $f(\tilde{c},\bar{\rho})$ is positive for a flame that propagates from left to right and is negative for a flame that moves in the opposite direction. The former flame is analyzed in the following text.

Thus, terms I, II, and III on the RHS of Equation 7.91 have been introduced (i) to allow for all basic phenomena (turbulent diffusion, pressure-driven transport, chemical reactions, development of turbulent diffusivity and mean reaction rate) that control the propagation of a developing premixed turbulent flame, (ii) to subsume as many currently available models of premixed turbulent combustion as possible, and (iii) to keep the problem analytically solvable. To do so, five unspecified functions, $f(\tilde{c},\bar{\rho})$, $\Omega(\tilde{c},\bar{\rho})$, $\chi(t)$, $\psi(t)$, and $\omega(t)$, have been invoked.

Although Equation 7.91 is rather general, it is less general than the unclosed Equation 7.36, because the unknown functions on the LHSs of Equations 7.92 and 7.93 may depend on both ξ and t in a general case, while these dependencies on the two variables are split into the products, $\psi(t)f[\tilde{c}(\xi),\bar{\rho}(\xi)] = \psi(t)f_1(\xi)$ and $\omega(t)\Omega[\tilde{c}(\xi),\bar{\rho}(\xi)] = \omega(t)\Omega_1(\xi)$, of two functions that depend on a single variable (either ξ or t) each. Thus, Equation 7.91 is a compromise between generality, solvability, and physical meaning.

Let us assume that the mean flame structure is self-similar, that is $\bar{\rho}(x,t) = \bar{\rho}(\xi)$ and $\tilde{c}(x,t) = \tilde{c}(\xi)$, in line with various experimental data discussed in the section "Mean Structure of a Premixed Turbulent Flame" in Chapter 4. Then, substitution of a self-similar mean profile of $\tilde{c}(x,t) = \tilde{c}(\xi)$, with the normalized distance ξ being determined by Equations 7.78 and 7.79, into the Favre-averaged continuity Equation 7.34 and Equation 7.91 yields

$$-\left(S + \xi \frac{d\delta_t}{dt}\right)\frac{d\bar{\rho}}{d\xi} + \frac{\partial}{\partial\xi}(\bar{\rho}\tilde{u}) = 0, \qquad (7.94)$$

and

$$-\left(S + \xi \frac{d\delta_t}{dt}\right)\bar{\rho}\frac{d\tilde{c}}{d\xi} + \bar{\rho}\tilde{u}\frac{d\tilde{c}}{d\xi} = \frac{\chi D_{t,\infty}}{\delta_t}\frac{d}{d\xi}\left(\bar{\rho}\frac{d\tilde{c}}{d\xi}\right) + \rho_u \psi V \frac{df}{d\xi} + \rho_u \frac{\delta_t \omega \Omega}{\tau_f}, \qquad (7.95)$$

where $S = dx_f/dt$ is the flame speed. Integrating Equation 7.94 from ξ to ∞, we obtain

$$\bar{\rho}\tilde{u} = -S(\rho_u - \bar{\rho}) - \frac{d\delta_t}{dt}\int_{\xi}^{\infty}\zeta\frac{d\bar{\rho}}{d\zeta}d\zeta \qquad (7.96)$$

for a flame that moves from left to right in the coordinate framework such that $\tilde{u}(\xi \to \infty) = 0$. Substitution of Equation 7.96 into Equation 7.95 results in

$$-\left(S+\xi\frac{d\delta_t}{dt}\right)\bar{\rho}\frac{d\tilde{c}}{d\xi}-\left[S(\rho_u-\bar{\rho})+\frac{d\delta_t}{dt}\int\limits_{\xi}^{\infty}\zeta\frac{d\bar{\rho}}{d\zeta}d\zeta\right]\frac{d\tilde{c}}{d\xi}$$

$$=\frac{\chi D_{t,\infty}}{\delta_t}\frac{d}{d\xi}\left(\bar{\rho}\frac{d\tilde{c}}{d\xi}\right)+\rho_u\psi V\frac{df}{d\xi}+\rho_u\frac{\delta_t\omega\Omega}{\tau_f}. \tag{7.97}$$

Therefore,

$$-\left(\rho_u S+\Psi\frac{d\delta_t}{dt}\right)\frac{d\tilde{c}}{d\xi}=\frac{\chi D_{t,\infty}}{\delta_t}\frac{d}{d\xi}\left(\bar{\rho}\frac{d\tilde{c}}{d\xi}\right)+\rho_u\psi V\frac{df}{d\xi}+\rho_u\frac{\delta_t\omega\Omega}{\tau_f}. \tag{7.98}$$

Where

$$\Psi \equiv \bar{\rho}\xi+\int\limits_{\xi}^{\infty}\zeta\frac{d\bar{\rho}}{d\zeta}d\zeta. \tag{7.99}$$

Let us discuss Equation 7.98. First, because the mean flame coordinate $x_f(t)$ in Equation 7.78 may be associated with any reference value c_0 of the combustion progress variable, one can always choose x_f so that (Lipatnikov and Chomiak, 2002c)

$$\int\limits_0^1 \Psi d\tilde{c} = 0. \tag{7.100}$$

Then, the integration of Equation 7.98 from $\xi = -\infty$ to $\xi = \infty$ yields

$$S(t)=\frac{\delta_t(t)\omega(t)}{\tau_f}\int\limits_{-\infty}^{\infty}\Omega d\zeta=B\frac{\delta_t(t)\omega(t)}{\tau_f}, \tag{7.101}$$

where $B = \int_{-\infty}^{\infty}\Omega d\zeta$ is a constant, because $\Omega[\tilde{c}(\xi),\bar{\rho}(\xi)]=\Omega_1(\xi)$. Equation 7.101 clarifies the physical meaning of the function $\omega(t)$; it has been introduced in order for the development of the turbulent burning velocity to differ qualitatively from the development of the mean flame brush thickness, in line with the discussion in the section "Development of Premixed Turbulent Flame in the Flamelet Regime" in Chapter 5. Moreover, Equation 7.101 clarifies the physical meaning of Equation 7.100, such that a definition of the mean flame coordinate x_f results in the equality of the speed of the mean flame surface and turbulent burning velocity.

Second, if we assume that the studied flame tends to a fully developed flame as $t \to \infty$ (such an assumption is justified at least in the case of a constant density, when the hydrodynamic instability of the entire flame brush vanishes), then Equation 7.101 results in

$$S_\infty = B \frac{\delta_{t,\infty}}{\tau_f}. \tag{7.102}$$

Accordingly, Equation 7.101 may be rewritten as

$$s(t) = \omega(t) d_t(t) \tag{7.103}$$

using the fully developed flame speed $S_\infty \equiv S(t \to \infty)$ and thickness $\delta_{t,\infty} \equiv \delta_t(t \to \infty)$ to normalize the developing flame speed $s(t) \equiv S(t)/S_\infty$ and thickness $d_t(t) \equiv \delta_t(t)/\delta_{t,\infty}$.

Third, at $t \to \infty$, Equation 7.98 reads

$$-\frac{\delta_{t,\infty}^2}{D_{t,\infty}\tau_f}\left(B\frac{d\tilde{c}}{d\xi}+\Omega\right)-\frac{\delta_{t,\infty}V}{D_{t,\infty}}\frac{df}{d\xi}=\frac{d}{d\xi}\left(\frac{\bar{\rho}}{\rho_u}\frac{d\tilde{c}}{d\xi}\right) \tag{7.104}$$

by virtue of Equation 7.102.

Fourth, using Equation 7.104 to exclude the diffusion term from Equation 7.98 and invoking Equation 7.102, we arrive at

$$\left(\frac{1}{2D_{t,\infty}}\frac{d\delta_t^2}{dt}\right)\left[\Psi\frac{d\tilde{c}}{d\xi}\right]-\frac{\delta_{t,\infty}^2}{D_{t,\infty}\tau_f}(d_t^2\omega-\chi)\left[B\frac{d\tilde{c}}{d\xi}+\Omega\right]-(\psi d_t-\chi)\left[\frac{\delta_{t,\infty}V}{D_{t,\infty}}\frac{df}{d\xi}\right]=0, \tag{7.105}$$

where the t- and ξ-dependent terms are written in round and square brackets, respectively. If time is normalized, $\vartheta = t/\tau_f$, using the flame timescale τ_f, then Equation 7.105 finally reads

$$\left(\frac{1}{2}\frac{dd_t^2}{d\vartheta}\right)\left[\Psi\frac{d\tilde{c}}{d\xi}\right]+(\chi-d_t^2\omega)\left[B\frac{d\tilde{c}}{d\xi}+\Omega\right]+(\chi-\psi d_t)\left[\frac{V\tau_f}{\delta_{t,\infty}}\frac{df}{d\xi}\right]=0. \tag{7.106}$$

Equation 7.106 has the following form:

$$\sum_{k=1}^{3}h_k(\vartheta)g_k(\xi)=0, \tag{7.107}$$

where at least $g_1(\xi)$, $g_3(\xi)$, and $h_1(\xi)$ are not identically equal to zero. Such an equation may have a solution only in the following five cases (Lipatnikov, 2007a,b):

1. $b_1 g_1 = b_2 g_2 = b_3 g_3$ and $\sum_{k=1}^{3} h_k / b_k = 0$,
2. $b_1 h_1 = b_2 h_2 = b_3 h_3$ and $\sum_{k=1}^{3} g_k / b_k = 0$,
3. $g_2 = 0$, $b_1 g_1 = b_3 g_3$, and $h_1 / b_1 + h_3 / b_3 = 0$,
4. $h_2 = 0$, $b_1 h_1 = b_3 h_3$, and $g_1 / b_1 + g_3 / b_3 = 0$, and
5. $h_3 = 0$, $b_1 h_1 = b_2 h_2$, and $g_1 / b_1 + g_2 / b_2 = 0$,

where b_k are constants. In case 1 (and case 3, to which case 1 is reduced if $|b_2| \to \infty$),

$$-\Psi \frac{d\tilde{c}}{d\xi} = b_2 \left[B \frac{d\tilde{c}}{d\xi} + \Omega \right] = b_3 \frac{V\tau_f}{\delta_{t,\infty}} \frac{df}{d\xi}. \tag{7.108}$$

Combining Equations 7.104 and 7.108, we obtain

$$\frac{\delta_{t,\infty}^2}{D_{t,\infty}\tau_f} \left(\frac{1}{b_2} + \frac{1}{b_3} \right) \Psi \frac{d\tilde{c}}{d\xi} = \frac{d}{d\xi} \left(\frac{\bar{\rho}}{\rho_u} \frac{d\tilde{c}}{d\xi} \right), \tag{7.109}$$

$$\Omega = -B \frac{d\tilde{c}}{d\xi} - \frac{b_3}{b_2 + b_3} \frac{D_{t,\infty}\tau_f}{\delta_{t,\infty}^2} \frac{d}{d\xi} \left(\frac{\bar{\rho}}{\rho_u} \frac{d\tilde{c}}{d\xi} \right), \tag{7.110}$$

and

$$V \frac{df}{d\xi} = -\frac{b_2}{b_2 + b_3} \frac{D_{t,\infty}}{\delta_{t,\infty}} \frac{d}{d\xi} \left(\frac{\bar{\rho}}{\rho_u} \frac{d\tilde{c}}{d\xi} \right). \tag{7.111}$$

Therefore, Equation 7.92 reads

$$-\overline{\rho u'' c''} = \frac{\chi D_{t,\infty}}{\delta_t} \bar{\rho} \frac{d\tilde{c}}{d\xi} - \frac{b_2}{b_2 + b_3} \frac{\Psi D_{t,\infty}}{\delta_{t,\infty}} \bar{\rho} \frac{d\tilde{c}}{d\xi} = \frac{D_{t,\infty}}{\delta_t} \left(\chi - \frac{b_2}{b_2 + b_3} \Psi d_t \right) \bar{\rho} \frac{d\tilde{c}}{d\xi} \tag{7.112}$$

and Equation 7.36 reduces to

$$\frac{\partial}{\partial t} (\bar{\rho}\tilde{c}) + \frac{\partial}{\partial x} (\bar{\rho}\tilde{u}\tilde{c}) = -\frac{\partial}{\partial x} \overline{\rho u'' c''} + \frac{\rho_u \omega \Omega}{\tau_f} =$$

$$= D_{t,\infty} \left(\chi - \frac{b_2}{b_2 + b_3} \Psi d_t \right) \frac{\partial}{\partial x} \left(\bar{\rho} \frac{\partial \tilde{c}}{\partial x} \right) - B \frac{\rho_u \omega}{\tau_f} \frac{d\tilde{c}}{d\xi}$$

$$- D_{t,\infty} \frac{b_3}{b_2 + b_3} \omega d_t^2 \frac{\partial}{\partial x} \left(\bar{\rho} \frac{\partial \tilde{c}}{\partial x} \right). \tag{7.113}$$

Finally, using Equations 7.101 and 7.103, we obtain

$$\frac{\partial}{\partial t}(\bar{\rho}\tilde{c}) + \frac{\partial}{\partial x}(\bar{\rho}\tilde{u}\tilde{c}) = D\frac{\partial}{\partial x}\left(\bar{\rho}\frac{\partial\tilde{c}}{\partial x}\right) - \rho_u S\frac{\partial\tilde{c}}{\partial x} \tag{7.114}$$

for a statistically planar, 1D, developing flame that moves from left to right. Here, the generalized diffusivity

$$D(t) = D_{t,\infty}\left[\chi(t) - (1+a)\psi(t)\frac{\delta_t(t)}{\delta_{t,\infty}} + a\omega(t)\left(\frac{\delta_t(t)}{\delta_{t,\infty}}\right)^2\right] \tag{7.115}$$

is introduced and $a = -b_3/(b_2 + b_3)$ is a constant. It may be shown that $0 \le a \le 1$ (Lipatnikov and Chomiak, 2005b).

It is also worth stressing that the flame speed S in Equation 7.114 is equal to the turbulent burning velocity $U_t \equiv \rho_u^{-1}\int_{-\infty}^{\infty}\bar{W}dx$ (see Equations 7.93 and 7.101), if the mean flame coordinate $x_f(t)$ is chosen so that Equation 7.100 holds.

Thus, in the statistically planar, 1D case, the key balance Equation 7.75 of the SSDF model has been derived from a rather general Equation 7.91 invoking the sole assumption of a self-similar mean structure of the flame considered. Recently, Pagnini and Bonomi (2011) developed another theoretical substantiation of the Prudnikov Equation 7.75.

The foregoing derivation has shown one more peculiarity of the SSDF model and has revealed at least one issue that needs further study. Equation 7.115 indicates that the generalized diffusion term on the RHS of Equation 7.114 is affected not only by turbulent diffusion but also by the pressure-driven transport and chemical reactions. Therefore (and in line with the discussion in the section "Mean Reaction Rate and Turbulent Diffusion"), the diffusion and source terms on the RHS of Equation 7.75 should not be considered to be two separate closure relations for the transport and mean reaction terms, respectively, on the RHS of Equation 7.36. A comparison of Equations 7.36 and 7.75 shows only that

$$-\frac{\partial}{\partial x_k}\overline{\rho u_k''c''} + \bar{W} = \frac{\partial}{\partial x_k}\left(\bar{\rho}D_t\frac{\partial\tilde{c}}{\partial x_k}\right) + \rho_u U_t|\nabla\tilde{c}|, \tag{7.116}$$

that is, the RHS of Equation 7.75 is a joint closure of the two terms on the RHS of Equation 7.36, as hypothesized by Karpov et al. (1994) and Zimont and Lipatnikov (1995). Accordingly, the SSDF model does not involve a closure of $\bar{W}$, and other physical reasoning should be invoked to evaluate the mean reaction rate.

Equation 7.115 also indicates that the use of turbulent diffusivity in Equation 7.75 is an oversimplification, which does not allow for the effects of heat release and chemical reactions on the turbulent scalar transport. Such effects make D dependent on the flame-development time. In particular, Equation 7.115 yields $D \to 0$ as $t \to \infty$, that is, the flame modeled by Equations 7.114 and 7.115 becomes fully developed due to the effects discussed earlier. Moreover, using Equation 7.103, one can rewrite Equation 7.115 as

$$D(t) = D_{t,\infty}\left[\chi(t) - (1+a)\psi(t)\frac{\delta_t(t)}{\delta_{t,\infty}} + as(t)\frac{\delta_t(t)}{\delta_{t,\infty}}\right]. \tag{7.117}$$

Because $0 \le a \le 1$ (Lipatnikov and Chomiak, 2005b) and $s(t) \le 1$, Equation 7.117 implies that the pressure-driven transport reduces the generalized diffusivity, while the influence of chemical reactions on D weakens this reduction effect but does not overwhelm it at least if $s(t) \le 2\psi(t)$. Accordingly, $D \le D_{t,\infty}\,\chi$ (cf. dotted–dashed and solid curves in Figure 7.7), and the pressure-driven transport reduces the rate of increase of the mean flame brush thickness.

Because Equation 7.117 involves one unknown constant a and at least one unknown function $\psi(t)$, evaluation of the diffusivity within the framework of the SSDF model is an issue that definitely needs further investigation. A few numerical results on the behavior of the generalized diffusivity defined by Equation 7.117 are reported by Lipatnikov (2007a,b).

It is also worth noting that even evaluation of the fully developed turbulent diffusivity $D_{t,\infty}$ is not simple. The following closure:

$$D_{t,\infty} = \frac{C_\mu}{Sc_t}\frac{\bar{k}^2}{\varepsilon} \tag{7.118}$$

is often invoked jointly with the $k - \varepsilon$ model when simulating turbulent mixing. However, contrary to the constant $C_\mu = 0.09$, there is no widely accepted turbulent Schmidt number, and significantly different values of Sc_t may be found in the literature. In particular, $Sc_t = 0.7$–0.9 in most RANS simulations, while Bilger et al. (1991)

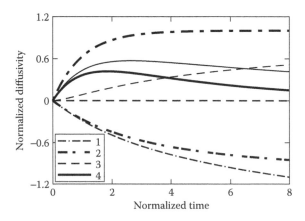

FIGURE 7.7 The normalized generalized diffusivity $D/D_{t,\infty}$ versus normalized time t/τ_L. Curves 1, 2, and 3 show turbulent diffusion, pressure-driven transport (with $\psi = 1$), and reaction terms, respectively, on the RHS of Equation 7.117. Curve 4 shows the sum of these three terms. The thin and bold curves have been calculated for $a = 1$ and $a = 0$, respectively. (Adapted from Lipatnikov, A.N., *Combust. Sci. Technol.*, 179, 91–115, 2007. With permission.)

obtained $Sc_t = 0.35$ by experimentally investigating a chemical reaction in a scalar mixing layer in grid-generated turbulence. By analyzing the DNS data computed in the constant-density case, Yeung (2002) has argued that the ratio of $\bar{k}/(\bar{\varepsilon}\tau_L)$ tends to 4.8 at large Reynolds numbers but is about 2.4 at $Re_\lambda = 40$ (see Figure 3 in the cited paper). Then, substitution of Equation 7.118 with $C_\mu = 0.09$ into $\tau_L = D_{t,\infty}/u'^2$ results in the estimate of $Sc_t \approx 0.32$ at $Re_\lambda = 40$, but $Sc_t \approx 0.64$ at $Re_\lambda \gg 1$.

Validation I: TFC Models

Contrary to the competitive RANS models of premixed turbulent combustion, the SSDF model has already been tested by several independent research groups against the experimental data obtained from substantially different laboratory premixed turbulent flames. It is worth remembering, however, that there are at least three different versions of the SSDF model.

The first release of the model invoked Equations 7.13 and 7.118 to close Equation 7.75. Accordingly, in the latter balance equation, $D_t = D_{t,\infty}$ given by Equation 7.118 and

$$U_t = U_{t,\infty} = Au'\mathrm{Da}^{1/4}\left(1 - \mathrm{P}_q\right). \tag{7.119}$$

This release of the SSDF model was implemented into commercial CFD codes such as Fluent and ANSYS CFX. Following Zimont (2000), we will call Equations 7.75, 7.118, and 7.119, the "TFC model." Here, $A = 0.5$ (Zimont and Lipatnikov, 1995; Karpov et al., 1996a) is a constant and P_q is the probability of local combustion quenching by turbulent eddies, which will be discussed in the section "Concept of Critical Stretch Rate (Flamelet Quenching by Strong Perturbations)" (see Equations 7.174 through 7.179). Because the available submodels for P_q involve unknown parameters and, therefore, offer an opportunity for tuning, the following discussion will mainly be restricted to the tests of the TFC model that were performed under well-defined conditions associated with $\mathrm{P}_q \ll 1$. Under such conditions, the model involves no tuning parameters if the same value $A = 0.5$ of a single model constant is set, as occurred in the tests summarized later.

The TFC model has been shown to predict:

- The effects of a mixture composition (the equivalence ratio and various fuels such as hydrogen, methane, ethane, and propane) on the turbulent burning velocities obtained from statistically spherical, premixed turbulent flames expanding in a fan-stirred bomb at various u' (Zimont and Lipatnikov, 1995; Karpov et al., 1996a), for example, see Figures 7.8 and 7.37 and note that many more results are reported in the cited papers. A comparison of the dashed and the solid curves in Figure 7.8, calculated for the same mixture setting $\mathrm{P}_q = 0$ and invoking a submodel for the probability P_q, given by Equation 7.178, respectively, indicates that the submodel for P_q affects the computed results only for high rms velocities, which are comparable with u'_m associated with the maximum burning velocity. The increasing branches of the solid curves are weakly affected by P_q. In certain cases (e.g., see the stoichiometric propane–air mixture in Figure 7.8), the results computed either setting $\mathrm{P}_q = 0$

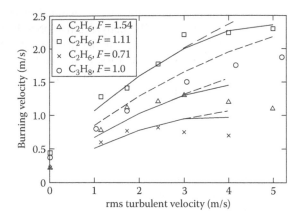

FIGURE 7.8 Dependencies of the burning velocity on the rms turbulent velocity measured (symbols) by Karpov and Severin (1980) and computed (curves) by Karpov et al. (1996a).

or using Equation 7.178 were practically the same. In such cases, solid curves are not shown. Therefore, the agreement between the measured and the computed data does not result from tuning, at least if u' is markedly lower than u'_m. The chemical timescale τ_c substituted into $Da = \tau_t/\tau_c$ in these simulations was equal to $\tau_{c0} = a_u/(S_L^0)^2$ if $Le = 1$ and depended on the Lewis number, as discussed in the section "Critically Strained Flamelets."

- The mean structure of a statistically stationary, oblique, confined, lean ($\Phi = 0.8$) methane–air turbulent flame stabilized by a hot jet (Zimont et al., 2001; Ghirelli, 2011) (e.g., see Figure 7.9).
- The mean structure of open, V-shaped, lean ($\Phi = 0.5$, 0.58, and 0.7) methane–air turbulent flames (Dinkelacker and Hölzler, 2000; Moreau, 2009; Ghirelli, 2011).

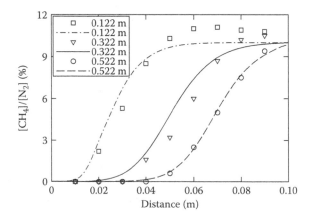

FIGURE 7.9 Transverse profiles of the mean volume fraction of CH_4 measured (symbols) by Moreau (1977) and computed (curves) by Zimont et al. (2001) at different distances from the flame holder, as specified in the legends.

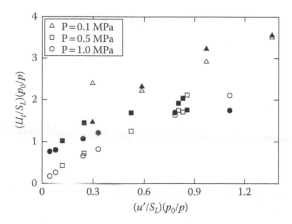

FIGURE 7.10 Normalized turbulent burning velocities measured (filled symbols) by Kobayashi et al. (1996) and computed (open symbols) by Muppala and Dinkelacker (2004) at three different pressures specified in the legends, $p_0 = 0.1$ MPa.

- The mean structure of an open, Bunsen, premixed turbulent flame (Dinkelacker, 2002).
- The profiles of the mean combustion progress variable in two open, swirl-stabilized flames (Dinkelacker, 2002).
- The overall burning velocities obtained from slightly lean ($\Phi = 0.9$) methane–air Bunsen flames under normal and elevated pressures (Muppala and Dinkelacker, 2004), provided that the ratio of u'/S_L^0 is markedly greater than unity, cf. the filled and open symbols at $(u'/S_L^0)(p_0/p) \geq 0.6$ in Figure 7.10 and note that both u'/S_L^0 and U_t/S_L^0 are multiplied by p_0/p, where $p_0 = 0.1$ MPa, in order for the scales of the data obtained at different pressures to be comparable.

It is also worth noting that the TFC model predicted well the mean structure of swirl-stabilized flames in a prototype of a gas turbine combustor (Zimont et al., 1998), but this result is not free from tuning, because the probability P_q was substantial in that simulation.

The author is aware of the sole test that was claimed to put the TFC model into question. Bray et al. (2001) invoked several turbulent combustion models and, in particular, the TFC model, to simulate five premixed flames stabilized in impinging jets (a sketch of such a flame is shown in Figure 4.8b). These authors claimed that the TFC model significantly underestimated the mean reaction rate. However, the model tested by Bray et al. (2001) differed substantially from the original TFC model, as discussed in detail by Lipatnikov (2002). In particular, Bray et al. (i) wrongly assumed that the last term on the RHS of Equation 7.75 was a closure of the mean reaction rate and (ii) neglected the diffusion term in that balance equation. However, a test of such an oversimplified model against the experimental data obtained from statistically stationary flames stabilized in impinging jets does not seem to be meaningful, because Equation 7.75 has no stationary solution if the diffusion term is neglected, but $\bar{\rho}\tilde{u}$ varies substantially along the normal to the mean flame brush as occurs in impinging jet flames. If the diffusion term is retained, a stationary solution exists (Lipatnikov and Chomiak, 2002b). Moreover, Lipatnikov (2011d) has shown that the experimental data obtained from the impinging jet flames addressed by Bray et al. (2001) are consistent with the TFC model. In these flames, $\rho_u U_t |\nabla \tilde{c}|$ differs substantially from $\overline{W}$, because the transport term on the LHS of Equation 7.75 is comparable with the two source terms.

Besides Bray et al. (2001), the TFC model was also criticized by Peters (2000, p. 161), who discussed a paper by Zimont and Lipatnikov (1995) and stressed that Equation 7.75 "has no solution for a steady planar flame" and leads "to an indefinite broadening of the region where $0 < \tilde{c} < 1$, which is unphysical." Unfortunately, Peters forgot to mention that, in the work criticized by him, it is clearly stated that Equation 7.75 "should be modified at long times" (Zimont and Lipatnikov, 1995, p. 998) and is valid under the condition of $t \ll t_{st}$, see Equation 12 in the cited paper, where t_{st} was "an estimated time needed for the average thickness of the turbulent flame zone to become steady" (Zimont and Lipatnikov, 1995, p. 999).

Nevertheless, the TFC model does have limitations, one of which is shown in Figure 7.10. While the agreement between the data measured by Kobayashi et al. (1996) and that computed by Muppala and Dinkelacker (2004) is reasonable at $u'/S_L^0 > 4$, the TFC model underestimates the burning velocity at weak turbulence. This observation is not surprising, because Equation 7.119 results in $U_{t,\infty} \to 0$, rather than in $U_{t,\infty} \to S_L^0$, as $u' \to 0$. Therefore, the TFC model in its present form does not allow us to predict the burning velocity in weakly turbulent premixed flames.

Moreover, when applying the TFC model to a flame, any $\tilde{c} = \text{const}$ satisfies Equation 7.75 if the boundary conditions involve only the derivatives of $\tilde{c}$ (Lipatnikov and Chomiak, 1997). This peculiarity of Equation 7.75 may cause difficulties in practical applications of the TFC model. Certainly, the problem may easily be resolved by setting $\tilde{c} = 1$ at one boundary and $\tilde{c} = 0$ at another, but such simple boundary conditions are not always feasible.

Furthermore, Equation 7.75 with $D_t = D_{t,\infty}$ given by Equation 7.118 cannot yield a fully developed flame in the statistically planar, 1D case, but such a flame should exist at least in the constant-density case.

From TFC to FSC Models

As a simple practical solution aimed at overcoming the aforementioned difficulties, Lipatnikov and Chomiak (1997, 2002a) proposed to insert an extra laminar-like source term into Equation 7.75. The extended equation reads

$$\frac{\partial}{\partial t}(\overline{\rho}\tilde{c}) + \frac{\partial}{\partial x_k}(\overline{\rho}\tilde{u}_k\tilde{c}) = \frac{\partial}{\partial x_k}\left[\overline{\rho}(a+D_t)\frac{\partial \tilde{c}}{\partial x_k}\right] + \rho_u V_t |\nabla \tilde{c}| + \frac{\overline{\rho}(1-\tilde{c})}{t_r\left(1+\dfrac{D_t}{a_b}\right)}\exp\left(-\frac{\Theta}{\tilde{T}}\right),$$

$$(7.120)$$

where $a = a(\tilde{T})$ is the molecular heat diffusivity evaluated at the Favre-averaged temperature, Θ is the activation temperature ($\Theta \gg T_b$), the combustion progress variable is equal to the normalized temperature, that is,

$$\tilde{c} = \frac{\tilde{T} - T_u}{T_b - T_u}, \tag{7.121}$$

and the reaction timescale t_r is set so that the following equation:

$$\frac{\partial}{\partial t}(\bar{\rho}\tilde{c}) + \frac{\partial}{\partial x_k}(\bar{\rho}\tilde{u}_k\tilde{c}) = \frac{\partial}{\partial x_k}\left(\bar{\rho}a\frac{\partial\tilde{c}}{\partial x_k}\right) + \frac{\bar{\rho}(1-\tilde{c})}{t_r}\exp\left(-\frac{\Theta}{\tilde{T}}\right), \tag{7.122}$$

yields the unperturbed laminar flame speed S_L^0 in the statistically planar, 1D case. Accordingly, if $u' \to 0$, $D_t \to 0$, and $V_t \to 0$, then Equation 7.120 is reduced to Equation 7.122, thereby yielding the correct limit $U_t(u' \to 0) \to S_L^0$ for the burning velocity in the statistically planar, 1D case.

Moreover, because the last term on the RHS of Equation 7.120 depends on $\tilde{c}$ rather than on its derivatives, $\tilde{c} = \text{const}$ is not a solution to Equation 7.120. Consequently, this equation is compatible with any boundary condition.

The extra source term introduced into Equation 7.120 looks like the reaction rate $\overline{W}(\tilde{T})$ calculated at the Favre-averaged temperature and multiplied by a factor of $(1 + D_t/a_b)^{-1} \ll 1$ This small factor is used in order that an increase in the burning velocity caused by the discussed source term is limited by the laminar flame speed.

Let us (i) consider the case of $u' > 0$, $D_t > 0$, $V_t > 0$ and (ii) rewrite Equation 7.120 as follows:

$$\frac{\partial}{\partial t}(\bar{\rho}\tilde{c}) + \frac{\partial}{\partial x}\left[(\bar{\rho}\tilde{u} - \rho_u V_t)\tilde{c}\right] = \frac{\partial}{\partial x}\left[\bar{\rho}(a + D_t)\frac{\partial\tilde{c}}{\partial x}\right] + \frac{\bar{\rho}(1-\tilde{c})}{t_r\left(1 + \dfrac{D_t}{a_b}\right)}\exp\left(-\frac{\Theta}{\tilde{T}}\right) \tag{7.123}$$

for a statistically planar, 1D flame that moves from right to left. Let us apply Equation 7.123 to a fully developed flame. In the coordinate framework attached to the latter flame, (i) the unsteady term vanishes on the LHS of Equation 7.123, (ii) $\bar{\rho}\tilde{u} = \rho_u U_{t,\infty}$ by virtue of the Favre-averaged continuity Equation 3.78, and, hence, (iii) Equation 7.123 is similar to the basic Equations 2.14 and 2.15 of the AEA theory of laminar premixed flames, provided that (i) $\rho_u U_{t,\infty} - \rho_u V_t$ is substituted with $\rho_u S_\infty$ in the convection term on the LHS of Equation 7.123 and (ii) $a(1 + D_t/a)$ and $t_r(1 + D_t/a)$ are substituted with D and t_r, respectively, on the LHS. Accordingly, we may invoke the solution already obtained in the section "AEA Theory of Unperturbed Laminar Premixed Flame" in Chapter 2, in particular, Equations 2.36 and 2.42, to determine S_∞, $U_{t,\infty} = S_\infty + V_t$, and the fully-developed flame thickness, $\Delta_{t,\infty}$. Consequently, at $t \to \infty$, we obtain $S_\infty = S_L^0$ and

$$\Delta_{t,\infty} = \frac{\rho_b(a_b + D_t)}{\rho_u S_\infty} = \frac{\rho_b a_b}{\rho_u S_L^0}\left(1 + \frac{D_t}{a_b}\right) = \Delta_L\left(1 + \frac{D_t}{a_b}\right) \propto \sigma^{-1}L\frac{u'}{S_L^0}. \tag{7.124}$$

Thus, Equation 7.120 yields a flame that (i) moves at a speed $V_t + f(t)S_L^0$ and (ii) may reach a fully developed stage with the fully developed mean flame brush thickness scaling according to Equation 7.124. Here, $f(t)$ is a nondimensional positive function such that $f(t \to \infty) \to 1$. The time required for the flame described by Equation 7.120 to be fully developed scales as $\Delta_{t,\infty}/S_L^0 \propto \sigma^{-1}\tau_t(u'/S_L^0)^2 \gg \tau_t$. Note that this time estimate and Equation 7.124 are consistent with the estimates made within the framework of the Zimont theory of turbulent burning velocity in the constant-density ($\sigma = \rho_u/\rho_b = 1$) case (see Equations 7.19 through 7.21).

According to the aforementioned time estimate, the last term on the RHS of Equation 7.120 weakly affects the increase in the mean flame brush thickness if $t = O(\tau_t)$ and $u' \gg S_L^0$. Furthermore, because this term (i) is of importance only in the high-temperature zone $T_b - \tilde{T} = O(T_b^2/\Theta)$ (see the section "Reaction Zone Thickness" in Chapter 2) and (ii) is multiplied by a factor of $(1 + D_t/a_b)^{-1} \propto Re_t^{-1} \ll 1$, this laminar-like source term weakly affects the spatial profiles of $\tilde{c}$ computed under typical conditions (e.g., see Figure 20 in Lipatnikov and Chomiak [2002a]).

In summary, the modification of Equation 7.75 discussed earlier, that is, the introduction of the laminar-like source term into Equation 7.120:

- Involves neither new constants nor unknown parameters
- Overcomes the boundary condition problem emphasized earlier
- Offers an opportunity to obtain a fully developed mean flame brush thickness
- Yields a correct equation in the limit case of $u' \to 0$

Thus, this modification extends the domain of the model applicability, but the validity of Equation 7.120 at $u' < S_L^0$ or/and $t \to \infty$ has not yet been tested, due to lack of experimental data.

Another important extension of the TFC model, proposed by Lipatnikov and Chomiak (1997, 2002a), consists of introducing developing turbulent diffusivity (see Equation 3.122 or 7.89) and burning velocity (see Equation 7.29) into Equation 7.75 and/or Equation 7.120. This method extends the domain of the model applicability and, in particular, offers an opportunity to simulate not only flames characterized by $U_t = $ const and $\delta_t \propto \sqrt{D_{t,\infty}t}$, which the TFC model is aimed at, but also flames characterized by a slowly increasing burning velocity and a linearly (with time) increasing mean flame brush thickness (see Figures 4.6 and 4.7). Indeed, Equation 7.90, which results from Equations 7.80 and 7.89, shows that the mean flame brush thickness $\delta_t \propto u't$ if the diffusivity is modeled invoking the Taylor theory and $t \ll \tau_t$.

Thus, the diffusivity in Equation 7.120 is calculated using Equations 7.89 and 7.118 and

$$V_t = Bu'\text{Da}^{1/4}\left(1 - P_q\right)\left\{1 + \frac{\tau_L}{t}\left[\exp\left(-\frac{t}{\tau_L}\right) - 1\right]\right\}^{1/2}. \tag{7.125}$$

The constant B in Equation 7.125 may differ from the constant $A = 0.5$ in Equation 7.119 due to (i) the time-dependent term on the RHS of the former equation and

(ii) the laminar-like source term on the RHS of Equation 7.120. On the one hand, in order for Equations 7.75 and 7.119 and Equations 7.120 and 7.125 to yield the same fully developed turbulent burning velocity in the statistically planar, 1D case, the following equality:

$$Au'\mathrm{Da}^{1/4}\left(1-\mathrm{P}_q\right) = Bu'\mathrm{Da}^{1/4}\left(1-\mathrm{P}_q\right) + S_L^0 \tag{7.126}$$

should hold, that is, $B < A$. On the other hand, because the constant $A = 0.5$ was tuned by applying Equation 7.119 to the developing flames, the tuned value of B in Equation 7.125 may be greater than 0.5.

In the following text, Equation 7.120 supplemented with Equations 7.89, 7.118, 7.121, and 7.125 will be called the ZFK-FSC model in order to stress that the algebraic source term was introduced on the RHS of Equation 7.120 via analogy with the theory of a laminar premixed flame by Zel'dovich and Frank-Kamenetskii (1938).

In summary, the ZFK-FSC model involves two modifications of the TFC model: (i) the use of developing turbulent diffusivity and burning velocity and (ii) the insertion of the laminar-like source term into Equation 7.120. While the former modification is both of importance for applications and consistent with the foregoing basic justification of Equation 7.75, discussed in the section "Derivation," the latter modification is a simple practical solution, which, however, changes the form of Equation 7.75 and, therefore, is not consistent with the derivation of the latter balance equation, reported in the section "Derivation." Moreover, although Equation 7.120 is correct in the limit case of $u' \to 0$, this feature of the ZFK-FSC model is not sufficient to claim that it is a proper tool for simulating weakly turbulent flames. The point is that the Zimont model of burning velocity, that is, Equation 7.125, was developed for large $\mathrm{Re}_t \gg 1$, as discussed in the section "Zimont Model of Burning Velocity in Developing Turbulent Premixed Flames." Accordingly, the validity of Equation 7.125 in weakly turbulent flames may be put into question.

Therefore, for the sake of consistency, one may wish to use the model without the laminar-like source term in Equation 7.120. We will call such a model the "truncated FSC model" (or the TFC model extended to address developing diffusivity and burning velocity).

Finally, because the last source term on the RHS of Equation 7.120 depends strongly nonlinearly on the mean combustion progress variable, the use of such a model in multidimensional numerical simulations of internal combustion engines may be computationally demanding. As a practical remedy, the following alternative balance equation:

$$\frac{\partial}{\partial t}\left(\bar{\rho}\tilde{c}\right) + \frac{\partial}{\partial x_k}\left(\bar{\rho}\tilde{u}_k\tilde{c}\right) = \frac{\partial}{\partial x_k}\left[\bar{\rho}\left(a+D_t\right)\frac{\partial\tilde{c}}{\partial x_k}\right] + \rho_u V_t\left|\nabla\tilde{c}\right| + \frac{\rho_u\left(S_L^0\right)^2}{4\left(a_u+D_t\right)}\tilde{c}\left(1-\tilde{c}\right), \tag{7.127}$$

was also proposed to be used by Lipatnikov and Chomiak (2002a, 2003, 2004b). In the statistically planar, 1D case and in the coordinate framework that moves at the speed V_t, Equation 7.127 is a particular example of the well-known

Kolmogorov–Petrovsky–Piskounov (KPP) Equation 7.201, which will be addressed in the section "Leading Points and the KPP Problem." Equation 7.127 supplemented with Equations 7.89, 7.118, 7.121, and 7.125 will be called the KPP-FSC model in the following text in order to stress that the algebraic source term was introduced into the RHS of Equation 7.127 via analogy with the KPP equation.

Using the seminal mathematical results by Kolmogorov et al. (1937), one can show that both Equation 7.120 and Equation 7.127 can yield statistically stationary, planar, 1D flames that move at the same speed $V_t + S_L^0$ with respect to the unburned mixture. However, the thickness of the flame modeled by Equation 7.127 is much greater than the thickness of the flame modeled by Equation 7.120. Because the last source term on the RHS of Equation 7.127 should be integrated over a large flame thickness to obtain S_L^0, the overall burning rate associated with this source term is significantly less than S_L^0 in a typical developing flame characterized by much less thickness. Accordingly, when invoking Equation 7.127, the constant B on the RHS of Equation 7.125 should be larger than the constant B tuned for Equation 7.120. Moreover, under typical conditions, a solution to Equation 7.127 is very close to the counterpart solution to Equation 7.75, because the last source term on the RHS of the former equation plays a minor role in the largest part of the turbulent flame brush but results in $\tilde{c} = 1$ in combustion products. Thus, the KPP-FSC model is closer to the truncated FSC model than the ZFK-FSC model.

Like Equation 7.120, the introduction of the algebraic source term into Equation 7.127:

- Involves neither new constants nor unknown parameters
- Overcomes the boundary condition problem
- Offers an opportunity to obtain a fully developed mean flame brush thickness

But, contrary to the former equation, Equation 7.127 does not model a typical premixed flame in the limit case of $u' \to 0$. This limitation of the KPP-FSC model does not seem to impede its application to practical turbulent premixed flames.

Validation II: FSC Models

The ZFK-FSC model has been tested by the author using a wide set of experimental data obtained by seven research groups from dozens of expanding, statistically spherical, premixed turbulent flames. In all these simulations, the probability P_q was equal to zero, that is, the model involved a single empirical constant $B = 0.4$, which was the same for all the flames. The results of these tests reported by Lipatnikov and Chomiak (1997, 2000, 2002a) and Lipatnikov et al. (1998) indicate the capabilities of the ZFK-FSC model for quantitatively predicting:

- The dependence of the flame radius increase on the mixture composition (e.g., see Figure 7.11)
- The dependence of the flame radius increase on the rms turbulent velocity (e.g., see Figure 7.11)
- An increase in the turbulent flame speed as the flame kernel grows (e.g., see computed curves in Figure 4.17)

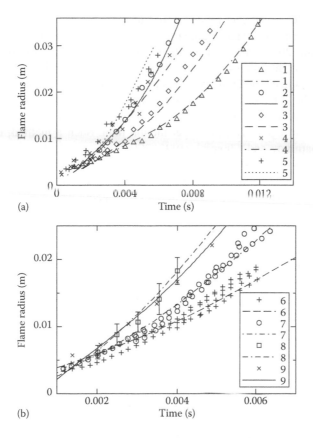

(a)

(b)

FIGURE 7.11 Increase in the radii of expanding statistically spherical (a) methane–air and (b) propane–air turbulent flames, computed (curves) using the ZFK-FSC model by Lipatnikov and Chomiak (1997) and Lipatnikov et al. (1998) and measured (symbols) by (1–3) Bradley et al. (1994c), (4, 5) Hainsworth (1985), (6–8) Mouqalid et al. (1994), and (9) Groff (1987). 1: $\Phi = 0.83$, $u' = 1.73$ m/s, $L = 20$ mm, $T_0 = 328$ K; 2: $\Phi = 1.0$, $u' = 3$ m/s, $L = 20$ mm, $T_0 = 328$ K; 3: $\Phi = 1.1$, $u' = 1.73$ m/s, $L = 20$ mm, $T_0 = 328$ K; 4: $\Phi = 0.8$, $u' = 1.93$ m/s, $L = 3$ mm; 5: $\Phi = 1.1$, $u' = 1.93$ m/s, $L = 3$ mm; 6: $\Phi = 0.75$, $u' = 0.8$ m/s, $L = 5$ mm; 7: $\Phi = 0.85$, $u' = 0.8$ m/s, $L = 5$ mm; 8: $\Phi = 1.0$, $u' = 1.0$ m/s, $L = 5$ mm; 9: $\Phi = 1.0$, $u' = 2$ m/s, $L = 25$ mm; $p_0 = 0.2$ MPa. At the ignition instant, temperature T_0 and pressure p_0 correspond to room conditions if the opposite is not previously specified.

- The opposite effects of pressure on the laminar and turbulent flame speeds (e.g., see Figure 4.17)
- The dependence of the burning velocity on the rms turbulent velocity and mixture composition (see Figure 31 in a review paper by Lipatnikov and Chomiak [2002a])

The ZFK-FSC model was successfully applied by Wallesten et al. (2002a,b) to multidimensional numerical simulations of turbulent combustion in conventional and stratified charge spark ignition engines.

The KPP-FSC model was validated by Lipatnikov and Chomiak (2003, 2004b) against the experimental data obtained from expanding statistically spherical flames by Renou et al. (2002), Atashkari et al. (1999), Nwagwe et al. (2000), and Bradley et al. (1994b, 2003b). In all these simulations, the probability P_q was equal to zero, that is, the considered model involved a single constant $B = A = 0.5$, which was the same for all the studied flames. However, the parameters of the invoked ignition model were tuned in order for the flame radius, computed at 1 ms after the start of spark ignition, to agree with the experimental data. Such a method, discussed in detail by Lipatnikov and Chomiak (1997, 2003), was invoked to yield reasonable initial conditions for testing the KPP-FSC model.

As summarized in Figures 7.12 through 7.16, the results computed using the KPP-FSC model agree with the experimental data reasonably well. Certain experiments

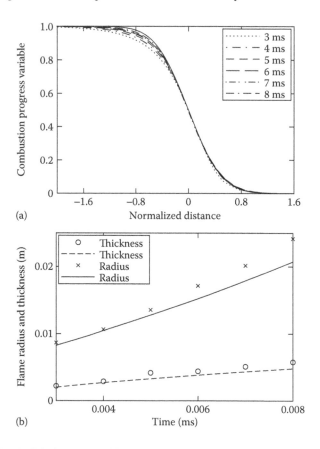

(a)

(b)

FIGURE 7.12 Validation of the KPP-TFC model against experimental data (symbols) obtained by Renou et al. (2002) from expanding, statistically spherical, stoichiometric CH_4– air flames. The solid curve in (a) has been calculated using Equation 4.34, which well approximates the experimental data obtained by Renou et al. (2002) (see Figure 4.22a). Other curves have been computed by Lipatnikov and Chomiak (2003). $u' \approx S_L^0$. (a) Self-similar mean flame structure calculated using Equations 4.1 and 4.33 at different instants specified in the legends. (b) Increase in flame radius $R_f = r(\bar{c} = 0.5)$ and thickness evaluated using Equation 4.1.

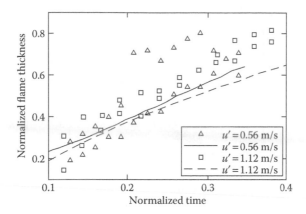

FIGURE 7.13 Increase in the mean flame brush thickness evaluated using Equation 4.1. The symbols show the experimental data obtained by Atashkari et al. (1999) from expanding, statistically spherical, stoichiometric methane–air flames at different initial rms turbulent velocities, as specified in the legends. The curves have been computed by Lipatnikov and Chomiak (2003).

were simulated by Lipatnikov and Chomiak (2003) invoking both the ZFK-FSC and KPP-FSC models, with all other things being equal. These computations have shown that the latter model predicts the observed flame speed and mean flame brush thickness in the case of $u'/S_L^0 = O(1)$ in a better way.

The truncated FSC model, that is, Equations 7.75, 7.89, 7.118, and $U_t = V_t$ given by Equation 7.125 with $B = 0.5$, was validated by Sathiah and Lipatnikov (2007) by computing the transverse temperature profiles in two ($T_u = 298$ and 600 K) lean ($\Phi = 0.61$), confined propane–air flames stabilized behind a bluff body. The computed

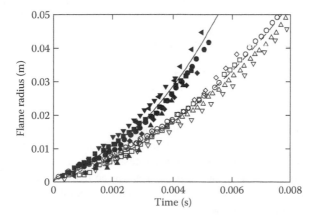

FIGURE 7.14 Validation of the KPP-FSC model against the experimental data obtained by Nwagwe et al. (2000) from expanding, statistically spherical, stoichiometric propane–air flames at $u' = 2.36$ (open symbols and dashed curve) and 4.72 m/s (filled symbols and solid curve). Different symbols show data obtained from different runs. The curves have been computed by Lipatnikov and Chomiak (2004b).

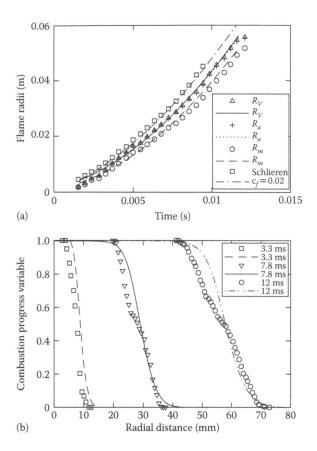

FIGURE 7.15 Increase in differently defined flame radii (a) and radial distributions of the Reynolds-averaged combustion progress variable (b). The symbols show the data measured by Bradley et al. (2003b). The curves have been computed invoking the KPP-FSC model with $B = 0.5$. A propane–air flame F; $\Phi = 1.32$, $u' = 1$ m/s, and $p = 5$ bar. $R_V^3 = 3\int_0^\infty \bar{c}r^2 dr$, $R_a^2 = 2\int_0^\infty \bar{c}r\,dr$, and $\int_0^{R_m} \bar{\rho}r^2 dr = \int_0^\infty \rho_b \bar{c}r^2 dr$. (Adapted from Lipatnikov, A.N. and Chomiak, J., *Proceedings of the Sixth International Symposium on Diagnostics and Modeling of Combustion in Internal Combustion Engines COMODIA2004*, 583–590, 2004. With permission.)

results indicate that the development of turbulent diffusivity and burning velocity, modeled by Equations 7.89 and 7.29, respectively, plays an important role even in such statistically stationary, premixed turbulent flames. Indeed, the thick curves in Figure 7.17, computed using the truncated FSC model, agree better with the experimental data obtained by Sjunnesson et al. (1992), shown as symbols, than the thin curves obtained invoking the TFC model.

It is also worth noting that the validation of the TFC model, discussed in the section "Validation I: TFC Model"), qualitatively supports the foregoing FSC models too, because the latter is an extension of the former. Nevertheless, this support should not be overestimated, and the tests of the TFC model do not validate the FSC models quantitatively.

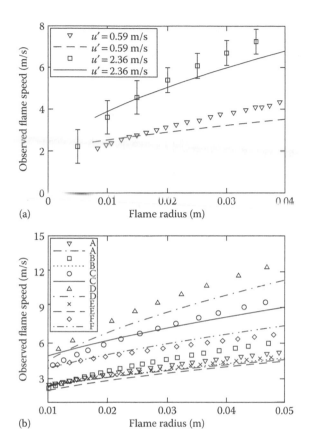

(a)

(b)

FIGURE 7.16 Validation of the KPP-FSC model against the experimental data obtained by (a) Bradley et al. (1994b) from preheated ($T_u = 358$ K) stoichiometric *iso*-octane–air flames at two different initial u' specified in the legends and (b) by Bradley et al. (2003b) from propane–air flames under the following conditions: A: $\Phi = 0.8$, $u' = 1$ m/s, $p = 0.1$ MPa; B: $\Phi = 1.32$, $u' = 1$ m/s, $p = 0.1$ MPa; C: $\Phi = 0.8$, $u' = 3$ m/s, $p = 0.1$ MPa; D: $\Phi = 1.32$, $u' = 3$ m/s, $p = 0.1$ MPa; E: $\Phi = 0.8$, $u' = 1$ m/s, $p = 0.5$ MPa; F: $\Phi = 1.32$, $u' = 1$ m/s, $p = 0.5$ MPa. The curves have been computed by Lipatnikov and Chomiak (2003 and 2004b). (Figure 7.16b is adapted from Lipatnikov, A.N. and Chomiak, J., *Proceedings of the Sixth International Symposium on Diagnostics and Modeling of Combustion in Internal Combustion Engines COMODIA2004*, 583–590, 2004. With permission.)

All in all, as compared with competitive approaches, a substantially wider quantitative and straightforward validation of both the TFC and the FSC models was already performed using the experimental data obtained from various laboratory flames under well-defined simple conditions (e.g., cf. validation of various models in the review papers by Lipatnikov and Chomiak [2002a] and Veynante and Vervisch [2002], published in the same journal in the same year). Here, the word "straightforward" means that the computed flame configurations and methods used to evaluate the output quantities were as close to the experimental techniques as possible. For instance, when simulating the experiments by Karpov and Severin (1978, 1980)

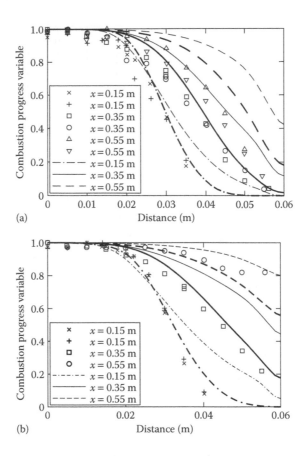

(a)

(b)

FIGURE 7.17 Profiles of the normalized Reynolds-averaged temperature, obtained numerically by Sathiah and Lipatnikov (2007) using the TFC (thin curves) and truncated FSC (thick curves) models and experimentally (symbols) by Sjunnesson et al. (1992) from lean ($\Phi = 0.61$) propane–air flames at different distances from a flame holder, as specified in the legends. Data measured in the up and down halves of the channel are shown as different symbols. (a) $T_u = 298$ K, (b) $T_u = 600$ K.

or Bradley et al. (2003b), expanding, statistically spherical, confined flames were addressed and (i) either the flame speed and radius were calculated by processing computed pressure curves $p(t)$, as done in the former measurements, or (ii) the radii R_V, R_a, and R_m were evaluated using the equations reported in the caption to Figure 7.15, as done in the latter measurements. On the contrary, comparison of a fully developed turbulent burning velocity yielded by a model in a statistically planar, 1D case with the speeds of developing nonplanar flames (e.g., see the papers by Duclos et al. [2003], Peters [1999], Bychkov [2003], Kolla et al. [2010], Lindstedt et al. [2011], and Chaudhuri et al. [2011] as examples of such tests) is not a straightforward quantitative validation of a model for the reasons discussed in the section "Determination of Turbulent Flame Speed and Burning Velocity" in Chapter 4, in the following subsection, and in Section 5.2 of a review paper by Lipatnikov and Chomiak (2010).

Moreover, both the TFC and the FSC models are well tailored for simulating developing premixed turbulent flames that are dominant in nature. For these reasons, the use of these models in RANS simulations of premixed turbulent combustion is strongly recommended. As far as the three different versions of the FSC model are concerned, the truncated FSC model is more consistent with the basic arguments discussed in the sections "General Features" and "Derivation," but the arbitrary $\tilde{c} = const$ satisfies Equation 7.75. The latter problem is resolved by the ZFK-FSC and KPP-FSC models, which have been validated against substantially wider sets of experimental data as compared with the truncated FSC model. Currently, the KPP-FSC model appears to be a reasonable compromise between basic and practical reasoning, because the algebraic source term on the RHS of Equation 7.127 is weakly nonlinear and plays a substantially less important role as compared with the algebraic source term on the RHS of Equation 7.120. Further tests of the models will allow us to rank them better.

Burning Velocity and Speed of Expanding Spherical Flame

To conclude the foregoing discussion, let us consider a single particular test of the truncated FSC model in more detail, because the test can be done semianalytically and offers an opportunity to clearly show the strong effect of flame development and mean flow nonuniformities on the turbulent burning velocity, thereby supplementing the discussion in the section "Determination of Turbulent Flame Speed and Burning Velocity" in Chapter 4.

Following Lipatnikov and Chomiak (2007), let us apply Equation 7.75 to an expanding, statistically spherical, premixed turbulent flame and assume that the same V_t may be substituted into the last term on the RHS to simulate both statistically planar and spherical flames, with all other things being equal. Here, V_t is a model expression for the turbulent burning velocity (e.g., Equation 7.29), while our goals are (i) to determine the difference between V_t and differently defined burning velocities U_t yielded by Equation 7.75 in the spherical case and (ii) to quantitatively test the truncated FSC model by comparing the obtained results with the experimental data.

In the statistically spherical case, Equation 7.75 reads

$$\frac{\partial}{\partial t}(\bar{\rho}\tilde{c}) + \frac{1}{r^2}\frac{\partial}{\partial r}(r^2\bar{\rho}\tilde{u}\tilde{c}) = \frac{1}{r^2}\frac{\partial}{\partial r}\left(r^2\bar{\rho}D\frac{\partial\tilde{c}}{\partial r}\right) - \rho_u V_t\frac{\partial\tilde{c}}{\partial r}. \tag{7.128}$$

Multiplying the LHS and RHS of this equation by r^2 and integrating the result from $r = 0$ to $r = \infty$, we obtain

$$\frac{dm_b}{dt} = -4\pi\rho_u V_t\int_0^\infty \frac{\partial\tilde{c}}{\partial r}r^2dr, \tag{7.129}$$

where

$$m_b \equiv 4\pi\rho_b\int_0^\infty \bar{c}r^2dr = 4\pi\int_0^\infty \bar{\rho}\tilde{c}r^2dr \tag{7.130}$$

is the total mass of combustion products in the flame and the integrals were transformed using the BML Equations 5.25 and 5.26. Then, integrating the RHS of Equation 7.129 by parts, we arrive at

$$\frac{1}{4\pi\rho_b}\frac{dm_b}{dt} = 2\sigma V_t \int_0^\infty \tilde{c} r \, dr. \tag{7.131}$$

To find a relation between V_t and U_t, Equation 7.131 could be supplemented with Equation 4.9, but, to do so, the radius R_f in the latter expression should be defined in the case of a turbulent flame of substantial and increasing thickness. There are different ways of resolving the problem.

First, if we invoke Equation 4.7, then the radius R_f in Equations 4.6 and 4.9 is equal to the radius R_V determined as

$$R_V^3 = 3 \int_0^\infty \bar{c} r^2 \, dr. \tag{7.132}$$

Subsequently, Equations 4.6, 4.9, 7.131, and 7.132 yield

$$U_t = \sigma^{-1}\frac{dR_V}{dt} = 2V_t \left(\int_0^\infty \tilde{c} r \, dr \right) \left(3 \int_0^\infty \bar{c} r^2 \, dr \right)^{-2/3}, \tag{7.133}$$

that is, the turbulent burning velocity differs from V_t in the considered spherical case, whereas U_t yielded by Equation 7.75 is always equal to V_t in a statistically planar, 1D flame.

The radius R_V is so defined that a sphere of this radius envelopes only the combustion products and the mass of the products inside the sphere is equal to the mass of the products in the entire flame brush. Karpov and Severin (1978, 1980) measured the radius R_V, but they approximated the measured $R_V(t)$ dependencies with straight lines and did not investigate the variations in dR_V/dt as the flame kernels grew. Nonlinear $R_V(t)$ dependencies were reported by Bradley et al. (2003b), who recorded a flame radius "so chosen that the volume of unburned gas inside a sphere of this radius is equal to the volume of burned gas outside it." The cited definition results in

$$\int_0^{R_V} P_u r^2 \, dr = \int_{R_V}^\infty P_b r^2 \, dr. \tag{7.134}$$

Considering that, within the framework of the BML approach, the probabilities of finding unburned, P_u, and burned, P_b, mixtures are equal to $1 - \bar{c}$ and $\bar{c}$, respectively,

Equations 7.132 and 7.134 are identical to each other (Lipatnikov and Chomiak, 2004c).

Second, an alternative approach to defining the mean radius of an expanding, statistically spherical, premixed turbulent flame was put forward by Lipatnikov and Chomiak (2002c). They hypothesized that the total burning rate in such a flame should not be directly affected by the rate of increase of the flame thickness, because the latter process was associated with turbulent transport, which redistributed the products in space but did not form them. This constraint offers an opportunity to determine the radius r_f in Equation 4.33, provided that the mean flame structure is self-similar. Indeed, if $\bar{c}(r,t) = \bar{c}(\xi)$, then

$$\frac{1}{4\pi\rho_b}\frac{dm_b}{dt} = \frac{d}{dt}\int_0^\infty \bar{c}(r,t)r^2 dr = \int_0^\infty \frac{\partial \bar{c}}{\partial t}r^2 dr$$

$$= -\int_0^\infty \left(\frac{dr_f}{dt} + \xi\frac{d\delta_t}{dt}\right)\frac{d\bar{c}}{d\xi}r^2\frac{dr}{\delta_t}$$

$$= -\int_0^\infty \left(\frac{dr_f}{dt} + \xi\frac{d\delta_t}{dt}\right)\frac{\partial \bar{c}}{\partial r}r^2 dr$$

$$= 2\frac{dr_f}{dt}\int_0^\infty \bar{c}r dr + \frac{2}{\delta_t}\frac{d\delta_t}{dt}\int_0^\infty \bar{c}\frac{\partial}{\partial r}\left[(r-r_f)r^2\right]dr. \qquad (7.135)$$

Therefore, in order for the last integral to vanish, the radius r_f should be defined as

$$r_f = \left(3\int_0^\infty \bar{c}r^2 dr\right)\left(2\int_0^\infty \bar{c}r dr\right)^{-1} = R_V^3\left(2\int_0^\infty \bar{c}r dr\right)^{-1}. \qquad (7.136)$$

Consequently, Equation 7.135 reads

$$\frac{1}{4\pi\rho_b}\frac{dm_b}{dt} = 2\frac{dr_f}{dt}\int_0^\infty \bar{c}r dr \qquad (7.137)$$

and Equations 7.131 and 7.137 result in

$$\sigma^{-1}\frac{dr_f}{dt} = V_t\left(\int_0^\infty \tilde{c}r dr\right)\left(\int_0^\infty \bar{c}r dr\right)^{-1}. \qquad (7.138)$$

Because the mean flow velocity at $r = 0$, that is, in combustion products, vanishes for symmetry reasons, the flame speed with respect to the burned gas is equal to the observed flame speed dr_f/dt in the considered case. Furthermore, because $\sigma S_{L,u} \approx S_{L,b}$ in a typical laminar flame and $\sigma S_{t,u} = S_{t,b}$ in a statistically planar, 1D, fully developed turbulent premixed flame, it is reasonable to associate the LHS of Equation 7.138 with the turbulent flame speed S_t with respect to the unburned mixture. Then, Equation 7.138 shows that the flame speed is less than V_t, because the Favre-averaged combustion progress variable cannot be larger than the Reynolds-averaged one by definition: $\tilde{c} \equiv \overline{\rho c}/\overline{\rho} = \overline{c} + \overline{\rho'c'}/\overline{\rho} \leq \overline{c}$. Here, $\overline{\rho'c'} \leq 0$, because an increase in c is always accompanied by a decrease in the density of a premixed flame.

If the turbulent burning velocity is defined by Equation 4.9, then $U_t = S_t$, provided that the flame speed is equal to the LHS of Equation 7.138 and the radius R_f in the former equation is equal to the radius R_a defined as

$$R_a^2 = 2 \int_0^\infty \overline{c} r \, dr = \frac{R_V^3}{r_f}.$$

(7.139)

In this case, the turbulent burning velocity does not depend straightforwardly on the rate of increase of the mean flame brush thickness.

To complete this analogy between laminar and turbulent expanding spherical flames, it is worth noting that, like Equation 5.96,

$$\frac{dr_f}{dt} = U_t + u_f = S_t + u_f,$$

(7.140)

provided that the reference mean velocity u_f of the unburned mixture, with respect to which the turbulent flame speed S_t is determined, is equal to the mean velocity of a fresh gas extrapolated to $r = R_a$ (Lipatnikov and Chomiak, 2002c). Indeed, substitution of $\overline{\rho}(r,t) = \overline{\rho}(\xi)$ into the Favre-averaged continuity Equation 3.78 written in the spherical coordinate framework yields

$$\frac{1}{r^2}\frac{\partial}{\partial r}\left(r^2 \overline{\rho}\tilde{u}\right) = \frac{1}{\delta_t}\left(\frac{dr_f}{dt} + \xi\frac{d\delta_t}{dt}\right)\frac{d\overline{\rho}}{d\xi} = \left(\frac{dr_f}{dt} + \xi\frac{d\delta_t}{dt}\right)\frac{\partial\overline{\rho}}{\partial r}.$$

(7.141)

Integrating this equation from 0 to r, we obtain

$$r^2\overline{\rho}\tilde{u} = \frac{dr_f}{dt}\int_0^r y^2 \frac{\partial\overline{\rho}}{\partial y}\,dy + \frac{1}{\delta_t}\frac{d\delta_t}{dt}\int_0^r y^2\left(y - r_f\right)\frac{\partial\overline{\rho}}{\partial y}\,dy.$$

(7.142)

In the unburned mixture, $\overline{\rho} = \rho_u = \text{const}$, the integration upper limit may be extended to infinity, and Equation 7.142 reads

$$r^2 \rho_u \bar{u} = \frac{dr_f}{dt} \int_0^\infty y^2 \frac{\partial \bar{\rho}}{\partial y} dy + \frac{1}{\delta_t} \frac{d\delta_t}{dt} \int_0^\infty y^2 (y - r_f) \frac{\partial \bar{\rho}}{\partial y} dy. \qquad (7.143)$$

Using the BML Equations 5.25 and 5.26, we arrive at

$$r^2 \rho_u \bar{u} = -\rho_u \frac{\sigma-1}{\sigma} \left[\frac{dr_f}{dt} \int_0^\infty y^2 \frac{\partial \bar{c}}{\partial y} dy + \frac{1}{\delta_t} \frac{d\delta_t}{dt} \int_0^\infty y^2 (y - r_f) \frac{\partial \bar{c}}{\partial y} dy \right]$$

$$= \rho_u \frac{\sigma-1}{\sigma} \left\{ \frac{dr_f}{dt} 2 \int_0^\infty y\bar{c}dy + \frac{1}{\delta_t} \frac{d\delta_t}{dt} \left[3 \int_0^\infty y^2 \bar{c}dy - 2r_f \int_0^\infty y\bar{c}dy \right] \right\}$$

$$= R_a^2 \rho_u \frac{\sigma-1}{\sigma} \frac{dr_f}{dt} \qquad (7.144)$$

by virtue of Equations 7.136 and 7.139. Therefore, if the velocity u_f is determined using Equation 7.144 with $r = R_a$, then Equation 7.140 holds.

The radius R_a was measured by Bradley et al. (2003b), who introduced a radius so chosen that "the area of unburned gas on the flame image inside the circumference with this radius is equal to that of burned gas outside it." The cited definition reads

$$\int_0^{R_a} P_u r dr = \int_{R_a}^\infty P_b r dr \qquad (7.145)$$

and, within the framework of the BML approach, Equations 7.139 and 7.145 are identical to each other (Lipatnikov and Chomiak, 2004c).

Bradley et al. (2003b) showed that the radii R_V and R_a were almost equal to each other in all flames studied by them (e.g., cf. the triangles and pluses in Figure 7.15a). Numerical simulations by Lipatnikov and Chomiak (2002c, 2004b) also indicated that $R_V \approx R_a$ (e.g., cf. the solid and dotted lines in Figure 7.15a). Substitution of this approximate equality into Equation 7.139 results in $r_f \approx R_V \approx R_a$. Then, Equations 7.133 and 7.138 may be rewritten as

$$\sigma^{-1} \frac{dR_V}{dt} \approx \sigma^{-1} \frac{dR_a}{dt} \approx V_t \left(\int_0^\infty \tilde{c}rdr \right) \left(\int_0^\infty \bar{c}rdr \right)^{-1}. \qquad (7.146)$$

This result may straightforwardly be tested against the experimental data obtained by Bradley et al. (2003b). To do so, Lipatnikov and Chomiak (2007) evaluated the integrals on the RHS using Equations 4.1, 4.2, 4.33, and 4.34 and invoked the RHS of Equation 7.29 to model the development of V_t, with the fully developed unperturbed burning velocity $U_{t,\infty}$ being tuned in order to obtain the best agreement between the measured and the calculated data. In other words, the ability of the RHS of the following equation was tested,

$$\frac{1}{\sigma U_{t,\infty}} \frac{dR_V}{dt} = \left\{ 1 + \frac{\tau_L}{t} \left[\exp\left(-\frac{t}{\tau_L} \right) - 1 \right] \right\}^{1/2} \left(\int_0^\infty \tilde{c} r dr \right) \left(\int_0^\infty \bar{c} r dr \right)^{-1}, \quad (7.147)$$

to predict the measured $R_V(U_{t,\infty} t)$ curves for a tuned $U_{t,\infty}$.

Figure 7.18 shows that the solid curves calculated using Equations 4.1, 4.2, 4.33, 4.34, and 7.147, which are fully consistent with the truncated FSC model, agree well with the experimental data by Bradley et al. (2003b), shown as symbols. These results not only further validate the model but also offer an opportunity to clearly

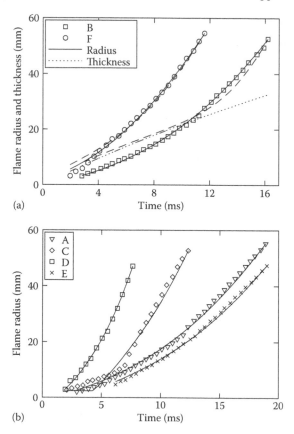

(a)

(b)

FIGURE 7.18 (a,b) Increase in the mean radii of spherical turbulent flame kernels. The solid and dashed curves have been calculated by Lipatnikov and Chomiak (2007) using Equations 7.147 and 7.150 through 7.151, respectively. The dotted curve shows the increase in the mean turbulent flame brush thickness evaluated using Equation 4.2. Experimental data obtained by Bradley et al. (2003b) are shown as symbols. The conditions of the experiments are specified in the caption to Figure 7.16b. (Figure 7.18a is reprinted from Lipatnikov, A.N. and Chomiak, J., *Proc. Combust. Inst.*, 31, 1361–1368, 2006. With permission.)

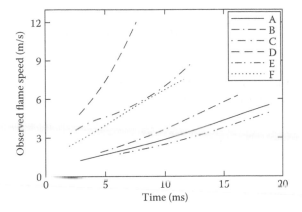

FIGURE 7.19 Increase in the observed flame speed dR_V/dt calculated using Equation 7.147 for six flames investigated by Bradley et al. (2003b). The conditions of the experiments are specified in the caption to Figure 7.16b.

illustrate the strong influence of flame development, curvature, and nonuniformities of the mean flow on the turbulent burning velocities obtained in a typical experimental study of expanding, statistically spherical, premixed turbulent flames.

For instance, Figure 7.19 indicates that the observed flame speeds calculated using Equation 7.147 increase with time and show no signs of leveling off. This trend implies that a flame speed or a burning velocity evaluated on the basis of the slope of a measured $R_V(t)$ curve, as done in a typical experimental study (Karpov and Severin, 1978, 1980; Bradley et al., 1994b, 2003a) of the discussed flames, may significantly underestimate the fully developed unperturbed burning velocity. Indeed, for the flames investigated by Bradley et al. (2003b), the tuned fully developed $U_{t,\infty}$ are larger than $\langle \sigma^{-1}dR_V/dt \rangle$ averaged at $R_V > L = 20$ mm by a factor ranging from about 3 (flames A, B, and F) to about 4 (flames C and D).

It is worth stressing that, as discussed by Lipatnikov and Chomiak (2007), such a strong effect is associated not only with the flame development and curvature but also and mainly with the nonuniformity of the mean flow. As already shown for a laminar spherical flame (see Equations 5.91 through 5.97), the stretch rate caused by the nonuniformity of the flow induced by the flame expansion is significantly larger than the stretch rate associated with the flame curvature, and a similar analysis is applicable to a turbulent spherical flame. Moreover, in the case of a constant density, Equation 7.146 yields $U_t = V_r$, that is, the effect of the mean flame curvature on the burning velocity vanishes due to the lack of mean flow nonuniformities.

Finally, invoking the BML Equations 5.25 and 5.26, one may rewrite Equation 7.138 as

$$\frac{1}{\sigma V_t}\frac{dr_f}{dt} = 1-(\sigma-1)\int_0^\infty \bar{\rho}\tilde{c}(1-\tilde{c})rdr\left(\int_0^\infty \bar{c}rdr\right)^{-1}. \qquad (7.148)$$

In the asymptotic case of $\delta_t \ll R_V$, the second term on the RHS is small, because the former and latter integrals are of the order of $\delta_t r_f$ and r_f^2, respectively. Therefore, we have

$$\frac{1}{\sigma V_t}\frac{dr_f}{dt} = 1 - 2(\sigma-1)\frac{\delta_t}{r_f}\int_{-\infty}^{\infty}\bar{\rho}\tilde{c}(1-\tilde{c})d\xi \qquad (7.149)$$

to the leading order with respect to $\delta_t/r_f \ll 1$ Here, the normalized distance ξ is determined by Equation 4.33. If we introduce a turbulent Markstein number

$$\mathrm{Ma}_t = (\sigma-1)\int_{-\infty}^{\infty}\bar{\rho}\tilde{c}(1-\tilde{c})d\xi, \qquad (7.150)$$

then, Equation 7.149 reads

$$\frac{1}{\sigma V_t}\frac{dr_f}{dt} = 1 - \frac{2\mathrm{Ma}_t\delta_t}{r_f} \qquad (7.151)$$

and becomes very similar to Equations 5.99 through 5.101.

It is worth stressing, however, that although Lipatnikov and Chomiak (2004a, 2007) introduced the number Ma_t via analogy with the classical Markstein number, the two quantities are related to different physical mechanisms of turbulent combustion. The classical Markstein number characterizes the response of the local burning rate in laminar flamelets to perturbations provided by turbulent eddies. The number Ma_t characterizes variations in U_t under the influence of global stretching of the entire turbulent flame by the mean flow. In other words, as far as a turbulent flame is concerned, the classical Markstein number and Ma_t are related to local turbulent and global deterministic perturbations, respectively.

Because Equations 7.150 and 7.151 have been derived from Equation 7.138 in the asymptotic case of $\delta_t/r_f \to 0$, they should be tested under proper conditions. Among the six flames investigated by Bradley et al. (2003b), only flame F is characterized by a sufficiently low ratio of δ_t/r_f and the dashed curve calculated using Equations 7.150 and 7.151 agrees well with the experimental data shown as circles in Figure 7.18a. For the other five flames, the foregoing ratio is of unity order or even larger than unity. Therefore, it is not surprising that the asymptotic Equations 7.150 and 7.151 cannot predict the experimental data (e.g., see flame B in Figure 7.18a).

Thus, to evaluate the fully developed, unperturbed turbulent burning velocity $U_{t,\infty}$ by processing the $r_f(t)$ curves obtained from expanding, statistically spherical, turbulent premixed flames, Equation 7.146, which results straightforwardly from the truncated FSC model and is validated by the experimental data by Bradley et al. (2003b), is proposed to be used. Under typical conditions, the difference between $U_{t,\infty}$ and the measured $\sigma^{-1}dr_f/dt$ may be as large as 400%.

G-Equation

If the thickness of the instantaneous flame front in a turbulent flow is substantially less than the Kolmogorov length scale, then the front may be considered to be an infinitely thin flame sheet that (i) separates the unburned and burned gases and

(ii) moves at a speed S_L^0 relative to the unburned gas (or at a speed $S_{L,b} \equiv \sigma S_L^0$ relative to the burned mixture). In this approach, flame propagation can be described by (i) assigning an arbitrary value G_0 to a scalar function $G(\mathbf{x},t)$ to define the flame sheet as

$$G(\mathbf{x},t) = G_0 \tag{7.152}$$

and (ii) solving the so-called G-equation (Markstein, 1964; Williams, 1985b):

$$\frac{\partial G}{\partial t} + \mathbf{v} \cdot \nabla G = S_L^0 |\nabla G|, \tag{7.153}$$

where $\mathbf{v}$ is the unburned gas velocity at the flame sheet and the function $G(\mathbf{x},t)$ is assumed to increase toward the burned gas region. Kinematic Equation 7.153 is derived by taking the material derivative of Equation 7.152:

$$0 = \left. \frac{dG}{dt} \right|_{G=G_0} = \frac{\partial G}{\partial t} + \nabla G \cdot \left. \frac{d\mathbf{x}}{dt} \right|_{G=G_0} = \frac{\partial G}{\partial t} + \nabla G \cdot \left. \left(\mathbf{v} + S_L^0 \mathbf{n} \right) \right|_{G=G_0}, \tag{7.154}$$

where the unit vector $\mathbf{n} = -\nabla G / |\nabla G|$ is normal to the flame sheet and points to the unburned gas. It is clear from this derivation that G-Equation 7.153 is valid only at the flame sheet.

Although Equation 7.153 appears to be very simple, numerically solving it is difficult due to (i) the density jump at the flame sheet and (ii) formation of cusps at the surface defined by Equation 7.152 even in the laminar case. In a turbulent flow, moreover, the straightforward application of the G-Equation 7.153 requires the resolution of small-scale turbulent eddies (e.g., see a recent paper by Treurniet et al. [2006]). The RANS simulations, which deal with the Reynolds- (or Favre-) averaged quantities, call for a kinematic G_t-equation to track a mean flame surface in the mean velocity field. Here, the subscript t is used to highlight what we consider a mean turbulent flame surface.

There are two approaches to arriving at such a kinematic equation. On the one hand, one may apply Equation 7.154 to a surface

$$G_t \left[\bar{c}(\mathbf{x},t) = c_0 \right] = G_0 \tag{7.155}$$

characterized by a reference value, c_0, of the Reynolds-averaged combustion progress variable. Then, Equations 7.154 and 7.155 yield

$$\frac{\partial G_t}{\partial t} + \mathbf{v}(\bar{c} = c_0) \cdot \nabla G_t = S_t(\bar{c} = c_0) |\nabla G_t|, \tag{7.156}$$

where $S_t(\bar{c} = c_0)$ is the speed of the self-propagation of the reference flame surface with respect to the mean flow at the surface. However, such an approach faces the following basic problems:

First, in the best case, Equation 7.156 would allow us to find solely a mean flame position but neither the flame brush thickness nor the mean flame structure, for example, the fields of the mean temperature and density. Second, because Equation 7.156 does not allow us to simulate the mean density field, the mean velocity $\bar{\mathbf{v}}(\bar{c} = c_0)$ in Equation 7.156 cannot be evaluated within the framework of the G_t-equation approach. These challenges could, in principle, be overcome by combining Equation 7.156 with a submodel for the mean flame structure, for example, Equations 7.77, 7.78, and 7.80 or 7.90, which are well supported by experimental data (see Figures 4.22 and 4.6, respectively). However, in such a case, the use of the TFC Equation 7.75 seems to be much more logical, because (i) the solution to it involves Equations 7.77, 7.78, and 7.80 or 7.90 straightforwardly in the statistically planar, 1D case and (ii) the numerical integration of Equation 7.75 is much simpler than that of Equation 7.156. Indeed, from the mathematical (and numerical) viewpoint, the major difference between Equations 7.75 and 7.156 is the diffusion term in the former equation. This term smoothes out spatial nonuniformities and makes the numerical integration of Equation 7.75 easy, whereas the problem of cusp formation needs sophisticated treatment when numerically solving Equation 7.156.

Furthermore, the speed $S_t(\bar{c} = c_0)$ is also difficult to evaluate. While the focus of many models of premixed turbulent combustion is on determining the turbulent burning velocity U_t, in a general case, $S_t(\bar{c} = c_0) \neq U_t$ for a number of reasons that were discussed in the section "Determination of Turbulent Flame Speed and Burning Velocity" in Chapter 4 and in Section 5.2 of a recent review paper by Lipatnikov and Chomiak (2010), in particular, due to the increase in the mean flame brush thickness. For instance, Lipatnikov and Chomiak (2002c) have proved analytically that the speed of any isoscalar surface determined by Equation 7.155 is affected not only by the burning rate but also by the rate of increase of the mean thickness of an expanding, statistically spherical flame with a self-similar mean structure, whereas the turbulent burning velocity is controlled by the burning rate but is not affected straightforwardly by $d\delta_t/dt$.

In summary, Equation 7.156 does not allow us to simulate the mean flame structure, contains unknown quantities $\bar{\mathbf{v}}(\bar{c} = c_0)$ and $S_t(\bar{c} = c_0)$, and is difficult to integrate numerically.

On the other hand, attempts were made to obtain a kinematic equation for tracking a mean flame surface in a mean turbulent flow by averaging the instantaneous G-Equation 7.153. A model developed by Peters (1992, 1999, 2000) is the most known approach of that kind. The model is aimed at determining not only the mean flame surface but also the mean flame brush thickness. The governing equations of the model were formally obtained as follows:

First, the G-equation was "considered to be valid everywhere in the flow field" (see Peters, 1999) and the scalar G was split into mean and fluctuating parts, $\bar{G}$ and $G' = G - \bar{G}$, respectively. Subsequently, Equation 7.153 was averaged to obtain the following equation

$$\bar{\rho}\frac{\partial \tilde{G}}{\partial t} + \bar{\rho}\tilde{\mathbf{v}} \cdot \nabla \tilde{G} = \bar{\rho}S_t \left| \nabla \tilde{G} \right| - 2\bar{\rho}D_t \tilde{h}_m \left| \nabla \tilde{G} \right|, \tag{7.157}$$

which "is valid at $\overline{G}(\mathbf{x},t) = G_0$ only" (Peters, 2000, p. 118). Here, $2h_m = \nabla\mathbf{n}$ is the flame curvature. Second, Equation 7.157 was subtracted from Equation 7.153 to obtain an equation for G'. Third, the following equation

$$\frac{\partial \overline{\rho G'^2}}{\partial t} + \nabla\left(\tilde{\mathbf{v}}\overline{\rho G'^2}\right) = \nabla_t\left(D_t\nabla_t\overline{\rho G'^2}\right) + 2\overline{\rho}D_t\left|\nabla\tilde{G}\right|^2 - c_s\frac{\tilde{\varepsilon}}{k}\overline{\rho G'^2} \tag{7.158}$$

for $\overline{\rho G'^2}(\mathbf{x},t)$ was obtained by averaging the G'-equation multiplied by G'. Here, D_t is the turbulent diffusivity, k and ε are the turbulent kinetic energy and its dissipation rate, respectively, c_s is a constant, and ∇_t is a gradient tangential to the flame. Note that Equation 7.158 results straightforwardly from Equation 2.152 in Peters (2000) and the continuity Equation 7.34.

Finally, Peters (2000) used Equations 7.157 and 7.158 as a basis for evaluating the mean flame brush thickness (see p. 118 in the cited paper)

$$\delta_t = \left(\frac{\overline{\rho G'^2}}{\overline{\rho\nabla\tilde{G}\cdot\nabla\tilde{G}}}\right)^{1/2}_{G=G_0}. \tag{7.159}$$

As discussed in detail by Lipatnikov and Sabel'nikov (2008), the aforementioned model is basically flawed for a number of reasons. Briefly speaking, first, the basic difficulty in averaging the G-field stems from the fact that the G-equation is physically meaningful only at the flame surface, as is clear from the foregoing derivation (see Equations 7.152 through 7.154). For instance, Peters (2000, p. 92) has stressed that "the quantity G is a scalar, defined at the flame surface only, while the surrounding G-field is not uniquely defined." However, if $G = G_0$ at the flame surface and "is not uniquely defined" outside it, the Reynolds-averaged value of G is equal either to G_0 if only the physically meaningful value of $G = G_0$ is averaged or to an arbitrary value if the entire G-field is processed.

As pointed out by Oberlack et al. (2001), propagation of the same flame may be modeled by Equations 7.152 and 7.153 with $G_0 = 0$ and by the same equations written for $F = \exp(G) - 1$, but the mean flame surfaces ($\overline{G} = 0$ or $\overline{F} = 0$) will be different in the two cases. This simple example clearly shows that the application of Reynolds averaging to the G-equation is a basically flawed method.

Second, Equations 7.157 through 7.159 do not resolve the problem of evaluating δ_t because the 3D $\nabla\overline{G}$, calculated using a solution to Equation 7.157, is an ill-defined quantity. If the latter equation is valid solely at a 2D flame surface, as claimed by Peters (2000, p. 118), then only the 2D tangential gradient $\nabla_t\tilde{G}$ is well defined, but $\nabla_t\tilde{G} = 0$.

Third, even if a correct kinematic equation for $\tilde{G}$ (valid at the mean flame surface only) is assumed to be known, a physically meaningful equation for $\overline{\rho G'^2}(\mathbf{x},t)$ cannot be obtained by subtracting the $\tilde{G}$-equation from Equation 7.153, as Peters did. If both the scalar G and Equation 7.153 are well defined at the instantaneous flame surface only, and $\tilde{G}$ and the $\tilde{G}$-equation are physically meaningful at the mean flame

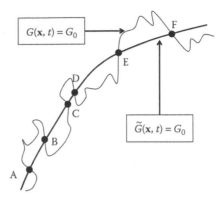

FIGURE 7.20   Instantaneous (thin curve) and mean (thick curve) flame surfaces.

surface only, then any partial differential equation written in terms of $G'' = G - \tilde{G}$ is ill defined. Indeed, the difference $G'' = G - \tilde{G}$ is well defined only at points A, B, C, etc., in Figure 7.20, where the thin and bold curves represent the instantaneous and mean flame surfaces, respectively. The evolution of a quantity defined in such a way is hardly amenable to modeling by any conventional partial differential equation.

Finally, in the simplest statistically stationary, planar, 1D case, Equation 7.157 reads $\bar{\rho}\tilde{\mathbf{v}} \cdot \nabla\tilde{G} = \bar{\rho}S_t \left|\nabla\tilde{G}\right|$. Therefore, for a flame that moves from right to left, Equation 7.157 and the continuity Equation 7.35 lead to the equality $\rho_u\bar{u}(x \to -\infty) = \bar{\rho}S_t$. Consequently, S_t in Equation 7.157 depends on the mean density at the reference flame surface and is not equal to the turbulent burning velocity commonly defined using Equation 4.4.

Unless these issues are resolved, the G-equation model by Peters (1992, 1999, 2000) should be considered to be basically flawed and the physical relevance of the results that are obtained by using it is unclear.

Oberlack et al. (2001) revised the foregoing model and (i) proposed to average the following characteristic equation:

$$\frac{d\mathbf{x}_f}{dt} \equiv \frac{d\mathbf{x}}{dt}\bigg|_{G=G_0} = \left(\mathbf{v} + S_L^0\mathbf{n}\right)\bigg|_{G=G_0}, \qquad (7.160)$$

which describes the motion of a point at the flame surface defined by Equation 7.152 and was used to derive the G-equation (see Equation 7.154), and (ii) considered the obtained equation for $d\bar{\mathbf{x}}_f/dt$ to be the characteristic equation for the mean flame surface.

However, the foregoing hypothesis that $d\bar{\mathbf{x}}_f/dt$ characterizes the motion of the mean flame surface is wrong in a general case, as shown in Figure 7.21. If instantaneous (thin lines) and mean (bold line) flame fronts are not locally parallel to each other, then the displacement $dx_A = S_L^0 dt\cos\varphi$ of a point A at the instantaneous front in the x-direction during the time interval dt is smaller than the displacement $dx_A = S_L^0 dt/\cos\varphi$ of this front in the x-direction during the same

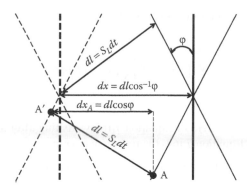

FIGURE 7.21 Propagation of instantaneous (thin solid and dashed lines) and mean (bold solid and dashed lines) flame fronts.

time interval dt. As a result, in the simple case shown in Figure 7.21, the mean flame front propagates in the x-direction faster than any point A at the instantaneous flame front, that is $d\bar{x}_f/dt$ underestimates the speed of the mean flame in the x-direction in this case.

For readers who would like to see a more rigorous mathematical discussion, let us consider the following 2D problem (Sabel'nikov and Lipatnikov, 2010). A planar laminar constant-density flame is embedded in a random shear flow $u = \{yf,0\}$ at $t = 0$. Here, $f = f(t)$ is a dimensional (s^{-1}) random function with zero mean value $\bar{f} = 0$, the y-axis is tangential to the flame surface at $t = 0$, and the origin of the coordinate framework lies on the flame surface at $t = 0$. By substitution, one can easily prove that Equation 7.153 with an initial condition of $G = x$ has the following solution:

$$G(x,y,t) = -x + yF + S_L^0 \int_0^t \sqrt{1 + F^2(\vartheta)}\,d\vartheta, \qquad (7.161)$$

where

$$F(t) \equiv \int_0^t f(\vartheta)\,d\vartheta. \qquad (7.162)$$

Therefore, the x-coordinate of the flame, associated with the level set at $G = 0$, is equal to

$$X_f(y,t) = yF + S_L^0 \int_0^t \sqrt{1 + F^2(\vartheta)}\,d\vartheta. \qquad (7.163)$$

Subsequently, averaging Equation 7.163 yields

$$\bar{X}_f(y,t) = S_L^0 \overline{\int_0^t \sqrt{1+F^2(\vartheta)}d\vartheta},$$

(7.164)

with the first term on the RHS of Equation 7.163 vanishing after averaging, because $\bar{f} = 0$. According to Equation 7.164, the turbulent burning velocity is proportional to an average increase $\overline{\sqrt{1+F^2}}$ in the instantaneous flame surface area, in line with the Damköhler (1940) hypothesis.

Let us consider the same problem within the framework of the approach by Oberlack et al. (2001), that is, by averaging the characteristic Equation 7.160:

$$\frac{dx_f}{dt} = fy_f + S_L^0 n_x = fy_f + \frac{S_L^0}{\sqrt{1+F^2}},$$

$$\frac{dy_f}{dt} = S_L^0 n_x = -\frac{S_L^0 F}{\sqrt{1+F^2}}$$

(7.165)

for a point at the flame surface with the initial coordinates $\{x_f(0) = 0, y_f(0) = y_0\}$. Here,

$$\mathbf{n} = -\frac{\nabla G}{|\nabla G|} = \left\{ \frac{1}{\sqrt{1+F^2}}, -\frac{F}{\sqrt{1+F^2}} \right\}$$

(7.166)

is the unit vector normal to the flame surface. The solution to Equation 7.165 with the foregoing initial conditions is

$$x_f(y_0,t) = y_0 F - S_L^0 YF + S_L^0 \int_0^t \sqrt{1+F^2(\vartheta)}d\vartheta,$$

$$y_f(y_0,t) = y_0 - S_L^0 Y,$$

(7.167)

where

$$Y = \int_0^t \frac{F(\vartheta)}{\sqrt{1+F^2(\vartheta)}}d\vartheta.$$

(7.168)

By averaging Equation 7.167, we obtain

$$\overline{x}_f\left(y_0,t\right) = S_L^0 \overline{YF} + S_L^0 \int_0^t \overline{\sqrt{1+F^2\left(\vartheta\right)}}\,d\vartheta$$

$$= -S_L^0 \int_0^t \overline{\frac{F(t)F(\vartheta)}{\sqrt{1+F^2(\vartheta)}}}\,d\vartheta + \overline{X}_f\left(t\right) \neq \overline{X}_f\left(t\right), \qquad (7.169)$$

$$\overline{y}_f\left(y_0,t\right) = y_0 - S_L^0 \int_0^t \overline{\frac{F(\vartheta)}{\sqrt{1+F^2(\vartheta)}}}\,d\vartheta = y_0 - S_L^0 \overline{Y}.$$

A comparison of Equations 7.164 and 7.169 clearly shows that the point with coordinates $\{\overline{x}_f(y_0,t), \overline{y}_f(y_0,t)\}$ does not belong to the mean flame surface $\overline{X}_f(t)$, contrary to the assumption invoked by Oberlack et al. (2001). Consequently, averaging Equation 7.160 to derive a kinematic equation for tracking the mean flame surface is a basically flawed approach. Therefore, the analysis by Oberlack et al. (2001) did not resolve the basic problems specific to the G-equation model by Peters (1992, 1999, 2000).

Recently, Sabel'nikov and Lipatnikov (2010) developed a rigorous method of averaging the instantaneous G-Equation 7.153 in the statistically 1D case. This rigorous analysis has shown that closing the mean flow velocity that advects the mean flame surface is an issue even in such a simple case. The mean G_t-equation derived in the cited paper has not yet been used in numerical simulations.

The author is aware of a few RANS computations of laboratory premixed turbulent flames performed by invoking the mean G_t-equation approach. Nilsson and Bai (2002) used Equation 7.156 to numerically simulate a confined, preheated, lean ($\Phi = 0.61$) propane–air flame stabilized by a bluff body and reported reasonable agreement with the experimental data obtained by Sjunnesson et al. (1992). Note that Sathiah and Lipatnikov (2007) studied the same flame using the TFC and FSC models, as discussed in the previous section (see Figure 7.17b). Peters' G-equation approach was invoked (i) by Herrmann (2006) to simulate three Bunsen flames F1, F2, and F3, experimentally investigated by Chen et al. (1996), and (ii) by Schneider et al. (2008) to numerically model two open, partially premixed, swirl-stabilized flames (such a flame is sketched in Figure 4.8c). Herrmann (2006) reported acceptable agreement between the measured and the computed radial profiles of the Favre-averaged temperature, but the mean flame brush thickness was underpredicted "by a factor of roughly 2" in all the three flames (see Figure 14 in the cited paper). In the simulations by Schneider et al. (2008), the measured mean flame brush thickness was also significantly underpredicted.

In summary, as compared with the SSDF (TFC or FSC) model, the mean G_t-equation approach does not seem to have any advantage but faces both basic and numerical problems and suffers from scant validation.

MODELS OF MOLECULAR TRANSPORT EFFECTS IN PREMIXED TURBULENT FLAMES

The models discussed so far in this chapter do not address the effects of differences in the molecular transfer coefficients on the turbulent burning rate, while such effects are well pronounced in premixed turbulent flames, as discussed in the section "Dependence of Turbulent Burning Velocity on Differences in Molecular Transport Coefficients" in Chapter 4. Various approaches to evaluating such effects are briefly discussed in this section. For a more detailed discussion, interested readers are referred to a review paper by Lipatnikov and Chomiak (2005a).

CONCEPT OF MARKSTEIN NUMBER (WEAKLY STRETCHED FLAMELETS)

As already discussed in the section "Variations in Flamelet Structure Due to Turbulent Stretching" in Chapter 5, the burning velocity and speed of stretched (i.e., strained and/or curved) laminar flames differ from S_L^0, with the effect depending substantially on the difference in the molecular diffusivities of the fuel, D_F, and the oxidizer, D_O, and on the difference between the Lewis number, Le, and unity. Theories of weakly ($\tau_c \dot{s} \ll 1$) stretched laminar flames yield linear dependencies of the burning rate integrated across the flame (consumption velocity u_c) and of the flame speed (displacement speed S_d) on the stretch rate $\dot{s}$ defined by Equation 5.50. Such a linear relation (see Equations 5.99 through 5.101) involves a nondimensional coefficient, which is called the Markstein number, Ma, and depends substantially on (Le $-$ 1) (see Equations 5.102 through 5.104).

The linear relation between the laminar flame speed and the stretch rate has been confirmed at least for weakly perturbed flames in a number of experimental and numerical studies (e.g., see Figure 5.29). The substantial dependence of the Markstein number on the Lewis number has also been confirmed in a number of studies. For instance, the aforementioned Figure 5.29 shows that $S_L > S_L^0$, that is, $Ma_d < 0$, in a lean ($\Phi = 0.6$, Le < 1) hydrogen–air flame, while the opposite inequalities hold in a rich ($\Phi = 4.5$, Le > 1) flame.

In a turbulent flow, a laminar flamelet is perturbed by various turbulent eddies that provide different stretch rates (both positive and negative). Therefore, the effect of flamelet stretching on the mean consumption velocity $\bar{u}_c$ could be modeled (Abdel-Gayed et al., 1984; Bray and Cant, 1991) by invoking a PDF for the flamelet stretch rate to average Equation 5.99. Subsequently, the effect of flamelet stretching on the turbulent burning velocity can be evaluated by replacing S_L^0 with $\bar{u}_c$ in an expression for U_t or in a RANS model of premixed turbulent combustion.

If the probability of finding a flamelet stretched with a rate $\dot{s} > 0$ were always equal to the probability of finding a flamelet compressed with a rate $-\dot{s} < 0$, then the linear Equations 5.99 through 5.101 would predict no mean effect of local flamelet stretching on the local consumption velocity and, thus, no effect of D_F/D_O and Le on U_t, because a local increase in u_c in stretched flamelets (e.g., in lean hydrogen–air mixtures) would be fully cancelled by a local decrease in u_c in compressed flamelets. This appears to be the case as far as flamelet curvature within the middle of a turbulent flame brush is concerned. Numerous experimental and numerical studies

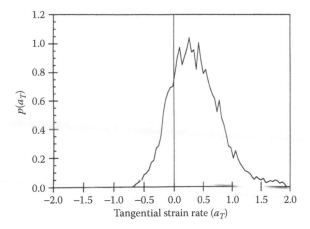

FIGURE 7.22 Probability density function of the normalized tangential strain rate, $\tau_c a_t$, on the flamelet surface, obtained by DNS. (Reprinted from Rutland, C.J. and Trouvé, A., *Combust. Flame*, 94, 41–57, 1993. With permission.)

reviewed by Bradley et al. (2003a, 2005) have shown that PDFs for flamelet curvature in turbulent flows look like the Gaussian function with a zero mean value. Accordingly, the effect of flamelet curvature on $\bar{u}_c$ and U_t is commonly neglected when averaging Equation 5.99 in turbulent flows (Bray and Cant, 1991; Bradley et al., 2003a, 2005).

This is not the case as far as flamelet straining is concerned. PDFs for the local flamelet strain rate, obtained in DNS and reviewed by Bray and Cant (1991) and Bradley et al. (2003a), look like the Gaussian function with a positive mean value, $\bar{a}_t$, which is inversely proportional to the Kolmogorov timescale (e.g., see Figure 7.22). For instance, Bray and Cant (1991) have suggested the following approximation of the DNS data available in the early 1990s:

$$P\left(a_t\right) = \frac{1}{\sqrt{2\pi \overline{a_t'^2}}} \exp\left[-\frac{1}{2} \frac{\left(a_t - \bar{a}_t\right)^2}{\overline{a_t'^2}}\right], \tag{7.170}$$

where

$$\overline{a_t'^2} = \frac{0.45}{\tau_\eta^2} \tag{7.171}$$

and

$$\bar{a}_t = \frac{0.28}{\tau_\eta} \min\left\{1; \exp\left[0.25\left(1 - \frac{S_L^0}{u_\eta}\right)\right]\right\}. \tag{7.172}$$

Recently, Bradley et al. (2003a, 2005) discussed other approximations of the PDF for the local flamelet strain rate.

Invoking Equations 5.58, 7.170 through 7.172 to average Equation 5.99, Bray and Cant (1991) obtained

$$\frac{\bar{u}_c - S_L^0}{S_L^0} = -0.28 \text{ Ma}_c \text{Ka} \min\left\{1; \exp\left[0.25\left(1 - \text{Ka}^{-1/2}\right)\right]\right\}. \tag{7.173}$$

If $\text{Ma}_c < 0$, for example, in lean hydrogen–air mixtures, Equation 7.173 predicts an increase in the mean consumption velocity and, hence, in the turbulent burning velocity due to local straining of the laminar flamelets in a turbulent flow. If $\text{Ma}_c > 0$, for example, in rich hydrogen–air mixtures, Equation 7.173 yields the opposite trend. Thus, the approach outlined earlier is capable of yielding a higher turbulent burning velocity in a lean hydrogen–air mixture as compared with a typical hydrocarbon–air mixture if the two mixtures are characterized by the same S_L^0.

Higher (lower) mean values $\bar{u}_c$ of the consumption velocity have been documented in many experimental and DNS studies of premixed turbulent flames characterized by high (low) diffusivity of the deficient reactant (e.g., see Figures 5.32 and 5.33). These and numerous similar data reviewed in Section 5.2 of a paper by Lipatnikov and Chomiak (2005a) qualitatively support the concept of Markstein number and Equation 7.173, in particular. However, the concept has not yet yielded a predictive tool for modeling the dependence of $\bar{u}_c$ on D_F/D_O and Le. For instance, no model based on the concept discussed earlier has been demonstrated to be capable of predicting the strong effect of Le on U_t (shown in Figures 4.20 and 4.21). Moreover, the concept suffers from the following weak points:

First, it characterizes the effect of using the ratio of $\bar{u}_c/S_L^0$, which depends on $\dot{s}\tau_c\text{Ma} \propto \text{KaMa}$. In order to yield an increase in U_t by the diffusivity D_d of the deficient reactant, the values of u_c in a mixture 2 with a higher $D_{d,2}$ must be larger than in a mixture 1 with a lower $D_{d,1}$. If the laminar flame speeds show the opposite behavior, for example, the ratio of $S_L^0(D_{d,1}) / S_L^0(D_{d,2})$ is as large as 16 for rich hydrogen–air and very lean $H_2/O_2/N_2$ flames in Figure 4.21, then the values of u_c must vary by several times as compared with S_L^0, to describe the experimental data. However, a simple linear relation between u_c and $\dot{s}$, which the model is based on, is only substantiated in weakly stretched flames, characterized by $|u_c - S_L^0| \ll S_L^0$. Thus, to yield a strong increase in U_t by D_d, documented in various measurements discussed in the section "Dependence of Turbulent Burning Velocity on Differences in Molecular Transport Coefficients" in Chapter 4 and Section 3.1 of a review paper by Lipatnikov and Chomiak (2005a), Equations 5.99 through 5.101 should be used well beyond the domain of their validity.

Second, the following qualitative discrepancy between the concept of Markstein number and the experimental data obtained from turbulent flames is worth stressing. The concept associates the effect of Le on U_t with the linear dependence of u_c on $\text{Ma}_c\tau_c\dot{s}$, which, after averaging, leads to a linear dependence of U_t on KaMa (e.g., see Equation 7.173). Because the Karlovitz number $\text{Ka} \propto u'^{3/2}$ depends on u', such a model predicts a stronger effect of Le on U_t at a higher u'. In particular, if the model yields a strong increase in U_t with decreasing Le (or Ma), this implies a substantial dependence of U_t on Ka, because the Lewis number affects the flame speed through KaMa $\propto$ Ka(Le − 1). Consequently, a decrease in u' and, hence, in Ka, should markedly reduce the effect of Le (or Ma) on U_t, because dU_t/dLe $\propto$ Ka. However, the experimental data shown in Figures 4.20 and 4.21 exhibit no dependence of the effect on u' at moderate turbulence.

Even if we forget about the strong effect of D_F/D_O and Le on U_t, a number of other basic issues associated with the concept of Markstein number should be resolved. In particular, the approach discussed earlier not only yields an increase in $\bar{u}_c$ by the diffusivity D_d of the deficient reactant but also assumes that the local consumption velocity depends linearly on the local stretch rate (see Equation 5.99). Contrary to the laminar flames, the linear dependence is not supported by the experimental and DNS data obtained from a turbulent flow. Even the best pronounced correlations between u_c and a_t obtained in DNS (e.g., see Figure 7.23 where the DNS data obtained by Chen and Im [2000] are reported) indicate that the linear Equation 5.99 oversimplifies the dependence of the local consumption velocity on the local stretch rate in a turbulent flow. Many other examples supporting this claim are discussed in Sections 5.1 and 5.2.1 in a review paper by Lipatnikov and Chomiak (2005a).

In turbulent flows, significant deviations of the dependencies of $u_c(\dot{s})$ from the linear Equation 5.99 can be caused by various phenomena, with the following two being the most important:

First, as discussed in the section "Local Quenching of Premixed Combustion in Turbulent Flows" in Chapter 5, the response of a laminar flame to unsteady

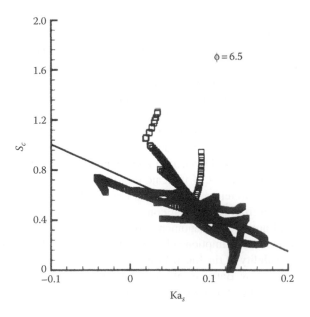

FIGURE 7.23 Correlation of the normalized consumption speed $S_c \equiv u_c/S_L^0$ with the normalized local strain rate Ka$_s \equiv a_t\delta_L/S_L^0$. (Reprinted from Chen, J.H. and Im, H.G., *Proc. Combust. Inst.*, 28, 211–218, 2000, With permission.)

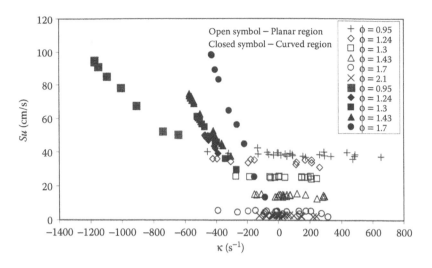

FIGURE 7.24 Reaction zone speed, measured for planar (open symbols) and strongly curved (filled symbols) reaction zones in laminar Bunsen methane–air flames versus stretch rate. (Reprinted from Choi, C.W. and Puri, I.K., *Combust. Flame*, 126, 1640–1654, 2001. With permission.)

perturbations is sensitive to transient effects if the timescale of the perturbation variations is comparable with the chemical timescale (Clavin and Joulin, 1997).

Second, Equations 5.99 through 5.101 have been derived theoretically in the limit of $\tau_c \dot{s} \ll 1$. Therefore, the response of the laminar flame speed (or consumption velocity) to perturbations can be substantially nonlinear if the perturbations are sufficiently strong (Mikolaitis, 1984; Zel'dovich et al., 1985). The nonlinear dependencies of the flame speed (or consumption velocity) on the flame stretch rate (or its constituents such as the strain rate and curvature) have been predicted theoretically, obtained numerically (e.g., see Figure 7.30), and documented experimentally. For instance, Choi and Puri (2001) measured the local stretch rate and the speed of the reaction zone in two different parts of the laminar Bunsen methane–air flames, in the vicinity of the flame tip, where the reaction zone was strongly negatively curved, and well outside the tip, where the reaction zone appeared to be planar. The experimental results shown in Figure 7.24 indicate that the effects of the positive (planar reaction zone) and negative (curved reaction zone) stretch rates on the flame speed are drastically different from each other, contrary to the linear Equation 5.100.

The transient and nonlinear phenomena emphasized earlier can strongly affect the response of the local consumption velocity to the local stretch rate and, thus, can strongly reduce the validity of the linear Equations 5.99 through 5.101. For instance, the DNS data obtained by Baum et al. (1994) from lean $H_2/O_2/N_2$, 2D turbulent flames do not indicate any simple dependence of the consumption velocity on the local strain rate (see Figure 7.25).

This brief review of the experimental and numerical results unambiguously shows that differences in the molecular transport coefficients substantially affect the local consumption velocity in premixed turbulent flames, with the effects being

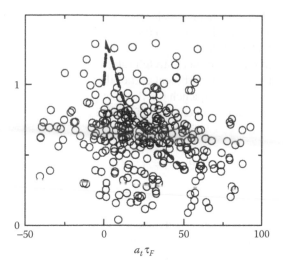

FIGURE 7.25 Normalized consumption speed u_c/S_L^0 versus normalized local strain rate $\tau_c a_t$. The symbols show the DNS data, and the dashed curve has been calculated for a steadily strained laminar flame. Lean ($\Phi = 0.5$) hydrogen flame, case 9. (Reprinted from Baum, M., Poinsot, T.J., Haworth, D.C., and Darabiha, N., *J. Fluid Mech.*, 281, 1–32, 1994. With permission.)

in qualitative agreement with the Markstein number concept of weakly perturbed laminar flames. However, such a qualitative agreement does not mean that premixed turbulent combustion models that utilize the theoretical results in the form of the linear Equations 5.199 through 5.101, supplemented with measured or computed values of variously defined Markstein numbers, provide a well-developed tool for predicting the dependence of U_t on D_F/D_O and Le.

Certainly, turbulent combustion models that utilize Ma can yield a reduction in U_t by Le. However, although the concept of Markstein number yields the qualitatively correct dependence of U_t on Le, the concept does not seem to be able to quantitatively predict the strong effect of Le on U_t, which is well documented in lean hydrogen mixtures. This limitation results from a number of assumptions, for example, weak and steady perturbations, that the concept is based on but that do not hold in a typical turbulent flame.

No model of premixed turbulent combustion, based on the concept of Markstein number, has yet been shown to predict the strong dependence of U_t on D_F/D_O and Le.

CONCEPT OF CRITICAL STRETCH RATE (FLAMELET QUENCHING BY STRONG PERTURBATIONS)

As discussed in the section "Variations in Flamelet Structure Due to Turbulent Stretching" in Chapter 5, differences in the molecular transport coefficients can affect the turbulent burning velocity, because, in particular, the quenching of laminar flamelets stretched by turbulent eddies depends substantially on D_F/D_O and Le.

The flamelet-quenching mechanism is highlighted by the concept of critical stretch rate, which was put forward by Bray (1987), Abdel-Gayed et al. (1988), Bradley (1992), and Bradley et al. (1992).

The concept introduces a critical stretch rate, $\dot{s}_q$, such that the flamelet structure is assumed to be unaffected by turbulent eddies that provide $\dot{s} < \dot{s}_q$, whereas stronger eddies ($\dot{s} > \dot{s}_q$) are assumed to quench the flamelet instantly. Subsequently, the probability $P_q \leq 1$ of local flamelet quenching is calculated as

$$P_q = 1 - \int\limits_{\dot{s}_{q,-}}^{\dot{s}_{q,+}} P(\dot{s}) d\dot{s} \tag{7.174}$$

invoking a PDF $P(\dot{s})$ for the local stretch rate, and the mean reaction rate is reduced by a factor of $(1 - P_q)$. Here, $\dot{s}_{q,+}$ and $\dot{s}_{q,-}$ are associated with the quenching of strained and compressed flamelets, respectively (Bradley et al., 2003a, 2005), but only the former effect was addressed in earlier papers, where $\dot{s}_{q,-} = -\infty$.

The concepts of Markstein number and critical stretch rate are compared in Figure 7.26. The former concept highlights weakly stretched flamelets and averages the linear Equation 5.99. In other words, the real dependence of $u_c(\dot{s})$ (see solid line) is replaced with a straight dashed line, the slope of which is controlled by the Markstein number, Ma_c. The latter concept highlights strongly stretched flamelets and replaces $u_c(\dot{s})$ with a step function (see the dotted-dashed lines). Recently, Bradley et al. (2003a, 2005) developed a more complicated approach that combines both the concept of Markstein number and the concept of critical stretch rate to evaluate the mean consumption velocity in a turbulent flow by averaging Equation 5.99 and invoking Equation 7.174.

To close Equation 7.174, the critical stretch rates $\dot{s}_{q,+}$ and $\dot{s}_{q,-}$ and the PDF $P(\dot{s})$ should be specified. For a particular flame configuration and mixture

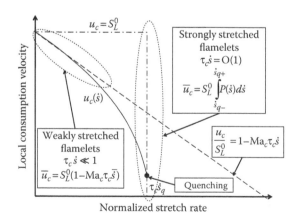

FIGURE 7.26 Concepts of Markstein number and critical stretch rate.

composition, the critical steady stretch rate can be either measured or calculated using an available CFD code (e.g., www.softpredict.com) and a detailed chemical mechanism.

Different parameterizations of $P(\dot{s})$ can be found in the literature (Bray, 1987; Bray and Cant, 1991; Bradley et al., 2003a, 2005). For instance, to evaluate the quenching probability, Bray (1987) utilized the well-known log-normal PDF (Frisch, 1995)

$$P(\varepsilon_l) = \frac{1}{\sigma_{\varepsilon,l}\varepsilon_l\sqrt{2\pi}} \exp\left\{-\frac{1}{2\sigma_{\varepsilon,l}^2}\left[\ln\left(\frac{\varepsilon_l}{\bar{\varepsilon}}\right) + \frac{\sigma_{\varepsilon,l}^2}{2}\right]^2\right\}, \tag{7.175}$$

where

$$\sigma_{\varepsilon,l}^2 = C_\varepsilon \ln\left(\frac{l}{\eta}\right) \tag{7.176}$$

and

$$\varepsilon_l = \iiint \frac{\nu}{2}\sum_{i,j=1}^{3}\left(\frac{\partial u_i}{\partial x_j} + \frac{\partial u_j}{\partial x_i}\right)^2 d\mathbf{x} \tag{7.177}$$

is the viscous dissipation of turbulent energy, locally averaged over a cube of size l. Substitution Equation 7.175 into Equation 7.174 yields (Bray, 1987):

$$1 - P_q = \frac{1}{2}\mathrm{erfc}\left\{\frac{1}{\sqrt{2}\sigma_{\varepsilon,L}}\left[\ln\left(\frac{\varepsilon_q}{\bar{\varepsilon}}\right) + \frac{\sigma_{\varepsilon,L}^2}{2}\right]\right\}, \tag{7.178}$$

where erfc is the complementary error function, $\sigma_{\varepsilon,L}$ is associated with the integral length scale, that is, $\sigma_{\varepsilon,L} = \sigma_{\varepsilon,l}(l = L)$, and the critical dissipation rate, ε_q, is controlled by the critical stretch rate:

$$\varepsilon_q = 15\nu_u\dot{s}_q^2. \tag{7.179}$$

If the critical stretch rate is reduced, for example, by increasing the Lewis number, then the probability of flamelet quenching is increased, leading to a decrease in the turbulent burning velocity. For instance, Figure 7.27 shows the probability $(1 - P_q)$ and turbulent burning velocity

$$U_t = u'(1 - P_q), \tag{7.180}$$

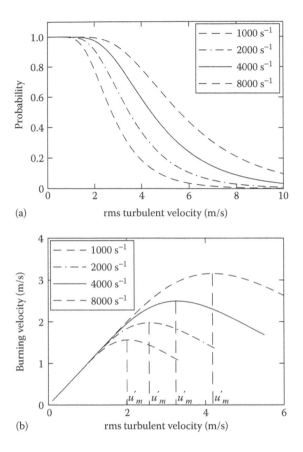

(a)

(b)

FIGURE 7.27 Probability $(1 - P_q)$ (a) and burning velocity (b) calculated using Equations 7.176, 7.178 through 7.180 for different critical stretch rates specified in the legends versus the rms turbulent velocity.

calculated using Equations 7.176, 7.178, and 7.179 for $L = 10$ mm and different critical stretch rates specified in the legends. A decrease in $\dot{s}_q$ reduces u'_m and U_t when the rms turbulent velocity is close to u'_m.

The dependence of the critical stretch rates on the differences in the molecular transport coefficients is well known from studies of laminar premixed flames (Zel'dovich et al., 1985). In particular, $\dot{s}_q$ is higher for mixtures with a lower Le. Therefore, the approach discussed earlier can yield a lower probability of flamelet quenching and, hence, a higher value of U_t in a lean hydrogen–air mixture as compared with a hydrocarbon–air mixture characterized by the same S_L^0, because the factor $(1 - P_q)$ is higher in the former mixture, with u' being the same for both flames.

The rapid increase in the amount of the numerical data obtained from DNS of premixed turbulent flames allows one to approximate the PDF for the local stretch rate reasonably well (e.g., see recent papers by Bradley et al. [2003a, 2005]).

Although the key ideas and blocks of the concept of critical stretch rate appear to be consistent with the contemporary knowledge on premixed turbulent combustion, this claim does not mean that a predictive tool for modeling the local flamelet quenching has already been developed and validated using the considered concept. To do so, a number of basic issues discussed in the section "Local Quenching of Premixed Combustion in Turbulent Flows" in Chapter 5 should be resolved.

Moreover, the concept of critical stretch rate is unlikely to be able to predict the strong effect of Le on U_t (shown in Figures 4.20 and 4.21). First, the key role of local combustion extinction by strong turbulent stretching implies that the effect of Le on U_t should increase with u'. For instance, the dependence of the quenching probability P_q, calculated using Equations 7.175 through 7.179, on u' is well pronounced (see Figure 7.27a). The model yields $P_q \ll 1$ in a certain, mixture-dependent range of u', but a further increase in u' results in a drastic increase in the probability of flamelet quenching. Based on such a behavior of P_q, one could expect a lack of influence of Le on U_t at relatively weak turbulence (e.g., all the curves in Figure 7.27b coincide with one another at $u' < 1.5$ m/s), whereas the effect of Le on U_t should be accompanied by bending of the $U_t(u')$ curves due to an increase in P_q. However, experiments show different trends (see Figures 4.20 and 4.21):

- The effect of Le on U_t is well pronounced in the entire range of moderate turbulence, where the slope of dU_t/du' is roughly constant.
- The data exhibit no dependence of the effect on u' in this range.

Second, the local flamelet quenching manifests itself in the bending of $U_t(u')$ curves. If the maxima in the curves are attributed to increasing probabilities of local combustion quenching, then the values of u'_m associated with these maxima must be related to other effects explained by local quenching, including the effects of D_F/D_O and Le on U_t. In particular, the dependencies of U_t and dU_t/du' on D_F/D_O and Le at $u' < u'_m$ must correlate with the dependence of u'_m on Le and D_F/D_O, that is, a weaker slope or a lower burning velocity must be associated with a lower u'_m.

Burning velocities obtained from typical hydrocarbon–air premixed turbulent flames often show this behavior (e.g., see Figure 4.13). However, the experimental data plotted in Figure 7.28 show the opposite behavior: A lower u'_m corresponds to a higher slope dU_t/du' (cf. lean hydrogen mixtures shown as filled symbols and solid curves with hydrocarbon–air mixtures shown as open symbols and dashed curves) and, sometimes (cf. lean $H_2/O_2/Ar$ mixture with hydrocarbon–air mixtures), to a higher U_t. This trend is unlikely to be predicted by a model that highlights the local quenching of flamelets in a turbulent flow.

In summary, differences in the molecular transport coefficients can affect the probability of local flamelet quenching, mean consumption velocity, and burning velocity in premixed turbulent flames, with the expected effects being in qualitative agreement with the influence of D_F/D_O and Le on U_t, as documented in experiments (see Figures 4.20 and 4.21). However, such a qualitative agreement does not mean that premixed turbulent combustion models that utilize a critical stretch rate determined for a stationary laminar flame are a well-developed tool for predicting the dependence of U_t on D_F/D_O and Le.

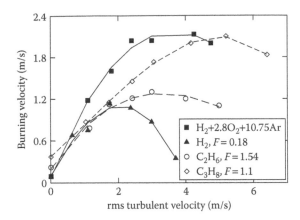

FIGURE 7.28 The burning velocity versus the rms turbulent velocity in hydrocarbon–air mixtures (open symbols and dashed curves) and in mixtures that contain H_2 (filled symbols and solid curves). The symbols show the data obtained by Karpov and Severin (1978 and 1980). The curves approximate the data.

Certainly, turbulent combustion models that utilize $\dot{s}_q$ can yield a reduction in U_t by Le, because $\dot{s}_q$ depends substantially on the Lewis number. However, although the concept of critical stretch rate yields the qualitatively correct dependence of U_t on Le, the concept does not seem to be able to quantitatively predict the strong effect of Le on U_t, which is independent of u' at moderate turbulence, as is well documented in lean hydrogen mixtures. No model of premixed turbulent combustion, based on the discussed concept, has yet been shown to predict the strong dependence of U_t on D_F/D_O and Le.

CONCEPT OF UNSTABLE FLAMELETS

As discussed in the section "Flame Instabilities" in Chapter 5, differences in the molecular transport coefficients can also affect the turbulent burning velocity by either promoting (Le < 1 and/or $D_e < D_d$) or impeding (Le > 1 and/or $D_e > D_d$) the development of the DL instability of laminar flamelets in a turbulent flow. Because of the instability, the flamelet surface area and, hence, the turbulent burning velocity could increase, with the increase being more pronounced in mixtures with a larger D_d.

A few models allow for the effects of the DL instability on premixed turbulent combustion (Kuznetsov and Sabel'nikov, 1990; Paul and Bray, 1996; Peters et al., 2000; Bychkov, 2003; Bradley et al., 2003a, 2005). Expressions for the turbulent burning velocity, reported by Paul and Bray (1996) and Bychkov (2003), indirectly allow the effect of D_d on U_t via the dependence of the so-called neutral wavelength Λ_n (see the section "Hydrodynamic Instability" in Chapter 5) on D_d. For instance, Bychkov (2003) has obtained the following expression:

$$U_t^2 = \left(S_L^0\right)^2 \left(\frac{l_m}{\Lambda_n}\right)^{2/3} + \frac{4}{3}C_t u'^2 \ln\left(\frac{L}{\eta}\right) \tag{7.181}$$

in the case of $\eta < \Lambda_n$ (see Equation 47 in the cited paper). Here, C_t is a constant of unity order, η is the Kolmogorov length scale, and l_m is the length scale of the largest possible flamelet perturbations controlled by the length scale of the problem considered.

The neutral wavelength characterizes such perturbations that a laminar flame is unstable (stable) if $\kappa < 2\pi/\Lambda_n$ ($\kappa > 2\pi/\Lambda_n$). The value of Λ_n can be specified using the results of the theoretical and numerical studies of laminar flames, as reviewed by Lipatnikov and Chomiak (2005a, Section 4.1). For instance, Figure 9 from the cited review paper shows that $\Lambda_n \approx 20\delta_L$ at Le = 1 and the neutral wavelength decreases when the Lewis number decreases. Substitution of such a dependence of Λ_n on Le into Equation 7.181 can yield an increase in the turbulent burning velocity when the Lewis number decreases.

The model developed by Bradley et al. (2003a, 2005) allows for the effect of D_d on the response of laminar flamelets to the DL instability via the dependence of a length scale l_c on D_d. The latter length scale characterizes the smallest cells that appear on the surface of an unstable, expanding, spherical laminar flame. Al-Shahrany et al. (2005) reported a method for evaluating l_c.

Although certain experimental and DNS data discussed in the section "Instabilities of Laminar Flamelets in Turbulent Flows" in Chapter 5 imply an important role played by the DL instability in weakly turbulent premixed combustion, models that highlight this phenomenon do not seem to be able to predict the strong effect of D_d on U_t shown in Figures 4.20, 4.21, and 7.28. For instance, in order for Equation 7.181 to predict roughly equal burning velocities for the rich and leanest H_2 mixtures in Figure 4.21, the ratios of $S_{L0}^3 \Lambda_n^{-1} = v_u^{-1}(S_L^0)^4(\delta_L/\Lambda_n)$ should also be roughly equal for the two mixtures (see the first term on the RHS of Equation 7.181 and note that the second term does not involve mixture characteristics). Consequently, the ratios of δ_L/Λ_n in the two mixtures should differ by a factor of roughly 16^4, because S_L^0 is 16 times higher in the rich mixture, while the difference in the viscosities of the two mixtures is much less than the difference in S_L^0. However, the results of studies of unstable laminar flames, reviewed recently by Lipatnikov and Chomiak (2005a), show a substantially weaker effect of Le on δ_L/Λ_n (see Figure 9 in the cited paper). Moreover, if Equation 7.181 predicts an increase in U_t when Le is decreased and/or the ratio of D_d/D_e is increased, then by differentiating this equation, one obtains a decrease in dU_t/du', contrary to numerous experimental data.

Moreover, the majority of the discussed models (Paul and Bray, 1996; Peters et al., 2000; Bradley et al., 2003a, 2005) yield a marked effect of the instabilities on U_t only at weak turbulence, when u' is of the order of S_L^0 or less; for example, the total burning rate shown in Figure 5.44 begins to rapidly increase with time after the initial turbulence decays substantially so that $u' < S_L^0$. On the contrary, experimental data indicate a strong dependence of U_t on D_d even for u' as large as $40 S_L^0$ (see lean $H_2/O_2/N_2$ and $H_2/O_2/Ar$ mixtures in Figures 4.21 and 7.28, respectively).

Furthermore, in sufficiently intense turbulence, the instabilities are likely to be suppressed due to the stretching of flamelets by turbulent eddies, as discussed in the section "Instabilities of Laminar Flamelets in Turbulent Flows" in Chapter 5.

Thus, although the experimental data obtained from turbulent flames (see the section "Instabilities of Laminar Flamelets in Turbulent Flows" in Chapter 5) indirectly support the concept of an important role played by the DL instabilities in weakly turbulent premixed combustion, decisive experimental evidence has not yet been reported. The majority of turbulent combustion models that allow for the DL instabilities, as well as numerical simulations, indicate that the effects of the instabilities on turbulent flames are substantially reduced by u'/S_L^0. Models based on the instability concept have not yet been shown to be able to predict the strong dependence of U_t on D_F/D_O and Le, observed in experiments.

CONCEPT OF LEADING POINTS

For a fully developed, statistically planar, 1D, turbulent premixed flame, the burning velocity, U_t, should be equal to the speed, S_{le}, of the leading edge of the flame brush relative to the unburned gas. Under this assumption, one can either model the processes inside a mean flame brush to predict U_t or develop a model for predicting S_{le} based on leading-edge characteristics, and the final result must be the same, $U_t = S_{le}$. For instance, if the local heat release is confined to thin flamelets, the flamelet surface production inside the mean flame brush must provide a balance between U_t and S_{le}.

If we assume that the strong effect of D_d/D_e on U_t, which is documented in developing turbulent flames, also exists in fully developed flames, then the constraint of $U_t = S_{le}$ in a steadily propagating, statistically 1D, planar turbulent flame strongly challenges the three concepts discussed earlier. If $U_t = S_{le}$, a consistent model of premixed turbulent combustion should predict or, at least, explain (i) the values of S_{le} of the order of u', because U_t and u' are of the same order at moderate turbulence (see Figures 4.20, 4.21, and 7.28) and (ii) the substantial dependence of S_{le} on D_F/D_O and Le. The models of premixed turbulent combustion, discussed so far, provide no clue to explaining both features together.[*] For instance, a model based on the concept of Markstein number can yield $S_{le} > S_L^0$ if Ma < 0 (see Equation 5.100). However, the increase in S_{le} as compared with S_L^0 seams to be of the order of S_L^0, while a strong effect of D_d/D_e on U_t is observed in the case of $S_{le} \approx U_t > 30 S_L^0$ (see lean $H_2/O_2/N_2$ and $H_2/O_2/Ar$ mixtures in Figures 4.21 and 7.28, respectively). Similarly, a flamelet-instability model could be used to predict an increase in S_{le} due to the local acceleration of unstable laminar flamelets. However, again, the increase appears to be insufficient (of the order of S_L^0, because even unstable flamelets retain a laminar structure) for explaining the aforementioned experimental data. As far as the concept of critical stretch rate is concerned, it cannot yield even $S_{le} > S_L^0$, not to mention $S_{le} \approx U_t > 30 S_L^0$. In this section, we will discuss models of premixed turbulent combustion that place the focus of consideration on the leading edge of the flame brush and allow for the strong influence of D_F/D_O and Le on the local burning rate at the leading edge.

---

[*] Note that the experimental data plotted in Figures 4.20, 4.21, and 7.28 show the burning velocities, U_t, obtained from developing turbulent flames. It is not yet clear whether or not the effect of D_F/D_O and Le on the speed of the leading edge of a developing turbulent flame is so strong (see Section 7.2 in a review paper by Lipatnikov and Chomiak [2005a]). It is admissible that this speed, which is larger than U_t, is mainly controlled by turbulence characteristics. In such a case, the feature (ii) would not be relevant to the developing flames. Target-directed experimental research on the effects of D_F/D_O and Le on the speed of the leading edge of a developing turbulent flame is definitely necessary to clarify the issue.

Zel'dovich and Frank-Kamenetskii (1947) highlighted the crucial role played by the leading points in turbulent flame propagation, and the idea was developed by several groups in Russia, as reviewed by Kuznetsov and Sabel'nikov (1990) and Lipatnikov and Chomiak (2005a, Section 6.2). Here, we restrict ourselves to a brief discussion on the concept of leading points and refer interested readers to the afore-mentioned book and review paper.

Kuznetsov and Sabel'nikov (1990) have thoroughly analyzed the physical basis of the concept and have argued that (i) the turbulent flame speed is controlled by the flamelets that advance furthest into the unburned mixture (leading points, see Figure 7.29) and (ii) the structure of the leading kernels is universal, that is, independent of turbulence characteristics, as hypothesized earlier by Baev and Tretjakov (1969).

If u' and U_t are substantially larger than S_L^0, a flamelet element may become a leading point due to the convection of the element by a strong turbulent eddy. At the same time, due to the highly uneven structure of a turbulent flow, the convected element may enter a spatial region characterized by high local stretch rates produced by other small-scale turbulent eddies. If the stretch rate were steady, the flamelet would be locally quenched. However, in an unsteady and nonuniform turbulent flow, the flamelet seems to be ousted from the aforementioned spatial regions, rather than quenched locally by the high stretch rates. Therefore, convection of a flamelet element toward the leading edge of a turbulent flame brush may be limited by the local stretching of the flamelet by turbulent eddies. Owing to the balance between the convection and the stretching, critically perturbed (under near-extinction conditions) flamelets may be hypothesized to dominate at the leading edge. Consequently, the structure of the leading points is assumed to be universal and determined by the strongest possible nonquenching perturbation of flamelets by turbulent eddies.

Such a concept offers an opportunity to substantially simplify the modeling of premixed turbulent combustion by reducing the flamelet library (a collection of the basic characteristics of perturbed laminar flames that allow for a number of possible perturbations) to a single "flamelet page" (parameters of a critically perturbed leading kernel). Then, the following issue is of paramount importance. What is the structure of the leading kernels?

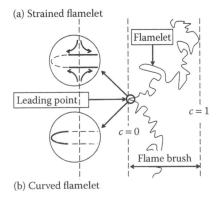

(a) Strained flamelet

Flamelet

Leading point

$c = 1$

$c = 0$

Flame brush

(b) Curved flamelet

FIGURE 7.29 (a,b) A sketch of a leading point and its possible structures.

From geometrical reasoning (see Figure 7.29), the flamelet surface in the vicinity of a leading point looks like a cone. Accordingly, there are two approaches to modeling the structure of the leading points. One approach, developed by Kuznetsov and Sabel'nikov (1977), places the focus of consideration on two almost parallel flamelets just behind the leading point (see circle a in Figure 7.29) and considers these flamelets to be critically strained, that is, a further increase in the strain rate would quench the flamelets. Another approach, developed later by the author (Karpov and Lipatnikov, 1994, 1995; Betev et al., 1995; Karpov et al., 1996a,b, 1997; Lipatnikov, 1997; Lipatnikov and Chomiak, 1998), highlights the tip of the conical flamelet surface (see circle b in Figure 7.29) and considers the flamelet structure at the tip to be the critically curved flamelet, that is, flamelets with a higher curvature cannot sustain themselves, because conductive heat losses from the reaction zone exceed the heat release within the zone.

Critically Strained Flamelets

Kuznetsov and Sabel'nikov (1977) considered a critically strained, steady, planar, symmetric, counterflow laminar flame (see Figure 5.20) to be a model of the leading point. By extending the preceding study of the same problem by Klimov (1953) to the case of $D_F \neq D_O \neq a$, Kuznetsov and Sabel'nikov (1977) theoretically analyzed such a flame in the limit case of single-step, two-reactant chemistry with $\Theta/T_b \to \infty$. They obtained the following solutions for the temperature, T_r, and mass fraction of the excess reactant in the critically strained reaction zone:

$$\left.\begin{aligned} T_r &= T_u + Y_{F,u}\left(T_s - T_u\right)\left(1 + St\right)\sqrt{D_F/a} \\ Y_{O,r} &= Y_{O,u} - St Y_{F,u}\sqrt{D_F/D_O} \end{aligned}\right\} \Phi_r \leq 1 \qquad (7.182)$$

if the mixture is lean in the reaction zone and

$$\left.\begin{aligned} T_r &= T_u + Y_{O,u}\left(T_s - T_u\right)\left(1 + St^{-1}\right)\sqrt{D_O/a} \\ Y_{F,r} &= Y_{F,u} - St^{-1}Y_{O,u}\sqrt{D_O/D_F} \end{aligned}\right\} \Phi_r \geq 1 \qquad (7.183)$$

if the mixture is rich in the reaction zone. Here, St is the mass stoichiometric coefficient, indexes u and r designate the unburned mixture and the reaction zone, respectively, and T_s is the adiabatic combustion temperature of the stoichiometric mixture. Even if $\Phi = 1$, $T_r \neq T_s$ in a general case of $Le \neq 1$ due to local variations in the mixture enthalpy in the critically strained reaction zone. Similarly, the local composition in the zone ($Y_{F,r}$ and $Y_{O,r}$) differs from the composition of the unburned mixture far ahead of the flame ($Y_{F,u}$ and $Y_{O,u}$), because the molecular diffusion fluxes of the fuel and oxidizer into the zone are not equal to one another if $D_F \neq D_O$.

By noting that $D_O \approx a$ in a typical fuel–air mixture (with the exception of rich H_2–air mixtures), Kuznetsov and Sabel'nikov (1977) neglected the difference between $T_b = T_u + (T_s - T_u)(1 + St)Y_{F,u}$ and T_r. Subsequently, they proposed to model premixed turbulent combustion using the unperturbed laminar flame speed $S_L^0(\Phi_r)$ associated with the local mixture composition within the reaction zone of the critically strained laminar flamelets, instead of the unperturbed laminar flame speed $S_L^0(\Phi_u)$ associated with the composition of a fresh mixture at far distances from the flamelet. In the limit case of $St \gg 1$ and $|D_F - D_O| \ll D_O$, Equations 7.182 and 7.183 yield

$$\Phi_r = \Phi_u \sqrt{D_F/D_O} \tag{7.184}$$

if the local equivalence ratio $\Phi_r \leq 1$ and

$$\Phi_r = \Phi_u - \sqrt{D_O/D_F} + 1 \tag{7.185}$$

if $\Phi_r \geq 1$.

By using this method, Kuznetsov and Sabel'nikov (1990) succeeded in explaining the following well-documented phenomenon: the dependencies of the burning velocity on the equivalence ratio reach maxima in leaner or richer hydrogen–air (Karpov and Severin, 1977) or gasoline–air (Talantov, 1978) mixtures in turbulent flows as compared with laminar flows.

Zimont and Lipatnikov (1993, 1995) and Karpov et al. (1994) incorporated the approach discussed earlier into the TFC model of premixed turbulent combustion, by replacing the standard chemical timescale $\tau_c^0 = a_u/[S_L^0(\Phi_u,T_u)]^2$ in Equation 7.13 with the modified timescale $\tau_c' = a_u/[S_L^0(\Phi_r,T_r)]^2$, where the leading point speed $S_L^0(\Phi_r,T_r)$ was associated with the temperature and equivalence ratio determined using Equations 7.182 and 7.183. At moderate turbulence, the model was able to quantitatively predict a considerable amount of data on $U_t(u')$ obtained by Karpov and Severin (1978, 1980) from lean, stoichiometric, and rich hydrocarbon–air flames. However, the model failed to predict the strongest effect of D_d on U_t, obtained by Karpov and Severin (1977, 1980) from the leanest hydrogen flames.

We may also note that the model by Kuznetsov and Sabel'nikov (1990) highlights critically strained, stationary laminar flamelets and, thus, does not address transient phenomena, which can substantially affect the influence of turbulent eddies on the flamelet local structure and burning rate.

Critically Curved Flamelets

In order to predict the high values of U_t and dU_t/du' in lean hydrogen mixtures characterized by high diffusivity of the deficient reactant, Lipatnikov and Chomiak (1998, 2005a) proposed laminar flame perturbations that gave rise to the highest local burning rate to be a model of the leading points. This hypothesis is based not only on the strong effect of D_d on U_t but also on the following very simple rationale; in order for an element of a flamelet to become the leading point, the local burning rate in this element should be as high as possible.

Then, to select a proper model of the leading points, Lipatnikov and Chomiak (1998) performed numerical simulations of various perturbed laminar flames: (i) expanding spherical flames, (ii) converging spherical flames, (iii) expanding cylindrical flames, (iv) strained expanding planar flames (see Figure 5.20), and (v) strained expanding cylindrical flames (see Figure 5.20 and imagine that it shows the intersection of a plane with a cylindrical flame surface). In all these computations with the exception of case (ii), a planar (iv) or cylindrical (iii and v) or spherical (i) kernel of a small radius r_i, filled with equilibrium adiabatic combustion products, was set at $t = 0$ to simulate ignition. Subsequently, the time history of the consumption velocity was calculated at various r_i. Numerical results showed that, if Le<; then the maximum (for each particular flame configuration) consumption velocities were obtained when the ignition radius was equal to its critical value, r_{cr}, such that the initial kernel shrunk if $r_i < r_{cr}$.

The results obtained in the case of Le < 1 are reported in Figure 7.30. The triangles and circles show the maximum values of $u_c(t)$, computed at $r_i = r_{cr}$ and different strain rates (each symbol corresponds to its own strain rate). Curves 3–5 have been drawn by rewriting the computed dependencies of u_c (see Figure 7.31) and $\dot{s}$ on time in the form of u_c versus $\dot{s}$. Curves 6 and 7 have been calculated for stationary strained flames at different stain rates (each point on a curve corresponds to its own strain rate). Figure 7.30 clearly indicates that the highest (Le < 1) burning rate is reached in highly curved, expanding spherical flames. For this reason, the highly curved, expanding spherical laminar flame kernel may be proposed to be used as a model of the structure of the leading points in turbulent flames if Le < 1. Historically, Karpov and Lipatnikov (1994, 1995) put forward such a proposal based on other reasons discussed in Section 6.2.2 of a review paper by Lipatnikov and Chomiak (2005a). As far as Figure 7.29 is concerned, a highly curved, spherical laminar flame is associated with a model of the highly curved tip of the leading flamelet cone (see circle *b*).

Based on the foregoing ideas, the following method has been proposed to characterize the burning rate in the leading points for mixtures with Le < 1 (Betev et al., 1995; Karpov et al., 1996b; Lipatnikov, 1997; Lipatnikov and Chomiak, 2005).

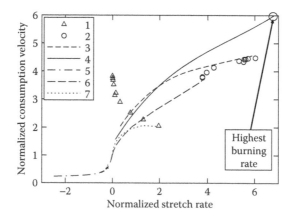

FIGURE 7.30 Normalized consumption velocity u_c/S_L^0 versus normalized flame stretch rate $\tau_c^0 \dot{s}$, computed for Le = 0.44 and $r_i = r_{cr}$. 1: Strained planar expanding flame; 2: strained cylindrical expanding flames; 3: cylindrical flame expanding in quiescent mixture; 4: expanding spherical flame; 5: converging spherical flame; 6: critically strained stationary cylindrical flame; 7: critically strained stationary planar flame. (Adapted from Lipatnikov, A.N. and Chomiak, J., *Combust. Sci. Technol.*, 137, 277–298, 1998, With permission.)

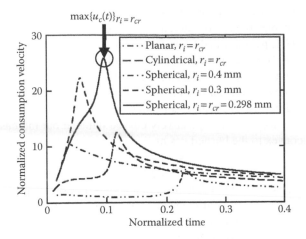

FIGURE 7.31 Normalized consumption velocity u_c/S_L^0 versus normalized time, computed for planar, cylindrical, and spherical expanding hydrogen–air flames. $\Phi = 0.26$. The circle and arrow show the consumption velocity used for modeling the burning rate at the leading edge of the premixed turbulent flame brush.

- First, the expansion of an adiabatic, spherical laminar flame from an initial pocket of radius r_i, filled with the equilibrium adiabatic combustion products, is numerically simulated to obtain the dependencies of the consumption velocity on time at different radii r_i (e.g., see the dashed and double dotted–dashed curves in Figure 7.31).
- Second, the critical radius of the pocket is determined so that the initial kernel expands if $r_i \geq r_{cr}$, but the kernel shrinks if $r_i < r_{cr}$.
- Third, $u_c(t)$ is calculated at $r_i = r_{cr}$ (see the solid curve in Figure 7.31).
- Fourth, the maximum of the latter $u_c(t)$ curve is determined (see the circle and arrow in Figure 7.31) and is used to evaluate a chemical timescale

$$\frac{\tau_{lp}}{\tau_c^0} = \frac{u_c\left(t \to \infty\right)}{\max\left\{u_c\left(t\right)\right\}_{r_i = r_{cr}}} \tag{7.186}$$

that characterizes the leading points. Note that the correct numerical model must yield $u_c(t \to \infty) \to S_L^0$ in Equation 7.186.

For flames with Le < 1, the highest instantaneous burning rate, $\max\{u_c(t)\}_{r_i = r_{cr}}$, is substantially larger than $u_c(t \to \infty) = S_L^0$ (see Figure 7.31). On the contrary, if Le > 1, the consumption velocity in a highly curved spherical flame is markedly less than S_L^0 and $\max\{u_c(t)\}_{r_i = r_{cr}} = u_c(t \to \infty) = S_L^0$, that is, Equation 7.186 yields $\tau_{lp} = \tau_c^0$ in the case of Le > 1. Therefore, this model disregards the effects of differences in the molecular transport coefficients on the local burning rate at the leading edge if Le > 1. This simplification requires discussion.

Two different physical mechanisms, convection of a flamelet by a large-scale eddy and an increase in the local burning rate due to diffusive-thermal effects, contribute to the formation of a leading point. If the latter mechanism is emphasized, strongly curved laminar flamelets are considered to be the leading points only for Le < 1 (Lipatnikov, 1997; Lipatnikov and Chomiak, 1998). In such mixtures, the local burning rate in a positively curved flamelet is substantially increased due to a higher flux of chemical energy into the flamelet as compared with the heat flux from it. As a result, the flamelet propagates faster and moves to the leading edge of the flame brush.

In the case of Le > 1, a similar scenario does not seem to be realistic because the local burning rate decreases in a positively curved flamelet but increases in a negatively curved (convex to the burned gas as in a collapsing spherical flame) flamelet. However, the latter flamelet cannot be a leading point due to geometrical consideration. At the leading points, if Le > 1, the highest local burning rate is associated with planar flamelets compressed by turbulent eddies, but such a planar structure is characterized by the lowest local flamelet surface area A per unit area of the mean flame surface. Therefore, a balance between an increase in the area due to flamelet wrinkling by small-scale turbulent eddies and a decrease (Le > 1) in the local burning rate due to the same wrinkling should be reached at the leading points.

For instance, let us consider the flame surface in the vicinity of the leading point to be the tip of a cone. The more acute the tip, the larger the aforementioned area ratio, but lower is the consumption velocity, u_c, for Le > 1. Consequently, the leading point is associated with the tip that has a specific acuteness such that $u_c A$ reaches the maximum (with respect to various acutenesses) value.

Such a leading "superflamelet" (this term was introduced by A.S. Betev) may be described in terms of a chemical timescale, which is larger than τ_c^0, but substantially lower than the chemical timescale that characterizes the highly curved, expanding spherical flame. Because no reliable model is currently available, τ_c^0 may be used as a higher estimate. Based on the aforementioned reasoning, $\max\{u_c(t)\}_{r_i=r_{cr}}$ is substituted in the denominator on the RHS of Equation 7.186. Certainly, such a simplified method underestimates the effects of D_d/D_e on U_t in the case of Le > 1, but errors due to this simplification appear to be acceptable at the present stage of research into the discussed problem, as shown in the section "Validation."

Flame Ball

The foregoing method (i.e., Equation 7.186) for characterizing the burning rate in the leading points requires numerical simulations of highly curved, expanding spherical laminar flames. Although this problem is 1D, such simulations may be time consuming, as a number of runs have to be performed to evaluate the critical radius of the ignition pocket. To quickly estimate the effect of D_d on U_t, a simpler method for characterizing the burning rate in critically curved leading points would be useful. Such a method was developed by drawing an analogy (Betev et al., 1995; Karpov et al., 1997) between the leading points and the so-called flame ball.

The latter phenomenon, predicted theoretically by Zel'dovich (1944), is the adiabatic burning of a motionless mixture at the surface of a small ball (see Figure 7.32), with the chemical energy (heat) being transported from infinity to the ball (from the ball to infinity) solely by molecular diffusion (heat conduction). Such a regime of premixed combustion received plenty of attention recently. Here, we will restrict ourselves to a brief introduction to the problem, relevant to the main subject of this

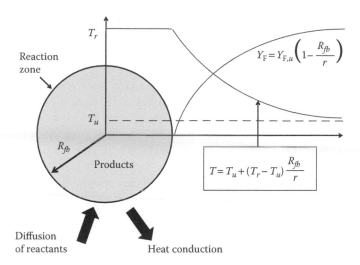

FIGURE 7.32 Flame ball in the case of a lean mixture.

section. Interested readers are referred to a book by Zel'dovich et al. (1985) and to review papers by Buckmaster (1993) and Ronney (1998).

The flame ball constitutes the asymptotically $(\Theta/T_b \to \infty)$ exact solution of stationary, 1D heat transfer and mass diffusion equations for the temperature and mass fraction of the deficient reactant, written in the spherical coordinate system.

If the activation temperature is very high, then the chemical reaction is confined to a thin reaction zone, outside which the stationary equations read

$$\frac{d}{dr}\left(r^m \rho a \frac{dT}{dr}\right) = 0 \tag{7.187}$$

and

$$\frac{d}{dr}\left(r^m \rho D_d \frac{dY_d}{dr}\right) = 0, \tag{7.188}$$

where the values 0, 1, and 2 of the power exponent m correspond to planar, cylindrical, and spherical flames, respectively. The convection terms vanish in Equations 7.187 and 7.188, because we consider a motionless mixture.

The solution to Equation 7.188 is as follows:

$$Y_d = Y_{d,1} + g_1 r \tag{7.189}$$

in the planar case ($m = 0$),

$$Y_d = Y_{d,2} + g_2 \ln r \tag{7.190}$$

in the cylindrical case ($m = 1$), and

$$Y_d = Y_{d,u}\left(1 - \frac{R_{fb}}{r}\right) \tag{7.191}$$

in the spherical case ($m = 2$). Here, $Y_{d,1}$, g_1, $Y_{d,2}$, and g_2 are constants and R_{fb} is the radius of the flame ball. Equations 7.189 and 7.190 yield $Y_d \to \infty$ if $g_k \neq 0$ and $r \to \infty$. Therefore, stationary combustion in a motionless mixture is impossible in the planar and cylindrical cases. However, in the spherical case, such a regime is theoretically admissible, as Equation 7.191 yields the physically correct result: $Y_d = Y_{d,u}$ at infinity and $Y_d = 0$ at the surface of the flame ball.

The solution to Equation 7.187 in the spherical case is as follows:

$$T = T_u + (T_r - T_u)\frac{R_{fb}}{r}. \tag{7.192}$$

Because the flux of chemical energy toward the thin reaction zone located at the surface of the flame ball should be equal to the heat flux from the zone, we obtain

$$Q\rho D_d \frac{Y_{d,u}}{R_{fb}} = Q\left(\rho D_d \frac{dY_d}{dr}\right)_{r=R_{fb}} = -\left(\rho a \frac{dT}{dr}\right)_{r=R_{fb}} = \rho a \frac{T_r - T_u}{R_{fb}} \tag{7.193}$$

by differentiating Equations 7.191 and 7.192. Here, Q is the heat of the reaction per unit mass of the deficient reactant. If $Le = 1$, Equation 7.193 yields $T_r = T_b$. If the Lewis number is lesser (greater) than unity, Equation 7.193 yields

$$T_r = T_u + \frac{T_b - T_u}{Le}. \tag{7.194}$$

Therefore, the temperature of the flame ball is higher (lower) than the adiabatic combustion temperature $T_b = T_u + QY_{d,u}$, because the flux of chemical energy toward the reaction zone is stronger (weaker) than the heat flux from it.

Thus, the temperature of the flame ball may be substantially different from the adiabatic combustion temperature if the Lewis number differs markedly from unity. Because the activation temperature of the combustion reaction is assumed to be high, such temperature variations strongly affect the burning rate at the ball surface. To evaluate the burning rate, the reaction–diffusion equation should be integrated over the reaction zone, as discussed elsewhere (Zel'dovich et al., 1985; Champion et al., 1986). If $\rho a = \rho_u a_u$ and the rate of reaction is proportional to the concentration of the deficient reactant, but does not depend on the concentration of the excess reactant, then we have

$$R_{fb} = \delta_L Le^{-1}\left(\frac{T_b}{T_r}\right)^{3/2}\exp\left(\frac{\Theta}{2T_b}\frac{T_b - T_r}{T_r}\right). \tag{7.195}$$

In the case of a small Lewis number, $T_r > T_b$ and the flame ball radius may be substantially smaller than the thickness of the unperturbed laminar flame.

Because the mass flux of the deficient reactant through the unit area of the surface of the flame ball scales as R_{fb}^{-1} (see Equation 7.193), the burning rate per unit surface of the flame ball, which is equal to the mass flux in the stationary case, is also proportional to R_{fb}^{-1}. Considering that $R_{fb} = \delta_L$ and $u_c = S_L^0$ in the case of Le = 1, the effect of Le on the burning rate has been proposed to be characterized using the following ratio (Betev et al., 1995; Karpov et al., 1997):

$$\frac{\tau_{fb}}{\tau_c^0} = \frac{S_L^0}{u_{fb}} = \frac{R_{fb}}{\delta_L} = \mathrm{Le}^{-1}\left(\frac{T_b}{T_r}\right)^{3/2}\exp\left(\frac{\Theta}{2T_b}\frac{T_b - T_r}{T_r}\right) \tag{7.196}$$

of a chemical timescale, τ_{fb}, associated with the flame ball, to the chemical timescale, $\tau_c^0 = a_u/(S_L^0)^2$, associated with the unperturbed laminar flame.

Equation 7.193 shows that the total (multiplied by $4\pi R_{fb}^2$) heat flux from the flame ball scales as R_{fb}. The total heat release in the reaction zone is proportional to the ball surface area, that is, it scales as R_{fb}^2, because the ball temperature and, hence, the reaction rate do not depend on its radius (see Equation 7.194). Therefore, the flame ball solution previously discussed is unstable (Zel'dovich et al., 1985; Champion et al., 1986). If the ball radius is increased, then the total heat release rate will be higher than the conductive heat losses, the conductivity will not be able to transfer all of the heat released in the reaction zone and the flame kernel will expand. If the ball radius is decreased, then the total rate of the heat release will be lower than the conductive heat losses, the ball will begin to get cold and will rapidly shrink.

Thus, the flame ball solution is associated with the highest possible curvature of a laminar flame that is not extinguished. Because the burning rate is increased by curvature if Le < 1, the flame ball is also associated with the highest possible local laminar burning rate for mixtures characterized by a small Lewis number. Consequently, it appears to be physically justified to use the chemical timescale τ_{fb} given by Equation 7.196 for characterizing the burning rate in the leading points. This approach offers an opportunity to estimate easily and quickly the magnitude of the effect discussed earlier using analytical Equations 7.194 and 7.196.

In summary, Equations 7.186 and 7.196 address the same phenomenon (a high local burning rate in critically curved leading flamelets in the case of Le < 1) but involve different methods to evaluate the magnitude of the effect. While Equation 7.196 is an asymptotically exact analytical expression resulting from a theory of a simplified phenomenon, Equation 7.186 requires numerical simulations of more realistic flames. Numerically, the values of the two chemical timescales, τ_{lp} and τ_{fb}, introduced earlier (see Equations 7.186 and 7.196, respectively) to characterize the local burning rate in the leading points, are of the same order if the two scales are calculated using the same physical–chemical parameters for the same reaction in the same mixture. For example, $\tau_{fb}/\tau_c^0 \approx 0.05$ and $\tau_{lp}/\tau_c^0 \approx 0.03$ were calculated by Betev et al. (1995) for the leanest hydrogen mixtures investigated by Karpov and Severin (1980).

Validation

Because the timescales introduced earlier to characterize the burning rate in the leading points depend very strongly (e.g., exponentially in Equations 7.194 and 7.196) on the difference between the Lewis number and unity, the use of either of the two timescales offers an opportunity to explain the strong effects of D_d on U_t, shown in Figures 4.20, 4.21, and 7.28.

For instance, Betev et al. (1995) and Karpov et al. (1996b, 1997) analyzed the entire experimental database obtained by Karpov and Severin (1977, 1978, 1980) and selected two reduced databases. The former ("standard") database encompasses burning velocities measured at moderate turbulence ($u' \leq u'_m$) in mixtures characterized by $D_F \approx D_O \approx a$ (e.g., ethane–air mixtures) and is aimed at showing the dependence of U_t on u'/S_L^0 and L/δ_L. The latter ("challenging") database, aimed at highlighting the strong effect of D_d on U_t, encompasses about 10 pairs of burning mixtures selected as follows: (i) mixture A is characterized by a higher S_L^0 and a lower D_d as compared with mixture B; (ii) Le > 1 in mixture A, while Le < 1 in mixture B; and (iii) the slope dU_t/du', measured at moderate turbulence, is markedly higher in mixture B than in mixture A, for example, compare the stoichiometric propane–air mixture (A) with a lean ($\Phi = 0.26$) hydrogen mixture (B) in Figure 4.21.

Both databases were processed using the same method reported in the section "Empirical Parameterizations for Turbulent Burning Velocity and Flame Speed" in Chapter 4 (see Equations 4.16 and 4.17). The standard database is reasonably well approximated by either Equation 4.18 or Equation 4.25 (see Figure 4.16). On the contrary, the results of an approximation of the challenging database using Equation 4.16 were unsatisfactory when the Damköhler number was evaluated using the unperturbed chemical timescale τ_c^0. In this case, Equation 4.17 yielded the minimum scatter for $m = 0$, $q = 2.1$, $b = 0.2 \pm 0.03$, and $d = 0.4 \pm 0.3$. However, even in this "best" case, the experimental data are strongly scattered around the line calculated using Equation 4.16 and show no clear trend (see Figure 7.33).

Substitution of the standard chemical timescale τ_c^0 in Da by the leading-point chemical timescale, τ_{lp}, determined by Equation 7.186, offers an opportunity to substantially improve the approximation of the challenging database by Equation 4.16 (see Figure 7.34). The scatter of the experimental points around the line ($m = 0$, $q = 0.6$, $b = -0.2 \pm 0.03$, and $d = 0.25 \pm 0.02$) is substantially reduced as compared with Figure 7.33. Moreover, the best approximation,

$$U_t \propto u'\mathrm{Da}_{lp}^{0.25}\,\mathrm{Re}_t^{-0.15} \propto u'\mathrm{Ka}_{lp}^{-0.25}\,\mathrm{Re}_t^{-0.025}, \tag{7.197}$$

where $\mathrm{Da}_{lp} = \tau_t/\tau_{lp}$ and $\mathrm{Ka}_{lp} = \sqrt{\mathrm{Re}_t}/\mathrm{Da}_{lp}$, is sufficiently close to Equations 4.18 and 4.25, which approximate the standard database. The similarity of the equations that approximate the two different databases indirectly supports both of these approximations.

Betev et al. (1995), Karpov et al. (1996b), and Lipatnikov and Chomiak (1998) also showed that the slopes dU_t/du' measured by Karpov and Severin (1980) correlated with the chemical timescale τ_{lp} that had been evaluated for highly curved

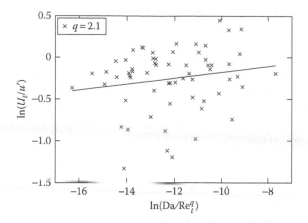

FIGURE 7.33 Approximation of the challenging database by Karpov and Severin (1980) using Equation 4.16 with the standard chemical timescale τ_c^0, $m = 0$ and $q = 2.1$. The symbols show the experimental data. The straight line approximates the data using least square fit.

leading flamelets using Equation 7.186 for mixtures included in the challenging database. The lower the chemical timescale τ_{lp}, the higher is the slope. For instance, let us consider the three mixtures shown in Figure 4.21. For the rich H_2–air and stoichiometric propane–air mixtures, $\tau_c^0 = 0.08$ and 0.15 ms, respectively. For a lean hydrogen–air mixture with $\Phi = 0.26$, the standard timescale is higher, $\tau_c^0 = 0.74$ ms. Therefore, a typical correlation associated with Equations 4.11 through 4.25, that is, the lower the chemical timescale, the higher is the slope, failed for the lean mixture, characterized by the steepest slope. On the contrary, the chemical timescale τ_{lp}, evaluated using Equation 7.186, is equal to 0.03 ms for the lean mixture,

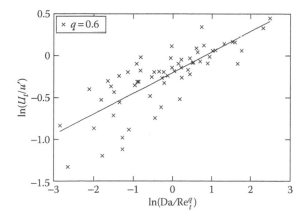

FIGURE 7.34 Approximation of the challenging database by Karpov and Severin (1980) invoking the timescale τ_{lp} computed for critically curved, expanding, spherical flames using Equation 7.186. The symbols show the experimental data. The straight line approximates the data using least square fit and $m = 0$.

while $\tau_{lp} = \tau_c^0$ for the other two mixtures. Thus, the aforementioned typical correlation works well if τ_c^0 is replaced with τ_{lp} for the lean hydrogen mixture, in line with the concept of leading points.

Finally, the entire database by Karpov and Severin (1977, 1978, 1980) was also processed using the previously discussed method. Equation 4.16 with τ_c^0 failed in satisfactorily approximating the database. Even in the best case of $m = 0$, $q = 0.9$, $b = 0.28 \pm 0.06$, and $d = 0.15 \pm 0.02$, the experimental data are substantially scattered (see Figure 7.35), with two clouds of the experimental points having two well-pronounced wings at low values of $Da/Re_t^{0.9}$. The wings are associated with burning velocities of the order of and much less than the rms turbulent velocity.

The use of the chemical timescale determined by Equation 7.186 has allowed us to approximate the database substantially better (see Figure 7.36). The scatter of the experimental data is significantly reduced and the aforementioned effect of the two wings has disappeared in Figure 7.36, where the data show a clear trend.

The minimum scatter Ξ (see Equation 4.17) has been obtained for $m = 0$, $q = 0.5$, $b = 0.34 \pm 0.02$, and $d = 0.32 \pm 0.01$, that is, the best approximation of the database is as follows:

$$U_t \propto u' Ka_{lp}^{-0.32}, \tag{7.198}$$

(see the squares and solid line in Figure 7.36). The triangles and dashed line in Figure 7.36 indicate that the analyzed database is also well approximated by a function of the Damköhler number, $Da_{lp} = \tau_t/\tau_{lp}$, calculated using Equation 7.186. When $m = 0$ and $q = 0$, the least square fit yields $b = -1.47 \pm 0.05$ and $d = 0.32 \pm 0.01$. Therefore, Equation 4.16 reduces itself to

$$U_t \propto u' Da_{lp}^{0.32}. \tag{7.199}$$

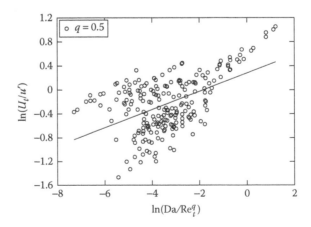

FIGURE 7.35 Approximation of the entire database by Karpov and Severin (1977, 1978, 1980) using Equation 4.16 with $m = 0$. The symbols show the experimental data. The straight line approximates the data using least square fit.

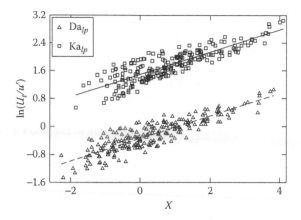

FIGURE 7.36 Approximation of the entire database by Karpov and Severin (1977, 1978, 1980) using Equation 4.16 with $m = 0$ and τ_{lp} computed for critically curved, expanding spherical flames using Equation 7.186. The symbols show the experimental data. The straight lines approximate the data using least square fit. The triangles and dashed line are associated with $q = 0$ in Equation 4.16 and $X = \ln(\text{Da}) - 3$. The squares and solid line are associated with $q = 0.5$ in Equation 4.16, that is, $X = -\ln(\text{Ka})$. The circles and solid line are shifted along the ordinate axis and correspond to $\ln(U_t/u') + 2$. (Reprinted from *Progress in Energy and Combustion Science*, 31, Lipatnikov, A.N. and Chomiak, J., Molecular transport effects on turbulent flame propagation and structure, 1–73, Copyright 2005, with permission from Elsevier.)

Similar results with a slightly larger scatter were obtained when invoking the timescale τ_{fb} calculated using Equation 7.196 for the flame ball (Betev et al., 1995; Karpov et al., 1997).

The scatter of the experimental data is still substantial in Figure 7.36 due to a number of causes:

- The chemical timescale τ_{lp} is assumed to be equal to τ_c^0 if Lc > 1.
- Combustion chemistry has been strongly simplified (a single-step chemistry with the activation temperature being independent of the equivalence ratio) in the discussed simulations of highly curved, expanding spherical flames.
- The approach does not address eventual local flamelet quenching by turbulent eddies.
- Equations 7.198 and 7.199 are not valid at low u'/S_L^0, that is, for mixtures with a high S_L^0, for example, rich hydrogen–air mixtures, because these equations lead to $U_t \to 0$ as $u' \to 0$.
- The raw experimental data are scattered.
- The values of U_t, reported by Karpov and Severin (1977, 1978, 1980), were obtained by approximating raw $R_f(t)$ curves with straight lines in the range of $15 < R_f < 40$ mm, that is, the transient behavior of U_t was disregarded.
- The values of S_L^0, reported by Karpov and Severin (1977, 1978, 1980), were obtained without any consideration of the effect of the stretch rates on the observed speed of expanding spherical laminar flames (see the sections

"Strained Laminar Premixed Flames" "Curved Laminar Premixed Flames," and "A Brief Summary of Theory of Weakly Stretched Laminar Premixed Flames" in Chapter 5).

Nevertheless, Figures 7.33 through 7.36 seem to be sufficient in highlighting the crucial role played by strongly curved flamelets in premixed turbulent flame propagation.

It is worth remembering that an expression similar to Equation 7.198 in the case of Le ≈ 1 has been obtained by Bradley et al. (1992) and a similar Equation 4.18 is highlighted in the section "Empirical Parameterizations for Turbulent Burning Velocity and Flame Speed" in Chapter 4 for mixtures associated with $D_F \approx D_O \approx a$. The universal applicability of Equation 4.18 or 7.198 has been achieved by using the chemical timescale τ_{lp}, which is 40–50 times smaller than τ_c^0 in very lean hydrogen mixtures. Because no tuning parameters have been invoked to simulate this very strong effect, the correlations obtained are unlikely to be fortuitous. Even considering the scatter of the experimental data in Figure 7.36, such results should not be underestimated, especially as alternative models discussed in the previous sections do not seem to be able to yield the strong effect of D_d on U_t.

Finally, a recent study by Venkateswaran et al. (2011) lent further support to the leading point concept. In the cited work, global turbulent burning velocities were obtained from lean H_2/CO–air and methane–air Bunsen flames under a wide range of conditions ($1 < u'/S_L^0 < 100$, the volume percentage of hydrogen in the fuel blend was changed from 30 to 90, various equivalence ratios, etc.). The burning velocities measured by keeping both S_L^0 and u' constant indicated a significant increase in U_t by the volume percentage of hydrogen in H_2–CO mixtures, with these effects clearly persisting "over the entire range of turbulence intensities used in measurements." The authors tried to parameterize the experimental data either by using the unperturbed laminar flame speed as the main mixture characteristic or by substituting S_L^0 with the maximum laminar flame speed $\max\{S_L(\dot{s})\}$ computed for two symmetrical counterflow flames, sketched in Figure 5.20, at various stretch rates $\dot{s}$. The substitution of S_L^0 with $\max\{S_L(\dot{s})\}$ was based on the leading point concept and allowed the authors to substantially improve the parameterization of the entire experimental database. Even if the maximum burning rate reached in stretched counterflow laminar flames is lower than the highest burning rate reached in critically curved laminar flames (cf. the upper edges of the solid and long-dashed lines in Figure 7.30), the use of $\max\{S_L(\dot{s})\}$ is a reasonable simplification that is conceptually in line with the leading point concept. Accordingly, the parameterization of the extensive database, achieved by Venkateswaran et al. (2011), seems to support the concept.

Concept of Critically Curved Leading Points in RANS Simulations

The chemical timescales previously introduced for characterizing critically curved flamelets at the leading edge can be used straightforwardly in numerical simulations of premixed turbulent combustion by extending the TFC or FSC models. For this purpose, the unperturbed chemical timescale τ_c^0 in the submodel for turbulent burning velocity (e.g., Equation 7.13 or 7.29) should be replaced with either τ_{fb} calculated using Equation 7.196 or τ_{lp} evaluated using Equation 7.186.

For instance, to simulate the analyzed experiments, Karpov et al. (1996a) extended the Zimont model by invoking the following expression:

$$U_t = Au' Da_{lp}^{1/4}\left(1-P_q\right) = Au'\left(\frac{\tau_t}{\tau_{lp}}\right)^{1/4}\left(1-P_q\right) = Au' Da^{1/4}\left(\frac{\tau_c^0}{\tau_{lp}}\right)^{1/4}\left(1-P_q\right), \quad (7.200)$$

where A is a constant, the timescale τ_{lp} is evaluated using Equation 7.186, and the probability of flamelet quenching, P_q, is calculated using Equation 7.178.

For each particular mixture and each particular value of the rms turbulent velocity ($L = 10$ mm was not varied in the discussed measurements), three simulations were performed by Karpov et al. (1996a). First, the same value of the constant $A = 0.5$ and the same value of the quenching probability $P_q = 0$ were used for all mixtures studied. It is worth emphasizing that, because no adjustable parameters were invoked to calculate the chemical timescale in Equation 7.186, the computed burning velocities were affected by a single model constant A, which was the real constant, that is, it was the same for all mixtures.

The results of the simulations are shown as dashed curves in Figures 7.8 and 7.37. For most mixtures (with the exception of the lean $H_2/O_2/N_2$ mixture), the numerical results agree very well with the experimental data (see symbols) at moderate turbulence, that is, at values of u' such that an increase in U_t by u' is well pronounced. However, these simulations have not yielded a decrease in U_t by u' at strong turbulence (e.g., see the circles at $u' > 2$ m/s and the crosses at $u' > 4$ m/s in Figure 7.37).

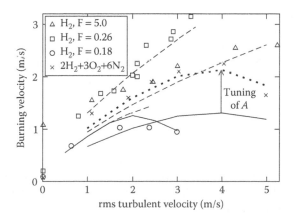

FIGURE 7.37 Turbulent burning velocities measured (symbols) by Karpov and Severin (1980) and computed (curves) by Karpov et al. (1996a) using the extended TFC model, that is, Equations 7.186 and 7.200 with either $P_q = 0$ (dashed curves) or P_q determined using Equation 7.178 (solid and dotted curves). All curves have been computed using $A = 0.5$ with the exception of a dotted curve, which has been obtained for $A = 0.67$.

Second, to obtain a decrease in U_t by u' at strong turbulence, simulations were also performed using $A = 0.5$ and P_q evaluated from Equations 7.178 and 7.179, with ε_q in the latter equation being adjusted in order for the maxima of the measured and computed $U_t(u')$ curves to be observed at approximately the same rms turbulent velocity u'_m. The results of these simulations are shown as solid curves. The adjustment of ε_q notably affected the computed burning velocities only at u' comparable with u'_m, whereas the adjustment did not affect the $U_t(u')$ curves at less intense turbulence. In certain cases, for example, the rich mixture in Figure 7.37, the results computed either setting $P_q = 0$ or using Equations 7.178 and 7.179 were practically the same. In such cases, solid curves are not shown in Figure 7.37.

Third, although almost all of the experimental data used in these numerical tests were quantitatively predicted for $u' < u'_m$ by using τ_{lp} and the same value of $A = 0.5$ (cf. the dashed curves with symbols in Figure 7.37), in a few mixtures, it was necessary to tune A to obtain a quantitative agreement between the computed and the measured turbulent burning velocities (see the dotted curve and arrow in Figure 7.37).

Figure 7.38 shows the best fitted values of A, associated with either the standard chemical timescale τ_c^0 (a) or the chemical timescale τ_{lp} (b) that characterizes critically curved leading flamelets. When employing τ_c^0, the constant A increases strongly with decreasing Le. By contrast, the adjusted values of A are slightly scattered around 0.5 if τ_{lp} is invoked.

In summary, the leading point concept, supplemented with the aforementioned submodel of τ_{lp} for critically curved flamelets, offers an opportunity not only to explain the strong effect of D_d on U_t at moderate turbulence but also to quantitatively predict it in RANS simulations.

Highly Curved Leading Points and Flamelet Tip in a Vortex Tube

To the best of the author's knowledge, the predominance of strongly curved flamelets at the leading edge of a turbulent flame brush has not yet been confirmed experimentally or in DNS, probably due to (i) the small size of the strongly curved parts of flamelets and (ii) the difficulties in encompassing appropriate statistics at $\tilde{c} \to 0$. The only relevant, indirect, well-documented evidence consists of the predominance of positive curvature at low $\tilde{c}$ (Trouvé and Poinsot, 1994; Veynante et al., 1994; Ashurst and Shepherd, 1997; Kostiuk et al., 1999), but this trend seems to be obvious for purely geometrical reasons. Certainly, a lack of direct experimental evidence impedes the recognition of an important role played by highly curved flamelets in the propagation of premixed turbulent flames. We may note, however, that almost all DNS and experimental studies of premixed turbulent combustion have yielded data obtained well inside the flame brush, but have not yet paid particular attention to the leading edge of the flame brush.

Moreover, the physical mechanism responsible for the predominance of highly curved flamelets at the leading edge of a turbulent flame brush is not clear. Because (i) U_t is of the order of u' and much higher than S_L^0 in many of the foregoing experiments (see Figures 4.20, 4.21, and 7.28) and (ii) the speed of the leading edge is equal to (in a fully developed flame) or even higher than (in a developing flame with increasing thickness) U_t, the leading points should move at a speed much higher

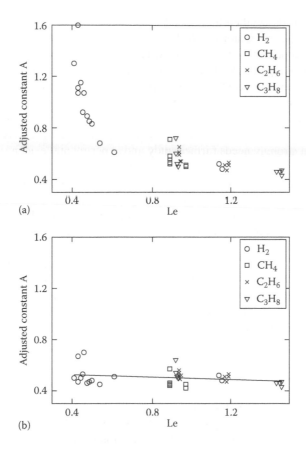

FIGURE 7.38 Adjusted values of the constant A in the extended TFC model, associated with either (a) Equation 7.119 or (b) Equation 7.200. (Adapted from *Progress in Energy and Combustion Science*, 31, Lipatnikov, A.N. and Chomiak, J., Molecular transport effects on turbulent flame propagation and structure, 1–73, Copyright 2005, with permission from Elsevier.)

than S_L^0 but retain a laminar structure in order to be affected by D_F/D_O and Le. Consequently, strongly curved burning zones having an inherently laminar structure and propagating at a high speed mainly controlled by turbulence should be introduced in order to construct a self-consistent leading-point model capable of predicting the strong effect of D_d on U_t.

The author is aware of the sole candidate for a laminar flame zone submodel of the kind outlined earlier—a laminar flame propagating along a vortex tube. Such flames were discussed in the section "Flame Propagation along a Laminar Vortex" in Chapter 5. Subsequently, a physical mechanism of premixed turbulent combustion that (i) highlights laminar flamelets propagating along vortex tubes and (ii) is capable of explaining the strong effect of D_d on U_t in turbulent flames in a consistent manner was hypothesized in the section "Vortex Tubes in Turbulent Flows and Premixed Turbulent Flames" in Chapter 5. Flamelets in a vortex tube may easily be associated with the leading points, because such flamelets may advance furthest into the unburned gas due to their high speeds. The ability to explain the strong effect of D_d on U_t in a consistent manner is the key merit of this hypothesis as compared with the alternative approaches available today. However, a model that highlights flamelet propagation along vortex tubes and yields an expression for the turbulent burning velocity that is consistent with the experimental data addressed in the section "Turbulent Flame Speed and Burning Velocity" in Chapter 4 has not yet been developed. Thus, the discussed association between the structure of the leading points and the highly curved tip of a flamelet propagating in a vortex tube is an attractive hypothesis that strongly needs further study and experimental evidence.

Certainly, the behavior of such a coarse characteristic as the turbulent burning velocity is insufficient to establish the governing physical mechanism of premixed turbulent combustion, but the existence of a paradox not resolved by most known approaches appears to be quite sufficient to point out a certain physical mechanism as the potentially governing one, based on the ability of this mechanism to resolve the paradox.

The strong effect of D_d on U_t should definitely be associated with an important role played by some thin, inherently laminar, reaction zones in the propagation of turbulent flames. Because the characteristics of the different types of such zones are substantially different and depend strongly on D_F/D_O and Le, the magnitude of the effect of D_d on U_t, predicted by different models, should substantially depend on the type of zone that the model highlights. This distinction offers an opportunity to select from a range of potentially important zones and to gain insight into the basic mechanism of premixed turbulent combustion.

Leading Points and KPP Problem

So far, we have discussed the physical arguments that highlight the leading points. This hypothesis also has a rigorous mathematical foundation. In their seminar paper, Kolmogorov et al. (1937) studied a biological problem and proved that the balance equation

$$S\frac{dc}{dx} = D\frac{d^2c}{dx^2} + \Omega \tag{7.201}$$

with (i) a constant positive diffusivity D; (ii) a nonlinear source term $\Omega = \Omega(c) \geq 0$, such that

$$\Omega(c = 0) = \Omega(c = 1) = 0, \tag{7.202}$$

$$\Omega(0 < c < 1) > 0, \tag{7.203}$$

and

$$0 < \frac{d\Omega}{dc}(0 < c \leq 1) < \frac{d\Omega}{dc}(c = 0); \tag{7.204}$$

and (iii) the boundary conditions:

$$c(-\infty) = 0,$$
$$c(\infty) = 1$$

(7.205)

has a solution, provided that the speed S satisfies the following constraint:

$$S \geq 2\sqrt{D\frac{d\Omega}{dc}\bigg|_{c=0}}.$$

(7.206)

If a nonsteady problem, modeled by Equations 7.202 through 7.205 and the balance equation

$$\frac{\partial c}{\partial t} + S\frac{\partial c}{\partial x} = D\frac{\partial^2 c}{\partial x^2} + \Omega$$

(7.207)

is considered, then for a wide set of initial conditions, for example, $c(x,t = 0) = 0$ if $x \leq x_1$, $0 < c(x,t = 0) < 1$ if $x_1 < x < x_2$, and $c(x,t = 0) = 1$ if $x \geq x_2$, the solution tends to $c = c(x + St)$ (Kolmogorov et al., 1937; Ebert and van Saarloos, 2000) with

$$S = 2\sqrt{D\frac{d\Omega}{dc}\bigg|_{c=0}}.$$

(7.208)

This outstanding mathematical result is widely used in different fields of contemporary science and is called the KPP problem in honor of Kolmogorov, Petrovsky, and Piskounov. Zel'dovich (1980) drew the attention of the combustion community to the KPP problem, while Hakberg and Gosman (1984) first applied Equation 7.208 to premixed turbulent combustion. Indeed, if c, S, D, and Ω are associated with the mean combustion progress variable, the turbulent burning velocity, turbulent diffusivity, and normalized (with ρ_u) mean reaction rate, respectively, then Equation 7.208 yields an exact analytical expression for evaluating the fully developed turbulent burning velocity in the statistically planar, 1D case, provided that Equations 7.202 through 7.204 hold. This is so for the most popular algebraic closure relations for the mean reaction rate, such as $\tau_f\bar{W} \propto \rho_u\bar{c}(1-\bar{c})$ or $\tau_f\bar{W} \propto \rho_u\tilde{c}(1-\tilde{c})$.

In the turbulent combustion literature, Equation 7.208 is commonly used to test various closure relations for $\bar{W}$, by substituting them into the RHS and comparing the calculated turbulent burning velocities with the experimental data obtained from laboratory flames (Hakberg and Gosman, 1984; Bray, 1990; Duclos et al., 1993; Kolla et al., 2010). Although such an assessment of the closure relations appears to be useful as far as qualitative trends are concerned, a quantitative comparison of the turbulent burning velocities calculated for fully developed, statistically planar, 1D flames with measured speeds of the developing and curved flames in divergent mean flows does not seem to be a correct test for a number of reasons discussed in the section "Determination of Turbulent Flame Speed and Burning Velocity" in Chapter 4 and in Section 5.2 of a recent review paper by Lipatnikov and Chomiak (2010), as well as in other papers (e.g., Lipatnikov and Chomiak, 2004a, 2007).

Here, it is worth stressing another feature of the KPP problem; Equation 7.208 straightforwardly shows that, at least for a fully developed turbulent flame with gradient scalar transport, the burning velocity is controlled by the processes at the leading edge of the flame brush ($\bar{c} \to 0$). Thus, the KPP solution supports the concept of leading points from the mathematical standpoint.

CHEMISTRY IN RANS SIMULATIONS OF PREMIXED TURBULENT COMBUSTION: FLAMELET LIBRARY

All the models discussed so far in this chapter are restricted to simulating the burning rate and mean structure of a premixed turbulent flame, that is, the spatial distributions of the mean temperature, density, and reactant mass fractions. The models do not care about chemical kinetics but assume that the influence of chemistry on a premixed turbulent flame may, in a first approximation, be addressed using the correct laminar flame speed, which depends on the chemistry.

Such a simplification was experimentally supported by Burluka et al. (2009), who investigated the expansion of statistically spherical turbulent flames in a Leeds fan-stirred bomb. These authors studied not only hydrocarbon–air flames but also flames of di-t-butyl-peroxide (DTBP) decomposition, with the latter flames being associated with a much simpler chemistry and a lower heat release as compared with the former flames. Despite significant differences in chemical kinetics and exothermicity, similar $U_t(u')$ curves were reported for the two kinds of flames, provided that they were characterized by approximately the same laminar flame speeds and approximately the same Lewis numbers.

However, because the contemporary society pays paramount attention to environmental issues and, in particular, to clean and effective combustion technologies, the industry demands much more from the turbulent combustion community. To develop future combustion engines, the formations of pollutants, for example, NO, CO, unburned hydrocarbons, and soot, in turbulent flames should be well understood and predictable. Accordingly, the modeling of turbulent combustion should not be limited to the burning rate and mean flame structure, but the influence of turbulence on the rates of various reactions must also be the focus of model development.

Unfortunately, our abilities to predict the mean rate of pollutant formation in a premixed turbulent flame are still very limited, at least as far as combustion in the flamelet regime is concerned. In this regime, a local value of temperature randomly jumps from T_u to T_b and then back to T_u (see Figure 5.7), and such strong temperature fluctuations substantially impede the evaluation of a mean reaction rate if the activation temperatures of the relevant reactions are markedly higher than T_b (see the section "A Key Problem").

The so-called flamelet library concept (Peters, 1986) appears to be the most attractive approach to modeling combustion chemistry in the flamelet regime of premixed turbulent burning. Very briefly, the key idea of the concept may be summarized as follows. If the chemical reactions are assumed to be confined to thin zones that retain the structure of stretched laminar flames, then for any species S, one can

- Preliminarily simulate a representative set of variously stretched laminar flames using detailed chemistry.
- Process the computed results to obtain a library of the various characteristics of these flames, for example, the spatial profiles $W_S(x, \dot{s})$ of the rate of production/consumption of species S or the integrated rate $Q_S(\dot{s}) \equiv \int_{-\infty}^{\infty} W_S(x, \dot{s}) dx$ versus the stretch rate.
- Average these characteristics using a parameterization of the PDF $P(\dot{s})$ for the stretch rate within the turbulent flame brush, for example,

$$\left(\bar{Q}_S\right)_f \left(\text{fuel}, \Phi, T_u, p\right) = \int_0^1 \int_{\dot{s}_{q-}}^{\dot{s}_{q+}} W_S\left(\text{fuel}, \Phi, T_u, p, c, \dot{s}\right) P\left(c, \dot{s}\right) d\dot{s} dc \qquad (7.209)$$

or

$$\left(\bar{Q}_S\right)_f \left(\text{fuel}, \Phi, T_u, p\right) \approx \int_{\dot{s}_{q-}}^{\dot{s}_{q+}} Q_S\left(\text{fuel}, \Phi, T_u, p, \dot{s}\right) P_s\left(\dot{s}\right) d\dot{s}. \qquad (7.210)$$

Here, integration is performed over the range $\dot{s}_{q-} < \dot{s}_q < \dot{s}_{q+}$ of compression ($\dot{s}_{q-} < \dot{s} < 0$) and stretch ($0 < \dot{s} < \dot{s}_{q+}$) rates that do not quench the flamelet.

- Finally, within the framework of the BML approach (see the section "Bray–Moss–Libby Approach" in Chapter 5),

$$\bar{W}_S = \left(1 - \bar{c}\right) \bar{W}_{S,u} + \bar{c} \bar{W}_{S,b} + \gamma \left(\bar{Q}_S\right)_f. \qquad (7.211)$$

Equation 7.211 is still not closed, because it involves four unknown quantities: the production/consumption rates $\bar{W}_{S,u}$ and $\bar{W}_{S,b}$ conditioned on the unburned and burned mixtures, respectively, the probability γ of finding flamelets, and the PDF $P(c, \dot{s})$. For many species, $\bar{W}_{S,u} = \bar{W}_{S,b} = 0$. If, however, either $\bar{W}_{S,b} \neq 0$, for example, the formation of thermal NO according to the Zel'dovich mechanism (see reactions XX and XXI in the section "Chemical Reactions in Flames" in Chapter 1), or $\bar{W}_{S,u} \neq 0$, for example, knock in car engines due to autoignition of the unburned gas, caused by chemical reactions responsible for the low-temperature oxidation of hydrocarbons, then in a first approximation, the rates $\bar{W}_{S,u}$ and $\bar{W}_{S,b}$ may be estimated using T_b or T_u, ρ_b or ρ_u, and the mean values of the relevant mass fractions, conditioned on the unburned or burned mixtures, respectively. In such a case, the turbulent fluctuations in T, ρ, and Y_k in the burned and/or unburned mixtures are neglected, but such a simplification seems to be justified, because these fluctuations are much lower than the fluctuations in T, ρ, and Y_k, caused by the flamelet motion (unburned–burned intermittency).

As regards the probability γ, it may be estimated by applying Equation 7.211 to the combustion progress variable. Since $\bar{W}_{c,u} = \bar{W}_{c,b} = 0$, Equation 7.211 applied to c yields

$$\gamma = \frac{\bar{W}}{\left(\bar{Q}_c\right)_f},\tag{7.212}$$

where the integrated rate $(\bar{Q}_c)_f$ is preliminarily calculated using the flamelet library and Equation 7.209 or 7.210. Thus, if a model of premixed turbulent combustion that yields the mean rate $\bar{W}$ of product formation is selected, then a mean rate $\bar{W}_S$ could be evaluated (i) using Equations 7.209 through 7.212, (ii) using the flamelet library for $(\bar{Q}_c)_f$ and $(\bar{Q}_S)_f$, and (iii) neglecting the influence of the turbulent fluctuations on $\bar{W}_{S,u}$ and $\bar{W}_{S,b}$ if these conditioned rates do not vanish.

Finally, the PDF $P(c,\dot{s})$ is commonly modeled as $P(c,\dot{s}) \approx P_s(\dot{s})P_c(c)$. A simple parameterization of the PDF $P_s(\dot{s})$ is reported in the section "Concept of Critical Stretch Rate (Flamelet Quenching by Strong Perturbations)" (see Equation 7.175), and more sophisticated parameterizations are discussed by Bradley et al. (2003a, 2005). The simplest model of the PDF $P_c(c) = (dc/d\zeta)_L^{-1}$, proposed to be used by Bray and Moss (1977), is based on the assumption that flamelets statistically retain the structure of an unperturbed laminar flame in a turbulent flow. Here, ζ is the spatial distance normal to the unperturbed laminar flame and the derivative $(dc/d\zeta)_L^{-1}$ is computed in the latter flame. If such a submodel of $P_c(c)$ is invoked, then averaging of a quantity $\int_0^1 q(c)P_c(c)dc$ is reduced to integration of this quantity $\int_{-\infty}^{\infty} q_L(\zeta)d\zeta$ over the unperturbed laminar flame.

Certainly, the approach outlined earlier requires a lot of simulations, as (i) the flamelet library should be computed and saved in look-up tables for various fuels, equivalence ratios, pressures, temperatures, and stretch rates and (ii) such huge look-up tables should be used by a CFD code in many grid points. Therefore, for many practical applications, the approach could be simplified by neglecting the influence of flamelet stretching on $\bar{W}_S$, that is, by replacing $P(\dot{s})$ in Equation 7.210 with the Dirac delta function $\delta(\dot{s})$. Indeed, considering the strong sensitivity of the characteristics of stretched laminar premixed flames to the transient effects and flame geometry (see the section "Local Quenching of Premixed Combustion in Turbulent Flows" in Chapter 5), an attempt to model the dependence of $\bar{W}_S$ on the mean stretch rate by invoking the dependencies of $Q_S(\dot{s})$ computed for steadily stretched laminar flames of a single geometrical configuration does not seem to be justified. The use of values of Q_S calculated for unperturbed laminar flames with detailed chemistry appears to be an acceptable compromise between physics and numerics.

The foregoing discussion addresses the flamelet regime of premixed turbulent combustion, characterized by the strongest temperature fluctuations. In the opposite limit case of weak temperature fluctuations, the mean reaction rates could be evaluated using the mean values of temperature, density, and species mass fractions. Such a simplification appears to be reasonable, for example, when simulating autoignition in turbulent flows, because a computed ignition delay is much more sensitive to a chemical mechanism than to a turbulence submodel.

In an intermediate case of nonflamelet combustion, but still large fluctuations in Θ/T, the mean reaction rates are sometimes evaluated invoking a presumed PDF, for example, the beta function approximation given by Equation 7.53 or the sum of two Dirac delta functions (see Equation 7.57). However, such an approach, as discussed in the section "Presumed PDF Models," strongly needs basic substantiation, validation, and further development. In particular, a choice of the shape of a presumed PDF should be supported by physical and/or mathematical arguments.

Because combustion of a typical hydrocarbon fuel involves hundreds of species and thousands of reactions, reduction of combustion chemistry is a hot topic of contemporary research into turbulent flames. Reduction of chemical mechanisms is commonly performed invoking partial-equilibrium and steady-state approximations (see the section "Chemical Reactions" in Chapter 1). This can be done both analytically for simpler chemical schemes (Peters and Williams, 1987; Peters, 1994; Seshadri, 1996; Williams, 2000) and numerically using several advanced tools developed over the past two decades, for example, the computational singular perturbation (CSP) method (Lam and Goussis, 1991), intrinsic low-dimensional manifold (ILDM) approach (Maas and Pope, 1992), flame prolongation (FPI) of ILDM (Gicquel et al., 2000), flamelet-generated manifold (FGM) (van Oijen et al., 2001), phase-space ILDM (PS-ILDM) concept (Bongers et al., 2002), and self-similar FPI (Ribert et al., 2006). Because these methods are solely aimed at reducing combustion chemistry and do not contribute to modeling the influence of turbulence on the mean reaction rates, a discussion on the enumerated tools is beyond the scope of this book and interested readers are referred to the aforementioned papers. It is worth noting, however, that a tool chosen to reduce combustion chemistry should be consistent with a model used to evaluate the mean reaction rate in a turbulent flow. For instance, the FPI and FGM methods based on the flamelet concept do not seem to be consistent with the presumed PDF models given by Equations 7.53 and 7.57, because these models are flawed in the flamelet regime of premixed turbulent combustion, that is, if $\overline{c'^2} \to \overline{c}(1 - \overline{c})$, as discussed in the section "Presumed PDF Models."

LARGE EDDY SIMULATION OF PREMIXED TURBULENT COMBUSTION

The previous discussion in this chapter dealt with the so-called RANS models of premixed turbulent combustion, that is, models that address mean quantities (either spatially averaged or time-averaged or ensemble-averaged). Such an approach does not resolve turbulent eddies and, therefore, has to model flame–turbulence interaction at all length scales. However, even if the behavior of small-scale eddies in various turbulent flows is often considered to be universal and well described by the Kolmogorov theory, the statistical characteristics of the large-scale eddies depends substantially on the nonuniformities and nonstationarity of the mean flow, that is, on boundary and initial conditions. Consequently, it seems to be difficult to obtain a unified RANS model that describes well various substantially different turbulent flows without tuning. To develop a more unified method for simulating turbulent flows, one could try to resolve large-scale eddies in numerical simulations and to restrict modeling to more universal small-scale phenomena. The large eddy simulation (LES) approach is an attempt to realize the aforementioned idea, that is, to model small-scale turbulent eddies and to numerically resolve large-scale unsteady turbulent eddies.

Contrary to RANS methods, LES deals with filtered quantities $\hat{q}$ defined as follows:

$$\hat{q}(\mathbf{x},t) = \int q(\mathbf{x}-\mathbf{r},t)\Pi(\mathbf{x},\mathbf{r},t)d\mathbf{r}, \qquad (7.213)$$

where integration is performed over the entire flow domain and $\Pi(\mathbf{x},\mathbf{r},t)$ is a filter function such that

$$\int \Pi(\mathbf{x},\mathbf{r},t)d\mathbf{r} = 1. \qquad (7.214)$$

In order for filtering to commute with spatial derivatives, the filter function should be uniform (Pope, 2000), that is, $\Pi(\mathbf{x},\mathbf{r},t) = \Pi(\mathbf{r},t)$. In order to resolve large-scale phenomena, the filter function should be localized either in the physical or in the wave number space, with the filter size Δ being much smaller than the integral turbulence length scales. The so-called box filter $\Pi(r) = \Delta^{-1}H(0.5\Delta - |r|)$ and the sharp spectral filter $\Pi(r) = \sin(\pi r/\Delta)/(\pi r)$ are 1D examples of the two types of filters, respectively (Pope, 2000). Here, H is the Heaviside function.

Because filtering commutes with time and spatial derivatives in the case of a uniform filter, the basic Equations 1.49, 1.51, and 1.79 of the premixed combustion theory can easily be filtered. The filtered continuity

$$\frac{\partial\hat{\rho}}{\partial t} + \frac{\partial}{\partial x_k}\left(\hat{\rho}\breve{u}_k\right) = 0, \qquad (7.215)$$

Navier–Stokes

$$\frac{\partial}{\partial t}\left(\hat{\rho}\breve{u}_i\right) + \frac{\partial}{\partial x_k}\left(\hat{\rho}\breve{u}_k\breve{u}_i\right) + \frac{\partial\hat{p}}{\partial x_i} = -\frac{\partial}{\partial x_k}\left(\langle\rho u_k u_i\rangle - \hat{\rho}\breve{u}_k\breve{u}_i\right) + \frac{\partial\hat{\tau}_{ik}}{\partial x_k}, \qquad (7.216)$$

and combustion progress variable balance

$$\frac{\partial}{\partial t}\left(\hat{\rho}\breve{c}\right) + \frac{\partial}{\partial x_k}\left(\hat{\rho}\breve{u}_k\breve{c}\right) = -\frac{\partial}{\partial x_k}\left(\langle\rho u_k c\rangle - \hat{\rho}\breve{u}_k\breve{c}\right) + \frac{\partial}{\partial x_k}\left\langle\rho D\frac{\partial c}{\partial x_k}\right\rangle + \hat{W} \qquad (7.217)$$

equations look similar to their counterpart Favre-averaged Equations 7.34 through 7.36. Here, $\breve{q}$ is the density-weighted value of a quantity q, that is,

$$\breve{q} \equiv \frac{1}{\hat{\rho}} \int \rho q \Pi d\mathbf{r}, \qquad (7.218)$$

similar to the Favre-averaged $\tilde{q}$, and $\langle ab \rangle$ designates the filtered product ab, that is, $\langle ab \rangle \equiv \hat{q}$ if $q = ab$. Here and in the following text, the dependencies of q and Π on the time and spatial coordinates are not specified, for the sake of brevity.

Like Equations 7.35 and 7.36, the RHSs of Equations 7.216 and 7.217 are not closed. The transport terms on the RHSs of the two sets of equations are written in different forms for reasons that will be discussed later.

An analogy between averaging and filtering may be further extended by introducing conditionally filtered quantities

$$(1-\hat{c})\hat{q}_u \equiv \int (1-c) q \Pi d\mathbf{r} \qquad (7.219)$$

and

$$\hat{c}\hat{q}_u \equiv \int c q \Pi d\mathbf{r} \qquad (7.220)$$

and obtaining the following BML-like (see the section "Bray–Moss–Libby Approach" in Chapter 5) equations:

$$\hat{\rho}\breve{c} = \int \rho c \Pi d\mathbf{r} = \rho_b \int c \Pi d\mathbf{r} = \rho_b \hat{c}, \qquad (7.221)$$

$$\hat{\rho}(1-\breve{c}) = \int \rho(1-c)\Pi d\mathbf{r} = \rho_u \int (1-c)\Pi d\mathbf{r} = \rho_u (1-\hat{c}), \qquad (7.222)$$

$$\hat{\rho} = \frac{\rho_u \hat{c}}{\sigma \breve{c}} = \frac{\rho_u}{1+(\sigma-1)\breve{c}}, \qquad (7.223)$$

$$\hat{\mathbf{u}} = \int (1-c+c)\mathbf{u}\Pi d\mathbf{r} = (1-\hat{c})\hat{\mathbf{u}}_u + \hat{c}\hat{\mathbf{u}}_b, \qquad (7.224)$$

$$\hat{\rho}\breve{\mathbf{u}} = \int (1-c+c)\rho\mathbf{u}\Pi d\mathbf{r} = \int (1-c)\rho\mathbf{u}\Pi d\mathbf{r} + \int c\rho\mathbf{u}\Pi d\mathbf{r}$$

$$= \rho_u(1-\hat{c})\hat{\mathbf{u}}_u + \rho_b\hat{c}\hat{\mathbf{u}}_b = \hat{\rho}\left[(1-\breve{c})\hat{\mathbf{u}}_u + \breve{c}\hat{\mathbf{u}}_b\right], \qquad (7.225)$$

$$\hat{\mathbf{u}} - \breve{\mathbf{u}} = (1 - \hat{c})\hat{\mathbf{u}}_u + \hat{c}\hat{\mathbf{u}}_b - (1 - \breve{c})\hat{\mathbf{u}}_u - \breve{c}\hat{\mathbf{u}}_b = -(\hat{\mathbf{u}}_b - \hat{\mathbf{u}}_u)(\breve{c} - \hat{c})$$

$$= \frac{(\sigma - 1)\hat{\rho}}{\rho_u}\breve{c}(1 - \breve{c})(\hat{\mathbf{u}}_b - \hat{\mathbf{u}}_u), \tag{7.226}$$

and

$$\langle \rho \mathbf{u}c \rangle = \int \rho \mathbf{u}c \Pi d\mathbf{r} = \rho_b \int \mathbf{u}c \Pi d\mathbf{r} = \rho_b \hat{c}\hat{\mathbf{u}}_b = \hat{\rho}\breve{\mathbf{u}} - \hat{\rho}(1 - \breve{c})\hat{\mathbf{u}}_u$$

$$= \hat{\rho}\breve{c}\breve{\mathbf{u}} + \hat{\rho}(1 - \breve{c})(\breve{\mathbf{u}} - \hat{\mathbf{u}}_u) = \hat{\rho}\breve{c}\breve{\mathbf{u}} + \hat{\rho}\breve{c}(1 - \breve{c})(\hat{\mathbf{u}}_b - \hat{\mathbf{u}}_u). \tag{7.227}$$

Nevertheless, despite the foregoing similarities between filtering and averaging (especially spatial averaging in the case of a flow statistically uniform in the y–z plane), there are important differences, some of which are illustrated in Figures 7.39 and 7.40 in the 2D case.

First, the length scale Λ of spatial averaging is much larger than the filter size Δ. Second, when evaluating a spatially averaged quantity, integration is performed over a strongly elongated rectangle shown in gray in Figure 7.39a, with the x-side (normal to the mean flame brush) of the rectangle being much shorter than the Kolmogorov length scale η and the y-side being much longer than an integral length scale L. When determining a filtered quantity, spatial integration is performed over the rectangles shown in gray in Figure 7.39b, with all the sides being of the same order Δ. Accordingly, spatial averaging is performed along a line parallel to the mean flame brush, while filtering is performed not only in the tangential but also in the normal direction. In other words, in the considered simple case, spatial averaging is a strongly anisotropic operation, whereas filtering is an isotropic method.

Due to these differences, the spatially averaged combustion progress variable does not depend on y and the mean flame position associated with $\overline{c}(x,t) = 0.5$ is a straight line normal to the x-axis (see Figure 7.39a), while the filtered combustion progress variable depends on both x and y (cf. the values of $\hat{c} = 1$, 0, and 0.5 in three points in Figure 7.39b, characterized by the same x but different $y_3 < y_2 < y_1$) and the filtered flamelet surface associated with $\hat{c}(x,y,t) = 0.5$ is wrinkled, as shown in the right down corner of Figure 7.39b. In other words, spatial averaging smoothes all the flamelet wrinkles produced by turbulent eddies, whereas filtering smoothes the small-scale wrinkles but retains the large-scale wrinkles.

Figure 7.39b shows one more peculiarity of filtering. Repeated filtering of an already filtered field $\hat{c}(x,y,t)$ does not yield $\hat{c}(x,y,t)$, that is, $\hat{\hat{q}} \neq \hat{q}$ in a general case, while $\overline{\overline{q}} = \overline{q}$ for a Reynolds-averaged quantity q. Indeed, let us consider the isosurface characterized by $\hat{c}(x,y,t) = 0.5$ (see the solid curve in the right down corner of Figure 7.39b). Application of the box filter, shown in gray and centered to a trailing point of the isosurface, to the filtered field $\hat{c}(x,y,t)$ yields $\hat{\hat{c}} > 0.5$ and $\hat{c} = 0.5$ at the trailing point.

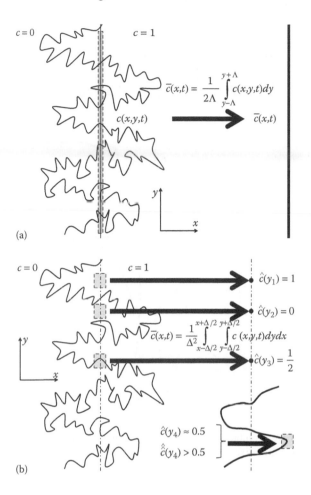

FIGURE 7.39 Spatial averaging (a) and filtering using a box filter (b).

Moreover, because $\hat{\hat{q}} \neq \hat{q}$, the filtered residual field $q' \equiv q - \hat{q}$ does not vanish, contrary to the mean rms field $\overline{q'} = 0$. Indeed,

$$\langle q' \rangle = \langle q - \hat{q} \rangle = \hat{q} - \hat{\hat{q}} \neq 0. \tag{7.228}$$

Equation 7.228 explains why the residual fields are not used when writing the filtered Equations 7.216 through 7.217, contrary to the Favre-averaged Equations 7.35 and 7.36.

Thus, although there are similarities between the filtered and the spatially averaged fields of the combustion progress variable, there are also important differences. To discuss one more difference, let us consider Figure 7.40.

In LES of a nonreacting turbulent flow, the filter size and the length scale of a numerical grid are typically of the same order. However, as stressed by Pitsch (2006), such a method of simulation faces serious problems when applied to

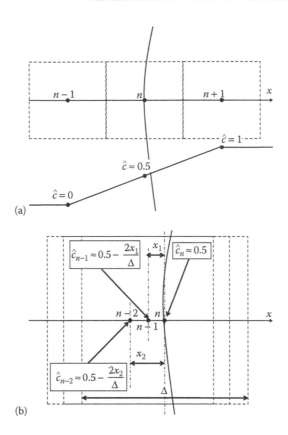

FIGURE 7.40 Filtered combustion progress variable field on coarse (a) and fine (b) grids in the flamelet regime of premixed turbulent combustion. A solid weakly wrinkled curve shows the flamelet surface.

the flamelet regime of premixed turbulent combustion. Indeed, if unburned and burned mixtures are separated by a thin flamelet ($\delta_L \ll \eta \ll \Delta$) and, for simplicity, the flamelet is weakly curved locally and instantly, then the filtered field $\hat{c}(x,y,t)$ is ill-resolved on the grid shown in Figure 7.41a, that is, if $\hat{c} \approx 0.5$ in the n-th grid point, while $\hat{c} = 0$ and $\hat{c} = 1$ in the $(n-1)$-th and $(n+1)$-th grid points, respectively. The simplest way of solving this problem is to use a finer grid, with Δx being significantly smaller than Δ. In such a case, the values of $\hat{c}$ in two neighboring grid points will be close to one another. However, such a solution raises both technical and basic problems.

From the technical viewpoint, the use of such a fine grid is expensive (especially for industrial applications), while an increase in Δ may reduce the accuracy of the LES models.

From the basic viewpoint, the physical meaning of the filtered field $\hat{c}(x,y,t)$ obtained on a fine grid (here, the word "fine" means that the grid size is significantly less than the filter size Δ) appears to differ substantially from the physical meaning of the Reynolds-averaged combustion progress variable $\bar{c}(x,t)$, which, in

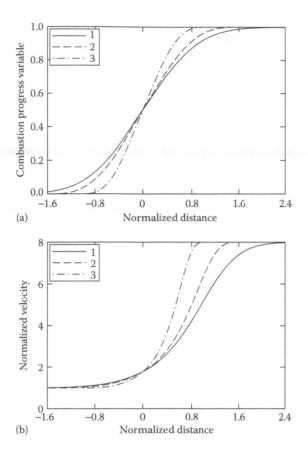

(a)

(b)

FIGURE 7.41 Filtered combustion progress variable $\check{c}$ (a) and normalized velocity $\check{u}/S_L^0$ (b) computed for an unperturbed laminar flame using the following filters: 1: $\Pi(\xi) = e^{-\xi^2}/(\Delta\sqrt{\pi})$, where $\xi = x/\Delta$; 2: $\Pi(\xi) = 2\cos^2\xi/(\pi\Delta)$ if $-\pi/2 < \xi < \pi/2$ and $\Pi(\xi) = 0$ otherwise; 3: $\Pi(\xi) = 15(1-\xi^2)^2/(16\Delta)$ if $-1 < \xi < 1$ and $\Pi(\xi) = 0$ otherwise.

the flamelet regime of premixed turbulent combustion, is equal to the probability of finding burned mixture on line $x = \text{const}$ at instant t (henceforth in this section, we address a statistically planar, 1D, premixed turbulent flame if the opposite is not stated). Indeed, if grid point number n lies on the instantaneous flamelet (see Figure 7.40b), then the difference $|\hat{c}_{n-m} - \hat{c}_n|$ is controlled by the distance between the two grid points. In other words, the dependence of $\hat{c}$ on x is controlled by the distance between the point x and the flamelet in the simple case considered.

In more general terms, the filtered combustion progress variable $\hat{c}$ characterizes the topology of the flamelet surface in the vicinity of a particular point at a particular instant, whereas the Reynolds-averaged progress variable $\bar{c}$ is equal to the probability of finding combustion products. Therefore, the physical meaning of the filtered $\hat{c}$ differs from the physical meaning of the Reynolds-averaged $\bar{c}$ in a general case.

The emphasized difference in the physical meanings of $\hat{c}$ and $\bar{c}$ should not be ignored even if they have common features. For instance, both $\rho_b\,\partial\bar{c}/\partial t$ and $\rho_b\,\partial\hat{c}/\partial t$ characterize an increase in the mass of combustion products within the corresponding control volumes, although the control volumes are very different in the two cases (see Figure 7.39).

Let us consider one more very simple example: an infinitely thin, planar, 1D, laminar flame expanding in a quiescent mixture from right to left. In the coordinate framework attached to the flame, the continuity

$$\frac{d}{dx}(\rho u) = 0 \tag{7.229}$$

and combustion progress variable balance

$$\frac{d}{dx}(\rho u c) = W = \rho_u S_L^0 \delta(x) \tag{7.230}$$

equations hold, where δ is the Dirac delta function. The solution to Equations 7.229 and 7.230 is simple:

$$\rho(x < 0) = \rho_u,$$

$$u(x < 0) = u_u = S_L^0,$$

$$c(x < 0) = 0,$$

$$\rho(x > 0) = \rho_b, \tag{7.231}$$

$$u(x > 0) = u_b = \sigma S_L^0,$$

$$c(x > 0) = 1.$$

The mean values are equal to the instantaneous values in the selected coordinate framework, and the Reynolds-averaged continuity and combustion progress variable balance equations are reduced to Equations 7.229 and 7.230, respectively.

However, the behavior of the filtered quantities and the filtered balance equations are very different from Equation 7.231 and Equations 7.229 through 7.230, respectively. Indeed, applying any continuous and uniform filter $\Pi(\zeta)$ to Equation 7.231, one can easily obtain

$$\hat{\rho}\breve{u} = \rho_u S_L^0,$$

$$\hat{c} = \int_{-\infty}^{\infty} c(x - \zeta)\Pi(\zeta)d\zeta = \int_{-\infty}^{x} \Pi(\zeta)d\zeta,$$

$$\hat{\rho}\breve{c} = \frac{\rho_u}{\sigma}\hat{c},$$

$$\langle \rho u c \rangle = \rho_u S_L^0 \hat{c},$$

$$(1-\hat{c})\hat{u}_u = \int\limits_{-\infty}^{\infty} u(x-\zeta)\big[1-c(x-\zeta)\big]\Pi(\zeta)\,d\zeta = S_L^0\int\limits_{x}^{\infty}\Pi(\zeta)\,d\zeta = (1-\hat{c})S_L^0,$$

$$\hat{c}\hat{u}_b = \int\limits_{-\infty}^{\infty} u(x-\zeta)c(x-\zeta)\Pi(\zeta)\,d\zeta = \sigma S_L^0\int\limits_{-\infty}^{x}\Pi(\zeta)\,d\zeta = \hat{c}\sigma S_L^0. \tag{7.232}$$

Straightforward filtering of Equation 7.230 is also very simple and yields the following trivial result:

$$\rho_u S_L^0\frac{d\hat{c}}{dx} = \rho_u S_L^0\Pi(x) = \hat{W}(x). \tag{7.233}$$

However, if the same equation is rewritten in the form commonly used in numerical simulations (see Equation 7.217), then we have

$$\rho_u S_L^0\frac{d\breve{c}}{dx} = -\rho_u S_L^0\frac{d}{dx}(\hat{c}-\breve{c}) + \hat{W}. \tag{7.234}$$

Subsequently, if, as commonly assumed, the first term on the RHS of Equation 7.234 is associated with the derivative of a subfilter scalar flux, which is equal to $\langle\rho\mathbf{u}c\rangle - \hat{\rho}\breve{\mathbf{u}}\breve{c}$ in a general case, then this flux $\rho_u S_L^0(\hat{c}-\breve{c})$ points to the combustion products (Poinsot and Veynante, 2005). It is worth stressing that such a countergradient behavior results solely from filtering and has no relation with the physical mechanisms of countergradient scalar transport in premixed flames, which are discussed in the section "Countergradient Scalar Transport in Turbulent Premixed Flames" in Chapter 6.

Moreover, Figure 7.41 shows that the spatial profiles of $\breve{c}(x)$ and $\breve{u}(x)$ determined by Equation 7.232 are sensitive to the filter used. The filtered reaction rate $\hat{W}(x)$ is also sensitive to the filter used (see Equation 7.233), but the closure of $\hat{W} = \rho_u S_L^0\nabla\hat{c}$ holds for any continuous filter in the considered case.

The foregoing simple examples clearly show that any analogy between the behavior of a mean combustion progress variable $\bar{c}$ or $\tilde{c}$ in a premixed turbulent flame and the behavior of filtered $\hat{c}$ or $\breve{c}$ should be drawn with great care. In particular, the widely accepted practice of closing Equation 7.217 by recasting RANS models into a filtered form does not seem to be justified in a general case. The validity of such recast models should be thoroughly investigated from the basic viewpoint in each particular case. Unfortunately, this issue is on the periphery of LES research on turbulent combustion, which is focused on numerical and applied aspects, for example, simulating increasingly complicated problems using models that still strongly need basic substantiation and thorough validation in a representative set of well-defined simple cases.

Recast of already developed combustion models into a filtered form should be performed with great care not only for balance equations, as previously discussed, but also for much simpler algebraic expressions for the turbulent burning velocity. For instance, the TFC model developed for RANS applications and discussed in the section "Models of a Self-Similarly Developing Premixed Turbulent Flame" was applied by Flohr and Pitsch (2000), Fureby (2005), Zimont and Bataglia (2006), and Chakraborty and Klein (2008) to LES. In the cited papers, Equation 7.13 has been recast in the following form:

$$U_{t,\Delta} \propto u'^{3/4}_{\Delta} \Delta^{1/4} S_L^{1/2} a_u^{-1/4}, \tag{7.235}$$

where $U_{t,\Delta}$ is a subgrid burning velocity.

However, this expression does not seem to be consistent with Zimont's theory, which results in Equation 7.13 (see the section "Zimont Model of Burning Velocity in Developing Turbulent Premixed Flames"). The point is that the theory is based on different scaling of the influence of small-scale and large-scale eddies on the turbulent burning velocity (see Equations 7.5 and 7.12, respectively). Therefore, if the subgrid burning velocity is assumed to be solely controlled by turbulent eddies smaller than a filter size Δ and if the filter size is approximately equal to the Zimont length scale λ_Z determined by Equation 7.4, then the consistent recast of the Zimont theory into a filtered form appears to be as follows:

$$U_{t,\Delta} \propto u_Z \propto \left(\overline{\varepsilon} \tau_c \right)^{1/2} \propto u'^{3/2}_{\Delta} \Delta^{-1/2} S_L^{-1} a_u^{1/2}, \tag{7.236}$$

(see Equation 7.5).

If $\lambda_f \ll \Delta \ll L$, where λ_f is the length microscale introduced by Zimont in Equation 7.7, then the closure

$$U_{t,\Delta} \propto u_Z \left(\frac{A_t}{A_0} \right)_{\Delta} \propto u_Z \frac{\Delta}{\lambda_f} \propto u_Z \frac{\Delta}{(u'\lambda_{Z}t)^{1/2}} \propto \left[\frac{\overline{\varepsilon}\tau_c \Delta^2}{(\overline{\varepsilon}L)^{1/3} \left(\overline{\varepsilon}\tau_c^3 \right)^{1/2} t} \right]^{1/2}$$

$$\propto \overline{\varepsilon}^{1/12} \tau_c^{-1/4} L^{-1/6} t^{-1/2} \Delta \propto u'_{\Delta} \left(\frac{\Delta}{u'_{\Delta}t} \right)^{1/2} \left(\frac{\Delta}{u'_{\Delta}\tau_c} \right)^{1/4} \left(\frac{\Delta}{L} \right)^{1/6}$$

$$\propto u'^{1/4}_{\Delta} L^{-1/6} \Delta^{11/12} S_L^{1/2} a_u^{-1/4} t^{-1/2} \tag{7.237}$$

could be proposed to be used by recasting Equations 7.4 through 7.7 and Equation 7.10 with $\Psi_4(Da) = $ const into a filtered form. Obviously, Equations 7.236 and 7.237 are very different from Equation 7.235. Moreover, because Equations 7.7 and 7.10 were obtained by Zimont (1979) based on certain global properties of the entire turbulent flame brush, it is not clear whether or not the recast of these equations into a filtered form is a correct operation.

Furthermore, the Zimont theory does not result straightforwardly in a scaling of $U_{t,\Delta}$ if either $\lambda_Z \ll \Delta \ll \lambda_f$ or $\Delta < \lambda_Z$. This simple example indicates that the subgrid burning velocity may follow different scaling for different filter sizes and may straightforwardly depend on the flame-development time, t.

The author is aware of only a few attempts to allow for the peculiarities of filtering techniques when closing Equation 7.217. For instance, Duwig and Fuchs (2005) proposed to close the reaction rate $\hat{W}$ by applying the Gaussian filter to a planar laminar flame, as discussed earlier. For unknown reasons, these authors restricted themselves to using the filter-dependent closure relation $\hat{W} = \rho_u S_L^0 \Pi(x)$, rather than the filter-independent expression $\hat{W} = \rho_u S_L^0 \nabla \hat{c}$.

Colin et al. (2000) proposed to use the so-called thickened flame model for LES of premixed flames. This approach, put forward by Butler and O'Rourke (1977) and O'Rourke and Bracco (1979) at the dawn of RANS simulations of turbulent combustion, is based on the fact that the laminar flame speed is not affected by an increase in molecular diffusivity, D, by a factor f, provided that the reaction rate, W, is simultaneously reduced by the same factor f (see Equation 2.37), whereas the flame thickness is increased by such a transformation ($D \to fD$ and $W \to W/f$, where $f > 1$) (see Equations 2.42 through 2.44). Thus, one can make a planar laminar flame thicker, retaining the same burning rate. For LES applications, the factor f should be so large that the thickened flame is well resolved in the simulations.

Certainly, the burning rate in a turbulent flow is affected by the foregoing transformation, for example, because the small-scale wrinkles on a flamelet surface are smoothed out when the flamelet thickness is increased. To allow for the contribution of the small-scale wrinkles on the burning rate, Colin et al. (2000) proposed to use an efficiency function $E(u'_\Delta/S_L^0, \Delta/\delta_L, f)$, which parameterized the results of DNS of the interaction between a laminar flame and a vortex pair (Poinsot et al., 1990, 1991; Meneveau and Poinsot, 1991). Here, u'_Δ is the "subgrid scale turbulent velocity ... that is, the square root of the subgrid scale kinetic energy." According to Colin et al. (2000), the second transformation, $W/f \to EW/f$, models an increase in the burning rate due to small-scale wrinkles of the flamelet surface, which are smoothed out by the first transformation, $D \to fD$ and $W \to W/f$.

Even if we admit that the efficiency function $E(u'_\Delta/S_L^0, \Delta/\delta_L, f)$, obtained by simulating a single type of flame–vortex interaction for a limited set of controlling parameters, is well suited for modeling the contribution of the smoothed small-scale wrinkles to the burning rate under substantially different conditions, the approach discussed earlier still substantially oversimplifies the physics of premixed turbulent combustion, because the foregoing transformations do not seem to allow for the influence of small-scale stretching on the local burning rate within laminar flamelets (e.g., see the section "Variations in Flamelet Structure Due to Turbulent Stretching" in Chapter 5). In particular, in its present form, the thickened flame model by Colin et al. (2000) does not allow us to investigate the strong influence of D_d on U_t. Nevertheless, the considered model appears to be an acceptable compromise between the applied needs and consistency with the underlying physics. The model is widely used by the French school and has already been successfully applied to engines.

The G-equation for tracking a filtered flame surface is another attempt to achieve a compromise of that kind. In principle, if the instantaneous field $c(\mathbf{x},t)$ is assumed to be known, then one may apply a uniform filter function and obtain a filtered field $\hat{c}(\mathbf{x},t)$, which depends on the filter selected. Accordingly, any isosurface $\hat{c}(\mathbf{x},t) = c_0$, for example, $\hat{c}(\mathbf{x},t) = 0.5$, may be associated with a filtered flame surface. Subsequently, taking the material derivative of the equality

$$G_f\left[\hat{c}(\mathbf{x},t) = c_0\right] = G_0, \tag{7.238}$$

we arrive at

$$\frac{\partial G_f}{\partial t} + \mathbf{v}_f \cdot \nabla G_f = S_f\left|\nabla G_f\right|, \tag{7.239}$$

similar to the derivation of Equation 7.153 from Equation 7.152 using Equation 7.154.

Equation 7.239 involves two unknown quantities: the velocity $\mathbf{v}_f$ that advects the filtered flame surface and the filtered flame speed S_f. The severe problems associated with modeling the counterpart terms in Equation 7.156 for tracking a mean flame surface were discussed in the section "G-Equation." Closing the filtered terms appears to be even more difficult. In most LES studies, these issues are ignored, $\mathbf{v}_f$ is associated with a filtered flow velocity $\hat{\mathbf{v}}$, and S_f is closed by recasting a model for the turbulent burning velocity into a filtered form.

An attempt to derive a kinematic equation for tracking a filtered flame surface was undertaken by Pitsch (2005), by filtering Equations 7.152 and 7.153. However, because the analysis by Pitsch (2005) for filtered flames is an extension of the analysis by Oberlack et al. (2001) for mean flames, the former inherits all the basic flaws of the latter, which were discussed in the section "G-Equation." Moreover, the filter introduced by Pitsch (2005) does not commute with time differentiation. This can easily be shown by applying the filter to a segment of an expanding spherical flame (see Appendix A in Sabel'nikov and Lipatnikov [2011b]).

The foregoing simple examples highlight issues specific to LES of a premixed turbulent flame. However, even if we consider a constant-density, nonreacting flow, certain basic issues still challenge the turbulence community. For instance, the experimental data by Cerutti and Meneveau (2000) and the DNS data analyzed recently by Yakhot and Wanderer (2012) show that the dependencies of structure functions $D_p(r) \equiv [u(x+r) - u(x)]^p$ on the distance r, computed for unfiltered and filtered velocities, differ substantially from one another at $r/\Delta < O(10)$, with the difference being strongly increased by the exponent p. These results indicate that "the simple-minded filtering procedure is intrinsically flawed," because "large and small-scale velocity fluctuations strongly interact and one cannot obtain a reasonable representation of one by simply filtering out the other" (Yakhot and Wanderer, 2012).

Vortex filaments (see the section "Internal Intermittency and Vortex Filaments" in Chapter 3), which may have a length of the order of an integral length scale L and a diameter of the order of the Kolmogorov length scale, are a particular example of such a strong interaction between large- and small-scale features of a turbulent flow.

Considering the eventual crucial role of vortex filaments in premixed combustion (see the section "Highly Curved Leading Points and Flamelet Tip in a Vortex Tube" and Section 7.3 in a review paper by Lipatnikov and Chomiak [2005a]), this issue is of paramount importance for LES of turbulent flames.

The foregoing comments are not aimed at disparaging LES of turbulent combustion. No doubt, the LES approach is very promising, appears to have become the dominant tool for multidimensional computations of various applied combustion problems, and, as reviewed elsewhere (Janicka and Sadiki, 2005; Pitsch, 2006), has already yielded a number of valuable results for flames that are substantially affected by the unsteady dynamics of large-scale vortices. Unfortunately, LES has not yet contributed much to the basic understanding of the governing physical mechanisms of premixed turbulent combustion and the earlier discussion is aimed at drawing the attention of the LES community to this issue also.

Further consideration of LES of premixed turbulent flames is beyond the scope of this book, which is focused on combustion physics. Interested readers are referred to recent review papers by Janicka and Sadiki (2005) and Pitsch (2006), where relevant numerical issues are discussed and a number of successful applications of LES to various flames are reported.

8 Introduction to Nonpremixed Combustion

If a fuel and an oxidizer are supplied to a burner by two separate streams, then combustion may occur in the nonpremixed mode. In such a case, transport processes play a crucial role, because the fuel and oxidizer must be mixed in order for their molecules to collide and react. Therefore, in a nonpremixed flame, the burning rate cannot be higher than the mixing rate and, in that sense, diffusion controls the rate of nonpremixed combustion. Accordingly, nonpremixed flames are commonly called diffusion flames.

As will be discussed in this chapter, the physics of nonpremixed combustion may, in the first approximation, be reduced to mixing followed by infinitely fast chemical reactions. As far as the total burning rate and the fields of mean temperature and mean mass fractions of major species are concerned, the predictive capabilities of such a simple approach developed more than 30 years ago are comparable with the predictive capabilities of the best models of premixed turbulent combustion available today. Over the past three decades, the efforts of the nonpremixed combustion community were focused on investigating much more subtle effects, such as mean concentrations of intermediate species (e.g., O and CO) in diffusion flames, formation of nitrogen oxide and soot, and extinction and reignition. As far as simulations of such effects are concerned, contemporary models of nonpremixed combustion are superior to the models of premixed turbulent flames. In that sense, the nonpremixed combustion community has advanced substantially further than the premixed combustion community, even if the models of premixed turbulent flames considered in Chapters 6 and 7 appear to be more sophisticated than simple approaches to simulating diffusion combustion, addressed in this chapter.

The following discussion on turbulent nonpremixed flames will be brief as compared with the previous consideration of premixed burning for a number of reasons. First of all, this book is mainly aimed at the major governing physical mechanisms of turbulent combustion and the simple approaches capture such mechanisms reasonably well as far as diffusion flames are concerned. Over the past three decades, substantial progress has been made in elaborating sophisticated numerical models of nonpremixed burning, but they still rely mainly on the same physical mechanisms. A detailed discussion on these models is beyond the scope of this introduction to the physics of turbulent combustion.

Moreover, combustion in engines developed today occurs in intermediate modes such as partially premixed turbulent flames, which are addressed in Chapter 9.

To analyze these flames, models of nonpremixed and premixed combustion should be combined. The predictive capabilities of such a combined approach will be limited by the predictive capabilities of the less accurate approach, that is, a model of premixed turbulent burning. Accordingly, a combination of the most advanced models of premixed and diffusion combustion appears to be an unnecessary overcomplication, which does not improve the predictive capabilities as compared with a combination of an advanced model of premixed burning with a simple model of diffusion flames. Therefore, this chapter is mainly restricted to discussing relatively simple concepts of nonpremixed combustion, the accuracy of which is comparable with the predictive capabilities of most advanced models of premixed turbulent flames.

LAMINAR DIFFUSION FLAMES

Because transport processes control the rate of nonpremixed combustion, an efficient approach to studying a diffusion flame consists of highlighting a linear diffusion equation for a conserved (nonreacting) scalar. The contemporary theory of nonpremixed combustion is based on such an approach and is simpler than the theory of premixed flames, which deals with the much more complicated reaction–diffusion Equation 1.58, which involves a highly nonlinear source term, W_l.

In a general case of complex combustion chemistry, a part of the governing balance equations may easily be rewritten in the form of a diffusion equation for a conserved scalar, thanks to the conservation of elements in chemical reactions (see Equation 1.62). Such diffusion equations hold in both premixed and nonpremixed flames, but they do not help much in the former case, because their solution is trivial in a fully premixed system where $Z_m = $ const for any chemical element. In the nonpremixed case, element mass fractions, Z_m, vary in space and time, for example, $Z_C = Z_H = 0$, but $Z_O \neq 0$ in an airstream, while $Z_H \neq 0$, $Z_C \neq 0$, but $Z_O = 0$ in the stream of a hydrocarbon fuel, C_xH_y, that does not contain oxygen. Accordingly, the diffusion Equation 1.62 offers an opportunity to study many key features of nonpremixed combustion.

Another cornerstone simplification of the discussed theory of nonpremixed combustion is linearizing the nonlinear Equation 1.62 by assuming that the binary diffusion coefficients, D_l, are equal to the same D for all species and the Lewis number is equal to unity. It is worth remembering that the original Equation 1.62 is a nonlinear equation, because D_l depends on species mass fractions, Y_l (Hirschfelder et al., 1954).

To further simplify our discussion, let us begin by considering burning that is controlled by a single, irreversible global reaction between a hydrocarbon fuel, C_xH_y, and oxygen:

$$C_xH_y + aO_2 \rightarrow \text{Products}.$$

Such a simplification is widely invoked when discussing the governing physical mechanisms of turbulent combustion.

Under the aforementioned assumptions, Equation 1.58 reads

$$\frac{\partial}{\partial t}(\rho Y_F) + \frac{\partial}{\partial x_k}(\rho u_k Y_F) = \frac{\partial}{\partial x_k}\left(\rho D \frac{\partial Y_F}{\partial x_k}\right) - M_F W_F \tag{8.1}$$

for fuel mass fraction Y_F and

$$\frac{\partial}{\partial t}(\rho Y_O) + \frac{\partial}{\partial x_k}(\rho u_k Y_O) = \frac{\partial}{\partial x_k}\left(\rho D \frac{\partial Y_O}{\partial x_k}\right) - M_O W_O \tag{8.2}$$

for oxygen mass fraction Y_O. Because $W_O = a W_F$, a linear combination of Equations 8.1 and 8.2 yields the diffusion equation (Burke and Schumann, 1928)

$$\frac{\partial}{\partial t}(\rho y) + \frac{\partial}{\partial x_k}(\rho u_k y) = \frac{\partial}{\partial x_k}\left(\rho D \frac{\partial y}{\partial x_k}\right) \tag{8.3}$$

for $y = a Y_F/M_F - Y_O/M_O$.

Equation 8.3 is the cornerstone equation of the theory of diffusion flames and the following discussion will be based on this equation, renormalized in the next section.

MIXTURE FRACTION

Let us normalize the aforementioned linear combination of mass fractions using its values y_1 and y_2 in the fuel and oxidizer streams, respectively. Such a normalized linear combination,

$$f = \frac{a Y_F/M_F - Y_O/M_O + Y_{O,2}/M_O}{a Y_{F,1}/M_F + Y_{O,2}/M_O} = \frac{St Y_F - Y_O + Y_{O,2}}{St Y_{F,1} + Y_{O,2}}, \tag{8.4}$$

is called the mixture fraction. Here, $St = a M_O/M_F$ is the mass stoichiometric coefficient (see Equation 1.17). It can be easily shown that $0 \leq f \leq 1$ and

$$f_{st} = \frac{Y_{O,2}}{St Y_{F,1} + Y_{O,2}} = \left(1 + \frac{St Y_{F,1}}{Y_{O,2}}\right)^{-1} \tag{8.5}$$

in the locally stoichiometric mixture characterized by $St Y_F = Y_O$, with $f > f_{st}$ (or $f < f_{st}$) in the locally rich (lean) mixture.

Obviously, the mixture fraction and its stoichiometric value depend on the mass fractions of the fuel and oxidizer in streams 1 and 2, respectively, and these mass fractions may be different in different cases, for example, $Y_O = 1$ in an oxygen stream, but $Y_O < 1$ in an airstream. Therefore, contrary to the combustion progress variable used to describe premixed flames, mixture fraction is a problem-dependent quantity, that is, f_{st} is controlled not only by the processes within a flame but also by the boundary conditions.

Moreover, Equations 8.4 and 8.5 hold only in the case of a double-feed flame, that is, when a fuel (oxidizer) is supplied by a uniform flow. Here, flow uniformity means that element mass fractions, Z_m, do not depend on $\mathbf{x}$ within the flow, while species mass fractions, $Y_i(\mathbf{x},t)$, may be nonuniform, for example, due to pyrolysis of the fuel in certain spatial regions. In the case of several flows that supply different fuels characterized by different Z_C and/or Z_H, the introduction of a mixture fraction and the determination of f_{st} are more difficult tasks.

It is also worth noting that, if the diffusivities of various species are equal to one another, then the diffusion Equation 8.3 holds not only for $y = aY_F/M_F - Y_O/M_O$ but also for many other linear combinations of Y_F, Y_O, and Y_P, as well as for element mass fractions Z_m (see Equation 1.62). Accordingly, the aforementioned definition of mixture fraction given by Equation 8.4 is not unique and there are a number of alternative expressions. For instance, the general definition

$$f_Z = \frac{\displaystyle\sum_{m=1}^{N}\alpha_m Z_m - \left(\sum_{m=1}^{N}\alpha_m Z_m\right)_2}{\left(\displaystyle\sum_{m=1}^{N}\alpha_m Z_m\right)_1 - \left(\sum_{m=1}^{N}\alpha_m Z_m\right)_2}, \tag{8.6}$$

where α_m are the arbitrary coefficients, may be used not only in the case of a single global reaction between the fuel and oxygen, but also if complex combustion chemistry is addressed. In particular,

$$f_Z = \frac{Z_C/(n_C M_C) + Z_H/(n_H M_H) + 2(Y_{O,2} - Z_O)/(aM_{O_2})}{Z_{C,1}/(n_C M_C) + Z_{H,1}/(n_H M_H) + 2Y_{O,2}/(aM_{O_2})} \tag{8.7}$$

is often used for hydrocarbon fuels, $C_x H_y$ (Bilger, 1988). Here, n_C and n_H are the numbers of carbon and hydrogen atoms, respectively, in one molecule of the fuel, and the subscript O designates either molecular or atomic oxygen as far as either the species mass fraction, Y_O, or the element mass fraction, Z_O, respectively, is concerned. Because

$$\frac{Z_C}{n_C M_C} = \frac{Z_H}{n_H M_H} = \frac{Y_F}{M_F} \tag{8.8}$$

by virtue of Equation 1.15 and $Z_O = Y_O$ in a mixture of air and a fuel that does not contain oxygen, the mixture fraction f_Z given by Equation 8.7 is equal to f given by Equation 8.4 in this case.

Certainly, mixture fraction may be introduced not only for nonpremixed but also for premixed flames. However, in the latter case, this quantity and the relevant balance equation are useless, because f or f_z is simply equal to a constant. In a diffusion flame, mixture fraction has a clear physical meaning, as it characterizes variations in the local mixture composition due to mixing. Furthermore, because mixing is the limiting process that controls the heat release rate in a typical nonpremixed flame, a single diffusion equation

$$\frac{\partial}{\partial t}(\rho f) + \frac{\partial}{\partial x_k}(\rho u_k f) = \frac{\partial}{\partial x_k}\left(\rho D \frac{\partial f}{\partial x_k}\right) \tag{8.9}$$

for a mixture fraction, supplemented with the simple boundary conditions

$$f_1 = 1 \quad \text{and} \quad f_2 = 0, \tag{8.10}$$

allows us to describe the (mean) burning rate and (mean) spatial structure of a laminar (turbulent) diffusion flame well by considering the asymptotic case of infinitely fast combustion reactions (Burke and Schumann, 1928).

Paradigm of Infinitely Fast Chemistry

Within the framework of this paradigm, fuel and oxygen cannot coexist, that is, as soon as a small amount of fuel (oxygen) supplied by one stream is mixed with oxygen (fuel) supplied by another stream, this amount of fuel (oxygen) and the proper amount of oxygen (fuel) are assumed to be instantly burned. Therefore, because all fuel should be burned in the locally lean $(f < f_{st})$ mixture, $Y_F(f < f_{st}) = 0$ and Equations 8.4 through 8.5 yield

$$Y_O(f < f_{st}) = Y_{O,2} - f(StY_{F,1} + Y_{O,2}) = Y_{O,2}\left(1 - \frac{f}{f_{st}}\right) \tag{8.11}$$

in the lean region of a diffusion flame. Similarly, $Y_O(f > f_{st}) = 0$ and

$$Y_F(f > f_{st}) = \frac{f(StY_{F,1} + Y_{O,2}) - Y_{O,2}}{St} = fY_{F,1} - \frac{1-f}{St}Y_{O,2} = Y_{F,1}\frac{f - f_{st}}{1 - f_{st}} \tag{8.12}$$

in the rich mixture. Both Equations 8.11 and 8.12 yield $Y_O(f = f_{st}) = 0$ and $Y_F(f = f_{st}) = 0$ at the stoichiometric surface $f(\mathbf{x},t) = f_{st}$. Differentiating Equations 8.11 and 8.12 and using Equation 8.5, one can easily show that, at the stoichiometric surface, $(\rho D dY_F/df)_{st} = St(\rho D dY_O/df)_{st}$, that is, the ratio of the fuel and oxygen diffusion fluxes to the surface is stoichiometric, thereby implying that the composition of the combustion products at the surface should also be stoichiometric.

The disappearance of both fuel and oxygen at the stoichiometric surface means that they react with each other at this surface, that is, the reaction rate is infinitely fast at the stoichiometric surface and vanishes outside it. Thus, within the framework of the discussed paradigm, fuel (oxygen) is transported by diffusion from the fuel (oxygen) stream to the stoichiometric surface, where the fuel and oxygen burn infinitely fast, followed by the diffusion of stoichiometric combustion products and heat from the surface to both the fuel and oxygen streams.

In order to determine the product mass fraction in the lean region of a diffusion flame, let us rewrite the diffusion equation

$$\frac{\partial}{\partial t}(\rho Z_O) + \frac{\partial}{\partial x_k}(\rho u_k Z_O) = \frac{\partial}{\partial x_k}\left(\rho D \frac{\partial Z_O}{\partial x_k}\right)$$ (8.13)

for the mass fraction of oxygen atoms in the normalized form

$$\frac{\partial}{\partial t}(\rho \zeta_O) + \frac{\partial}{\partial x_k}(\rho u_k \zeta_O) = \frac{\partial}{\partial x_k}\left(\rho D \frac{\partial \zeta_O}{\partial x_k}\right),$$ (8.14)

where $\zeta_O \equiv (Z_{O,2} - Z_O)/(Z_{O,2} - Z_{O,1})$, with the subscripts 1 and 2 designating the fuel and oxygen streams, respectively. The boundary conditions to Equation 8.14 read

$$\zeta_{O,1} = 1 \quad \text{and} \quad \zeta_{O,2} = 0$$ (8.15)

that is, the problem described by Equations 8.14 and 8.15 is mathematically equivalent to the problem described by Equations 8.9 and 8.10 and, consequently, $f(\mathbf{x},t) = \zeta(\mathbf{x},t)$. For a hydrocarbon fuel C_xH_y, $Z_{O,1} = 0$, and we have $Z_O = (1 - f)Z_{O,2}$. If stream 2 contains O atoms only in the form of molecular oxygen (e.g., airstream), then the mass fraction of oxygen atoms is equal to the mass fraction of oxygen molecules in stream 2, that is, $Z_{O,2} = Y_{O,2}$ and, hence, $Z_O = (1 - f)Y_{O,2}$. Because, within a diffusion flame, (i) Z_O is equal to the sum of the mass fraction of unburned molecular oxygen and the mass fraction of oxygen atoms bounded in combustion products (CO_2, H_2O, etc.) and (ii) burning of 1 g of O_2 yields $(St^{-1} + 1)$ g of products, the mass fraction of the products in the lean region of a diffusion flame is equal to

$$Y_P(f < f_{st}) = (St^{-1} + 1)(Z_O - Y_O) = (St^{-1} + 1)\left[(1 - f)Y_{O,2} - Y_O\right]$$

$$= (St^{-1} + 1)\left[(1 - f)Y_{O,2} - \left(1 - \frac{f}{f_{st}}\right)Y_{O,2}\right]$$

$$= (St^{-1} + 1)Y_{O,2}\frac{(1 - f_{st})f}{f_{st}} = Y_{P,st}\frac{f}{f_{st}},$$ (8.16)

where

$$Y_{P,st} = \left(St^{-1}+1\right)\left(1-f_{st}\right)Y_{O,2} = \left(1+St\right)f_{st}Y_{F,1}$$

$$= \frac{\left(1+St\right)Y_{F,1}Y_{O,2}}{StY_{F,1}+Y_{O,2}} = \left[\frac{St}{\left(1+St\right)Y_{O,2}}+\frac{1}{\left(1+St\right)Y_{F,1}}\right]^{-1} \tag{8.17}$$

is the mass fraction of the products at the stoichiometric surface $f(\mathbf{x},t)=f_{st}$. The second and third equalities in Equation 8.17 result from Equation 8.5.

Similarly, by considering the diffusion equation for C atoms and properly normalizing it, we have $Z_C = fZ_{C,1}$. If the fuel were not burned, then its mass fraction would be equal to $Z_C(M_F/n_C M_C)$. Therefore, because burning of 1 g of fuel yields $(St + 1)$ g of products, we obtain

$$Y_P\left(f > f_{st}\right) = \left(1+St\right)\left(\frac{M_F}{n_C M_C}Z_C - Y_F\right)$$

$$= \left(1+St\right)\left(\frac{M_F}{n_C M_C}fZ_{C,1} - Y_F\right)$$

$$= \left(1+St\right)\left(Y_{F,1}f - Y_F\right) = \left(1+St\right)Y_{F,1}\left(f - \frac{f-f_{st}}{1-f_{st}}\right)$$

$$= Y_{P,st}\frac{1-f}{1-f_{st}}. \tag{8.18}$$

Equations 8.16 through 8.18 have a clear physical meaning. First, in order to produce 1 g of products, we need $(1 + St)^{-1}$ g of fuel or $(1+St)^{-1}Y_{F,1}^{-1}$ g of the gas from stream 1 and $(St^{-1} + 1)^{-1}$ g of O_2 or $(1+St^{-1})^{-1}Y_{O,2}^{-1}$ g of the gas from stream 2, that is, in total, $(1+St^{-1})^{-1}Y_{O,2}^{-1}+(1+St)^{-1}Y_{F,1}^{-1}$ g of gas contains 1 g of the products, in line with the last equality on the right-hand side (RHS) of Equation 8.17. In the case of streams that contain either pure fuel or pure oxygen, $Y_{F,1} = Y_{O,2} = 1$ and Equation 8.17 reads $Y_{P,st} = 1$. Second, in the lean mixture, the product mass fraction is lower than $Y_{P,st}$, because 1 g of unburned oxygen has not yielded $(St^{-1} + 1)$ g of products (see Equation 8.16). Third, in rich mixture, the product mass fraction is lower than $Y_{P,st}$, because 1 g of unburned fuel has not yielded $(1 + St)$ g of products (see Equation 8.18).

Thus, within the framework of the discussed paradigm, we have (i) the maximum concentration of products and zero concentrations of fuel and oxygen at the stoichiometric surface $f(\mathbf{x},t)=f_{st}$, (ii) fuel and products if $f > f_{st}$ with Y_F (or Y_P) linearly increasing (decreasing) with an increase in the mixture fraction, and (iii) oxygen and products if $f < f_{st}$ with Y_O (or Y_P) linearly increasing (decreasing) with a decrease in f (see the solid lines in Figure 8.1).

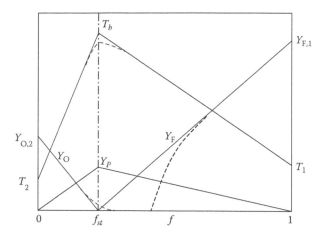

FIGURE 8.1 An asymptotic structure of a diffusion flame in the case of infinitely fast chemistry.

It is worth noting that, in a nonreacting case, the mass fraction balance Equation 1.58 and the relevant boundary conditions may easily be reduced to the problem described by Equations 8.9 and 8.10 and, therefore, $Y_l = Y_{l,2} + f(Y_{l,1} - Y_{l,2})$ depends linearly on the mixture fraction for any inert species. Thus, the fuel mass fraction decreases linearly with a decreasing f both in the nonreacting and burning mixing layers, but the decrease occurs in the entire layer in the former case and is bounded by the stoichiometric surface in the latter case.

If (i) Lewis numbers are equal to unity for all species, (ii) pressure is constant, and (iii) flame is adiabatic, then the enthalpy Equation 1.64 is reduced to

$$\frac{\partial}{\partial t}(\rho \zeta_h) + \frac{\partial}{\partial x_k}(\rho u_k \zeta_h) = \frac{\partial}{\partial x_k}\left(\rho D \frac{\partial \zeta_h}{\partial x_k}\right), \tag{8.19}$$

$$\zeta_{h,1} = 1 \quad \text{and} \quad \zeta_{h,2} = 0, \tag{8.20}$$

where $\zeta_h \equiv (h_2 - h)/(h_2 - h_1)$. A comparison of Equations 8.9 and 8.10 with Equations 8.19 and 8.20 shows that $\zeta_h = f$ and $h = h_2 - (h_2 - h_1)f$. The knowledge of the mixture composition and the specific enthalpy h allows us to evaluate the mixture temperature and density.

For instance, if partial specific heat capacities (i) are equal to the same c_p for all species and (ii) do not depend on temperature, then

$$c_p(T_b - T_0) + Y_F \Delta h_F^0 + Y_O \Delta h_O^0 + Y_P \Delta h_P^0$$

$$= h_2 - (h_2 - h_1)f$$

$$= c_p(T_2 - T_0) + Y_{O,2}\Delta h_O^0 - \left[c_p(T_2 - T_0) + Y_{O,2}\Delta h_O^0 - c_p(T_1 - T_0) - Y_{F,1}\Delta h_F^0\right]f, \tag{8.21}$$

where T_0 is a reference temperature used to determine Δh_i^0 (see the section "Energy, Enthalpy, and Heat Capacities" in Chapter 1). If, for the sake of simplicity, $T_2 = T_1 = T_u$, then

$$c_p\left(T_b - T_u\right) = Y_{O,2}\Delta h_O^0 - \left(Y_{O,2}\Delta h_O^0 - Y_{F,1}\Delta h_F^0\right)f - Y_F\Delta h_F^0 - Y_O\Delta h_O^0 - Y_P\Delta h_P^0 . \quad (8.22)$$

Consequently, using Equations 8.5, 8.11, 8.12, and 8.16 through 8.18, we obtain

$$c_p\left(T_b - T_u\right) = Y_{O,2}\Delta h_O^0\left(1 - f\right) + Y_{F,1}\Delta h_F^0 f - Y_{O,2}\Delta h_O^0 \frac{f_{st} - f}{f_{st}} - Y_{P,st}\Delta h_P^0 \frac{f}{f_{st}}$$

$$= \frac{f}{f_{st}}\left[Y_{O,2}\Delta h_O^0\left(1 - f_{st}\right) + Y_{F,1}\Delta h_F^0 f_{st} - Y_{P,st}\Delta h_P^0\right]$$

$$= \frac{f}{f_{st}}\frac{Y_{P,st}}{1 + St}\left[St\Delta h_O^0 + \Delta h_F^0 - \left(1 + St\right)\Delta h_P^0\right] \quad (8.23)$$

in the lean mixture and

$$c_p\left(T_b - T_u\right) = Y_{O,2}\Delta h_O^0\left(1 - f\right) + Y_{F,1}\Delta h_F^0 f - Y_{F,1}\Delta h_F^0 \frac{f - f_{st}}{1 - f_{st}} - Y_{P,st}\Delta h_P^0 \frac{1 - f}{1 - f_{st}}$$

$$= \frac{1 - f}{1 - f_{st}}\left[Y_{O,2}\Delta h_O^0\left(1 - f_{st}\right) + Y_{F,1}\Delta h_F^0 f_{st} - Y_{P,st}\Delta h_P^0\right]$$

$$= \frac{1 - f}{1 - f_{st}}\frac{Y_{P,st}}{1 + St}\left[St\Delta h_O^0 + \Delta h_F^0 - \left(1 + St\right)\Delta h_P^0\right] \quad (8.24)$$

in the rich mixture. At the stoichiometric surface,

$$c_p\left(T_{st} - T_u\right) = \frac{Y_{P,st}}{1 + St}\left[St\Delta h_O^0 + \Delta h_F^0 - \left(1 + St\right)\Delta h_P^0\right], \quad (8.25)$$

with T_{st} being equal to the adiabatic combustion temperature of the stoichiometric mixture, see Equation 1.35 and allow for $Y_{F,u} : Y_{O,u} : Y_P = 1 : St : (1 + St)$ in the stoichiometric mixture.

Finally, we have

$$\frac{T_b - T_u}{T_{st} - T_u} = \frac{f}{f_{st}} \quad (8.26)$$

if $f < f_{st}$ and

$$\frac{T_b - T_u}{T_{st} - T_u} = \frac{1 - f}{1 - f_{st}} \quad (8.27)$$

if $f > f_{st}$. Equations 8.26 and 8.27 indicate that, within a diffusion flame, the maximum temperature is reached at the stoichiometric surface and is equal to the adiabatic combustion temperature of the stoichiometric mixture. In the lean (rich) mixture, the temperature decreases linearly with a decrease (an increase) in the mixture fraction (see Figure 8.1).

Therefore, with the aforementioned simplifications (a single global reaction, $Le_l = Sc_l = 1$, infinitely fast chemistry, $c_{p,l} = c_p = const$, adiabatic burning at constant pressure), modeling of a diffusion flame is reduced to integration of Equations 8.9 and 8.10, followed by evaluation of the mass fractions of fuel, oxygen, and products and temperature using the partially linear Equations 8.11, 8.12, 8.16 through 8.18, 8.26 and 8.27. Thus, the physics of nonpremixed combustion is extremely simple within the framework of this first-order approximation.

Despite its simplicity, the previously discussed approach allows one to predict well the length and mean structure not only of laminar but also of turbulent diffusion flames, (e.g., see Figure 8.2), provided that a proper mixture fraction probability density function (PDF) is invoked to average Equations 8.11, 8.12, and 8.16 through 8.18,

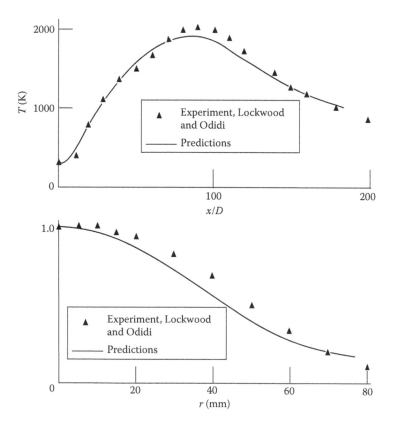

FIGURE 8.2 The axial distribution and radial profile of temperature in a jet diffusion flame. The symbols show the experimental data. The curves have been computed. (Reprinted from Lockwood, F.G. and Naguib, A.S., *Combust. Flame*, 24, 109–124, 1975. With permission.)

and $h = h_2 - (h_2 - h_1)f$. Lockwood and Naguib (1975) computed the curves plotted in Figure 8.2 by allowing for the temperature dependence of c_p in equilibrium combustion products, whereas the use of a constant c_p and Equations 8.26 and 8.27 would yield less satisfactory results (cf. Figures 1.1 and 1.2 in order to see the difference in the temperatures calculated using the two methods).

It is worth noting that the aforementioned simplification of adiabatic burning may result in a notable overestimation of the temperature within a diffusion flame, whereas the same simplification works better in a typical premixed flame. The point is that (i) the flux of thermal energy from the stoichiometric surface heats the rich zone of a nonpremixed flame, thereby promoting soot formation due to pyrolysis of the fuel and (ii) the thermal radiation from the soot particles results in substantial heat losses. Due to radiation from soot particles, many diffusion flames have a yellow color. On the contrary, in a typical premixed flow, soot formation and radiative heat losses are of much less importance due to oxidation of hydrocarbon fuels by oxygen, which is mixed with the fuel in this case but is separated from a fuel in a diffusion flame.

If radiative heat losses from a diffusion flame should be taken into account, then the simple linear relation of $h = h_2 - (h_2 - h_1)f$ does not hold and the temperature depends not only on the mixture fraction but also on the local value of h, which can be determined by solving Equation 8.19 with a radiative sink term introduced on the RHS. In such a case, Equations 8.11, 8.12, and 8.16 through 8.18 hold, whereas Equations 8.21 through 8.27 should be replaced with

$$c_p \left(T_b - T_0 \right) + Y_F \Delta h_F^0 + Y_O \Delta h_O^0 + Y_P \Delta h_P^0 = h. \tag{8.28}$$

Within the framework of the paradigm of infinitely fast reaction, complex combustion chemistry may be taken into account by replacing the linear Equations 8.11, 8.12, 8.16 through 8.18, and either Equations 8.21 through 8.27 or Equation 8.28 with the nonlinear equations

$$Y_l = Y_l^{eq} \left(f Z_{C,1}, f Z_{H,1}, (1 - f) Z_{O,2}, h \right) \quad \text{and} \quad T = T^{eq} \left(f Z_{C,1}, f Z_{H,1}, (1 - f) Z_{O,2}, h \right), \tag{8.29}$$

where the mixture temperature and composition are computed for given values of $Z_C = f Z_{C,1}$, $Z_H = f Z_{H,1}$, $Z_O = (1 - f) Z_{O,2}$, and h by utilizing the method described in the section "Calculation of Temperature and Composition of Adiabatic Combustion Products" in Chapter 1. In other words, the local mixture temperature and composition at a point $\mathbf{x}$ within a diffusion flame at instant t are associated with the temperature and composition of the equilibrium products formed during combustion of the fuel–air mixture characterized by $f(\mathbf{x},t)$ and $h(\mathbf{x},t)$ that have been evaluated by solving Equations 8.9 and 8.10 and Equations 8.19 and 8.20, respectively.

If Equation 8.29 is used, then the computed profiles of $Y_O(f)$, $Y_P(f)$, and $T(f)$ are continuous at $f = f_{st}$ (see the dashed curves in Figure 8.1), and small but finite concentrations of molecular oxygen can be calculated at $f > f_{st}$, because backward combustion reactions do not allow O_2 to vanish completely even in rich equilibrium products. Moreover, the equilibrium approach offers an opportunity (i) to evaluate

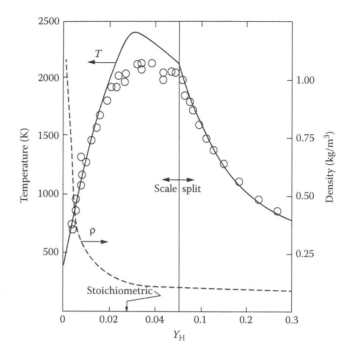

FIGURE 8.3 The correlation of a hydrogen element mass fraction and temperature in a turbulent jet diffusion flame. The symbols show the experimental data. The solid and dashed lines show the adiabatic equilibrium combustion temperature and density, respectively. (Reprinted from Kent, J.H. and Bilger, R.W., *Proc. Combust. Inst.*, 14, 615–625, 1973. With permission.)

not only Y_F and Y_O but also the mass fractions of CO, H_2, etc., and (ii) to improve the accuracy of temperature calculations. For instance, Figure 8.3 indicates that, in the lean and rich zones of a turbulent diffusion hydrogen flame, the mean temperature measured by Kent and Bilger (1973) agrees very well with the adiabatic combustion temperature calculated for the mean measured mass fraction of hydrogen element, whereas the difference in the temperatures measured and calculated in the near-stoichiometric zone is associated with the influence of turbulent fluctuations in the mixture fraction (if the mean mixture fraction is close to f_{st} when almost all fluctuations in f reduce the temperature).

As compared with Equations 8.11, 8.12, and 8.16 through 8.18, a limitation of the equilibrium approach lies in the fact that it leads to conversion of a hydrocarbon fuel to CO and H_2 in the rich zone of a diffusion flame (e.g., see the dashed curve for Y_F in Figure 8.1), but such an effect is contradicted by measurements.

DIFFUSION FLAMES WITH FINITE-RATE CHEMISTRY

Although the paradigm of infinitely fast chemistry is capable of predicting the mean length and structure of many diffusion flames reasonably well (e.g., see

Figures 8.2 and 8.3, as well as Figures 5.1 through 5.5 in a book by Kuznetsov and Sabel'nikov [1990]), it does not address a number of important phenomena, for example, ignition in the mixing layer of fuel and oxygen flows, quenching of diffusion combustion, and liftoff of turbulent jet diffusion flames. Moreover, there are processes that can be described qualitatively but cannot be modeled quantitatively if the chemical reactions that control heat release are assumed to be infinitely fast.

The formation of nitrogen oxide in flames is one of the most well-known processes of the latter kind. This problem received plenty of attention due to the growing interest in clean combustion technology and motivated the development of approaches aimed at allowing for finite rates of chemical reactions in nonpremixed combustion. As discussed in the section "Chemical Reactions in Flames" in Chapter 1, the formation of thermal NO in flames is a relatively slow process. Nevertheless, even within the framework of the paradigm of infinitely fast chemistry, this particular difficulty could easily be resolved by separating the modeling of the structure of a diffusion flame from the modeling of the formation of NO in the flame. Indeed, due to low concentrations of NO in combustion products, the influence of reactions XX–XXII discussed in the section "Chemical Reactions in Flames" in Chapter 1 on the structure of a diffusion flame may be neglected and the rate of nitrogen oxide formation may be calculated at a postprocessing stage after the flame structure has been modeled by assuming infinitely fast chemistry. A much greater challenge is that the rate of NO formation is proportional to the concentration of oxygen atoms. This concentration is not yielded by Equations 8.11, 8.12, and 8.16 through 8.18 but may be evaluated by invoking the equilibrium Equation 8.29. However, such models substantially underestimate the concentration of O and the rate of NO formation in diffusion flames, as reviewed by Kuznetsov and Sabel'nikov (1990). A finite rate of key chemical reactions should be taken into account to resolve the problem.

If the rates of the chemical reactions that control heat release are finite, then the knowledge of the mixture fraction field $f(\mathbf{x},t)$ obtained by integrating Equations 8.9 and 8.10 does not allow us to compute the fields of $Y_F(\mathbf{x},t)$, $Y_O(\mathbf{x},t)$, and $Y_P(\mathbf{x},t)$ even in the simplest case of a single combustion reaction and $Sc_l = 1$. Accordingly, Equation 1.58 should be solved in order to allow for the effects caused by finite reaction rates.

On account of the crucial role played by mixture fraction in diffusion combustion, it is tempting to rewrite Equation 1.58 in a coordinate framework attached to the stoichiometric surface $f(\mathbf{x},t) = f_{st}$ (Peters, 1983, 1984). For this purpose, Equations 8.9 and 8.10 should be solved to determine the surface, and an arbitrary point $\mathbf{x}_0$ on this surface could be considered. Then, the coordinate framework $\{x,y,z\}$ should be rotated so that the x'-axis is perpendicular to the stoichiometric surface and the plane (y',z') is tangential to the surface at the point $\mathbf{x}_0$ (see Figure 8.4). After doing so, let us transform the framework $\{x',y',z',t\}$ to a new framework $\{f(x',t),\eta,\zeta,\tau\}$, where $\eta = y'$, $\zeta = z'$ and $\tau = t$. For the sake of brevity, we will perform this coordinate transformation in a two-dimensional (2D) case in the following text. Because

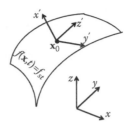

FIGURE 8.4 A stoichiometric surface.

$$\frac{\partial}{\partial t} = \frac{\partial}{\partial \tau} + \frac{\partial f}{\partial t}\frac{\partial}{\partial f}, \quad \frac{\partial}{\partial x'} = \frac{\partial f}{\partial x'}\frac{\partial}{\partial f}, \quad \text{and} \quad \frac{\partial}{\partial y'} = \frac{\partial}{\partial \eta} + \frac{\partial f}{\partial y'}\frac{\partial}{\partial f}, \tag{8.30}$$

we have

$$0 = \frac{\partial}{\partial t}(\rho Y_l) + \frac{\partial}{\partial x'}(\rho u Y_l) + \frac{\partial}{\partial y'}(\rho v Y_l) - \frac{\partial}{\partial x'}\left(\rho D\frac{\partial Y_l}{\partial x'}\right) - \frac{\partial}{\partial y'}\left(\rho D\frac{\partial Y_l}{\partial y'}\right) - M_l W_l$$

$$= \rho\frac{\partial Y_l}{\partial \tau} + \rho\frac{\partial Y_l}{\partial f}\frac{\partial f}{\partial t} + \rho u\frac{\partial Y_l}{\partial f}\frac{\partial f}{\partial x'} + \rho v\frac{\partial Y_l}{\partial \eta} + \rho v\frac{\partial f}{\partial y'}\frac{\partial Y_l}{\partial f}$$

$$\quad - \frac{\partial}{\partial x'}\left(\rho D\frac{\partial f}{\partial x'}\frac{\partial Y_l}{\partial f}\right) - \frac{\partial}{\partial y'}\left(\rho D\frac{\partial Y_l}{\partial \eta}\right) - \frac{\partial}{\partial y'}\left(\rho D\frac{\partial f}{\partial y'}\frac{\partial Y_l}{\partial f}\right) - M_l W_l$$

$$= \frac{\partial Y_l}{\partial f}\left(\rho\frac{\partial f}{\partial t} + \rho u\frac{\partial f}{\partial x'} + \rho v\frac{\partial f}{\partial y'}\right) + \rho\frac{\partial Y_l}{\partial \tau} + \rho v\frac{\partial Y_l}{\partial \eta} - \frac{\partial Y_l}{\partial f}\frac{\partial}{\partial x'}\left(\rho D\frac{\partial f}{\partial x'}\right)$$

$$\quad - \rho D\left(\frac{\partial f}{\partial x'}\right)^2\frac{\partial^2 Y_l}{\partial f^2} - \frac{\partial Y_l}{\partial \eta}\frac{\partial}{\partial y'}(\rho D) - \rho D\left(\frac{\partial^2 Y_l}{\partial \eta^2} + \frac{\partial f}{\partial y'}\frac{\partial^2 Y_l}{\partial f \partial \eta}\right)$$

$$\quad - \frac{\partial Y_l}{\partial f}\frac{\partial}{\partial y'}\left(\rho D\frac{\partial f}{\partial y'}\right) - \rho D\frac{\partial f}{\partial y'}\left(\frac{\partial^2 Y_l}{\partial f^2}\frac{\partial f}{\partial y'} + \frac{\partial^2 Y_l}{\partial f \partial \eta}\right) - M_l W_l$$

$$= \frac{\partial Y_l}{\partial f}\left[\rho\frac{\partial f}{\partial t} + \rho u\frac{\partial f}{\partial x'} + \rho v\frac{\partial f}{\partial y'} - \frac{\partial}{\partial x'}\left(\rho D\frac{\partial f}{\partial x'}\right) - \frac{\partial}{\partial y'}\left(\rho D\frac{\partial f}{\partial y'}\right)\right]$$

$$\quad + \rho\frac{\partial Y_l}{\partial \tau} + \rho v\frac{\partial Y_l}{\partial \eta} - \rho D\frac{\partial^2 Y_l}{\partial f^2}\left[\left(\frac{\partial f}{\partial x'}\right)^2 + \left(\frac{\partial f}{\partial y'}\right)^2\right] - \frac{\partial Y_l}{\partial \eta}\frac{\partial}{\partial y'}(\rho D)$$

$$\quad - \rho D\left(\frac{\partial^2 Y_l}{\partial \eta^2} + 2\frac{\partial f}{\partial y'}\frac{\partial^2 Y_l}{\partial f \partial \eta}\right) - M_l W_l.$$

Using Equation 8.9 and returning to a 3D case, we obtain

$$
\rho\frac{\partial Y_l}{\partial \tau} - \rho\chi\frac{\partial^2 Y_l}{\partial f^2} - M_l W_l = -\rho v\frac{\partial Y_l}{\partial \eta} - \rho w\frac{\partial Y_l}{\partial \zeta} + \frac{\partial Y_l}{\partial \eta}\frac{\partial}{\partial y'}(\rho D) + \frac{\partial Y_l}{\partial \zeta}\frac{\partial}{\partial z'}(\rho D)
$$

$$
+ \rho D\left(\frac{\partial^2 Y_l}{\partial \eta^2} + \frac{\partial^2 Y_l}{\partial \zeta^2} + 2\frac{\partial f}{\partial y'}\frac{\partial^2 Y_l}{\partial f\partial \eta} + 2\frac{\partial f}{\partial z'}\frac{\partial^2 Y_l}{\partial f\partial \zeta} \right), \qquad (8.31)
$$

where

$$
\chi \equiv D\frac{\partial f}{\partial x_k}\frac{\partial f}{\partial x_k} \qquad (8.32)
$$

is the scalar dissipation. The foregoing coordinate transformation has allowed us to arrive at Equation 8.31, which does not involve the normal (to the stoichiometric surface) convection term.

Furthermore, because (i) combustion reactions that control heat release are characterized by large activation energies and (ii) the highest temperatures are achieved in the vicinity of the stoichiometric surface, it is tempting to assume that the reaction rates drop sharply with distance from the surface of $f(\mathbf{x},t) = f_{st}$ or, in other words, the reactions are localized at thin (in the mixture fraction space) zones, which are somehow similar to the thin reaction zone within a premixed laminar flame. If, moreover, the spatial gradient $|\nabla f|$ (or $|\partial f/\partial x'|$) is sufficiently large in the vicinity of the stoichiometric surface, then the reaction zone will be thin (as compared with the length scales of flow nonuniformities) even in the physical space. In such a case, the reaction zone may be considered to be locally planar, that is, its thickness is much less than the length scales of the surface wrinkles caused by flow nonuniformities, and the tangential derivatives may be skipped on the RHS of Equation 8.31. Then, we arrive at (Peters, 1983, 1984)

$$
\rho\frac{\partial Y_l}{\partial t} - \rho\chi\frac{\partial^2 Y_l}{\partial f^2} - M_l W_l = 0. \qquad (8.33)
$$

If the scalar dissipation varies sufficiently slowly with time, then the unsteady term seems to be small as compared with the reaction term and Equation 8.33 is further simplified as

$$
-\chi\frac{\partial^2 Y_l}{\partial f^2} = \frac{M_l W_l}{\rho}. \qquad (8.34)
$$

A similar equation may be obtained for temperature in the case of $Le_l = 1$, $p = \text{const}$ and $c_p = \text{const}$.

It is interesting to note that the same equation was first obtained by Bilger (1976) for the reaction zone of any thickness, provided that $Y_l = Y_l(f)$ solely depends on the mixture fraction. The latter assumption appears to be justified, for example, if the differences between Y_l and $Y_l^{eq}(f)$ given by Equation 8.29 are small as compared with Y_l^{eq}. Under the aforementioned assumption, Equations 1.49 and 1.58 read

$$0 = \rho \frac{\partial Y_l}{\partial t} + \rho u_k \frac{\partial Y_l}{\partial x_k} - \frac{\partial}{\partial x_k}\left(\rho D_l \frac{\partial Y_l}{\partial x_k}\right) - M_l W_l$$

$$= \frac{dY_l}{df}\left(\rho \frac{\partial f}{\partial t} + \rho u_k \frac{\partial f}{\partial x_k}\right) - \frac{\partial}{\partial x_k}\left(\rho D_l \frac{dY_l}{df} \frac{\partial f}{\partial x_k}\right) - M_l W_l$$

$$= \frac{dY_l}{df}\left[\rho \frac{\partial f}{\partial t} + \rho u_k \frac{\partial f}{\partial x_k} - \frac{\partial}{\partial x_k}\left(\rho D_l \frac{\partial f}{\partial x_k}\right)\right] - \rho D_l \frac{d^2 Y_l}{df^2} \frac{\partial f}{\partial x_k} \frac{\partial f}{\partial x_k} - M_l W_l$$

and, using Equations 8.9 and 8.32, we obtain Equation 8.34.

It is worth stressing the following difference between Equation 8.34 derived by Bilger (1976) and the same equation obtained by Peters (1983) for a thin, stationary reaction zone. The former equation is valid in any point within a diffusion flame, provided that $Y_l = Y_l(f)$ solely depends on the mixture fraction. If the point is outside a reaction zone, that is, the reaction rates are negligibly small, then Equation 8.34 may be satisfied only if $d^2 Y_l/df^2 = 0$, that is, Y_l may solely depend on the mixture fraction only if the dependence is linear. Considering that the mass fraction of any nonreacting species varies linearly with the mixture fraction, we may consider Equation 8.34 to be valid in spatial regions where $W_l = 0$. Peters did not invoke the foregoing assumption about $Y_l = Y_l(f)$, but his derivation of Equation 8.34 is restricted to (i) a reaction zone that is much thinner than the smallest turbulence length scale and (ii) chemical reactions that are much faster than the shortest turbulence timescale.

Equation 8.34 is fully consistent with the paradigm of infinitely fast chemistry. Indeed, in this case, the reaction rates are proportional to the Dirac delta function, $\delta(f - f_{st})$, and the second derivatives, $\partial^2 Y_l/\partial f^2$, are also proportional to the same Dirac delta function if the partially linear Equations 8.11, 8.12, and 8.16 through 8.18 hold. If W_l is finite, then the dependence of Y_l on f should be nonlinear with negative (positive) $\partial^2 Y_l/\partial f^2$ for products (reactants) in order for Equation 8.34 to hold. A flame structure consistent with Equation 8.34 is shown in Figure 8.5, with the dependencies of $Y_F(f)$, $Y_O(f)$, etc., being nonlinear within a reaction zone of thickness $(\Delta f)_r$, which is bounded by two dashed lines in the figure. Outside the reaction zone, the dependencies are linear, as discussed earlier.

Equation 8.34 indicates a very important peculiarity of many diffusion flames. Because $W_l = W_l(Y_l, h)$ and both Y_l and h solely depend on f in the reaction zone of a constant-pressure adiabatic diffusion flame characterized by a given χ and $Le_l = 1$, the structure of the reaction zone is solely controlled by the scalar dissipation rate in the zone. Kuznetsov (1977) drew a similar conclusion regarding the structure of instantaneous reaction zones in turbulent nonpremixed flames, by analyzing a

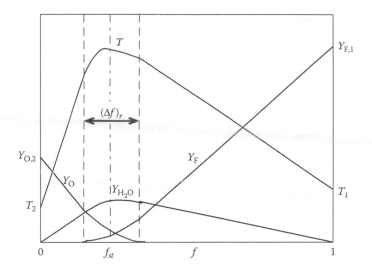

FIGURE 8.5 An asymptotic structure of a diffusion flame described by Equation 8.34.

balance equation for the PDF of the concentration of a reacting species in a turbulent flow.

Equation 8.34 highlights the scalar dissipation to be the key characteristic of non-premixed combustion, similar to the laminar flame speed, which is the key characteristic of premixed combustion. However, there is a basic difference between the two quantities. While S_L is solely controlled by the physical (molecular transport coefficients, heat capacity, and enthalpy) and chemical (reaction rates) characteristics of a burning gas mixture, χ depends on the boundary conditions and time in a general case.

Indeed, if we consider burning in a stationary, 2D mixing layer of fuel and oxygen laminar streams separated by a plane $\{x \leq 0; y = 0\}$ (see Figure 8.6a), then Equation 8.9 reads

$$u \frac{\partial f}{\partial x} = D \frac{\partial^2 f}{\partial x^2} + D \frac{\partial^2 f}{\partial y^2} \tag{8.35}$$

in the simplest case of a constant-density "flame." If the fuel is supplied at $y > 0$, that is, $Y_F(x,\infty) = 1$ and $Y_F(x,-\infty) = 0$, while $Y_O(x,\infty) = 0$ and $Y_O(x,-\infty) = 1$, then the boundary conditions are $f(x,\infty) = 1$ and $f(x,\infty) = 0$. If the diffusion in the x-direction is negligible as compared with the convection, then the problem has the approximate solution

$$f = 1 - \frac{1}{2} \operatorname{erfc}(\xi) = 1 - \frac{1}{\sqrt{\pi}} \int_{\xi}^{\infty} e^{-\eta^2} d\eta, \tag{8.36}$$

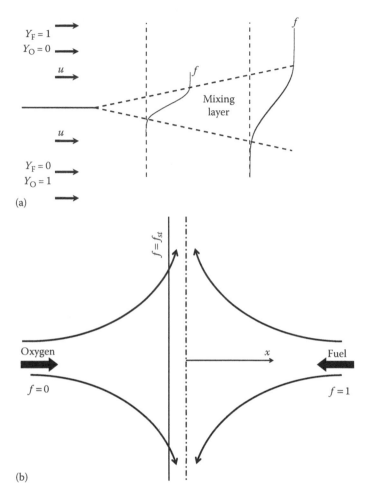

FIGURE 8.6 Two examples of laminar nonpremixed combustion: (a) a planar mixing layer and (b) two opposed jets.

where $\xi = y\sqrt{u/4Dx}$. Accordingly,

$$\frac{\partial f}{\partial y} = \frac{1}{2\sqrt{\pi\, Dx/u}}\, e^{-\xi^2} \tag{8.37}$$

and the scalar dissipation

$$\chi = \frac{u}{4\pi x}\, e^{-2\xi^2} \tag{8.38}$$

is controlled by the boundary conditions (flow velocity) and decreases with an increase in x, with $\chi(y)$ having a peak at $y = 0$. Therefore, different regions of such a diffusion flame are characterized by different magnitudes of the scalar dissipation, whereas different regions of an unperturbed laminar premixed flame are characterized by the same S_L.

Nonpremixed combustion of opposed fuel and oxygen streams is another well-known example of a diffusion flame (see Figure 8.6b). If we assume that the strain rate $\dot{s}$ is spatially uniform in the entire flame and $\rho^2 D = \rho_u^2 D_u$, then Equations 5.64, 5.65, 5.69, and 5.71 may be used to describe the axial profile of the mixture fraction, provided that $x_r = 0$, $S_{Lb} = 0$, and c is substituted with f in the foregoing equations. Accordingly,

$$\frac{\partial f}{\partial \zeta} = B \exp\left(-\frac{\dot{s}\zeta^2}{2\rho_u^2 D_u}\right), \tag{8.39}$$

where

$$\zeta = \int_x^0 \rho\, dx \tag{8.40}$$

and the constant $B = -\rho_u^{-1}\sqrt{\dot{s}/2\pi D_u}$ can be evaluated by integrating Equation 8.39 from $\zeta = -\infty$ to $\zeta = \infty$ and using the boundary conditions of $f(\zeta = -\infty) = f(x = \infty) = 1$ and $f(\zeta = \infty) = f(x = -\infty) = 0$. Therefore, the scalar dissipation rate

$$\chi = \frac{\dot{s}}{2\pi} \exp\left(-\frac{\dot{s}\zeta^2}{\rho_u^2 D_u}\right) \tag{8.41}$$

varies along the x-axis and is proportional to the strain rate, that is, it is controlled by the boundary conditions in an opposed-jet diffusion flame. Note that the scalar dissipation rates given by Equations 8.38 and 8.41 differ substantially from one another in the two simple flames discussed earlier.

Thus, similar to S_L, which controls the burning rate and the structure (together with the molecular diffusivity) of a laminar premixed flame, the scalar dissipation rate controls the structure of a diffusion flame in the mixture fraction space. However, contrary to the laminar flame speed, χ is controlled by the boundary conditions rather than by the mixture properties.

Equation 8.34 allows us to estimate the total mass rate, $\rho_u\Omega_p$, of formation (or consumption) of l-th species per unit area of the stoichiometric surface. Indeed, the integration of Equation 8.34 along the normal to the surface, that is, along the x'-axis in Figure 8.4, yields

$$\rho_u\Omega_p \equiv \int_{-\infty}^{\infty} M_l W\, d\xi = -\int_{-\infty}^{\infty} \rho\chi \frac{\partial^2 Y_l}{\partial f^2}\, d\xi = -\int_0^1 \rho\chi \frac{\partial^2 Y_l}{\partial f^2} \frac{df}{df/d\xi}, \tag{8.42}$$

provided that $f(-\infty) = 0$ and $f(\infty) = 1$. Although the second integral in Equation 8.42 is taken along an infinitely long line, the integral is finite, because $\partial^2 Y_l/\partial f^2 \neq 0$ only in a thin reaction zone. In the case of infinitely fast reactions localized to the stoichiometric surface, the second derivative $\partial^2 Y_l/\partial f^2$ is proportional to the Dirac delta function and

$$\rho_u \Omega_p \propto \rho_{st} \chi_{st} \left(\frac{\partial f}{\partial x'} \right)_{st}^{-1} \propto \left(\rho D \frac{\partial f}{\partial x'} \right)_{st}, \tag{8.43}$$

thereby supporting the hypothesis that the burning rate is controlled by the mixing rate in this limit case. In the case of finite-rate chemistry localized at a thin reaction zone, both χ and $\partial f/\partial x'$ seem to vary weakly in the zone due to its small thickness, and the scalar dissipation is mainly controlled by the normal gradient of the mixture fraction, that is, $\chi = D|\nabla f|^2 \approx D(\partial f/\partial x')^2$. Therefore,

$$\rho_u \Omega_p \approx -\left(\rho D \frac{\partial f}{\partial z'} \right)_{st} \int_0^1 \frac{\partial^2 Y_l}{\partial f^2} \, df = -\left(\rho D \frac{\partial f}{\partial z'} \right)_{st} \left[\frac{\partial Y_l}{\partial f} \right]_{f=0}^{f=1}, \tag{8.44}$$

again supporting the aforementioned hypothesis on the mixing-controlled burning rate.

EXTINCTION

Equations 8.43 and 8.44 imply that the specific, that is, per unit area of the stoichiometric surface, mass burning rate, $\rho_u \Omega_p$, is increased by the gradient of the mixture fraction in the vicinity of the stoichiometric surface, that is, $\rho_u \Omega_p$ is increased by χ. In the case of finite-rate chemistry, such a conclusion is valid only in the limited range of variations in the scalar dissipation. Indeed, because fast reactions occur under the stoichiometric conditions, the local heat release rate in an adiabatic diffusion flame cannot be higher than the maximum heat release rate reached in the adiabatic, stoichiometric, laminar premixed flame, provided that $Le_l = 1$ for all species. Accordingly, if the scalar dissipation rate is sufficiently high and is further increased, then the bounded reaction term on the RHS of Equation 8.34 can balance the term on the left-hand side (LHS) only if the magnitude of the second derivative, $\partial^2 Y_l/\partial f^2$, is decreased in the vicinity of the stoichiometric surface, $f(\mathbf{x},t) = f_{st}$. Therefore, the mass fractions of the fuel and oxygen are increased at the stoichiometric surface (see Figures 8.1 and 8.5 and note that $(\partial^2 Y_F/\partial f^2)_{st} \to -\infty$ and $(\partial^2 Y_O/\partial f^2)_{st} \to -\infty$ for infinitely fast reactions, while these second derivatives are finite if the reaction rates are finite), while the mass fractions of the products and temperature are decreased. As a result, the reaction rates are further decreased and the previously discussed laminar diffusion flame is quenched if the scalar dissipation rate is too high.

Zel'dovich (1949) developed a simple physical concept of quenching. If χ_{st} is sufficiently low, then the asymptotic $(\chi_{st}/|M_F W_F| \to 0)$ theory discussed in the section "Paradigm of Infinitely Fast Chemistry" works well; that is, (i) the heat release is localized to a very narrow reaction zone, (ii) the concentrations of the fuel and oxygen in this zone are very low (significantly lower than the counterpart concentrations in the reaction zone of the stoichiometric premixed laminar flame), but (iii) the temperature is very close to the adiabatic combustion temperature, T_{st}, for the stoichiometric mixture. Because of low Y_F and Y_O, the peak $|W_F|$ in such a diffusion flame is significantly lower than the peak fuel consumption rate $|W_F|_{max}$ reached in the stoichiometric premixed flame, with all other things being equal. Accordingly, an increase in χ_{st} can be balanced by an increase in $|W_F|$ due to an increase in the mass fractions of the fuel and oxygen within the reaction zone. The increase in Y_F and Y_O in the vicinity of the stoichiometric surface is accompanied by a decrease in the temperature in the reaction zone. If χ_{st} is sufficiently low and the reaction zone is sufficiently thin, then the former effect dominates (even in a laminar premixed flame, the peak $|W_F|$ is reached at $T < T_{st}$), and both $|W_F|$ and $\rho_u \Omega_p$ are increased by χ_{st}, as shown by Zel'dovich (1949) (see also Zel'dovich et al., 1985). For instance, if $Y_F \propto \varepsilon \ll 1$, $Y_O \propto St\varepsilon$, and $T_{st} - T \propto \varepsilon T_{st}$ in the reaction zone, then

$$|W_F| \propto St\varepsilon^2 \exp\left(-\frac{\Theta}{T_{st}}\right) \exp\left(-\frac{\Theta}{T_{st}}\varepsilon\right) \tag{8.45}$$

(see Equation 1.44 for a bimolecular reaction and Equation 2.45). The rate $|W_F|$ is increased by ε, provided that $\varepsilon\Theta \ll T_{st}$. However, because $|W_F|$ in a diffusion flame is bounded by $|W_F|_{max}$, the increase in $|W_F|$ by χ_{st} may occur in a bounded range of scalar dissipation rates. For instance, if $\varepsilon > 2T_{st}/\Theta \ll 1$, then Equation 8.45 yields a decrease in $|W_F|$ with an increase in ε. Thus, if χ_{st} exceeds a certain critical value χ_q, a further increase in the scalar dissipation rate rapidly decreases the reaction rates and quenches combustion, because the reaction term on the RHS of Equation 8.34 is not able to balance the term on the LHS if $\chi_{st} > \chi_q$.

In other words, an increase in the scalar dissipation rate is accompanied by an increase in the diffusion fluxes of oxygen and fuel to the reaction zone. Because the maximum reaction rate in the zone is bounded by $|W_F|_{max}$, the diffusion fluxes may be so intense that the fuel and oxygen are not fully burned. Under such conditions, the products are diluted by the unburned reactants, the product temperature is decreased, and the reaction rate drops, thereby further increasing the foregoing imbalance between the supply of the fuel and oxygen by diffusion and the consumption of them in combustion reactions. As a result, the flame is quenched. Because the most intense fuel and oxygen diffusion fluxes that can be fully burned are associated with the reaction zone of the adiabatic stoichiometric premixed flame, Zel'dovich (1949) argued that

$$\left| D\frac{df}{dx'} \right|_q \propto f_{st}(1 - f_{st}) S_{L,st} \tag{8.46}$$

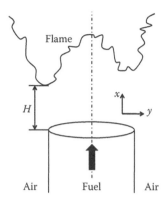

FIGURE 8.7 A lifted turbulent jet diffusion flame.

under quenching conditions, and, therefore,

$$\chi_q \propto \frac{f_{st}^2 \left(1 - f_{st}\right)^2}{\tau_{c,st}}, \tag{8.47}$$

that is, the critical scalar dissipation rate that quenches a diffusion flame is inversely proportional to the chemical timescale, $\tau_{c,st} = D_u/S_{L,st}^2$, that characterizes the counterpart stoichiometric premixed flame.

The local quenching of diffusion combustion because of high scalar dissipation rates was hypothesized to be a physical mechanism that controls such a well-known phenomenon as the liftoff of nonpremixed jet flames (Janicka and Peters, 1982; Peters, 1983, 1984; Peters and Williams, 1983). As sketched in Figure 8.7, such flames are stabilized at a certain distance H from the nozzle, depending on the jet velocity, nozzle diameter, etc. Because the scalar dissipation rate in the mixing layer in the vicinity of the nozzle is high and its maximum (for various y) value decreases with distance x from the nozzle, the flame liftoff may be caused by the quenching of diffusion combustion at small x associated with $\chi > \chi_q$. However, more recent experimental and direct numerical simulations (DNS) data support alternative explanations of the liftoff phenomenon, as discussed in detail in the review papers by Buckmaster (2002), Chung (2007), Lyons (2007), and Lawn (2009).

Another relevant phenomenon, well investigated in the opposed-jet diffusion flames sketched in Figure 8.6b, is a decrease in the peak product temperature with an increase in the scalar dissipation rate, followed by the extinction of nonpremixed combustion. Equation 8.34 describes the extinction conditions reasonably well (e.g., cf. the symbols and solid curve in Figure 8.8).

Certainly, contemporary computer hardware and software allows one to simulate such a flame by integrating Equation 1.58, which is more general than Equation 8.34, and by using a detailed chemical mechanism, $Le_l \neq 1$, etc., with the results of such simulations being in good agreement with the experimental data (e.g., cf. the symbols and dashed curve in Figure 8.8 or see Figure 8.9). Such simulations indicate,

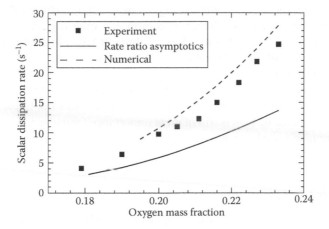

FIGURE 8.8 The scalar dissipation rate at extinction versus the mass fraction of oxygen in the oxidizer stream. (Adapted from Seshadri, K., *Proc. Combust. Inst.*, 26, 831–846, 1996. With permission.)

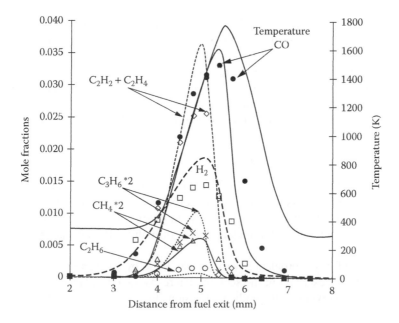

FIGURE 8.9 Mole fractions of various species versus distance from the exit of the fuel duct. The symbols show the experimental data. The curves have been computed. (Reprinted from Seiser, R., Truett, L., Trees, D., and Seshadri, K., *Proc. Combust. Inst.*, 27, 649–657, 1998. With permission.)

in particular, that the concentration of oxygen atoms in a diffusion flame may be much higher ("superequilibrium") than its equilibrium value (e.g., cf. the circles with crosses in Figure 8.15 or see Figures 3 and 5 through 8 in a paper by Barlow et al. [1990]). This effect substantially enhances the production of nitrogen oxide in non-premixed combustion.

Despite the progress made in numerical simulations of laminar diffusion flames by integrating Equation 1.58 with a detailed chemical mechanism, $Le_l \neq 1$, radiative heat losses etc., Equation 8.34 is still of great importance for understanding the physics of nonpremixed combustion, as well as for modeling turbulent diffusion flames, as discussed in the next section.

TURBULENT DIFFUSION FLAMES

Similar to premixed combustion, the major effects of turbulence on a diffusion flame consist of increasing the area of the stoichiometric surface and changing the structure of the reaction zone attached to the surface. The former mechanism increases the burning rate both in premixed and diffusion flames. The latter mechanism affects differently premixed and nonpremixed combustion. As far as a premixed flame with $Le_l = 1$ and finite reaction rates is concerned, the influence of turbulence on the inner structure of the local reaction zones decreases the local burning rate (Klimov, 1963) and, subsequently, can cause local combustion quenching. In a nonpremixed flame, a moderate increase in the scalar dissipation rate by turbulent eddies may increase the local burning rate, contrary to the premixed flame, but too high χ can cause local combustion quenching, similar to the premixed flame. Despite the foregoing analogies with the influence of turbulence on premixed and nonpremixed combustion, the latter phenomenon is much easier to model because of the dominant role played by mixing in a diffusion flame. In the latter case, the two effects discussed earlier consist of intensification of mixing by turbulence and, in the first approximation, may be described reasonably well within the framework of the concept of turbulent diffusion of a conserved scalar, that is, mixture fraction in the combustion case.

PARADIGM OF INFINITELY FAST CHEMISTRY

A simple, but efficient way of modeling a turbulent nonpremixed flame is to consider the behavior of a mixture fraction in a turbulent flow and invoke the paradigm of infinitely fast chemistry, which allows one to predict the mean temperature and the mean mass fractions of the major species in many diffusion flames (e.g., see Figure 8.10).

On the face of it, the problem is very simple. First, by (i) averaging Equation 8.9 using the method described in the section "Favre-Averaged Balance Equations" in Chapter 3 and (ii) invoking the gradient diffusion closure of the turbulent scalar flux $\overline{\rho u_i'' f''}$, we have

$$\frac{\partial}{\partial t}\left(\bar{\rho}\tilde{f}\right) + \frac{\partial}{\partial x_k}\left(\bar{\rho}\tilde{u}_k\tilde{f}\right) = \frac{\partial}{\partial x_k}\left(\bar{\rho}D_t \frac{\partial \tilde{f}}{\partial x_k}\right), \tag{8.48}$$

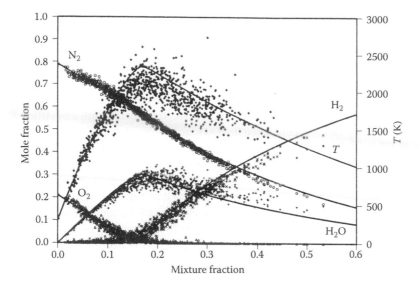

FIGURE 8.10 An ensemble of the Raman scattering measurements (symbols) of major species concentrations and temperature in a jet nonpremixed flame. The lines show the equilibrium concentrations and temperature calculated using Equation 8.29. (Reprinted from Barlow, R.S., Dibble, R.W., Chen, J.-Y., and Lucht, R.P., *Combust. Flame*, 82, 235–251, 1990. With permission.)

which is basically similar to the Reynolds-averaged Equations 3.102 through 3.104. Second, by solving Equation 8.48, one could determine the Favre-averaged mixture fraction field $\tilde{f}(\mathbf{x},t)$, followed by the calculation of $\tilde{Y}_F(\tilde{f})$, $\tilde{Y}_O(\tilde{f})$, $\tilde{Y}_P(\tilde{f})$, and $\tilde{T}(\tilde{f})$ using Equations 8.11, 8.12, 8.16 through 8.18, 8.26, and 8.27, based on the apparent linearity of these equations. It is worth remembering that $\overline{y(x)} = y(\bar{x})$ for a linear function $y(x) = a + bx$ due to the mutual cancellation of the effects of positive and negative fluctuations in x on y.

However, such an approach substantially oversimplifies the problem by neglecting the influence of turbulent pulsations in the mixture fraction in a diffusion flame. Indeed, first, although the foregoing equations are linear either at $f < f_{st}$ or at $f > f_{st}$, they are nonlinear at $f = f_{st}$. For instance, if we consider $T(f)$ given by Equations 8.26 and 8.27, then both positive ($f' \equiv f - \bar{f} > 0$) and negative ($f' < 0$) fluctuations in the mixture fraction around $\bar{f} = f_{st}$ reduce the temperature and make $\overline{T(f)}$ lower than $T(\bar{f})$ (e.g., cf. the symbols and solid curve in Figure 8.3).

Fluctuations in mixture fractions can also result in $\tilde{Y}_F(\mathbf{x})\tilde{Y}_O(\mathbf{x}) > 0$ in a statistically stationary turbulent diffusion flame with infinitely fast chemistry, despite $Y_F(\mathbf{x},t)$ $Y_O(\mathbf{x},t) = 0$, $\overline{Y_F(\mathbf{x},t)Y_O(\mathbf{x},t)} = 0$, and $Y_F(\tilde{f}(\mathbf{x}))Y_O(\tilde{f}(\mathbf{x})) = 0$ in this case. The problem is that even if the fuel and oxygen cannot be recorded in the same point at the same time, they may be recorded in this point at different instants due to fluctuations in f during the time interval of averaging.

Second, if the concentrations of intermediate species such as CO should be addressed or if the temperature should be calculated for equilibrium combustion products by allowing for the temperature dependence of heat capacities, then nonlinear

Equation 8.29 should be used. In such a case, the knowledge of $\tilde{f}$ is not sufficient to accurately evaluate $\tilde{T}$ even in a lean or rich region of a nonpremixed flame.

PRESUMED PROBABILITY DENSITY FUNCTIONS

The aforementioned problems may be easily resolved by invoking a PDF $P(f)$ for a mixture fraction. When $P(f)$ was studied properly, the mean length and structure of certain laboratory turbulent diffusion flames were predicted well (e.g., see Figures 5.1 through 5.5 in a book by Kuznetsov and Sabel'nikov [1990]). In principle, such a PDF could be determined by solving a PDF balance equation and substantial progress was obtained in this direction, as discussed elsewhere (O'Brien, 1980; Pope, 1985, 1994; Kuznetsov and Sabel'nikov, 1990; Dopazo, 1994; Haworth, 2010). However, because this method strongly needs better elaborated closures of the molecular mixing term in the PDF balance equation and is computationally demanding, especially as far as unsteady, 3D computations of combustion in engines are concerned, another simpler approach is widely used in engineering applications.

The approach consists of invoking a presumed expression for $P(f, \overline{\rho f^n}/\overline{\rho})$, which involves its first moments $\overline{\rho f^n}/\overline{\rho}$, where $n = 1, 2, \dots$, as unknown parameters. Subsequently, the presumed PDF is used to close the balance equations for the foregoing moments, these equations are numerically solved, and the PDF is fully recovered based on the computed $\overline{\rho f^n}/\overline{\rho}$. Within the framework of such an approach, the beta-function PDF given by the RHS of Equation 7.53, with the combustion progress variable c being substituted with the mixture fraction f, was most widely used (Janicka and Kollmann, 1979; Peters, 1984; Williams, 1985a) because the beta-function PDF is determined only by two first moments and may take substantially different shapes (see Figure 8.11), depending on the values of these two moments.

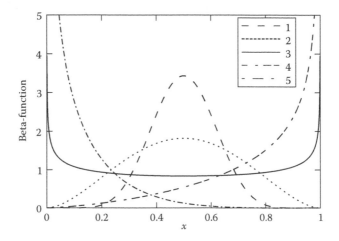

FIGURE 8.11 Beta-function PDFs $P(f, \overline{f}, \overline{f'^2})$ calculated for (1) $\overline{f} = 0.5$ and $\overline{f'^2} = 0.05\overline{f}(1 - \overline{f})$, (2) $\overline{f} = 0.5$ and $\overline{f'^2} = 0.15\overline{f}(1 - \overline{f})$, (3) $\overline{f} = 0.5$ and $\overline{f'^2} = 0.4\overline{f}(1 - \overline{f})$, (4) $\overline{f} = 0.1$ and $\overline{f'^2} = 0.15\overline{f}(1 - \overline{f})$, and (5) $\overline{f} = 0.8$ and $\overline{f'^2} = 0.25\overline{f}(1 - \overline{f})$.

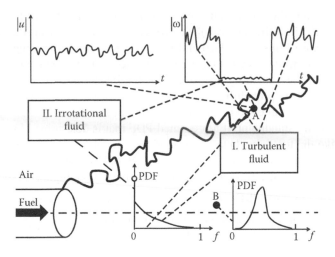

FIGURE 8.12 External intermittency.

Nevertheless, despite the variety of shapes taken by the beta-function PDF, there are many more PDFs that are not parameterized by Equations 7.53 through 7.55 (with $c = f$). For instance, nonpremixed combustion is often investigated by utilizing the open jet burners sketched in Figure 8.7, that is, fuel is supplied by a high-speed turbulent jet surrounded either by a low-speed airstream or by quiescent air. Such flows are well known to be intermittent, that is, the turbulent (rotational) fluid is separated from the ambient irrotational fluid by a thin interface, called a viscous superlayer (see Figure 8.12), with the instantaneous boundary of the turbulent mixing layer being very close (for gases that are burned in a typical diffusion flame) to the viscous superlayer (Bilger, 1980; Kuznetsov and Sabel'nikov, 1990; Pope, 2000). In such a flow, substantial pulsations of velocities are caused by pressure forces in wide spatial regions I and II on both sides of the viscous superlayer, while instantaneous velocity gradients and vorticity are large only on one side (region I) but almost vanish on the other side (region II). Similarly, the instantaneous concentration gradients are large in the turbulent fluid (I) but almost vanish in the ambient irrotational fluid (II). Accordingly, the time dependencies of the local vorticity and local scalar dissipation rate, measured at a point A in the vicinity of the mean boundary of the turbulent mixing layer, look very different depending on whether point A is in region I or II. The PDF $P(f)$ involves the Dirac delta function, $\delta(f)$, associated with a finite probability of finding pure air at this point. On the contrary, at a point B in the vicinity of the jet axis and at a sufficiently large distance from the jet exit (outside the potential core), there is neither pure fuel nor pure air and $P(f)$ contains neither $\delta(f)$ nor $\delta(f - 1)$. If the mixing of two half-space turbulent streams of fuel and oxygen is considered (see Figure 8.6a), then $P(f)$ may involve not only $\delta(f)$ at a sufficiently large negative y but also $\delta(f - 1)$ at a sufficiently large positive y.

Although the beta-function PDF may exhibit high peaks at $f = 0$ and $f = 1$ if the second moment $\overline{f'^2}$ is large (see the solid curve in Figure 8.11), these peaks differ substantially from the Dirac delta function. In particular, if $a \ll 1$, substitution of Equation 7.53 into the integral $\int_0^\varepsilon P(x)dx$ yields ε^a, considering that $\Gamma(a) \propto a^{-1}$ as $a \to 0$, because $\Gamma(a + 1) = a\Gamma(a)$ and $\Gamma(1) = 1$. Therefore, the integral vanishes as $\varepsilon \to 0$ for any finite $a \ll 1$, that is, contrary to $\delta(f)$, the beta-function PDF is associated with the vanishing probability of finding pure air even in the region of $\bar{f} \ll 1$, where this probability should be about unity. Consequently, the beta-function PDF is not suited to modeling intermittent flows.

From a general standpoint, if a presumed PDF that is determined by its first two moments, like the foregoing beta-function PDF, is invoked, then the mean value of $y(x)$ will be weakly sensitive to the choice of PDF only if the dependence of $y(f)$ is weakly nonlinear. When averaging a highly nonlinear $y(f)$, higher moments play a significant role, and two PDFs characterized by the same $\bar{x}$, the same $\overline{x^2}$, but different $\overline{x^n}$ ($n \geq 3$) may yield substantially different $\overline{y(x)}$.

Williams (1985a) noted that the PDF

$$P(f) = \alpha\delta(f) + \beta\delta(1 - f) + \gamma \exp\left[-\frac{(f - \mu)^2}{2\sigma^2}\right], \tag{8.49}$$

where

$$\gamma = (1 - \alpha - \beta)\left\{\sigma\sqrt{\frac{\pi}{2}}\left[\text{erf}\left(\frac{1 - \mu}{\sqrt{2}\sigma}\right) + \text{erf}\left(\frac{\mu}{\sqrt{2}\sigma}\right)\right]\right\}^{-1} \tag{8.50}$$

due to the normalization constraint $\int_0^1 P(f)df = 1$, is much more flexible than the beta-function PDF and is applicable to intermittent flows. However, the use of Equations 8.49 and 8.50, which involve four unknown parameters α, β, μ, and σ, requires the knowledge of four first moments $\overline{f^n}$ ($n \leq 4$) and makes simulations too complicated. For further discussion on the choice of a presumed PDF, interested readers are referred to a review paper by Bilger (1980) and to books by Williams (1985a), Kuznetsov and Sabel'nikov (1990), and Peters (2000).

To recover a presumed PDF, the balance equation

$$\frac{\partial}{\partial t}\overline{\rho f''^2} + \frac{\partial}{\partial x_k}\left(\tilde{u}_k\overline{\rho f''^2}\right) = -2\overline{\rho u_k''f''}\frac{\partial \tilde{f}}{\partial x_k} - \frac{\partial}{\partial x_k}\overline{\rho u_k''f''^2} + 2\overline{f''\frac{\partial}{\partial x_k}\left(\rho D\frac{\partial f}{\partial x_k}\right)}, \tag{8.51}$$

for the second moment $\overline{\rho f''^2}$ can be obtained using the method utilized to derive Equation 7.42. Balance equations for higher moments $\overline{\rho(f'')^n}/\bar{\rho}$ ($n \geq 3$) may be obtained in a similar manner. The last term on the RHS of Equation 8.51 is asymptotically equal to $-2\bar{\rho}\tilde{\chi}$ at high Reynolds numbers. Indeed, if $\rho D = \text{const}$ for simplicity, then

$$\overline{f'' \frac{\partial}{\partial x_k}\left(\rho D \frac{\partial f}{\partial x_k}\right)} = \overline{\rho D f'' \frac{\partial^2 f}{\partial x_k^2}} = \overline{\rho D f \frac{\partial^2 f}{\partial x_k^2}} - \overline{\rho D \tilde{f} \frac{\partial^2 \overline{f}}{\partial x_k^2}}$$

$$= \overline{\rho D \frac{\partial}{\partial x_k}\left(f \frac{\partial f}{\partial x_k}\right)} - \overline{\rho D \frac{\partial f}{\partial x_k}\frac{\partial f}{\partial x_k}} - \rho D\tilde{f}\frac{\partial^2 \overline{f}}{\partial x_k^2}$$

$$= -\overline{\rho\chi} + \frac{1}{2}\rho D \frac{\partial^2 \overline{f^2}}{\partial x_k^2} - \rho D\tilde{f}\frac{\partial^2 \overline{f}}{\partial x_k^2}$$

and the last two terms, which contain the gradients of the mean quantities multiplied with small molecular transport coefficients, are negligible at $\text{Re}_t \to \infty$. On the contrary, terms that contain molecular transport coefficients and gradients of instantaneous quantities, such as the scalar dissipation term, do not vanish because the latter gradients may be infinitely large as $\text{Re}_t \to \infty$.

Finally, if the transport term, that is, the second term on the RHS of Equation 8.51, is closed, invoking a gradient diffusion approximation, then we arrive at (Williams, 1985a; Peters, 2000)

$$\frac{\partial}{\partial t}\left(\overline{\rho}\frac{\overline{\rho f''^2}}{\overline{\rho}}\right) + \frac{\partial}{\partial x_k}\left(\overline{\rho}\tilde{u}_k \frac{\overline{\rho f''^2}}{\overline{\rho}}\right) = -2\overline{\rho u_k'' f''}\frac{\partial \tilde{f}}{\partial x_k} + \frac{\partial}{\partial x_k}\left[\overline{\rho D_t}\frac{\partial}{\partial x_k}\left(\frac{\overline{\rho f''^2}}{\overline{\rho}}\right)\right] - 2\overline{\rho}\tilde{\chi}.$$

(8.52)

The Favre-averaged scalar dissipation rate on the RHS of Equation 8.52 is commonly closed on the basis of the Favre-averaged turbulent kinetic energy $\tilde{k}$ and its dissipation rate $\tilde{\varepsilon}$ as

$$2\overline{\rho}\tilde{\chi} = C_\chi \frac{\tilde{\varepsilon}}{\tilde{k}}\overline{\rho f''^2},$$

(8.53)

where different values of the constant C_χ, varying from 1 to 3, may be found in the literature, as reviewed by Peters (2000). A more sophisticated approach deals with a balance equation for $\tilde{\chi}$ (e.g., see a review paper by Jones [1994]).

Because the balance Equations 8.48 and 8.52 are written for the Favre-averaged moments, it is the Favre-averaged presumed PDF

$$\tilde{P}(f) \equiv \frac{\rho}{\overline{\rho}}P(f)$$

(8.54)

that is widely used in the combustion literature. If $\tilde{P}(f)$ is parameterized with a beta-function (see Equation 7.53) and the Favre-averaged fields $\tilde{f}(\mathbf{x}, t)$ and $\overline{\rho f''^2}(\mathbf{x}, t)$ are computed by numerically solving Equations 8.48 and 8.52, respectively, then the parameters a and b of the PDF can be easily calculated using Equation 7.55, with c being substituted with f. If a beta-function is invoked to model $P(f)$, then the

Reynolds-averaged moments $\bar{f}(\mathbf{x},t)$ and $\overline{f'^2}(\mathbf{x},t)$ have to be evaluated on the basis of $\tilde{f}(\mathbf{x},t)$ and $\widetilde{\rho f''^2}(\mathbf{x},t)$ in order to use Equation 7.55. If the Favre-averaged PDF is presumed, then the mean density is evaluated as

$$\frac{1}{\bar{\rho}} = \frac{\overline{\rho\rho^{-1}}}{\bar{\rho}} = \int_0^1 \rho^{-1}\tilde{P}df. \qquad (8.55)$$

LAMINAR FLAMELET CONCEPT OF TURBULENT NONPREMIXED COMBUSTION

To study such issues as pollutant formation in diffusion flames or quenching of them, the paradigm of infinitely fast chemistry is not sufficient and more advanced approaches that deal with finite-rate chemistry are required. The flamelet library concept (Liew et al., 1981, 1984; Peters, 1984, 1986) is widely used to resolve the problem.

Within the framework of the concept, a turbulent nonpremixed flame is considered to be an ensemble of thin, inherently laminar, diffusion flamelets (Carrier et al., 1975; Williams, 1975; Bilger, 1976; Kuznetsov, 1977). Then, based on Equations 8.33 and 8.34, the inner structure of a flamelet is assumed to be solely controlled by the local scalar dissipation rate, χ_{st}, at the stoichiometric surface. Accordingly, turbulent nonpremixed combustion is divided into two subproblems: (i) turbulent mixing, which controls the fields of $\tilde{f}(\mathbf{x},t)$, $\widetilde{\rho f''^2}(\mathbf{x},t)/\bar{\rho}(\mathbf{x},t)$, and $\tilde{\chi}(\mathbf{x},t)$ and (ii) laminar burning in a diffusion flamelet subject to a given scalar dissipation rate. Therefore, one may simulate nonpremixed turbulent combustion by computing a library of laminar diffusion flames subject to various given χ_{st}, followed by averaging this library by invoking a proper joint PDF, $P(f, \chi_{st})$, the shape of which is determined on the basis of the calculated $\tilde{f}(\mathbf{x},t)$, $\widetilde{\rho f''^2}(\mathbf{x},t)/\bar{\rho}(\mathbf{x},t)$, and $\tilde{\chi}(\mathbf{x},t)$.

Note that although Equations 8.33 and 8.34 substantiate such an approach, the diffusion flamelet library may be computed by integrating the more general Equations 1.58 and 1.64 with $Le_l \neq 1$, while the former equations were obtained in the case of $Le_l = 1$. The stationary opposed-jet laminar diffusion flames sketched in Figure 8.6b are commonly addressed in such flamelet library simulations, which are performed for different strain rates, with the computed χ_{st} being used to parameterize the computed flamelet library. The flamelet library concept of nonpremixed turbulent combustion is sketched in Figure 8.13.

The considered approach, called the stationary laminar flamelet model (SLFM), is based on Equation 8.34, that is, the unsteady term in Equation 8.33 is neglected. Such a simplification may be justified if variations in χ are much slower than the burning in diffusion flamelets. Because the burning rate is controlled by the mixing rate, as discussed earlier (see Equations 8.43 and 8.44), SLFM appears to work well if $\chi_{st} t_x \gg 1$, where t_x is the timescale characterizing the rate of variations in χ. Therefore, in regions of low mean scalar dissipation rate, for example, at far distances from the nozzle exit in a jet nonpremixed flame, sketched in Figure 8.7, transient effects may be of great importance (e.g., cf. the symbols with solid and dashed curves in Figure 8.15 discussed later in the text), and the unsteady term in Equation 8.33 should be retained (Haworth et al., 1988; Mauss et al., 1990; Buriko et al., 1994; Pitsch et al., 1998; Peters, 2000).

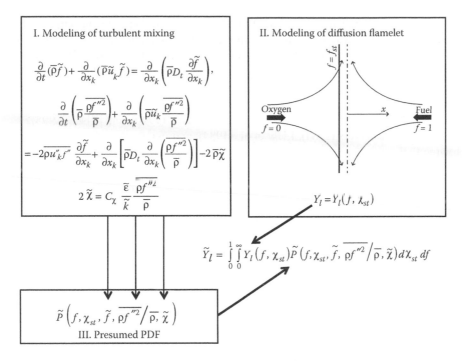

FIGURE 8.13 The flamelet library concept of nonpremixed turbulent combustion.

In order to use the flamelet library concept, one has to model the joint PDF $P(f, \chi_{st})$. Little is known about this function and an assumption on the statistical independence of f and χ_{st} at high Reynolds numbers is widely invoked, that is,

$$\tilde{P}(f,\chi_{st}) = \frac{\rho(f,\chi_{st})}{\overline{\rho}} P(f,\chi_{st}) = \frac{\rho(f,\chi_{st})}{\overline{\rho}} P_f(f) P_\chi(\chi_{st})$$

$$= \frac{\overline{\rho}}{\rho(f,\chi_{st})} \left[\frac{\rho(f,\chi_{st})}{\overline{\rho}} P_f(f) \right] \cdot \left[\frac{\rho(f,\chi_{st})}{\overline{\rho}} P_\chi(\chi_{st}) \right] = \frac{\overline{\rho}}{\rho(f,\chi_{st})} \tilde{P}_f \tilde{P}_\chi. \quad (8.56)$$

This assumption, discussed in detail by Kuznetsov and Sabel'nikov (1990), relies on the fact that the mixture fraction and scalar dissipation rate are controlled by large-scale and small-scale turbulent eddies, respectively. It is worth stressing that, in intermittent flows such as jets, this assumption may hold only for the fields $f_t(\mathbf{x},t)$ and $\chi_t(\mathbf{x},t)$ conditioned on turbulent fluid (Kuznetsov, 1972).

If Equation 8.56 is used, then the presumed PDFs, $\tilde{P}_f$ and $\tilde{P}_\chi$, are commonly invoked.[*] The former PDF is commonly parameterized invoking a beta function,

---

[*] Note that both $\tilde{P}_f(f) = P_f(f)\rho(f,\chi_{st})/\overline{\rho}$ and $\tilde{P}_\chi(\chi_{st}) = P_\chi(\chi_{st})\rho(f,\chi_{st})/\overline{\rho}$ depend both on f and χ_{st}. Therefore, (i) these two PDFs are not associated with two statistically independent phenomena and (ii) a widely invoked assumption (e.g., see Equation 8.57) that the latter (former) PDF depends only on χ_{st} (f) needs substantiation.

as discussed earlier, while the latter PDF (i) is either reduced to the Dirac delta function $\delta(\chi_{st} - \tilde{\chi})$, that is, fluctuations in the scalar dissipation rate are neglected, because solutions to Equation 8.34 depend weakly on χ_{st} (Kuznetsov, 1982; Buriko et al., 1994), or (ii) is assumed to be lognormal by referring to the third hypothesis of Kolmogorov (1962). For further discussion on this hypothesis and on the use of the lognormal PDF in turbulence theory, interested readers are referred to books by Monin and Yaglom (1975), Kuznetsov and Sabel'nikov (1990), and Frisch (1995).

In particular, Peters (1984), Liew et al. (1984), and many others invoked the following PDF:

$$\tilde{P}_\chi(\chi_{st}) = \frac{1}{\sqrt{2\pi}\sigma\chi_{st}} \exp\left[-\frac{(\ln\chi_{st}-\mu)^2}{2\sigma^2}\right]. \tag{8.57}$$

In this case,

$$\tilde{\chi} = \frac{1}{\bar{\rho}}\int_0^1\int_0^\infty \rho\chi_{st}P(f,\chi_{st})d\chi_{st}df = \int_0^1\left(\int_0^\infty \frac{\rho}{\bar{\rho}}\chi_{st}P_\chi(\chi_{st})d\chi_{st}\right)P_f(f)df$$

$$= \int_0^1\left(\int_0^\infty \chi_{st}\tilde{P}_\chi(\chi_{st})d\chi_{st}\right)P_f(f)df = \int_0^1 \exp\left(\mu+\frac{\sigma^2}{2}\right)P_f(f)df = \exp\left(\mu+\frac{\sigma^2}{2}\right), \tag{8.58}$$

with $\tilde{\chi}$ being modeled using Equation 8.53, and

$$\frac{\overline{\rho\chi''^2}}{\bar{\rho}\tilde{\chi}^2} = \frac{\overline{\rho\chi^2}-\bar{\rho}\tilde{\chi}^2}{\bar{\rho}\tilde{\chi}^2} = \exp(\sigma^2)-1. \tag{8.59}$$

In practical simulations, neither a balance equation for $\overline{\rho\chi''^2}$ nor Equation 8.59 is used, but σ^2 is assumed to be either constant (Peters, 1984) or proportional to $\ln(L/l)$, where L is an integral length scale of turbulence and l is associated with either the Taylor (Liew et al., 1984) or the Kolmogorov (Kuznetsov and Sabel'nikov, 1990) turbulence length scale. Liew et al. (1984) claimed that the probability $P_q = \int_{\chi_q}^\infty \tilde{P}_\chi d\chi$ of flamelet quenching, evaluated using Equations 8.57 and 8.58, is weakly sensitive to variations in σ^2 for values of $\ln(L/\lambda)$, $\tilde{\chi}$, and χ_q typical for turbulent nonpremixed combustion.

SLFM supplemented with a proper PDF, $\tilde{P}(f,\chi_{st})$, improves predictions of mass fractions of intermediate species in turbulent diffusion flames (e.g., cf. the symbols with solid and dashed curves in Figure 8.14 and note that the results are presented in the logarithmic scale). Worse agreement between the measured and the computed data obtained at large distances from the nozzle, that is, in the region characterized by a low mean scalar dissipation rate, is associated with the transient effects neglected in Equation 8.34. Figure 8.15 also indicates the importance of such effects

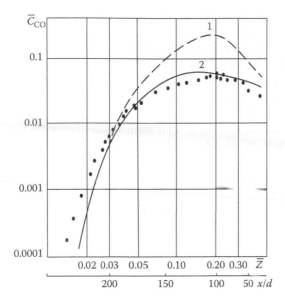

FIGURE 8.14 An axial profile of the mean mass fraction of CO in a jet diffusion flame of propane (Z designates the mixture fraction). The symbols show the experimental data. The solid and dashed lines have been computed by averaging the results obtained using SLFM and equilibrium $Y_{CO}^{eq}(f)$, respectively. (Reprinted from Buriko, Yu.Ya., Kuznetsov, V.R., Volkov, D.V., Zaitsev, S.A., and Uryvsky, A.F., *Combust. Flame*, 96, 104–120, 1994. With permission.)

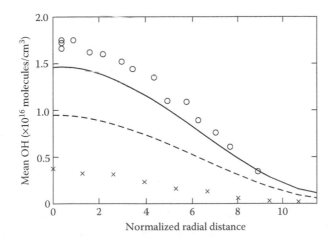

FIGURE 8.15 The radial profiles of the mean OH concentration measured in a $CO/H_2/N_2$–air jet diffusion flame at $x/D = 50$. The circles and crosses show the measured OH concentration and equilibrium OH concentration calculated on the basis of major species measurements. The solid and dashed lines have been computed by averaging the results obtained using an unsteady flamelet model and SLFM, respectively. (Adapted from Haworth, D.C., Drake, M.C., Pope, S.B., and Blint, R.J., *Proc. Combust. Inst.*, 22, 589–597, 1988. With permission.)

(cf. the symbols with solid and dashed curves) and shows superequilibrium concentrations of OH in the studied flame (cf. the circles and crosses).

Certainly, the domain of validity of SLFM is bounded. First, as mentioned earlier, the transient effects neglected in Equation 8.34 may be of importance as far as the regions characterized by a low mean $\tilde{\chi}$ or rapidly varying $\chi(t)$ are concerned. Even if unsteady diffusion flamelet models were proposed to be used (Haworth et al., 1988; Mauss et al., 1990; Pitsch et al., 1998; Peters, 2000), they strongly need thorough testing. Second, the assumption of a uniform scalar dissipation rate within a diffusion flamelet, that is, the use of a single χ_{st} to parameterize the flamelet library, may be correct only if the thickness of the flamelet reaction zone is much less than the smallest length scale of spatial variations in $\chi(\mathbf{x},t)$. Third, due to the substantial probability of finding $\chi_{st} \gg \tilde{\chi}$, the far tail of $\tilde{P}(\chi_{st})$ should be modeled very accurately in order to predict local extinction (Buriko et al., 1994). The lack of a well-elaborated approximation of $\tilde{P}_\chi(\chi_{st})$ in the range of $\chi_{st} \gg \tilde{\chi}$ (limitations of the lognormal PDF approach are discussed by Frisch, 1995) impedes the use of SLFM for the investigation of the extinction of nonpremixed combustion, even if Equation 8.34 could predict χ_q reasonably well for a slowly varying (both in space and time) field $\chi(\mathbf{x},t)$. Fourth, if this field varies rapidly such that the criterion of $\chi(\mathbf{x},t) > \chi_q$ holds locally in a small (as compared with the reaction zone thickness) spatial region or/and during a short (as compared with the chemical timescale) time interval, extinction may not occur (Darabiha, 1992). Therefore, transient and nonuniform effects may play a substantial role as far as extinction due to a large scalar dissipation rate in a highly uneven turbulent flow is concerned. Fifth, the structure of a curved diffusion flamelet may differ substantially from the structure of the counterpart planar diffusion flamelet if the curvature radius is of the order of the flamelet thickness (Cuenot and Poinsot, 1994).

Different governing physical mechanisms of turbulent combustion are commonly discussed with the help of combustion regime diagrams both in premixed (see Figure 5.46) and nonpremixed flames. However, in the latter case, the combustion regime diagrams published by different authors are very different (e.g., cf. Figure 6 by Borghi [1988], Figure 11 by Libby and Williams [1994], Figure 4 by Bray and Peters [1994], Figure 6 by Cuenot and Poinsot [1994], Figure 12 by Veynante and Vervisch [2002], and Figure 3.9 by Peters [2000]).

One difficulty in constructing combustion regime diagrams for nonpremixed burning is the lack of inherent characteristics of diffusion flames that are similar to the laminar flame speed and thickness in the premixed case. As previously discussed, the burning rate and, hence, the timescale of a diffusion flame are controlled by the local scalar dissipation rate, that is, they are different in different flows. Nevertheless, the shortest chemical timescale associated with near-quenching conditions appears to be a proper parameter for drawing combustion regime diagrams, because it characterizes the mixture, similar to $\tau_c = \kappa_u/S_L^2$ in the premixed case. Considering the straightforward relation between this timescale and the critical (quenching) scalar dissipation rate, given by Equation 8.47, χ_q may be used for this purpose. Then, a length scale that characterizes the mixture similar to the premixed laminar flame thickness is $\delta_f \propto \sqrt{\kappa_{st}/\chi_q}$, where κ_{st} is the molecular heat diffusivity at the stoichiometric surface. Subsequently, if χ_q and δ_f are invoked to characterize the timescale

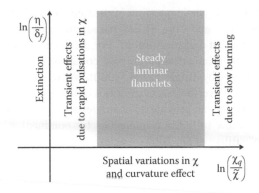

FIGURE 8.16   A combustion regime diagram for nonpremixed turbulent flows.

and length scale of the mixture, then the mean value of the scalar dissipation rate and the Kolmogorov length scale may be used to characterize the timescale and length scale of the flow.

It is worth noting that other scales and dimensional numbers are also highlighted to construct diagrams of turbulent diffusion combustion, for example, Bray and Peters (1994) used the ratio of the thickness Δf of a laminar diffusion flame, determined in the mixture fraction space, to the Favre-averaged variance $\overline{\rho f''^2}/\bar{\rho}$, while Peters (2000) pointed out that the reaction zone thickness $(\Delta f)_r$ is a more proper quantity than Δf for drawing combustion regime diagrams.

A combustion regime diagram for nonpremixed turbulent flows is sketched in Figure 8.16. The domain of validity of SLFM is associated with a gray rectangle in the center. If the Kolmogorov length scale is comparable with the thickness of the reaction zone of a laminar diffusion flamelet, then SLFM oversimplifies the problem (i) by assuming that the spatially nonuniform field, $\chi(\mathbf{x})$, may be substituted by a single scalar dissipation rate, χ_{st}, evaluated at the stoichiometric surface and (ii) by disregarding curvature effects, that is, by neglecting the tangential derivatives in Equation 8.31. If the mean scalar dissipation rate is too low, as happens in a far region of a jet flame, then SLFM oversimplifies the problem by disregarding the transient effects, that is, by neglecting the unsteady term in Equation 8.33. In this case, the transient effects may play a substantial role, because mixing-controlled burning is too slow. These effects may also play a substantial role in the case of moderately large $\tilde{\chi}$ if the local scalar dissipation rate varies rapidly and the use of a constant χ_{st} oversimplifies the problem. Finally, if the mean scalar dissipation rate is too high, as happens in the vicinity of a nozzle in a jet nonpremixed flame, then laminar diffusion flamelets are quenched, because even the highest possible local reaction rate is not sufficient in order for intense fluxes of fuel and oxygen toward the reaction zone to be completely burned.

It is difficult to specify the precise criteria for the validity of SLFM, because the length scale and timescale of a laminar diffusion flamelet are not constant quantities but depend on the local field of the scalar dissipation rate. Moreover, it is not clear a priori whether or not the two kinds of transient effects discussed earlier may be negligible in some region of a particular nonpremixed turbulent flame. As claimed by Bilger et al. (2005), "the whole question of the range of validity of flamelet models remains controversial" and "the SLFM paradigm cannot remain valid in the presence of local extinction and re-ignition." For a critical discussion on the laminar flamelet concept of turbulent nonpremixed combustion, interested readers are referred to review papers by Bilger (1988, 2000), Veynante and Vervisch (2002), and Bilger et al. (2005). Nevertheless, even critics of the concept recognize that "the laminar flamelet paradigm can provide an accurate description for sufficiently large turbulence scales and low turbulence intensities for combustion chemistry that is close to irreversible" (Bilger et al., 2005).

OTHER APPROACHES

Over the past decades, a few alternative advanced approaches to modeling turbulent nonpremixed burning, which are not restricted by the assumptions invoked by the laminar flamelet paradigm, were developed and are listed in the following text.

The conditional moment closure (CMC) put forward independently by Klimenko (1990) and Bilger (1993) is based on a hypothesis that the fluctuations in temperature and mass fractions in a turbulent nonpremixed flame are mainly controlled by the fluctuations in the mixture fraction. In other words, if one measures the temperature (or mass fraction, or reaction rate, etc.) in a point $\mathbf{x}$ only at that instant (here, the statistically stationary flow is considered for simplicity) when the mixture fraction is approximately equal to a certain given value z, that is, $z \leq f \leq z + dz$, and, subsequently, averages this random set of temperatures, $T|_{f=z}$, in order to evaluate conditionally averaged $\langle T|f = z \rangle$, then the fluctuations in the conditioned temperature $T|_{f=z}$ around $\langle T|f = z \rangle$ will be much less than the fluctuations in T around $\bar{T}$. Mathematically, this elegant hypothesis reads

$$\frac{\left(q\big|_{f=z} - \langle q|f = z \rangle \right)^2}{\langle q|f = z \rangle^2} \ll \frac{\overline{q^2}}{\bar{q}^2}, \tag{8.60}$$

where $q = \{T, Y_l, \rho, W_l, \ldots\}$. Unclosed balance equations for conditionally averaged quantities such as $\langle Y_l|f = z \rangle$ can be derived from the basic balance equations discussed in the section "Balance Equations" in Chapter 1, and Equation 8.60 offers an opportunity to use the simple closure

$$\langle W_l|f = z \rangle = W_l \left(\langle \rho|f = z \rangle, \langle T|f = z \rangle, \langle Y_l|f = z \rangle \right) \tag{8.61}$$

of strongly nonlinear reaction rates in the balance equations for $\langle T|f = z \rangle$ and $\langle Y_l|f = z \rangle$. In other words, conditionally averaged reaction rates are evaluated using conditionally averaged temperature, density, and mass fractions and by neglecting the fluctuations in the conditioned quantities. This simplification may not be accurate in certain cases, for example, if such fluctuations are significant due to large probabilities of local extinction and reignition. In such cases, a more complicated

(i) second-order closure (Swaminathan and Bilger, 1998; Kronenburg et al., 1998; Klimenko and Bilger, 1999), which allows for the influence of conditional second moments $\langle q_l q_m | f = z \rangle$ on the conditional reaction rates, or (ii) multiple mapping closure (Klimenko and Pope, 2003), which is aimed at combining "the advantages of the PDF and the CMC methods," or (iii) double conditioning (Kronenburg, 2004; Kronenburg and Papoutsakis, 2005), which deals with quantities conditioned on both mixture fraction and sensible enthalpy, could be used.

For further discussions on CMC, interested readers are referred to review papers by Klimenko and Bilger (1999) and Bilger et al. (2005). It is of interest to note that, although the flamelet approach and CMC were developed on the basis of different hypotheses, the closure relations widely used for conditional balance equations in homogeneous turbulence yield balance equations similar to Equation 8.33 (Klimenko, 2001).

The PDF balance equation approach, reviewed, for example, by O'Brien (1980), Pope (1985), Borghi (1988), Dopazo (1994), Kuznetsov and Sabel'nikov (1990), Bray (1996), Peters (2000), Veynante and Vervisch (2002), Bilger et al. (2005), and Haworth (2010), and the linear eddy model (LEM) developed by Kerstein (1988, 1992a, 1992b, 2002) are two other advanced methods of modeling turbulent nonpremixed combustion, with the finite-rate chemistry effects being taken into account, but these methods are computationally demanding.

Similar to premixed combustion, a diffusion flame affects turbulence, but the physical mechanisms of such effects are basically similar to the physical mechanisms of the influence of premixed burning on turbulence, which are discussed in Chapter 6. For numerical modeling of such effects in nonpremixed turbulent flows, interested readers are referred to a review paper by Jones (1994).

Similar to premixed combustion, a diffusion flame is affected by the difference in the molecular transport coefficients of major reactants and temperature (e.g., see Peters [2000]). However, in the latter case, such effects appear to be less pronounced than in premixed turbulent combustion. The point is that a premixed turbulent flame propagates, with the leading points controlling the speed of the flame propagation. Accordingly, the Lewis number and the preferential diffusion phenomena may strongly affect the flame speed by changing the local structure and burning rate of the leading points. Because a nonpremixed flame does not propagate, such leading-point processes do not seem to play a crucial role in diffusion combustion, and even strong local fluctuations in the mixture temperature and composition due to $Le_l \neq 1$ appear to be significantly damped after averaging.

SUMMARY

In a first approximation, the physics of diffusion combustion is reduced to a slow mixing process followed by infinitely fast chemical reactions. Then, burning is localized at the stoichiometric surface, which separates the mixture of fuel and combustion products from the mixture of oxidizer and products. At the stoichiometric surface, (i) there is neither fuel nor oxidizer, (ii) the temperature and mixture composition correspond to the stoichiometric combustion products, and (iii) the burning rate is controlled by the diffusion fluxes of the fuel and oxidizer toward the surface and is typically much less than the burning rate in the stoichiometric premixed flame of the same fuel and the same oxidizer. Within the framework of such an approximation, the influence of turbulence on diffusion flames consists of the intensification of mixing by increasing the local scalar gradients and wrinkling the stoichiometric surface.

In numerical simulations of turbulent nonpremixed combustion, the aforementioned simple physical scenario is implemented (i) by solving a balance equation for the mean concentration of a conserved scalar (mixture fraction), with the intensification of mixing by turbulence being addressed using turbulent diffusivity, and (ii) by averaging the dependencies of the equilibrium mass fractions and temperature on the conserved scalar, invoking a presumed PDF in order to allow for fluctuations in the scalar. The presumed PDF is commonly constructed on the basis of its first and second Favre-averaged moments, which are evaluated by numerically integrating proper balance equations. Such a simple method allows one to predict well the length, total burning rate, mean temperature, and mean mass fractions of major reactants in diffusion flames that are weakly affected by ignition and extinction phenomena.

If such phenomena are of importance or if the mass fractions of radicals, for example, O or OH, or pollutants, for example, NO or CO or soot, must be predicted, then the finite rates of chemical reactions should be taken into account. Analysis of laminar diffusion flames shows that, in the case of finite reaction rates, the flame structure depends not only on the mixture fraction but also on the scalar dissipation rate, with the latter quantity controlling the burning rate. Too high scalar dissipation rates may cause extinction, because finite-rate reactions cannot completely consume the fluxes of fuel and oxidizer transported to the stoichiometric surface by diffusion. The local burning rate in an adiabatic diffusion flame is increased with an increase in the scalar dissipation rate and reaches its peak value under near-extinction conditions, with this peak value being controlled by the burning rate in the stoichiometric premixed flame.

SLFM is the simplest numerical tool for simulating nonpremixed turbulent combustion by allowing for finite-rate chemistry effects. Within the framework of this model, a turbulent diffusion flame is an ensemble of inherently laminar diffusion flamelets characterized solely by the local magnitude of the scalar dissipation rate at the stoichiometric surface. Accordingly, a library of laminar diffusion flames subject to various scalar dissipation rates, that is, each particular flame from the library is subject to its particular χ, may be computed and parameterized as a function of the mixture fraction and the scalar dissipation rate at the stoichiometric surface. Subsequently, such a library may be averaged by assuming a joint PDF for the mixture fraction and the scalar dissipation rate.

Because SLFM highlights locally planar diffusion flamelets subject to a scalar dissipation rate that varies slowly in space (on scales of the order of flamelet thickness) and time (as compared with the timescale of burning), the model is justified only if the diffusion flamelets are significantly thinner than the Kolmogorov length scale and the chemical timescale characterizing the local burning rate is significantly shorter than the Kolmogorov timescale. Over the past two decades, more sophisticated approaches that do not suffer from the foregoing restrictions have been developed to simulate extinction and reignition or pollutant formation in turbulent nonpremixed flames. Unsteady flamelets, CMC, PDF balance equation, and LEM are examples of such approaches.

9 Partially Premixed Turbulent Flames

In many burners and internal combustion engines, heat is released neither in the nonpremixed mode nor in the homogeneously premixed mode. For instance, in a lean, prevaporized, premixed gas turbine combustor, fuel and air are premixed, but the mixture composition is inhomogeneous and lean, that is, the mean equivalence ratio varies in space with $\Phi(\mathbf{x},t) < 1$. Such a burning regime characterized by spatial variations in the equivalence ratio, which occur only in the lean (or only in the rich) domain, is often called stratified combustion (Bilger et al., 2005). In other words, a stratified flame consumes spatially inhomogeneous fuel–air mixture, with the equivalence ratio never being equal to unity. A premixed flame that is characterized by a homogeneous mean field $\overline{\Phi}(\mathbf{x},t) = \text{const}$ but is subject to fluctuations in $\Phi(\mathbf{x},t)$ also belongs to the stratified combustion regime, provided that $\Phi(\mathbf{x},t) \neq 1$.

In a direct injection gasoline spark ignition engine, the time interval between injection of the fuel into the combustion chamber and spark ignition may be too short for the mixture composition to be homogeneous at the instant of ignition but is sufficiently long for most of the fuel to be mixed with air before burning. Accordingly, the flame kernel created by the spark propagates through a highly inhomogeneous mixture characterized by large fluctuations in the equivalence ratio, with the ensemble-averaged mixture composition being lean (and even beyond the lean flammability limit) in some spatial regions and rich (and even beyond the rich flammability limit) in other regions. Such a burning regime is sometimes called premixed/nonpremixed combustion (Bilger et al., 2005). The same regime is associated with the burning of a rich fuel–air jet surrounded by a lean fuel–air jet (see Figure 9.1). In this regime, lean and rich premixed turbulent flames coexist with a turbulent diffusion flame, where CO and H_2 from the rich combustion products react with O_2 from the lean products.

In the two aforementioned examples of premixed/nonpremixed burning, inhomogeneously premixed combustion plays a major role, that is, it controls a major part of the total heat release, while the afterburning of the lean and rich products in the diffusion mode contributes a little to the total heat release, but it may be of significant importance as far as pollutant, for example, soot, formation is concerned. In a diesel engine, the time interval between fuel injection and autoignition is so short that only a small amount of the fuel is mixed with air before autoignition of the mixture due to compression. In such a case, similar to a direct injection gasoline engine, lean and rich premixed turbulent flames coexist with diffusion flames, but, contrary to the gasoline engine, the total heat release is mainly controlled by the nonpremixed mode of burning. This regime could be called nonpremixed/premixed combustion in order to stress the dominant role played by the diffusion flames. A lifted jet flame sketched in Figure 8.7 also seems to belong to this regime, as far as heat release is concerned,

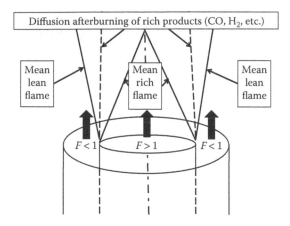

FIGURE 9.1 A sketch of premixed/nonpremixed jet flame.

but the premixed combustion mode may play a crucial role in flame stabilization in this case.

All of the aforementioned combustion modes (stratified, premixed/nonpremixed, and nonpremixed/premixed burning) are commonly subsumed under partially premixed flames. Contemporary understanding of the physics of partially premixed turbulent combustion is mainly based on the knowledge of premixed and diffusion turbulent flames, but there are physical mechanisms and phenomena that are specific to partially premixed burning. These mechanisms and phenomena may be divided into two groups: (i) effects taken somehow into account by contemporary combustion models for Reynolds-averaged Navier–Stokes (RANS) and large-eddy simulation (LES) applications and (ii) effects that have not yet been addressed by such models but were documented in experimental and direct numerical simulation (DNS) studies of laminar and turbulent partially premixed flames. These two groups of mechanisms and phenomena are considered in the sections "Effects Addressed in RANS or LES Studies of Partially Premixed Turbulent Flames" and "Effects Ignored in RANS or LES Studies of Partially Premixed Turbulent Flames," respectively. A multi-Dirac-delta-function approach that has been widely used for simulating partially premixed turbulent flames over the past few years is discussed in the section "Use of Presumed PDFs for Modeling both Premixed and Diffusion Burning Modes," followed by a brief summary.

EFFECTS ADDRESSED IN RANS OR LES STUDIES OF PARTIALLY PREMIXED TURBULENT FLAMES

INHOMOGENEITY OF MEAN MIXTURE COMPOSITION

Due to the inhomogeneity of the mean mixture composition, the average burning rate in the premixed mode of partially premixed combustion varies in space following variations in $\bar{\Phi}(\mathbf{x},t)$. In other words, a premixed turbulent flame propagates through a spatially nonuniform mixture. This process could be described by invoking a model of premixed turbulent combustion that is capable of well predicting the dependence of the total burning rate (or turbulent burning velocity U_t) on the equivalence ratio. For instance, if premixed turbulent combustion is simulated by solving the balance Equation 7.36 for the Favre-averaged combustion progress variable $\tilde{c}$, then the invoked closure of the mean rate $\overline{W}$ of product formation should yield a sufficiently accurate dependence of $\overline{W}$ on $\overline{\Phi}$ in the homogeneously premixed case. If either the turbulent flame closure (TFC) model or the flame speed closure (FSC) model of premixed turbulent combustion, both of which are discussed in the section "Models of a Self-Similarly Developing Premixed Turbulent Flame" in Chapter 7, is invoked, then, the dependence of $\overline{W}$ on $\overline{\Phi}$ is controlled by the dependence of U_t on $\overline{\Phi}$ (see Equation 7.116), and the use of Equation 7.13 allows one to well predict the latter dependence (see, e.g., Figures 4.17, 7.8, 7.11, 7.36, 7.37, and 7.38b).

Spatial variations in the mean mixture composition may be modeled by solving the balance Equation 8.48 for the Favre-averaged mixture fraction $\tilde{f}$ or another conserved scalar $\tilde{z}$, for example, the recovered mass fraction $Y_F^u = Z_C + Z_H$ of a hydrocarbon fuel. For the sake of brevity, both the symbol f and the term "mixture fraction" will designate not only a mixture fraction but also the aforementioned conserved scalar z in the following text.

The equivalence ratio may be easily calculated if the mixture fraction is known. For instance, bearing in mind that $(1 + St)$ kg of products is formed during the combustion of 1 kg of a fuel and St kg of an oxidizer, the recovered mass fraction of the fuel is equal to $Y_F + Y_P/(1 + St)$ and Equations 8.12, 8.17, and 8.18 yield

$$Y_F^u = Y_{F,1}\frac{f - f_{st}}{1 - f_{st}} + \frac{Y_{P,st}}{1 + St}\frac{1-f}{1 - f_{st}} = \frac{Y_{F,1}}{1 - f_{st}}\left[f - f_{st} + f_{st}\left(1 - f\right)\right] = Y_{F,1}f \qquad (9.1)$$

if $f > f_{st}$. If $f < f_{st}$, the same Equation 9.1 results from Equations 8.16 and 8.11. Similarly, Equations 8.11 and 8.16 through 8.18 yield

$$Y_O^u = Y_{O,2}\left(1 - f\right). \qquad (9.2)$$

Therefore, using Equation 8.5, we obtain

$$\Phi = \frac{StY_F^u}{Y_O^u} = \frac{StY_{F,1}}{Y_{O,2}}\frac{f}{1-f} = \left(\frac{1}{f_{st}} - 1\right)\frac{f}{1-f} = \frac{1 - f_{st}}{1 - f}\frac{f}{f_{st}}. \qquad (9.3)$$

Equation 9.3 is nonlinear, and averaging it is not simple. However, because the evaluation of the mean rate $\overline{W}$ (or turbulent burning velocity) using the Favre-averaged mixture fraction $\tilde{f}$ means, in fact, that the joint probability density function (PDF) $P(c,f,\mathbf{x},t)$ is reduced to $P_c(c,\mathbf{x},t)\,\delta[f - \tilde{f}(\mathbf{x},t)]$, we have

$$\overline{q(c,f,\mathbf{x},t)} = \int_0^1 q\left[c,\tilde{f}(\mathbf{x},t)\right]P_c\left(c,\mathbf{x},t\right)dc \qquad (9.4)$$

invoking the latter PDF. Here, q is an arbitrary continuous quantity, including the equivalence ratio. In particular, if the Bray–Moss–Libby (BML) PDF given by Equation 5.20 is invoked for the combustion progress variable, then

$$\overline{q(c,f,\mathbf{x},t)} = \left[1-\overline{c}(\mathbf{x},t)\right]\overline{q}_u\left(\tilde{f},\mathbf{x},t\right) + \overline{c}(\mathbf{x},t)\overline{q}_b\left(\tilde{f},\mathbf{x},t\right). \tag{9.5}$$

The approach outlined previously, that is, solving the two equations

$$\frac{\partial}{\partial t}\left(\overline{\rho}\tilde{c}\right) + \frac{\partial}{\partial x_k}\left(\overline{\rho}\tilde{u}_k\tilde{c}\right) = -\frac{\partial}{\partial x_k}\overline{\rho u_k''c''} + \overline{W} \tag{9.6}$$

and

$$\frac{\partial}{\partial t}\left(\overline{\rho}\tilde{f}\right) + \frac{\partial}{\partial x_k}\left(\overline{\rho}\tilde{u}_k\tilde{f}\right) = \frac{\partial}{\partial x_k}\left(\overline{\rho}D_t\frac{\partial\tilde{f}}{\partial x_k}\right) \tag{9.7}$$

supplemented with Equation 9.5, was used, for example, by Polifke et al. (2002) and Wallesten et al. (2002b) for simulating stratified turbulent combustion in a swirl-stabilized industrial burner and premixed/nonpremixed turbulent combustion in a direct injection spark ignition engine, respectively. In the two cited papers, Equation 9.6 was closed invoking the TFC and FSC models, respectively, with $U_t = U_t[\Phi(\tilde{f})]$.

It is worth noting that if the combustion progress variable is associated with the normalized mass fraction of the deficient reactant (henceforth in this chapter, for the sake of brevity, we will consider a fuel to be the deficient reactant), then the exact balance equation for c, derived from the exact balance equations for f and Y_F, involves extra source terms in the partially premixed case (Domingo et al. 2002; Bray et al., 2005). For instance, if

$$Y_F(\mathbf{x},t) = Y_F\left[c(\mathbf{x},t), f(\mathbf{x},t)\right], \tag{9.8}$$

then the equation

$$\frac{\partial}{\partial t}(\rho c) + \frac{\partial}{\partial x_k}(\rho u_k c)$$

$$= \frac{\partial}{\partial x_k}\left(\rho D\frac{\partial c}{\partial x_k}\right) + \left(\frac{\partial Y_F}{\partial c}\right)^{-1}W_F$$

$$+ \left(\frac{\partial Y_F}{\partial c}\right)^{-1}\left[\underbrace{\frac{\partial^2 Y_F}{\partial c^2}\rho D\frac{\partial c}{\partial x_k}\frac{\partial c}{\partial x_k}}_{\chi_c} + \underbrace{\frac{\partial^2 Y_F}{\partial f^2}\rho D\frac{\partial f}{\partial x_k}\frac{\partial f}{\partial x_k}}_{\chi_f} + \underbrace{\frac{\partial^2 Y_F}{\partial c\partial f}\rho D\frac{\partial c}{\partial x_k}\frac{\partial f}{\partial x_k}}_{\chi_{c,c}}\right] \tag{9.9}$$

results straightforwardly from

$$\frac{\partial}{\partial t}(\rho f) + \frac{\partial}{\partial x_k}(\rho u_k f) = \frac{\partial}{\partial x_k}\left(\rho D \frac{\partial f}{\partial x_k}\right) \tag{9.10}$$

and

$$\frac{\partial}{\partial t}(\rho Y_F) + \frac{\partial}{\partial x_k}(\rho u_k Y_F) = \frac{\partial}{\partial x_k}\left(\rho D \frac{\partial Y_F}{\partial x_k}\right) + W_F \tag{9.11}$$

(Bray et al., 2005). Because all the terms in square brackets on the right-hand side (RHS) of Equation 9.9 vanish in a premixed flame (f = const and Y_F depends linearly on c), the balance equations for c differ from one another in fully and partially premixed cases. Even if the first term in square brackets on the RHS of Equation 9.9 vanishes, provided that Y_F depends linearly on c, the second and third terms may be of importance in an inhomogeneous mixture, as discussed by Domingo et al. (2002, 2005a) and Bray et al. (2005). Accordingly, the straightforward use of a balance equation for $\tilde{c}$, closed for premixed turbulent combustion, for simulating a partially premixed flame may be put into question. Despite this clear disadvantage of the combustion progress variable concept, Bray et al. (2005) pointed out that $\tilde{c}$ is a sufficiently reasonable characteristic of partially premixed combustion, in particular, because the aforementioned extra terms are small in many practical cases.

It is also worth emphasizing that, in the flamelet regime of premixed turbulent burning, the Reynolds-averaged combustion progress variable $\bar{c}(\mathbf{x},t)$ is equal to the probability of finding equilibrium products in point $\mathbf{x}$ at instant t (see Equation 5.21), and certain RANS models of premixed turbulent combustion, for example, the TFC and FSC models discussed in the section "Models of a Self-Similarly Developing Premixed Turbulent Flame" in Chapter 7, consider $\bar{c}$ to be the probability, rather than the mean normalized mass fraction (or temperature or density). Accordingly, when applied to partially premixed flames, such RANS models do not need to address the balance equations for the instantaneous fields of $f(\mathbf{x},t)$ and $c(\mathbf{x},t)$, thus bypassing the problem of extra terms in Equation 9.9. Certainly, such a bypass could be justified only if $\bar{c}(\mathbf{x},t)$ is equal to the aforementioned probability, that is, if the probability $\gamma(\mathbf{x},t)$ of finding intermediate (between unburned and burned) states of the reacting mixture is much less than unity everywhere. If γ is of unity order, then the use of the combustion progress variable for simulating partially premixed flames does not seem to be the best choice, and the mass fraction of the deficient reactant appears to be a more appropriate characteristic of the mixture state.

TURBULENT PULSATIONS IN MIXTURE COMPOSITION

Equation 9.6 with $\overline{W(\Phi)} = W[\Phi(\tilde{f})]$ and Equation 7.75 with $\overline{U_t(\Phi)} = U_t[\Phi(\tilde{f})]$ do not allow for the influence of turbulent pulsations in the local equivalence ratio on

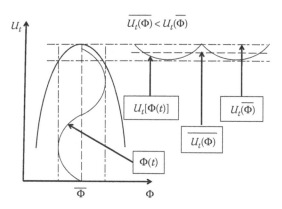

FIGURE 9.2 Difference between the averaged turbulent burning velocity and the turbulent burning velocity evaluated at the mean equivalence ratio.

the burning rate and other flame characteristics, but such an effect may be of substantial importance. For example, if $\tilde{\Phi} \approx 1$ and $U_t[\Phi(\tilde{f})] = \max\{U_t(\Phi)\}$, then both positive and negative pulsations in Φ will reduce $\overline{U_t(\Phi)}$ as compared with $U_t[\Phi(\tilde{f})]$, as sketched in Figure 9.2.

The simplest way to model the aforementioned effect consists of invoking a presumed PDF for the mixture fraction and evaluating the mean reaction rate and turbulent burning velocity as follows:

$$\overline{W\left[f\left(\mathbf{x},t\right)\right]} = \int_0^1 \langle W \rangle(f) P_f\left(f, \tilde{f}, \overline{\rho f''^2}\right) df, \tag{9.12}$$

$$\overline{U_t\left[f\left(\mathbf{x},t\right)\right]} = \int_0^1 U_t(f) P_f\left(f, \tilde{f}, \overline{\rho f''^2}\right) df. \tag{9.13}$$

Here, $\langle W \rangle = \langle W \rangle (L/u', u'/S_L, \ldots)$ is an expression for the mean reaction rate in a premixed turbulent flame. Both $\langle W \rangle$ and U_t depend on the mixture fraction f mainly due to the dependence of the laminar flame speed S_L on the equivalence ratio Φ. It is worth noting that Equations 9.12 and 9.13, in fact, imply the statistical independence of flame–turbulence interaction and turbulent pulsations in the mixture fraction. Simplicity appears to be the sole reason for invoking such an assumption, which is difficult to justify in a more solid manner.

As in the case of numerical models of turbulent diffusion flames, the beta-function approximation of $\tilde{P}_f(f, \tilde{f}, \overline{\rho f''^2})$ (see Figure 8.11 and Equations 7.53 through 7.55, with c being replaced by f) is widely invoked in the simulations of partially premixed turbulent flames and the second moment of the PDF is determined by solving the balance Equation 8.51.

If the combustion progress variable and mixture fraction are assumed to be statistically independent and the BML concept of premixed turbulent combustion is invoked, then

$$\bar{q}(\mathbf{x},t) = \left[1 - \bar{c}(\mathbf{x},t)\right] \int_0^1 q_u(f) P_f\left(f, \tilde{f}, \overline{\rho f''^2}\right) df$$

$$+ \bar{c}(\mathbf{x},t) \int_0^1 q_b(f) P_f\left(f, \tilde{f}, \overline{\rho f''^2}\right) df \qquad (9.14)$$

for an arbitrary continuous quantity q. When applying Equation 9.14 to the equivalence ratio, $\bar{\Phi}_u = \bar{\Phi}_b$ and we have

$$\bar{\Phi}(\mathbf{x},t) = \int_0^1 \frac{1 - f_{st}}{1 - f} \frac{f}{f_{st}} P_f\left(f, \tilde{f}, \overline{\rho f''^2}\right) df, \qquad (9.15)$$

using Equation 9.3. The approach outlined previously, that is, solving Equations 8.51, 9.6, 9.7, and 9.14 supplemented with the beta-function approximation of $P_f(f, \tilde{f}, \overline{\rho f''^2})$ and Equation 9.12, was used by Bigot et al. (2000) for simulating partially premixed propane–air flames stabilized by the recirculation zones in a channel with abrupt expansion. Two propane–air streams with different equivalence ratios were "injected in the two halves of the entrance part of the channel, separated by a plane." The model was successfully applied not only to a weakly stratified case (the equivalence ratios in the two streams were equal to 0.9 and 0.7) but also to highly stratified combustion, when the composition of one stream ($\Phi = 0.3$) was well beyond the lean flammability limit.

Bondi and Jones (2002) applied a basically similar approach with another submodel of $\langle W \rangle = \langle W \rangle$ $(L/u', u'/S_L, \ldots)$ "to the computation of a gas-fired burner representative of those found in industrial burner–boiler configuration." The equivalence ratio was varied from 0.5 to 1.6 when methane was fired in the burner. The computed results indicated that the model "is capable of predicting the main features" of the studied flame.

Note that Equations 9.13 through 9.15 involve the PDF P_f, rather than the Favre PDF $\tilde{P}_f$. On the contrary, Equations 7.53 through 7.55 yield the Favre PDF $\tilde{P}_f$, which may differ substantially from the counterpart PDF P_f. Similar equations could be invoked to presume $P_f(f, \bar{f}, f'^2)$, but, in a variable-density case, the balance equations for $\bar{f}$ and f'^2 involve extra unclosed terms as compared with widely used Equations 9.7 and 8.51, respectively. Thus, if a beta function is invoked to parameterize the Favre PDF $\tilde{P}_f$ in a partially premixed turbulent flame, then the calculation of P_f is an issue, which has not yet received proper attention in the literature. Moreover, if $\tilde{P}_f$ is independent of c, then the PDF $P_f = \tilde{P}_f \rho(c,f)/\bar{\rho}$ depends both on c and f, contrary to a widely invoked assumption on the statistical independence of fluctuations in the combustion progress variable and mixture fraction. Further research on this issue is definitely necessary.

MIXING-CONTROLLED AFTERBURNING

Partially premixed combustion involves one more specific effect, that is, mixing-controlled afterburning of lean and rich products, (e.g., see mean diffusion flames shown in dashed lines in Figure 9.1). This phenomenon is commonly considered to be basically similar to the classical nonpremixed combustion, at least if the chemical composition within the purely diffusion flame is assumed to be equal to the composition of the mixture in the chemically equilibrium state (see Equation 8.29). However, if the chemical composition within the purely diffusion flame is considered to be a mixture of stoichiometric combustion products and an unburned fuel if $f > f_{st}$, or oxygen if $f < f_{st}$ (see Equations 8.11, 8.12, 8.16 through 8.18), then the diffusion flamelets within a partially premixed turbulent flame may differ from the diffusion flamelets within a turbulent nonpremixed flame, because, in the former case, the fuel is decomposed into CO, CO_2, H_2, H_2O, etc., well before it reaches the stoichiometric surface (see Figure 9.3).

In principle, the models discussed earlier address mixing-controlled afterburning in a first approximation. Indeed, if Equation 9.5 or 9.14 is invoked and $\bar{c} = 1$, then the temperature and composition of the burned mixture are associated with the temperature and composition of the equilibrium combustion products at a given mixture fraction. Therefore, within the framework of such an approach, lean and rich combustion products are assumed to infinitely rapidly react with one another as soon as the mean (if Equation 9.5 is used) or the instantaneous (Equation 9.14) mixture fraction is equal to f_{st}.

The aforementioned approaches to simulating partially premixed flames are only a combination of a simple model of turbulent diffusion combustion, which was developed long time ago, with an advanced contemporary model of premixed turbulent burning. Such a combination of an old and a new model appears to be justified by the following reasoning. Because the predictive capabilities of the most advanced models of premixed turbulent flames are substantially poorer than the predictive capabilities of the contemporary models of turbulent diffusion combustion, an attempt to combine the most advanced models of both premixed and diffusion turbulent flames to simulate partially premixed burning appears to be unnecessary

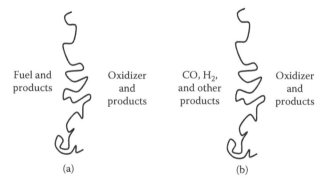

Fuel and products   Oxidizer and products   CO, H₂, and other products   Oxidizer and products

(a)   (b)

FIGURE 9.3 Diffusion flamelets within (a) nonpremixed and (b) partially premixed turbulent flames.

overcomplication, at least as far as stratified or premixed/nonpremixed combustion is concerned. In such a case, numerical predictions of a premixed/nonpremixed turbulent flame are limited by the former model and there is no need to invoke a significantly more precise advanced model of the diffusion burning regime. Accordingly, to reach a reasonable compromise between precision and computational efficiency of the simulations discussed, the use of a simple model of diffusion combustion would be sufficient, provided that its predictive capabilities are comparable with the predictive capabilities of the chosen advanced model of premixed turbulent flames. For this reason, the mixture fraction paradigm is the most widely used in numerical studies of mixing-controlled burning in stratified and premixed/nonpremixed turbulent flames, while different advanced models are invoked to describe the premixed burning mode.

The aforementioned models do not take combustion chemistry effects into account. In diffusion flamelets, such effects could be addressed by invoking the flamelet library approach, that is, by substituting the functions $\rho(f)$, $T(f)$, and $Y(f)$, calculated for laminar opposed-jet diffusion flames using a detailed chemical scheme, into the integrals on the RHS of Equation 9.14. Such an approach supplemented with the beta-function approximation of $\tilde{P}_f(f, \tilde{f}, \overline{\rho f''^2})$ was successfully applied by Müller et al. (1994) and Chen et al. (2000) to RANS simulations of jet flames, and the numerical results reported in the latter paper predicted the available experimental data on the liftoff heights of methane and propane jets reasonably well. In these simulations, the premixed combustion regime was modeled by solving the mean G-equation (Equation 7.157), with S_t being equal to U_t, evaluated using Equation 9.13. A basically similar approach, with Equation 7.157 being replaced by Equation 7.156, was utilized in the LES of various partially premixed and stratified turbulent flames by Nogenmyr et al. (2007, 2008, 2010) and Li et al. (2009).

Although the latter approach considers combustion chemistry effects in the diffusion combustion regime, it neglects similar effects in the premixed burning regime. On the one hand, as far as stratified or premixed/nonpremixed turbulent combustion is concerned, the use of flamelet libraries only for diffusion but not for premixed burning appears to be unnecessary overcomplication, which does not improve the predictive capabilities, as discussed earlier. On the other hand, the application of the flamelet library concept solely to diffusion flamelets may be useful, for example, to study soot formation in partially premixed combustion, because (i) the soot formation in premixed flamelets is negligible as compared with the soot formation in the mixing-controlled burning regime, with all other things being equal, and (ii) this process can only be described by invoking complex combustion chemistry. However, the eventual difference between the diffusion flamelets in nonpremixed and partially premixed turbulent flames, sketched in Figure 9.3, should be borne in mind when studying soot formation in partially premixed combustion. In a classical diffusion flame, soot production is associated with pyrolysis of the hydrocarbon fuel in the rich zone of the flame. However, as far as mixing-controlled afterburning of lean and rich products is concerned, the fuel has already been partially oxidized to CO, CO_2, H_2O, etc., in the products of rich premixed burning.

Another issue is as follows. A flamelet library of a diffusion flame is classically constructed on the basis of the local mixture fraction and the scalar dissipation rate at the stoichiometric surface, as discussed in the sections "Diffusion Flames with Finite Rate Chemistry" and "Laminar Flamelet Concept of Turbulent Nonpremixed Combustion" in Chapter 8. Accordingly, the library does not highlight the combustion progress variable. Therefore, if a diffusion flamelet library is invoked to simulate partially premixed turbulent combustion, then the second integral on the RHS of Equation 9.14 will be evaluated for diffusion flamelets even if combustion occurs locally in the inhomogeneously premixed mode. In other words, the straightforward substitution of a diffusion flamelet library into Equation 9.14 may result in simulating combustion chemistry effects both in the premixed and diffusion modes of partially premixed turbulent combustion, but the applicability of a diffusion flamelet library to modeling premixed flamelets may easily be put into question.

The following proposal by Domingo et al. (2002) may be useful in resolving the latter issue. These authors performed an LES study of lifted jet flames and introduced the following indicator of partial premixing:

$$\xi_p \equiv \frac{1}{2}\left(\frac{\mathbf{n}_F \cdot \mathbf{n}_O}{|\mathbf{n}_F \cdot \mathbf{n}_O|} + 1 \right), \tag{9.16}$$

where $\mathbf{n}_l = -\nabla Y_l / |\nabla Y_l|$ is the unit vector normal to an isoconcentration surface of the l-th reactant. In a planar diffusion flame, $\mathbf{n}_F \cdot \mathbf{n}_O = -1$ and $\xi_p = 0$, because the fuel and oxygen occupy two different half-spaces with respect to the heat release layer. In a planar premixed flame, $\mathbf{n}_F \cdot \mathbf{n}_O = 1$ and $\xi_p = 0$, because the fuel and oxygen occupy the same half-space with respect to the heat release layer. Accordingly, $\tilde{\xi}_p(\mathbf{x},t)$ and $1 - \tilde{\xi}_p(\mathbf{x},t)$ are associated with the probabilities of finding the mixtures undergoing premixed and diffusion burning, respectively, in point $\mathbf{x}$ at instant t. Thus, the discussed approach assumes the intermittency of premixed and diffusion flamelets and, therefore,

$$\tilde{Y}_l = \tilde{\xi}_P \tilde{Y}_{l,P} + \left(1 - \tilde{\xi}_P\right)\tilde{Y}_{l,D}, \tag{9.17}$$

where the species mass fractions $\tilde{Y}_{l,P}$ and $\tilde{Y}_{l,D}$ are associated with the premixed and diffusion burning modes, respectively (Domingo et al., 2002).

In the cited paper, Domingo et al. (2002) (i) modeled the influence of the mixture inhomogeneity on premixed turbulent burning by substituting

$$S_L(\mathbf{x},t) = \int_0^1 S_L(f)\tilde{P}_f\left(f, \tilde{f}, \overline{\rho f''^2}\right)df \tag{9.18}$$

into a closure of a filtered balance equation for the combustion progress variable, (ii) invoked a beta-function approximation of $\tilde{P}_f(f, \tilde{f}, \overline{\rho f''^2})$, and (iii) assumed that the diffusion burning is infinitely fast, that is, $Y_i(f)$ is equal to the mass fraction $Y_{i,b}(f)$ of the i-th reactant in the adiabatic equilibrium combustion products for the mixture fraction f. Unfortunately, to the best of the author's knowledge, Equations 9.16 and

9.17 have not yet been used jointly with a diffusion flamelet library in a RANS or an LES study of partially premixed turbulent combustion.

If not only inhomogeneously premixed turbulent combustion followed by mixing-controlled afterburning, similar to the case sketched in Figure 9.1, but also purely diffusion combustion due to autoignition occur in an engine, then a more complicated model of diffusion turbulent flames (unsteady flamelets, conditional moment closure, PDF balance equation, etc.) should be invoked to address the latter process. Definitely, neither the paradigm of infinitely fast chemistry nor a steady flamelet library can describe autoignition.

Alternatively, autoignition could be simulated using a detailed chemical scheme but neglecting the influence of turbulent fluctuations on the mean reaction rates, that is, by numerically integrating the balance equations

$$\frac{\partial}{\partial t}\left(\bar{\rho}\tilde{Y}_l\right) + \frac{\partial}{\partial x_k}\left(\bar{\rho}\tilde{u}_k\tilde{Y}_l\right) = -\frac{\partial}{\partial x_k}\overline{\rho u_k'' Y_l''} + M_l W_l\left(\tilde{T},\tilde{Y}_m\right), \tag{9.19}$$

$$\frac{\partial}{\partial t}\left(\bar{\rho}\tilde{T}\right) + \frac{\partial}{\partial x_k}\left(\bar{\rho}\tilde{u}_k\tilde{T}\right) = -\frac{\partial}{\partial x_k}\overline{\rho u_k'' T''} + \frac{W_T\left(\tilde{T},\tilde{Y}_m\right)}{c_p} + \frac{1}{c_p}\frac{\partial \bar{p}}{\partial t}, \tag{9.20}$$

supplemented the Favre-averaged continuity and Navier–Stokes equations and a proper turbulence model. Equations 9.19 and 9.20 result from averaging Equations 1.58 and 1.69, respectively, and neglecting the mean molecular transport terms as compared with the turbulent fluxes $\overline{\rho u_k'' Y_l''}$ and $\overline{\rho u_k'' T''}$ at high Reynolds numbers.

Contrary to the modeling of turbulent combustion, an assumption that $\overline{W_l(T,Y_m)} = W_l(\tilde{T},\tilde{Y}_m)$ in Equations 9.19 and 9.20 may be acceptable when simulating autoignition. The point is that such an assumption is inaccurate in turbulent flames mainly due to highly nonlinear (exponential) dependence of W_l on T and large turbulent pulsations in temperature caused by the intermittency of low-temperature unburned mixture and high-temperature combustion products (e.g., see Figure 5.4). However, because such intermittency is triggered by combustion, turbulent pulsations in temperature are much less before ignition than in the burning phase. Accordingly, the errors associated with the discussed simplification may be of significantly less importance than the errors associated with the invoked chemical mechanism of autoignition.

SUMMARY

The earlier brief review shows that contemporary RANS simulations of partially premixed turbulent flames deal with combinations of two combustion models developed independently for homogeneously premixed and nonpremixed burning regimes. Such a combination is capable of addressing the following three physical mechanisms specific to partially premixed combustion.

First, due to the inhomogeneity of the mean mixture composition, the mean burning rate in the premixed mode varies in space following variations in $\bar{\Phi}(x,t)$. This physical mechanism could, in the first approximation, be caught by invoking a sub-model of the mean rate $\overline{W}$ of product formation (or turbulent flame speed S_t or burning velocity U_t) that is capable of predicting the dependence of $\overline{W}$ (or S_t, or U_t) on the equivalence ratio.

Second, the local burning rate fluctuates due to turbulent pulsations in the mixture composition. This effect could, in the first approximation, be caught by invoking a presumed PDF for the mixture fraction. A beta-function PDF is widely used for this purpose, with the parameters of the PDF being evaluated on the basis of the local values of the first $\tilde{f}$ and second $\overline{\rho f''^2}/\overline{\rho}$ moments, which have been computed by solving the proper balance Equations 9.7 and 8.51, respectively.

Third, when products of lean combustion and rich combustion are mixed, the oxygen remaining in the former products reacts with the CO, H_2, and unburned hydrocarbons in the latter products. This physical mechanism could, in the first approximation, be caught either by assuming that the rate of afterburning is solely controlled by the rate of mixing or by invoking a diffusion flamelet library.

EFFECTS IGNORED IN RANS OR LES STUDIES OF PARTIALLY PREMIXED TURBULENT FLAMES

In addition to the three physical mechanisms discussed briefly in the previous section, there are effects specific to partially premixed combustion that have not yet been addressed in RANS or LES studies of turbulent flames. Such effects are discussed in this section.

BACK-SUPPORTED STRATIFIED COMBUSTION AND EXTENSION OF FLAMMABILITY LIMITS

One physical mechanism, specific to stratified combustion, is sketched in Figure 9.4, where the solid and dashed curves show spatial profiles of the equivalence ratio, temperature, and heat release rate in homogeneous and stratified, respectively, lean

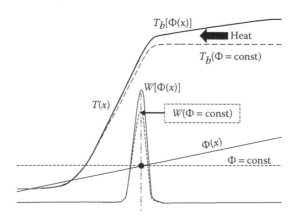

FIGURE 9.4 Homogeneous (*dashed curves*) and back-supported stratified (*solid curves*) laminar premixed flames.

laminar one-dimensional (1D) flames. In the two cases, the equivalence ratios conditioned on the maximum heat release rate are equal to one another (see filled circle). Due to an increase in the equivalence ratio behind the stratified flame, the adiabatic combustion temperature also increases in the stratified products with distance from the flame. Therefore, in the stratified case, there is an extra heat flux from the products to the reaction zone. As a result, both the temperature and heat release rate in the reaction zone of the stratified flame are higher than the temperature and heat release rate in the reaction zone of the homogeneous flame even if the local equivalence ratios are approximately the same in the two reaction zones. This mechanism, often called back-supported stratified combustion, enhances the local burning rate and extends the flammability limits. It is worth noting that the aforementioned scenario is simplified, and diffusion fluxes of the reactive species should be taken into account when modeling a stratified laminar flame.

An increase in flame speed and an extension of flammability limits due to mixture stratification were obtained in a couple of numerical simulations of laminar burning (Marzouk et al., 2000; Pirez da Cruz et al., 2000; Ra and Cheng, 2001; Richardson et al., 2010). Kang and Kyritsis (2007, 2009) experimentally investigated the propagation of laminar flames through steadily stratified lean or rich methane–air mixtures. Their data indicate that the laminar flame speeds obtained from stratified mixtures (e.g., see filled symbols in Figure 9.5) are typically higher than the speeds of their counterpart, that is, characterized by the same local Φ, homogeneously premixed flames (see open symbols), with the effect being increased by $|d\Phi/dx|$ (cf. triangles and circles, with the gradient of the equivalence ratio being specified in the legends). Moreover, back-supported stratified flames may sustain themselves in very lean or very rich mixtures, the composition of which is well beyond the flammability limits

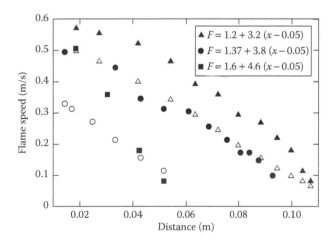

FIGURE 9.5 Laminar flame speeds obtained by Kang and Kyritsis (2009) from homogeneous (*open symbols*) and stratified (*filled symbols*) methane–air mixtures. The linear spatial profiles of the equivalence ratio measured in stratified cases in the range of $0.05 \leq x \leq 0.19$ m, specified in the legends. The open symbols show different speeds of the homogeneous flames at different distances, because these speeds were determined for the local equivalence ratio $\Phi(x)$.

of the counterpart homogeneously premixed flames. For instance, the filled squares in Figure 9.5 show that, in the discussed experiments, a laminar flame propagated through a stratified rich mixture even if the local equivalence ratio was significantly larger than $\Phi = 1.36$, whereas a homogeneously premixed mixture was flammable if $\Phi < 1.36$.

By simultaneously measuring the local flamelet speeds and local equivalence ratios in weakly turbulent propane–air flames, Pasquier et al. (2007) obtained lower flamelet speeds from a lean homogeneous mixture, as compared with a stratified ($\Phi = 0.6$–1.0) mixture characterized by the same local $\Phi = 0.8$. The mean local flamelet speeds were 0.4 and 0.5 m/s, respectively. This observation may be attributed to the back-supported burning in the stratified case.

To the best of the author's knowledge, the back support of stratified combustion was never addressed in RANS or LES studies of partially premixed turbulent flames. Development of a simple model of this physical mechanism appears to be a difficult task, because the response of the local burning velocity to mixture inhomogeneities depends on various parameters even for the simplest laminar flames. For instance, the experimental data obtained by Kang and Kyritsis (2007), shown in Figure 9.6, imply that a flame speed of about 15 cm/s for $\Phi \approx 0.55$ and two different spatial gradients of the equivalence ratio, 10 and 12.5 per m, may be obtained. Therefore, the knowledge of the local values of Φ and $|\nabla\Phi|$ is not sufficient to quantify the speeds of lean, stratified methane–air laminar flames even in the case of stationary stratification of the mixture.

In a turbulent flow, spatial gradients of the equivalence ratio vary rapidly with time. Numerical simulations of simple laminar flames indicate that the response of the local burning rate to fluctuations in the equivalence ratio is sensitive to transient effects (Marzouk et al., 2000; Sankaran and Im, 2002; Richardson et al., 2010). For

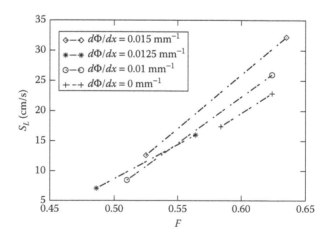

FIGURE 9.6 Laminar flame speeds measured in the stratified, lean methane–air mixtures subject to various spatial gradients of equivalence ratio, specified in the legends. (Adapted from Kang, T. and Kyritsis, D.C., *Proc. Combust. Inst.*, 31, 1075–1083, 2006. With permission.)

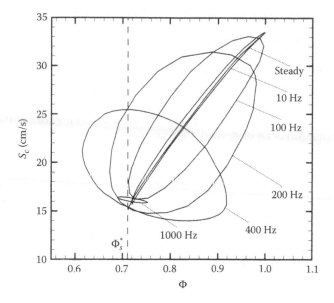

FIGURE 9.7 Response of the consumption speed to the harmonic oscillations of the equivalence ratio in two identical opposed methane–air jets. The curves have been computed for different frequencies of the oscillations, specified in the legends. (Reprinted from Sankaran, R. and Im, H.G., *Proc. Combust. Inst.*, 29, 77–84, 2002. With permission.)

instance, Figure 9.7 shows that the time-dependence of the burning rate computed for two identical opposed laminar methane–air jets depends on the frequency of the oscillations of the equivalence ratio at the jet nozzles. In this case, (i) the "flame can actually sustain itself, even if the equivalence ratio falls far below the steady flammability limit," but (ii) "the flame cannot be sustained if the mean value of the equivalence ratio oscillation" becomes lower than the lean flammability limit (Sankaran and Im, 2002). It is of interest to note that the largest extension of the instantaneous lean flammability limit is observed for an intermediate frequency of the oscillations (400 Hz), whereas too fast (1000 Hz) or too slow (10 Hz) oscillations weakly affect the lean flammability limit.

Thus, in order to quantify the response of the local burning rate to mixture inhomogeneities, we have to know not only the local value of the equivalence ratio but also its local spatial gradient and a timescale τ_Φ characterizing the turbulent pulsations in Φ. However, even the knowledge of these three parameters does not seem to be sufficient to accurately evaluate the local burning rate, as indicated in Figure 9.6. The problem of parameterizing $S_L(\Phi, |\nabla\Phi|, \tau_\Phi, \dots)$ and averaging such a parameterization in a turbulent flow has not yet been resolved.

In addition to back-supported stratified flames discussed earlier, tangentially supported burning (i.e., the heat flux from hotter products characterized by lower $|\Phi - 1|$ is tangential to the flamelet surface) may also play a role in partially premixed turbulent combustion, but this problem has not yet been investigated to the best of the author's knowledge.

Increase in Flamelet Surface Area Due to Mixture Inhomogeneity

If an initially planar laminar flame propagates through a quiescent, spatially uniform mixture, the flame surface remains planar, as sketched in Figure 9.8a. However, if the mixture composition varies in space, then different elements of the flame move at different speeds and an initially planar flame becomes curved (see Figure 9.8b), that is, mixture inhomogeneities may increase the flame surface area. Therefore, the flamelet surface may grow not only due to turbulent stretching (this is the primary physical mechanism of an increase in the burning rate by turbulence) but also due to turbulent pulsations in Φ.

It is worth stressing, however, that an increase in the flame surface area due to mixture inhomogeneities does not necessitate an increase in the overall burning rate. A simple example is sketched in Figure 9.9. The dotted line shows the initial position of a laminar flame, while the dashed line and solid curve show the flame fronts at an instant t_1 in the cases of homogeneous and inhomogeneous mixtures, respectively. The dashed and solid arrows show the laminar flame speeds in the two cases, respectively. Despite a larger surface of the inhomogeneous flame (cf. solid curve and dashed line), the homogeneous flame consumes more reactants (gray area) due to the higher burning rate per unit flame surface area (cf. dashed and solid arrows) in the considered cases.

In a turbulent flow, the overall effect of mixture inhomogeneities on the mean rate of product formation is controlled not only by an increase in the area, which accelerates combustion, but also by the spatial variations in $u_c(\Phi)$, caused by the spatial variations in the laminar flame speed due to the inhomogeneities. Indeed, in a stratified turbulent flame, Equation 7.65 may be generalized as

$$\overline{W}(\mathbf{x},t) = \rho_u \overline{u_c(\Phi)}(\mathbf{x},t)\overline{\Sigma}(\mathbf{x},t),\tag{9.21}$$

and, even if the mean flame surface density (FSD) $\overline{\Sigma}$ is increased by mixture inhomogeneities, fluctuations in Φ may reduce the mean rate $\overline{W}$ by decreasing the mean local consumption velocity $\overline{u_c(\Phi)}$.

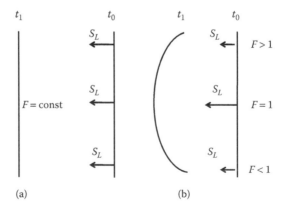

(a)   (b)

FIGURE 9.8 Increase in the instantaneous flame surface due to the mixture inhomogeneity. (a) homogeneous case and (b) inhomogeneous case.

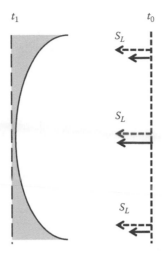

FIGURE 9.9 Mixture inhomogeneity may reduce the overall burning rate, despite an increase in the flame surface. The gray area shows the reactants consumed by a homogeneous flame (*dashed line*), but not consumed by an inhomogeneous flame (*solid curve*).

Various scenarios of the influence of the fluctuations in the mixture composition on $\overline{u_c(\Phi)}$ are sketched in Figure 9.10. If the amplitude of the fluctuations is sufficiently low so that the mixture composition is always lean (see rectangle (a)), or rich, then an increase (in the lean mixture) in $u_c(\Phi)$ due to positive fluctuations in Φ and a decrease in $u_c(\Phi)$ due to negative fluctuations mutually cancel one another after averaging in the first approximation, because the laminar flame speed depends quasi-linearly on the equivalence ratio in lean or rich mixtures (e.g., see Figure 2.8). As a result, $\overline{u_c(\Phi)} \approx u_c(\overline{\Phi})$. Moreover, the mechanism of back support, discussed in the previous section, may result in $\overline{u_c(\Phi)} > u_c(\overline{\Phi})$ (see dashed curve, arrow, and rectangle). When the amplitude of the fluctuations is increased, the probabilities of finding locally stoichiometric and locally inflammable mixtures become finite (see rectangle (b)). Therefore, if $\overline{\Phi} < 1$, too strong positive fluctuations in Φ make the mixture locally rich and do not further increase $u_c(\Phi)$, while too strong negative fluctuations in Φ yield vanishing local consumption velocity. As a result, $\overline{u_c(\Phi)}$ becomes lower than $u_c(\overline{\Phi})$ and is decreased when $\overline{\Phi'^2}$ is increased. The same trends, that is, $\overline{u_c(\Phi)} < u_c(\overline{\Phi})$ and a decrease in $\overline{u_c(\Phi)}$ with an increase in $\overline{\Phi'^2}$, are observed if the mean mixture composition is close to either stoichiometry (see rectangle (c)) or flammability limit (see rectangle (d)). In the former case, both positive and negative fluctuations in Φ reduce $u_c(\Phi)$. In the latter case, the drop in $u_c(\Phi)$ in a locally inflammable mixture overwhelms an increase in $u_c(\Phi)$ in a mixture with a composition that is more close to the stoichiometry.

As far as the increase in the flamelet surface area due to mixture inhomogeneities is concerned, the effect may be weakly pronounced in highly turbulent flows if the rate of growth of the flamelet surface due to turbulent stretching is much higher than the surface growth rate caused by mixture inhomogeneities. For instance, the results of a 2D DNS study of lean flames, shown in Figure 9.11, indicate a significant

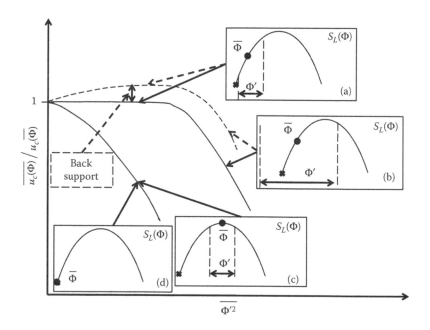

FIGURE 9.10 Influence of the mixture composition fluctuations on the ratio of the mean local consumption velocity to the local consumption velocity evaluated at the mean local equivalence ratio.

increase in the normalized flame surface area (more precisely, the flame length obtained in the 2D case and associated with the area in the 3D case) due to mixture fluctuations in the laminar case (see squares), but the effect is much less pronounced even in a weakly turbulent flow characterized by $u'/S_L = 0.58$ (see circles). Here and in the remainder of this chapter, a ratio of u'/S_L is evaluated at the mean equivalence ratio. Note that the dimensional averaged area of the flame surface was increased by u'/S_L for all mixture compositions studied in the discussed DNS.

It is of interest that Garrido-López and Sarkar (2005) simulated not only 2D single-step chemistry flames but also the progress of the same reaction in the case of a constant density, with all other things being equal. The results obtained in the latter case indicate a much weaker effect of mixture inhomogeneities on the overall burning rate and flamelet surface area. In the case of $\rho = \text{const}$, the instantaneous flame front looked like a weakly wrinkled, almost planar surface, whereas large-scale distortions of the front were well pronounced in the case of variable density, thus significantly increasing the front area (see Figure 7 in the cited paper). Based on these results, Garrido-López and Sarkar (2005) have concluded that "the hydrodynamic instability caused by gas expansion, catalyzed by the composition fluctuations interacting with the flame, is found to be responsible for the flame length enhancement." If the influence of mixture inhomogeneities on the flamelet surface area in a turbulent flow is mainly reduced to triggering out the Darrieus-Landau (DL) instability, a weak effect of the inhomogeneities on the area in sufficiently intense turbulence (e.g., see triangles in Figure 9.11) is associated with the suppression of

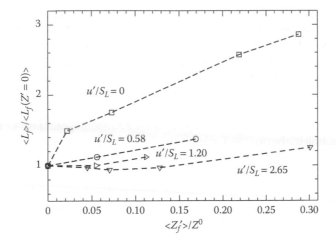

FIGURE 9.11 Average normalized flame length versus intensity $\langle Z_f' \rangle$ of mixture fraction fluctuations. The flame length is normalized using its value obtained for the same u'/S_L and $\langle Z_f' \rangle = 0$ (homogeneous mixture). (Adapted from Garrido-López, D. and Sarkar, S., *Proc. Combust. Inst.*, 30, 621–628, 2005. With permission.)

the instability by turbulent stretching, as discussed in the section "Instabilities of Laminar Flamelets in Turbulent Flows" in Chapter 5.

In summary, based on the aforementioned simple reasoning and DNS data, we may expect (i) an increase in the weakly turbulent burning rate of a lean or rich mixture due to an increase in the flamelet surface area caused by moderate mixture inhomogeneities, but (ii) a weak effect of moderate mixture inhomogeneities on the burning rate in highly turbulent lean or rich flames. Moreover, a decrease in the burning rate by intense mixture inhomogeneities appears to be possible due to a decrease in $u_c(\Phi)$ if there is a significant probability of finding inflammable reactants. A similar effect may be observed due to fluctuations in the equivalence ratio around $\bar{\Phi} = 1$, but, in this case, the effect may be damped by back-supported combustion.

Indeed, different effects of mixture inhomogeneities on the mean burning rate and flamelet surface area were reported in the literature. Zhou et al. (1997, 1998) performed very interesting measurements. In these experiments, propane was injected into a cylindrical vessel filled with air. After a short delay, turbulence was generated and the two gases were mixed during a few cycles of reciprocating movement of the two identical perforated plates in the vessel. Subsequently, the nonuniform propane–air mixture was ignited by a spark installed in the center of the vessel and the expansion of the flame kernel was visualized utilizing high-speed Schlieren photography technique. The rms turbulent velocity was varied by changing the time interval between the end of the reciprocating movement of the plates and the spark discharge, while the magnitude of the fluctuations in the mixture composition at the instant of ignition was independently varied by changing the number of cycles of the reciprocating movement. The Schlieren images reported by Zhou et al. (1997, 1998) clearly indicate wrinkling of the flame front due to mixture inhomogeneities, with the effect being documented in weakly turbulent ($u' \leq 0.185$ m/s) flows. This mechanism controlled an increase in the turbulent flame speed by a degree α of the mixture inhomogeneity, measured for $\overline{\Phi} = 0.7$ and 1.6 and moderate α (see the diamonds and crosses in Figure 9.12). In these two cases, the influence of moderate fluctuations in the mixture composition on the mean consumption velocity was of minor importance, that is, $u_c(\Phi) \approx u_c(\overline{\Phi})$, due to the linearity of $S_L(\Phi)$ in lean and rich flammable mixtures (e.g., see Figure 2.8). For a near-stoichiometric mixture ($\overline{\Phi} = 1.1$), fluctuations in Φ decrease the mean consumption velocity, as sketched in Figures 9.2 and 9.9. This mechanism counterbalances the increase in the flamelet surface area due to mixture inhomogeneities. As a result of the counteraction of the two mechanisms, the flame speed, shown in squares in Figure 9.12, depends weakly on the degree of the mixture inhomogeneity provided that α is sufficiently small. If the magnitude of the fluctuations in Φ is large, the former mechanism, that is, a decrease in $u_c(\Phi)$, overwhelms the aforementioned increase in the flamelet surface area, and the burning velocity is decreased when α is increased. In Figure 9.12, such a trend is observed for all mixtures, with the exception of the leanest mixture (diamonds). For the richest mixture (crosses), a decrease in $u_c(\Phi)$ when α is sufficiently large and is further increased is associated with strong fluctuations in the equivalence ratio that make the mixture locally inflammable.

A moderate increase in the mean FSD $\overline{\Sigma}$ due to mixture inhomogeneities was documented in V-shaped methane–air flames by Anselmo-Filho et al. (2009) and Sweeney et al. (2011) (e.g., see Figure 9.13). Vena et al. (2011) observed a similar effect in V-shaped *iso*-octane–air flames. It is worth stressing that turbulence was very weak in these measurements, for example, $Re_t = 38$ in methane–air flames. Böhm et al. (2011) experimentally investigated lean, stratified methane–air flames in more intense turbulence and did not report an increase in the burning rate due to mixture inhomogeneities.

Malkeson and Chakraborty (2010) did not observe any notable effect of random fluctuations in the equivalence ratio on the total burning rate obtained in the case of $\overline{\Phi} = 1$ in a 3D single-step chemistry DNS, but the fluctuations increased the burning rate when $\overline{\Phi}$ was set equal to 0.7. The former result may be attributed to mutual

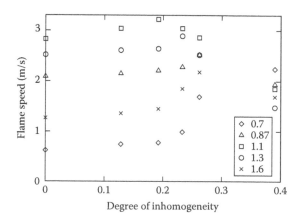

FIGURE 9.12 Flame propagation speeds measured by Zhou et al. (1998) at various degrees of the mixture inhomogeneity, different mean equivalence ratios specified in the legends, and $u' = 0.13$ m/s.

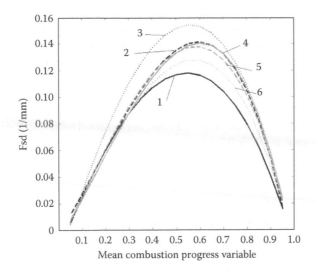

FIGURE 9.13 Measured mean flame surface density versus mean combustion progress variables. 1: premixed flame, $\Phi = 0.77$. 2–6: partially premixed flames with $\Phi_1/\Phi_2 = 1.2, 1.5, 1.8, 2.3,$ and 3.0, respectively. (Reprinted from Anselmo-Filho, P., Hochgreb, S., Barlow, R.S., and Cant, R.S., *Proc. Combust. Inst.*, 32, 1763–1770, 2009. With permission.)

cancellation of the effects of mixture inhomogeneities on the flamelet surface area and $\overline{u_c}(\Phi)$, while the latter result is associated with the increase in flamelet surface area due to mixture inhomogeneities. Unfortunately, Malkeson and Chakraborty (2010) did not report data that straightforwardly showed the influence of mixture inhomogeneities on the flamelet surface area.

Grout et al. (2009) performed a DNS study of premixed/nonpremixed ($\overline{\Phi} = 0.74$ and $\Phi = 0.4-1.2$), weakly turbulent ($u'/S_L = 0.7$) flames and reported that mixture inhomogeneities accelerated burning due to an increase in the flamelet surface area.

The results of DNS performed by Garrido-López and Sarkar (2005) indicated that the production of flamelet surface area due to the DL instability catalyzed by mixture inhomogeneities overwhelmed a decrease in $\overline{u_c}(\Phi)$ in laminar and weakly turbulent ($u'/S_L = 0.58$) cases and resulted in an increase in the averaged burning rate by fluctuations in Φ. At $u'/S_L = 1.2$, the two mechanisms counterbalanced each other and the total burning rate depended weakly on the magnitude of the fluctuations. At $u'/S_L = 2.65$, the latter mechanism overwhelmed the former mechanism and the averaged burning rate was decreased when $\overline{\Phi'^2}$ was moderately increased.

The production of flamelet surface area due to mixture inhomogeneities was not documented in a DNS performed in more intense turbulence. 2D complex-chemistry DNS by Haworth et al. (2000) indicated a weak influence of the harmonic spatial variations in the equivalence ratio (from 0.5 to 2.7 with $\overline{\Phi} = 1$) in the oncoming flow on the flamelet length in a moderately turbulent ($u'/S_L = 4.67$) flow. In 3D single-reaction DNS by Hélie and Trouvé (1998), fluctuations in the mixture composition around $\overline{\Phi} = 1$ reduced the highly turbulent ($u'/S_L = 7.5$) burning rate, with the effect being "a direct consequence of the nonlinear, laminar-like variations of mass burning rate with mixture composition," that is, fluctuations in the equivalence ratio around $\bar{\Phi} = 1$ reduced the mean consumption velocity in Equation 9.21 and, hence, the burning rate, as sketched in Figures 9.10 and 9.2, respectively. In 2D complex-chemistry DNS by Jiménez et al. (2002), mixture inhomogeneities ($\bar{\Phi} = 0.6$ and $\Phi = 0.03–2.06$) did not affect the flamelet length, but increased the total heat release rate in very intense ($u'/S_L = 11.6$) turbulence. The latter effect was attributed to a stronger resistance of the local burning rate $\bar{u}_c$ to straining in near-stoichiometric flamelets, which existed in the stratified case, as compared with lean flamelets. In the homogeneous case, $\bar{u}_c/S_L^0 \approx 0.7$ was lower than $\bar{u}_c/S_L^0 \approx 1$ obtained in the inhomogeneous case (see Figure 8 in the cited paper).

TRIBRACHIAL (TRIPLE) FLAMES

A tribrachial or triple flame, which was experimentally discovered by Philips (1965) and is sketched in Figure 9.14, is another local phenomenon that may play a role in partially premixed turbulent burning. In particular, as reviewed elsewhere (Lyons, 2007), such a flame is hypothesized to burn at the stabilization point of a lifted jet turbulent flame sketched in Figure 8.7.

In the case of a nonuniform field of the equivalence ratio, for example, in the vicinity of the aforementioned stabilization point, lean and rich premixed flames (see the bold, solid curve in Figure 9.14) and diffusion flames (bold, dashed line) may contact each other in a single point O. Such a tribrachial flame is a particular example of edge flames, that is, 2D (or 3D) structures burning in inhomogeneous flow of nonmixed or mixed reactants. Over the past decade, edge flames were the focus of theoretical, numerical, and experimental research on laminar burning, as reviewed elsewhere (Buckmaster, 2002; Buckmaster et al., 2005; Matalon, 2009). In particular, studies on laminar tribrachial flames were recently reviewed by Chung (2007).

The most known peculiarity of tribrachial flames is their ability to propagate at a speed U_f, which is significantly (e.g., by a factor of two) higher than the stoichiometric unperturbed laminar flame speed $S_L^0(\Phi = 1)$, provided that (i) the local transverse gradient $d\Phi/dy$ of the equivalence ratio is sufficiently small (too large a gradient may quench the flame; see Buckmaster, 2002), and (ii) U_f is measured with respect to the flow of the unburned gas at a large distance from the flame. However, it is not yet clear whether

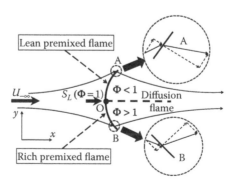

FIGURE 9.14   A sketch of a laminar tribrachial flame.

or not this effect plays an important role in turbulent combustion. The point is that $U_f > S_L^0(\Phi = 1)$ does not mean an increase in the local burning rate, but is controlled by the redirection of the flow ahead of a tribrachial flame (Ruetsch et al., 1995). Indeed, due to density drop across the premixed branches of the flame, the components of the flow velocity vector that are locally normal to the premixed branches are increased, that is, the flow velocity vectors change their directions at the premixed branches (see the insets A and B in Figure 9.14). Accordingly, as sketched in Figure 9.14, the area of a flow tube that is normal to a tribrachial flame is increased as the unburned mixture moves toward the flame. The increase in the tube area results in a decrease in the flow velocity as the unburned mixture moves toward the flame. Due to this purely hydrodynamic mechanism, which is also responsible for the DL instability of premixed flames (see the section "Hydrodynamic Instability" in Chapter 5 and Figure 5.39 in particular), the normal (i.e., parallel to the x-axis in Figure 9.14) flow velocity $U_{-\infty}$ at a large distance from the flame is significantly higher than the normal flow velocity of the unburned mixture in the vicinity of point O. If $D_F = D_O$ and Le = 1, then the two velocities are equal to U_f and $S_L^0(\Phi = 1)$, respectively, in the coordinate framework attached to the flame, that is, U_f is significantly higher than $S_L^0(\Phi = 1)$. Based on the previous reasoning, Ruetsch et al. (1995) have developed a simple model that yields $U_f/S_L^0 \approx \sqrt{\rho_u/\rho_{b,\Phi=1}}$ in the limit case of small but finite $|d\Phi/dy|$. This simple estimate is consistent with subsequent numerical simulations reviewed by Chung (2007).

Even if the discussed physical mechanism does not affect the local burning rate and, based on this, could be neglected when simulating partially premixed turbulent combustion, there are other phenomena associated with tribrachial laminar flames that could be of importance in turbulent flows. For instance, Ruetsch et al. (1995) have shown that the heat release in a tribrachial flame not only redirects the oncoming flow, as sketched in Figure 9.14, but also reduces $|d\Phi/dy|$ and, therefore, the scalar dissipation ahead of the flame. Such effects may be of importance when applying a diffusion flamelet model to simulate mixing-controlled afterburning behind a tribrachial flame. Moreover, the decrease in $|d\Phi/dy|$ impedes eventual quenching of the tribrachial flames by small-scale nonuniformities in the mixture composition.

Furthermore, because the premixed branches of the discussed flame are substantially curved in the vicinity of the leading point O, the influence of the differences in the molecular transport coefficients on the local burning rate may be significant.

To the best of the author's knowledge, such an eventual local small-scale burning structure as a tribrachial laminar flame has not yet been taken into account by RANS or LES models of turbulent partially premixed combustion.

SUMMARY

To summarize the earlier discussion, let us rearrange Equation 9.21 as

$$\frac{\overline{W}(\mathbf{x},t)}{\rho_u} = \underbrace{u_c\left[\overline{\Phi}(\mathbf{x},t)\right]\overline{\Sigma}\left[\overline{\Phi}(\mathbf{x},t)\right]}_{\text{I}} \underbrace{\frac{\overline{u_c(\Phi)}}{u_c(\overline{\Phi})}(\mathbf{x},t)}_{\text{II}} \underbrace{\frac{\overline{\Sigma(\Phi)}}{\overline{\Sigma}(\overline{\Phi})}(\mathbf{x},t)}_{\text{III}} \qquad (9.22)$$

in a stratified turbulent flame. Note that in a partially premixed flame where both locally lean and locally rich mixtures may be observed, Equation 9.22 should involve a term associated with mixing-controlled afterburning, as discussed in the section "Mixing-Controlled Afterburning."

Term I on the RHS describes the dependence of the mean rate of product formation on the mean mixture composition. This term may be evaluated by invoking a proper model of homogeneously premixed turbulent combustion, as discussed in the section "Inhomogeneity of Mean Mixture Composition."

Term II is associated with the influence of mixture fluctuations and back support on the mean consumption velocity. The former effect could be addressed by invoking a presumed PDF $\tilde{P}_f(f, \tilde{f}, \overline{\rho f''^2})$, as discussed in the section "Turbulent Pulsations in Mixture Composition," but modeling of back-supported combustion in a turbulent flow of inhomogeneous mixture, as well as the role played by this mechanism in partially premixed flames, is an issue.

Term III is associated with the production of flamelet surface area due to mixture inhomogeneities. This physical mechanism may result in a substantial increase in $\overline{\text{burning rate}}$, provided that (i) turbulence is sufficiently weak and (ii) the ratio of $\overline{u_c(\Phi)}/u_c(\overline{\Phi})$ is sufficiently close to or larger than unity. In more intense turbulence, the discussed increase in the flamelet surface area appears to be of minor importance as compared with an increase in the flamelet surface area due to turbulent stretching. The author is not aware of any model of term III for RANS applications.

USE OF PRESUMED PDFs FOR MODELING BOTH PREMIXED AND DIFFUSION BURNING MODES

The earlier discussion shows that contemporary combustion models invoked in RANS studies of partially premixed turbulent flames neglect certain physical mechanisms specific to turbulent burning of inhomogeneous gas mixtures. Moreover, such models do not address finite-reaction-rate effects in the premixed mode of burning. To bypass these issues, a simple mathematical method is sometimes used.

The method consists of invoking a joint PDF $P(c, f, \tilde{c}, \overline{\rho c''^2}, \tilde{f}, \overline{\rho f''^2})$ for the combustion progress variable and mixture fraction in order to straightforwardly average the reaction and heat release rates and, thus, bypass the problem of modeling the aforementioned physical mechanisms.

One may note that the models discussed in the section "Effects Addressed in RANS or LES Studies of Partially Premixed Turbulent Flames," in fact, invoked the following joint PDF:

$$\tilde{P}\left(c, f, \tilde{c}, \tilde{f}, \overline{\rho f''^2}\right) = \left\{\left[1 - \overline{c}\left(\mathbf{x}, t\right)\right]\delta(c) + \overline{c}\left(\mathbf{x}, t\right)\delta(1 - c)\right\}\tilde{P}_f\left(f, \tilde{f}, \overline{\rho f''^2}\right). \quad (9.23)$$

However, because this PDF does not model the statistics of intermediate values of $0 < c < 1$, Equation 9.23 does not allow us to average the reaction and heat release rates, but requires a model of the premixed mode of turbulent burning in order to close the problem. In this section, we will discuss the joint presumed PDFs that address the statistics of intermediate values of $0 < c < 1$ and allow us to average the aforementioned rates.

To the best of the author's knowledge, an approach of the latter kind was first developed by Bradley et al. (1990, 1998a) for computing the dependence of the liftoff height of a methane jet, for example, sketched in Figure 8.7, on the mean inlet velocity. These authors assumed that

$$\tilde{P}\left(f,c,\tilde{c},\overline{\rho c''^2},\tilde{f},\overline{\rho f''^2}\right)=\tilde{P}_c\langle c\mid f\rangle\tilde{P}_f\left(f,\tilde{f},\overline{\rho f''^2}\right)P_b\left(f\right), \qquad (9.24)$$

and invoked beta-function approximations for both the mixture fraction PDF $P_f(f,\tilde{f},\rho f''^2)$ and the conditioned PDF $\tilde{P}_c\langle c\mid f\rangle$ as a function of c. Here, $\tilde{P}_b(f)$ "expresses the statistical influence of strain rates on flamelets" (Bradley et al., 1998a).

An alternative presumed PDF

$$P\left(f,Y_F,\mathbf{x},t\right)=a_1\left(\mathbf{x},t\right)\delta\left(f-f^{(1)}\left(\mathbf{x},t\right)\right)\left(Y_F-Y_F^{(1)}\left(\mathbf{x},t\right)\right)$$

$$+a_2\left(\mathbf{x},t\right)\delta\left(f-f^{(2)}\left(\mathbf{x},t\right)\right)\delta\left(Y_F-Y_F^{(2)}\left(\mathbf{x},t\right)\right) \qquad (9.25)$$

was introduced by Libby and Williams (2000). Equation 9.25 characterizes the progress of premixed burning using fuel mass fraction Y_F, rather than c, and involves four Dirac delta functions $\delta(\underline{x})$ and six unknown parameters: $a_1(\mathbf{x},t)$, $a_2(\mathbf{x},t)$, $f^{(1)}(\mathbf{x},t)$, $f^{(2)}(\mathbf{x},t)$, $Y_F^{(1)}(\mathbf{x},t)$, and $Y_F^{(2)}(\mathbf{x},t)$. In the following text, dependencies of the aforementioned quantities on time t and spatial coordinates $\mathbf{x}$ will be skipped, for the sake of brevity.

The normalizing constraint of

$$\int_0^1 \int_{Y_{F,b}(f)}^{Y_{F,u}(f)} P\left(f,Y_F\right)dY_F df = 1 \qquad (9.26)$$

allows us to reduce the number of unknown parameters to five, because $a_1 = 1 - a_2$ by virtue of Equation 9.26. Libby and Williams (2000) proposed to determine these five parameters (i) by solving the balance equations for $\tilde{f}$, $\tilde{Y}_F$, $\rho f''^2$, $\rho Y_F''^2$, and $\rho f''^3$ or $\rho Y_F''^3$ and (ii) by equating the computed moments to the same moments resulting from Equation 9.25, for example,

$$\tilde{f}=\frac{1}{\rho}\int_0^1 \int_{Y_{F,b}(f)}^{Y_{F,u}(f)} \rho f P\left(f,Y_F\right)dY_F df = \frac{\left(1-a_2\right)\rho\left(f^{(1)},Y_F^{(1)}\right)f^{(1)}+a_2\rho\left(f^{(2)},Y_F^{(2)}\right)f^{(2)}}{\left(1-a_2\right)\rho\left(f^{(1)},Y_F^{(1)}\right)+a_2\rho\left(f^{(2)},Y_F^{(2)}\right)}. \qquad (9.27)$$

Thus, the five unknown parameters $(a_2, f^{(1)}, f^{(2)}, Y_F^{(1)}, \text{and } Y_F^{(2)})$ are evaluated by solving a system of five nonlinear equations

$$\Xi_l\left(\tilde{f}, \tilde{Y}_F, \overline{\rho f''^2}, \overline{\rho Y_F''^2}, \overline{\rho f''^3}, a_2, f^{(1)}, f^{(2)}, Y_F^{(1)}, Y_F^{(2)}\right) = 0, \qquad (9.28)$$

where (i) the five functions $\Xi_l(l = 1, ..., 5)$ are analytically determined using Equation 9.25, for example, Ξ_1 is equal to the difference between the left-hand side (LHS) and the RHS of Equation 9.27, while (ii) the quantities $\tilde{f}, \tilde{Y}_F, \overline{\rho f''^2},$ and $\overline{\rho Y_F''^2}$ and $\overline{\rho f''^3}$ (or $\overline{\rho Y_F''^3}$) are obtained by solving the proper balance equations. Note that if an expression for the local reaction rate $W = W(f, Y_F)$ is specified, then the closure of the reaction terms $\overline{W}, \overline{Y_F''W}$, and $\overline{f''W}$ in the balance equations for $\tilde{Y}_F, \overline{\rho Y_F''^2},$ and $\overline{\rho f''Y_F''}$, respectively, is simplified using Equation 9.25, for example,

$$\overline{W} = \int\limits_0^1 \int\limits_{Y_{F,b}(f)}^{Y_{F,u}(f)} WP(f, Y_F)dY_F df = (1 - a_2)W\left(f^{(1)}, Y_F^{(1)}\right) + a_2 W\left(f^{(2)}, Y_F^{(2)}\right). \qquad (9.29)$$

It is worth stressing, however, that obtaining a numerical solution to Equation 9.28 may be difficult due to the strongly nonlinear behavior of the functions Ξ_l, as discussed in detail by Libby and Williams (2000).

Although Libby and Williams (2000) solely analyzed the case of a single combustion reaction

$$CH_4 + 2O_2 \rightarrow CO_2 + 2H_2O$$

and used the analytical relations between the mass fractions of $O_2, CO_2, H_2O, f,$ and Y_F, their approach can easily be extended to more realistic cases, including flames with complex chemistry. To do so, the dependencies of mass fractions and the corresponding reaction rates on f and Y_F should be preliminarily approximated or tabulated by simulating proper laminar flames.

The aforementioned model is commonly called the Libby-Williams (LW) model and was used by several research groups. Ribert et al. (2004) simplified the LW model and obtained the expressions for $a_2, f^{(1)}, f^{(2)}, Y_F^{(1)},$ and $Y_F^{(2)}$ as functions of solely the first and second moments $\tilde{f}, \tilde{Y}_F, \overline{\rho f''^2},$ and $\overline{\rho Y_F''^2}$, (see Equations 15 through 17 in the cited paper) by invoking an arbitrary assumption about the joint fluctuations of f and Y_F. The simplified model was successfully applied to the numerical simulations of turbulent combustion of a lean ($\Phi = 0.8$) methane–air mixture ignited by hot products obtained from the burning of the stoichiometric methane–air mixture. The model by Ribert et al. (2004) is commonly called the LW-P model, that is, the LW model modified by the Poitiers group.

Darbyshire et al. (2010) augmented the LW-P model by invoking an advanced closure of the scalar dissipation rate in the balance equation for $\overline{\rho Y_F''^2}$ and simulated (i) stratified propane–air flames stabilized by the recirculation zones in a channel with abrupt expansion (these flames were also computed by Bigot et al. [2000], as mentioned in the section "Turbulent Pulsations in Mixture Composition") and (ii) partially premixed V-shaped methane–air flames in the case of moderate ($\Phi_1 = 0.52$ and $\Phi_2 = 0.95$) and strong ($\Phi_1 = 0.37$ and $\Phi_2 = 1.1$) stratifications. In all the simulated cases, the computed results agree reasonably well with the available experimental data.

Robin et al. (2006) extended the LW model by replacing the PDF given by Equation 9.25 with the following presumed PDF:

$$P(f, Y_F) = a_1 \left[\alpha \delta \left(Y_F - Y_F^{(11)} \right) + (1 - \alpha) \delta \left(Y_F - Y_F^{(12)} \right) \right] \delta \left(f - f^{(1)} \right)$$

$$+ a_2 \left[\beta \delta \left(Y_F - Y_F^{(21)} \right) + (1 - \beta) \delta \left(Y_F - Y_F^{(22)} \right) \right] \delta \left(f - f^{(2)} \right). \tag{9.30}$$

Because the latter PDF involves four more unknown parameters as compared with the former PDF, the model by Robin et al. (2006) requires more closure relations. In particular, Robin et al. (2006) solved not only the balance equations for $\tilde{f}$, $\tilde{Y}_F$, $\overline{\rho f''^2}$, and $\overline{\rho Y_F''^2}$, but also a balance equation for $\overline{\rho f'' Y_F''}$. Such a complication was justified as follows; while the model by Ribert et al. (2004) results in $\overline{\rho f'' Y_F''} = \sqrt{\overline{\rho f''^2} \cdot \overline{\rho Y_F''^2}}$, the use of Equation 9.30 yields $\overline{\rho f'' Y_F''} \leq \sqrt{\overline{\rho f''^2} \cdot \overline{\rho Y_F''^2}}$. Note that the former equality is contradicted by the recent DNS data obtained by Malkeson and Chakraborty (2010).

Robin et al. (2006) tested their model by simulating partially premixed propane–air flames stabilized by the recirculation zones in a channel with abrupt expansion (the same flames were also studied by Bigot et al. [2000] and Darbyshire et al. [2010]). The model yielded reasonable results not only in a weakly stratified case (equivalence ratios in the two streams were equal to 0.9 and 0.7) but also when the composition of one stream ($\Phi = 0.3$) was well beyond the lean flammability limit. It is worth noting that Robin et al. (2006) also applied the model by Ribert et al. (2004) to the same two cases and did not emphasize any substantial difference in the transverse profiles of the mean velocity computed using the two models.

In a subsequent paper, Robin et al. (2008) further developed their model by improving the closure relations for the dissipation terms in the balance equations for $\overline{\rho Y_F''^2}$ and $\overline{\rho f'' Y_F''}$. These authors applied the improved model to numerical simulations of four V-shaped methane–air flames; a lean ($\Phi = 0.6$) premixed case and three inhomogeneous cases with the equivalence ratio varying from either 0.8 or 1.0 or 1.2 to zero. A reasonable agreement with the experimental data on the mean velocity and mean combustion progress variable, presented in the same paper, was reported. It is of interest to note that no diffusion afterburning was computed in the richest case ($0 \leq \Phi \leq 1.2$), while a part of the fuel remained unburned (see Figures 27 and 28 in the cited paper).

It is worth emphasizing the following two issues associated with the joint presumed PDFs given either by a product of beta functions in Equation 9.24 or by Equation 9.25. First, such PDFs are not applicable to modeling the simplest case of homogeneously premixed turbulent combustion in the flamelet regime, as discussed in detail in the section "Presumed PDF Models" in Chapter 7. Briefly speaking, to determine all parameters of the aforementioned presumed PDFs, the first $\tilde{c}$ and the second $\overline{\rho c''^2}/\bar{\rho}$ moments are considered to be two independent input quantities. Accordingly, the two equations

$$\bar{\rho}\tilde{c} = \int_0^1 \int_0^1 \rho c P(c, f) \, df \, dc \tag{9.31}$$

and

$$\overline{\rho c''^2} = \int_0^1 \int_0^1 \rho c^2 P(c, f) \, df \, dc - \bar{\rho}\tilde{c}^2 \tag{9.32}$$

allow us to evaluate the parameters of two PDFs, provided that Equations 9.31 and 9.32 are independent. However, the number of independent input quantities is reduced by one as $\overline{\rho c''^2} \to \bar{\rho}\tilde{c}(1 - \tilde{c})$ in the flamelet regime (see Equation 5.29). Consequently, the number of independent equations becomes lower than the number of parameters of the unknown PDFs, thus making the problem ill-defined.

Second, both the beta function and the sum of the Dirac delta functions are totally arbitrary presumed PDFs, which are difficult to justify with solid physical reasoning. The flexibility of the shape of a beta function (see Figure 8.11) is not sufficient to justify its use, because there are a number of other PDFs that cannot be approximated with the beta function. For instance, as shown in the section "Presumed Probability Density Functions" in Chapter 8, the beta-function PDF $P(f)$ is associated with the vanishing probability of finding pure oxygen. Therefore, because there are no solid reasons for showing preference to a particular shape of $P(c)$ (or $P(f)$), weak sensitivity of the key averaged quantities (e.g., mean reaction and heat release rates) to a particular shape of a presumed PDF should be demonstrated to substantiate the discussed presumed joint PDF approach. However, the author does not know such results.

On the contrary, simulations by Huang and Lipatnikov (2012) indicate that the mean burning rate and flame speed are sensitive to the shape of a presumed PDF invoked to average them. These authors (i) computed laminar flame speeds and distributions of heat release rate in laminar gasoline–air flames by invoking a semi-detailed chemical mechanism developed and validated by Huang et al. (2010) (see Figure 2.10) and (ii) averaged these results by invoking either the LW PDF given by Equation 9.23 or the beta-function PDF either for the combustion progress variable or for the mixture fraction. Note that both the LW and beta-function PDFs were characterized not only by the same first and second moments, but also by the same third moments either for c in the premixed case or for f in the partially premixed case.

Figure 9.15 shows that both the mean normalized heat release rate and mean normalized turbulent burning velocity, that is, the normalized heat release rate integrated over the turbulent flame brush, depend substantially on the shape of a presumed PDF

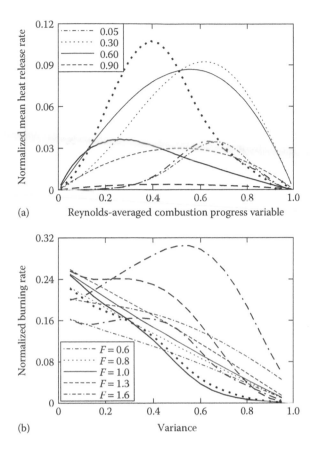

(a)

(b)

FIGURE 9.15 (a) Dependencies of the normalized heat release rate $\overline{\omega} g_c$ on $\overline{c}$, calculated for the stoichiometric gasoline–air mixture and various values of the segregation factor g_c, specified in the legends. (b) Dependencies of the normalized turbulent burning velocity on the segregation factor, calculated at various equivalence ratios, specified in the legends. In both cases, the thin and thick curves have been obtained invoking a beta-function PDF $P(c, \overline{c}, g_c)$ and Equation 9.23 (with $f^{(1)} = f^{(2)} = \delta(\overline{f})$), respectively. The two PDFs are characterized by the same $\overline{c}$, the same g_c, and the same $\overline{c'^3}$.

$P(c, \overline{c}, g_c)$ in homogeneous flames, with the exception of the limit case of weak fluctuations in c, that is, low values of the segregation factor $g_c \equiv \overline{c'^2} / \overline{c}(1 - \overline{c})$, which are not typical for premixed combustion. These results (i) are not surprising, considering the highly nonlinear dependence of the heat release rate on the combustion progress variable and (ii) further support the aforementioned criticism of the presumed PDF approach to modeling premixed turbulent combustion.

Figure 9.16a shows that the sensitivity of the mean normalized laminar flame speed $\overline{s}_L$ to the shape of a presumed PDF $P(f, \overline{f}, g_f)$, invoked to average it, is reasonably weak for values of the segregation factor $g_f \equiv \overline{g'^2} / \overline{g}(1 - \overline{g})$ typical for inhomogeneously premixed combustion, provided that the mixture composition is always within flammability limits. When the probability of finding an inflammable

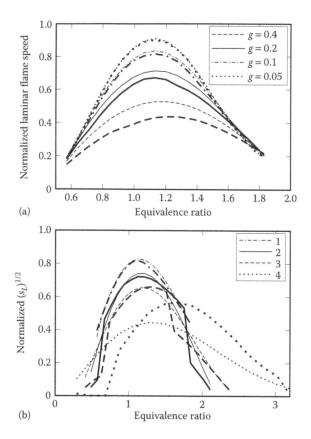

FIGURE 9.16 (a) Dependencies of the mean normalized laminar flame speed $\bar{s}_L$ on the mean equivalence ratio, calculated for various values of the segregation factor g_f, specified in the legends. Fluctuations in Φ are restricted by $\Phi_{min} = 0.5$ and $\Phi_{max} = 1.9$, with the lean and rich flammability limits being set equal to 0.5 and 2.0, respectively. (b) Dependencies of $\sqrt{\bar{s}_L}$ on $\bar{\Phi}$, calculated for $g_f = 0.2$ and (1) $\Phi_{min} = 0.5$, $\Phi_{max} = 1.9$, (2) $\Phi_{min} = 0.4$, $\Phi_{max} = 2.3$, (3) $\Phi_{min} = 0.3$, $\Phi_{max} = 2.5$, and (4) $\Phi_{min} = 0.1$, $\Phi_{max} = 4.0$. In both cases, the thin and thick curves have been obtained invoking a beta-function and the LW PDFs $P(f, \bar{f}, g_f)$, respectively. The two PDFs are characterized by the same $\bar{f}$, the same g_f, and the same $\overline{f'^3}$.

mixture is increased due to an increase in the amplitude of the fluctuations in the equivalence ratio, the sensitivity of $\bar{s}_L$ to the PDF shape is substantially increased; cf. thin and thick curves 3 and 4 in Figure 9.16b, and note that the effect is more pronounced for $\bar{s}_L$ than for $\sqrt{\bar{s}_L}$. The latter figure reports $\sqrt{\bar{s}_L}$ rather than $\bar{s}_L$, because the dependence of the turbulent burning velocity on the laminar flame speed is closer to $U_t \propto \sqrt{S_L}$ than to $U_t \propto S_L$, as discussed in the section "Empirical Parameterizations for Turbulent Burning Velocity and Flame Speed" in Chapter 4.

Finally, it is worth noting that recent DNS data obtained by Malkeson and Chakraborty (2010) do not support the use of a presumed PDF given by the sum of Dirac delta functions for modeling partially premixed turbulent combustion.

SUMMARY

As compared with flames that propagate through perfectly homogeneous mixtures, partial premixed turbulent combustion involves at least the following five specific effects:

1. Dependence of the burning rate on the inhomogeneity of the mean mixture composition
2. Dependence of the burning rate on the turbulent pulsations in the local mixture composition
3. Mixture-controlled afterburning of lean and rich combustion products
4. An increase in the flame surface area due to mixture inhomogeneities
5. Variations in the local burning rate due to the heat and reactant fluxes from/ to neighboring leaner or richer combustion products, for example, back-supported stratified burning

In contemporary RANS studies, the first and second mechanisms are often modeled by assuming that the influence of turbulence on premixed combustion is statistically independent of the turbulent pulsations in the mixture composition. The third mechanism is commonly addressed by invoking a model of nonpremixed turbulent flames, with the accuracy of simulations being typically limited by another combustion submodel that deals with premixed burning. The fourth and fifth mechanisms are usually neglected without solid justification.

The use of a joint presumed PDF for the combustion progress variable and mixture fraction offers an opportunity to bypass modeling the aforementioned physical mechanisms, but such an approach appears to be flawed in the flamelet regime of inhomogeneously premixed turbulent combustion and suffers from the sensitivity of the averaged burning rate to the shape of a presumed PDF.

References

Abdel-Gayed, R.G. and D. Bradley. 1985. Criteria for turbulent propagation limits of premixed flames. *Combust. Flame* 62:61–8.

Abdel-Gayed, R.G., Bradley, D., and M. McMahon. 1979. Turbulent flame propagation in premixed gas: Theory and experiment. *Proc. Combust. Inst.* 17:245–54.

Abdel-Gayed, R.G., Al-Khishali, K.J., and D. Bradley. 1984. Turbulent burning velocities and flame straining in explosions. *Proc. R. Soc. Lond. A* 391:391–414.

Abdel-Gayed, R.G., Bradley, D., and A.K.C. Lau. 1988. The straining of premixed turbulent flames *Proc. Combust. Inst.* 22:731–8.

Abdel-Gayed, R.G., Bradley, D., and F.K.-K. Lung. 1989. Combustion regimes and the straining of turbulent premixed flames. *Combust. Flame* 76:213–18.

Akkerman, V., Bychkov, V., and L.E. Eriksson. 2007. Numerical study of turbulent flame velocity. *Combust. Flame* 151:452–71.

Al-Shahrany, A.S., Bradley, D., Lawes, M., and R. Woolley. 2005. Measurement of unstable burning velocities of *iso*-octane–air mixtures at high pressure and the derivation of laminar burning velocities. *Proc. Combust. Inst.* 28:225–32.

Al-Shahrany, A.S., Bradley, D., Lawes, M., Liu, K., and R. Woolley. 2006. Darrieus-Landau and thermo-acoustic instabilities in closed vessel explosions. *Combust. Sci. Tech.* 178:1771–802.

Aldredge, R.C. 2006. The speed of isothermal-front propagation in isotropic, weakly turbulent flows. *Combust. Sci. Tech.* 178:1201–15.

Aldredge, R.C., Vaezi, V., and P.D. Ronney. 1998. Premixed-flame propagation in turbulent Taylor-Couette flow. *Combust. Flame* 115:395–405.

Anselmo-Filho, P., Hochgreb, S., Barlow, R.S., and R.S. Cant. 2009. Experimental measurements of geometric properties of turbulent stratified flames. *Proc. Combust. Inst.* 32:1763–70.

Asato, K., Wada, H., Hiruma, T., and Y. Takeuchi. 1997. Characteristics of flame propagation in a vortex core: Validity of a model for flame propagation. *Combust. Flame* 110:418–28.

Ashurst, Wm.T. 1994. Modeling turbulent flame propagation. *Proc. Combust. Inst.* 25:1075–89.

Ashurst, Wm.T. and I.G. Shepherd. 1997. Flame front curvature distributions in turbulent premixed flame zone. *Combust. Sci. Tech.* 124:115–44.

Ashurst, Wm.T., Checkel, M.D., and D.S.-K. Ting. 1994. The eddy structure model of turbulent flamelet propagation, the expanding spherical and steady planar cases. *Combust. Sci. Tech.* 99:51–74.

Aspden, A.J., Bell, J.B., Day, M.S., Woosley, S.E., and M. Zingale. 2008. Turbulence–flame interactions in type Ia supernovae. *Astrophys. J.* 689:1173–85.

Aspden, A.J., Bell, J.B., and S.E. Woosley. 2010. Distributed flames in type Ia supernovae. *Astrophys. J.* 710:1654–63.

Aspden, A.J., Day, M.S., and J.B. Bell. 2011a. Characterization of low Lewis number flames. *Proc. Combust. Inst.* 33:1463–71.

Aspden, A.J., Day, M.S., and J.B. Bell. 2011b. Lewis number effects in distributed flames. *Proc. Combust. Inst.* 33:1473–80.

Aspden, A.J., Day, M.S., and J.B. Bell. 2011c. Turbulence–flame interactions in lean premixed hydrogen: Transition to the distributed burning regime. *J. Fluid Mech.* 680:287–320.

Atashkari, K., Lawes, M., Sheppard, C.G.W., and R. Woolley. 1999. Towards a general correlation of turbulent premixed flame wrinkling. In *Engineering Turbulence Modelling and Measurements 4*, eds. W. Rodi and D. Laurence, pp. 805–14. Amsterdam: Elsevier.

Aung, K.T., Hassan, M.L., and G.M. Faeth. 1997. Flame stretch interactions of laminar premixed hydrogen/air flames at normal temperature and pressure. *Combust. Flame* 109:1–24.

Aung, K.T., Hassan, M.L., and G.M. Faeth. 1998. Effects of pressure and nitrogen dilution on flame/stretch interactions of laminar premixed $H_2/O_2/N_2$ flames. *Combust. Flame* 112:1–15.

Ayoola, B.O., Balachandran, R., Frank, J.H., Mastorakos, E., and C.F. Kaminski. 2006. Spatially resolved heat release rate measurements in turbulent premixed flames. *Combust. Flame* 144:1–16.

Baev, V.K. and P.K. Tretjakov. 1969. Calculation of flame position in turbulent flow. *Izv. SO AN SSSR* 3(1):32–7 (in Russian).

Barlow, R.S., Dibble, R.W., Chen, J.-Y., and R.P. Lucht. 1990. Effect of Damköhler number on superequilibrium OH concentration in turbulent nonpremixed jet flames. *Combust. Flame* 82:235–51.

Batchelor, G.K. 1952. The effect of homogeneous turbulence on material lines and surfaces. *Proc. R. Soc. Lond. A* 213:349–66.

Baum, M., Poinsot, T.J., Haworth, D.C., and N. Darabiha. 1994. Direct numerical simulation of $H_2/O_2/N_2$ flames with complex chemistry in two-dimensional turbulent flows. *J. Fluid Mech.* 281:1–32.

Bechtold, J.K. and M. Matalon. 2001. The dependence of the Markstein length on stoichiometry. *Combust. Flame* 127:1906–13.

Betev, A.S., Karpov, V.P., Lipatnikov, A.N., and Z.P. Vardosanidze. 1995. Hydrogen combustion in engines and preferential diffusion effects in laminar and turbulent flames. *Arch. Combust.* 15:199–227.

Biagioli, F. 2004. Position, thickness and transport properties of turbulent premixed flames in stagnating flows. *Combust. Theory Model.* 8:533–54.

Bigot, P., Champion, M., and D. Garréton-Bruguieres. 2000. Modeling a turbulent reactive flow with variable equivalence ratio: Application to a flame stabilized by a two-dimensional sudden expansion. *Combust. Sci. Tech.* 158:299–320.

Bilger, R.W. 1976. Turbulent jet diffusion flames. *Prog. Energy Combust. Sci.* 1:87–109.

Bilger, R.W. 1980. Turbulent flows with nonpremixed reactions. In *Turbulent Reacting Flows*, eds. P.A. Libby and F.A. Williams, pp. 65–113. New York: Springer.

Bilger, R.W. 1988. The structure of turbulent nonpremixed flames. *Proc. Combust. Inst.* 22:475–88.

Bilger, R.W. 1993. Conditional moment closure for turbulent reacting flows. *Phys. Fluids A* 5:436–44.

Bilger, R.W. 2000. Future progress in turbulent combustion research. *Prog. Energy Combust. Sci.* 26:367–80.

Bilger, R.W., Saetran, L.R., and L.V. Krishnamoorthy. 1991. Reaction in a scalar mixing layer. *J. Fluid Mech.* 233:211–42.

Bilger, R.W., Pope, S.B., Bray, K.N.C., and J.F. Driscoll. 2005. Paradigms in turbulent combustion research. *Proc. Combust. Inst.* 30:21–42.

Bill, R.G., Naimer, I., Talbot, L., Cheng, R.K., and F. Robben. 1981. Flame propagation in grid-induced turbulence. *Combust. Flame* 43:229–42.

Bondi, S. and W.P. Jones. 2002. A combustion model for premixed flames with varying stoichiometry. *Proc. Combust. Inst.* 29:2123–9.

Bongers, H., van Oijen, J.A., and L.P.H. de Goey. 2002. Intrinsic low-dimensional manifold method extended with diffusion. *Proc. Combust. Inst.* 29:1371–8.

Borghi, R. 1988. Turbulent combustion modeling. *Prog. Energy Combust. Sci.* 14:245–92.

Borghi, R. 1990. Turbulent premixed combustion: Further discussions of the scales of fluctuations. *Combust. Flame* 80:304–12.

Bosschaart, K.J. and L.P.H. de Goey. 2004. The laminar burning velocity of flames propagating in mixtures of hydrocarbons and air measured with the heat flux method. *Combust. Flame* 136:261–9.

Boudier, P., Henriot, S., Poinsot, T., and T. Baritaud. 1992. A model for turbulent flame ignition and propagation in spark ignition engines. *Proc. Combust. Inst.* 24:503–10.

Boughanem, H. and A. Trouvé. 1998. The domain of influence of flame instabilities in turbulent premixed combustion. *Proc. Combust. Inst.* 27:971–8.

Bourguignon, E., Kostiuk, L.W., Michou, Y., and I. Gökalp. 1996. Experimentally measured burning rates of premixed turbulent flames. *Proc. Combust. Inst.* 26:447–53.

Bradley, D. 1992. How fast can we burn? *Proc. Combust. Inst.* 24:247–62.

Bradley, D. 1999. Instabilities and flame speeds in large-scale premixed gaseous explosions. *Phil. Trans. R. Soc. Lond.* 357:3567–81.

Bradley, D., Gaskell, P.H., and A.K.C. Lau. 1990. A mixedness–reactedness flamelet model for turbulent diffusion flames. *Proc. Combust. Inst.* 23:685–92.

Bradley, D., Lau, A.K.C., and M. Lawes. 1992. Flame stretch rate as a determinant of turbulent burning velocity. *Phil. Trans. R. Soc. Lond. A* 338:359–87.

Bradley, D., Gaskell, P.H., and X.J. Gu. 1994a. Application of a Reynolds stress, stretched flamelet, mathematical model to computations of turbulent burning velocities: Comparisons with experiments and the predictions of other models. *Combust. Flame* 96:221–48.

Bradley, D., Lawes, M., and C.G.W. Sheppard. 1994b. Study of turbulence and combustion interaction: Measurement and prediction of the rate of turbulent burning. Periodic Report. Leeds: University of Leeds.

Bradley, D., Lawes, M., Scott, M.J., and E.M.J. Mushi. 1994c. Afterburning in spherical premixed turbulent explosions. *Combust. Flame* 99:581–90.

Bradley, D., Gaskell, P.H., and X.J. Gu. 1998a. The mathematical modeling of liftoff and blowoff of turbulent non-premixed methane jet flames at high strain rates. *Proc. Combust. Inst.* 27:1199–206.

Bradley, D., Hicks, R.A., Lawes, M., Sheppard, C.G.W., and R. Woolley. 1998b. The measurement of laminar burning velocities and Markstein numbers for *iso*-octane–air and *iso*-octane-*n*-heptane–air mixtures at elevated temperatures and pressures in an explosion bomb. *Combust. Flame* 115:126–44.

Bradley, D., Gaskell, P.H., Sedaghat, A., and X.J. Gu. 2003a. Generation of PDFs for flame curvature and for flame stretch rate in premixed turbulent combustion. *Combust. Flame* 135:503–23.

Bradley, D., Haq, M.Z., Hicks, R.A., Kitagawa, T., Lawes, M., Sheppard. C.G.W., and R. Woolley. 2003b. Turbulent burning velocity, burned gas distribution, and associated flame surface definition. *Combust. Flame* 133:415–30.

Bradley, D., Gaskell, P.H., Gu, X.J., and A. Sedaghat. 2005. Premixed flamelet modelling: Factors influencing the turbulent heat release rate source term and the turbulent burning velocity. *Combust. Flame* 143:227–45.

Bradley, D., Lawes, M., Liu, K., and R. Woolley. 2007. The quenching of premixed turbulent flames of *iso*-octane, methane and hydrogen at high pressures. *Proc. Combust. Inst.* 31:1393–400.

Bradshaw, P. 1971. *An Introduction to Turbulence and Its Measurements*. Oxford: Pergamon Press.

Bray, K.N.C. 1979. The interaction between turbulence and combustion. *Proc. Combust. Inst.* 17:223–33.

Bray, K.N.C. 1980. Turbulent flows with premixed reactants. In *Turbulent Reactive Flows*, eds. P.A. Libby and F.A. Williams, pp. 115–83. New York: Springer.

Bray, K.N.C. 1987. Methods of including realistic chemical reaction mechanisms in turbulent combustion models. In *Complex Chemical Reaction Systems: Mathematical Modelling and Simulation*, eds. J. Warnatz and W. Jager, pp. 356–75. Heidelberg: Springer.

Bray, K.N.C. 1990. Studies of the turbulent burning velocity. *Proc. R. Soc. Lond. A* 431:315–35.

Bray, K.N.C. 1995. Turbulent transport in flames. *Proc. R. Soc. Lond. A* 451:231–56.

Bray, K.N.C. 1996. The challenge of turbulent combustion. *Proc. Combust. Inst.* 26:1–26.

Bray, K.N.C. and R.S. Cant. 1991. Some applications of Kolmogorov's turbulence research in the field of combustion. *Proc. R. Soc. Lond. A* 434:217–40.

Bray, K.N.C. and J.B. Moss. 1977. A unified statistical model for the premixed turbulent flame. *Acta Astronaut.* 4:291–319.

Bray, K.N.C. and N. Peters. 1994. Laminar flamelets in turbulent flames. In *Turbulent Reacting Flows*, eds. P.A. Libby and F.A. Williams, pp. 63–113. London: Academic Press.

Bray, K.N.C., Libby, P.A., and J.B. Moss. 1985. Unified modeling approach for premixed turbulent combustion. Part I: General formulation. *Combust. Flame* 61:87–102.

Bray, K.N.C., Champion, M., and P.A. Libby. 2001. Premixed flames in stagnating turbulence. Part V: Evaluation of models for the chemical source term. *Combust. Flame* 127:2023–40.

Bray, K.N.C., Domingo, P., and L. Vervisch. 2005. Role of the progress variable in models for partially premixed turbulent combustion. *Combust. Flame* 141:431–7.

Bray, K.N.C., Champion, M., Libby, P.A., and N. Swaminathan. 2006. Finite rate chemistry and presumed PDF models for premixed turbulent combustion. *Combust. Flame* 146:665–73.

Bray, K.N.C., Champion, M., Libby, P.A., and N. Swaminathan. 2011. Scalar dissipation and mean reaction rates in premixed turbulent combustion. *Combust. Flame* 158:2017–22.

Brodkey, R.S. 1967. *The Phenomena of Fluid Motions*. London: Addison-Wesley.

Buckmaster, J.D. 1993. The structure and stability of laminar flames. *Ann. Rev. Fluid Mech.* 25:21–53.

Buckmaster, J.D. 2002. Edge flames. *Prog. Energy Combust. Sci.* 28:435–75.

Buckmaster, J.D. and G.S.S. Ludford. 1982. *Theory of Laminar Flames*. Cambridge: Cambridge University Press.

Buckmaster, J.D. and M. Short. 1999. Cellular instabilities, sublimit structures and edge-flames in premixed counterflows. *Combust. Theory Model.* 3:199–214.

Buckmaster, J., Clavin, P., Linán, A., Matalon, M., Peters, N., Sivashinsky, G., and F.A. Williams. 2005. Combustion theory and modeling. *Proc. Combust. Inst.* 30:1–19.

Buriko, Yu.Ya., Kuznetsov, V.R., Volkov, D.V., Zaitsev, S.A., and A.F. Uryvsky. 1994. Test of a flamelet model for turbulent non-premixed combustion. *Combust. Flame* 96:104–20.

Burke, S.P. and T.E. Schumann. 1928. Diffusion flames. *Ind. Eng. Chem.* 20:998–1004.

Burluka, A.A. 2010. Combustion Physics. In *Handbook of Combustion, Vol. 1: Fundamentals and Safety*, eds. M. Lackner, F. Winter, and A.K. Agrawal, pp. 53–83. Weinheim: Wiley.

Burluka, A.A., Griffiths, J.F., Liu, K., and M. Ormsby. 2009. Experimental studies of the role of chemical kinetics in turbulent flames. *Combust. Explos. Shock Waves* 45:383–91.

Burluka, A.A., El-Dein Hussin, A.M.T., Sheppard, C.G.W., Liu, K., and V. Sanderson. 2011. Turbulent combustion of hydrogen–CO mixtures. *Flow Turbul. Combust.* 86:735–49.

Buschmann, A., Dinkelacker, F., Schäfer, T., Schäfer, M., and J. Wolfrum. 1996. Measurement of the instantaneous detailed flame structure in turbulent premixed combustion. *Proc. Combust. Inst.* 26:437–45.

Bush, W.B. and F.E. Fendell. 1970. Asymptotic analysis of laminar flame propagation for general Lewis numbers. *Combust. Sci. Tech.* 1:421–8.

Butler, T.D. and P.J. O'Rourke. 1977. A numerical method for two-dimensional unsteady reacting flows. *Proc. Combust. Inst.* 16:1503–15.

Bychkov, V. 2003. Importance of the Darrieus-Landau instability for strongly corrugated turbulent flames. *Phys. Rev. E* 68:066304.

Böhm, B., Frank, J.H., and A. Dreizler. 2011. Temperature and mixing field measurements in stratified lean premixed turbulent flames. *Proc. Combust. Inst.* 33:1583–90.

Candel, S. and T. Poinsot. 1990. Flame stretch and the balance equation for the flame area. *Combust. Sci. Tech.* 170:1–15.

Carrier, G.F., Fendell, F.E., and F.E. Marble. 1975. The effect of strain rate on diffusion flames. *SIAM J. Appl. Math.* 28:463–500.

Cerutti, S. and C. Meneveau. 2000. Statistics of filtered velocity in grid and wake turbulence. *Phys. Fluids* 12:1143–65.

Chakraborty, N. and R.S. Cant. 2009. Effects of Lewis number on turbulent scalar transport and its modelling in turbulent premixed flames. *Combust. Flame* 156:1427–44.

Chakraborty, N. and M. Klein. 2008. A priori direct numerical simulation assessment of algebraic flame surface density models for turbulent premixed flames in the context of large eddy simulation. *Phys. Fluids* 20:085108.

Chakraborty, N. and A.N. Lipatnikov. 2011. Statistics of conditional fluid velocity in the corrugated flamelets regime of turbulent premixed combustion: A direct numerical simulation study. *J. Combust.* 2:628208.

Chakraborty, N. and N. Swaminathan. 2007a. Influence of the Damköhler number on turbulence–scalar interaction in premixed flames. I: Physical insight. *Phys. Fluids* 19:045103.

Chakraborty, N. and N. Swaminathan. 2007b. Influence of the Damköhler number on turbulence–scalar interaction in premixed flames. II: Model development. *Phys. Fluids* 19:045104.

Chakraborty, N., Rogerson, J.W., and N. Swaminathan. 2008. A priori assessment of closures for scalar dissipation rate transport in turbulent premixed flames using direct numerical simulation. *Phys. Fluids* 20:045106.

Champion, M., Deshaies, B., Joulin, G., and K. Kinoshita. 1986. Spherical flame initiation: Theory versus experiments for lean propane–air mixtures. *Combust. Flame* 65:319–37.

Chan, C.K., Lau, K.S., Chin, W.K., and R.K. Cheng. 1992. Freely propagating open premixed turbulent flames stabilized by swirl. *Proc. Combust. Inst.* 24:511–18.

Chaudhuri, S., Akkerman, V., and C.K. Law. 2011. Spectral formulation of turbulent flame speed with consideration of hydrodynamic instability. *Phys. Rev. E* 84:026322.

Chen, J.H. 2011. Petascale direct numerical simulation of turbulent combustion—Fundamental insights towards predictive models. *Proc. Combust. Inst.* 33:99–123.

Chen, J.H. and H.G. Im. 2000. Stretch effects on the burning velocity of turbulent premixed hydrogen–air flames. *Proc. Combust. Inst.* 28:211–18.

Chen, J.H., Lumley, J.L., and F.C. Gouldin. 1986. Modeling of wrinkled laminar flames with intermittency and conditional statistics. *Proc. Combust. Inst.* 21:1483–91.

Chen, M., Herrmann, M., and N. Peters. 2000. Flamelet modeling of lifted turbulent methane/air and propane/air jet diffusion flames. *Proc. Combust. Inst.* 28:167–74.

Chen, Y.-C. 2009. Measurements of three-dimensional mean flame surface area in turbulent premixed Bunsen flames. *Proc. Combust. Inst.* 32:1771–7.

Chen, Y.-C. and R.W. Bilger. 2000. Turbulence and scalar transport in premixed Bunsen flames of lean hydrogen/air mixtures. *Proc. Combust. Inst.* 28:521–8.

Chen, Y.-C. and R.W. Bilger. 2002. Experimental investigation of three-dimensional flame-front structure in premixed turbulent combustion. I: Hydrocarbon/air Bunsen flames. *Combust. Flame* 131:400–35.

Chen, Y.-C. and R.W. Bilger. 2004. Experimental investigation of three-dimensional flame-front structure in premixed turbulent combustion. II: Lean hydrogen/air Bunsen flames. *Combust. Flame* 138:155–74.

Chen, Y.-C. and R.W. Bilger. 2005. Detailed measurements of local scalar-front structures in stagnation-type turbulent premixed flames. *Proc. Combust. Inst.* 30:801–8.

Chen, Y.-C. and M.S. Mansour. 1999. Topology of turbulent premixed flame fronts resolved by simultaneous planar imaging of LIPF of OH radical and Rayleigh scattering. *Exp. Fluids* 26:277–87.

Chen, Y.-C. and M.S. Mansour. 2003. Geometric interpretation of fractal parameters measured in turbulent premixed Bunsen flames. *Exp. Thermal Fluid Sci.* 27:409–16.

Chen, Y.-C. Peters, N., Schneemann, G.A., Wruck, N., Renz, U., and M.S. Mansour. 1996. The detailed flame structure of highly stretched turbulent premixed methane–air flames. *Combust. Flame* 107:223–44.

Cheng, R.K. 1995. Velocity and scalar characteristics of premixed turbulent flames stabilized by weak swirl. *Combust. Flame* 101:1–14.

Cheng, R.K. and I.G. Shepherd. 1986. Interpretation of conditional statistics in open oblique premixed turbulent flames. *Combust. Sci. Tech.* 49:17–40.

Cheng, R.K. and I.G. Shepherd. 1991. The influence of burner geometry on premixed turbulent flame propagation. *Combust. Flame* 85:7–26.

Cheng, R.K., Shepherd, I.G., and L. Talbot. 1988. Reaction rates in premixed turbulent flames and their relevance to the turbulent burning speed. *Proc. Combust. Inst.* 22:771–80.

Cho, E.-S., Chung, S.H., and T.K. Oh. 2006. Local Karlovitz numbers at extinction for various fuels in counterflow premixed flames. *Combust. Sci. Tech.* 178:1559–84.

Cho, P., Law, C.K., Cheng, R.K., and I.G. Shepherd. 1988. Velocity and scalar fields of turbulent premixed flames in stagnation flow. *Proc. Combust. Inst.* 22:739–45.

Choi, C.R. and K.Y. Huh. 1998. Development of a coherent flamelet model for spark-ignited turbulent premixed flame in a closed vessel. *Combust. Flame* 114:336–48.

Choi, C.W. and I.K. Puri. 2001. Contribution of curvature to flame-stretch effects on premixed flames. *Combust. Flame* 126:1640–54.

Chomiak, J. 1970. A possible propagation mechanism of turbulent flames at high Reynolds numbers. *Combust. Flame* 15:319–21.

Chomiak, J. 1977. Dissipation fluctuations and the structure and propagation of turbulent flames in premixed gases at high Reynolds numbers. *Proc. Combust. Inst.* 16:1665–73.

Chomiak, J. 1979. Basic considerations of the turbulent flame propagation in premixed gases. *Prog. Energy Combust. Sci.* 5:207–21.

Chomiak, J. 1990. *Combustion: A Study in Theory, Fact and Application.* New York: Gordon and Breach Science.

Chung, S.H. 2007. Stabilization, propagation and instability of tribrachial triple flames. *Proc. Combust. Inst.* 31:877–92.

Chung, S.H. and C.K. Law. 1984. An invariant derivation of flame stretch. *Combust. Flame* 55:123–5.

Chung, S.H. and C.K. Law. 1988. An integral analysis of the structure and propagation of stretch premixed flames. *Combust. Flame* 72:325–36.

Chung, S.H., Chung, D.H., Fu, C., and P. Cho. 1996. Local extinction Karlovitz numbers for premixed flames. *Combust. Flame* 106:515–20.

Cintosun, E., Smallwood, G.J., and Ö.L. Gülder. 2007. Flame surface fractal characteristics in premixed turbulent combustion at high turbulence intensities. *AIAA J.* 45:2785–9.

Class, A.G., Matkowsky, B.J., and A.Y. Klimenko. 2003a. A unified model of flames as gasdynamic discontinuities. *J. Fluid Mech.* 491:11–49.

Class, A.G., Matkowsky, B.J., and A.Y. Klimenko. 2003b. Stability of planar flames as gasdynamic discontinuities. *J. Fluid Mech.* 491:51–63.

Clavin, P. 1985. Dynamical behavior of premixed flame fronts in laminar and turbulent flows. *Prog. Energy Combust. Sci.* 11:1–59.

Clavin, P. and G. Joulin. 1997. High-frequency response of premixed flames to weak stretch and curvature: A variable density analysis. *Combust. Theory Model.* 1:429–46.

Clavin, P. and F.A. Williams. 1979. Theory of premixed-flame propagation in large-scale turbulence. *J. Fluid Mech.* 90:589–604.

Clavin, P. and F.A. Williams. 1982. Effects of molecular diffusion and of thermal expansion on the structure and dynamics of premixed flames in turbulent flows of large scale and low intensity. *J. Fluid Mech.* 116:251–82.

Cohé, C., Halter, F., Chauveau, C., Gökalp, I., and Ö.L. Gülder. 2007. Fractal characterization of high-pressure and hydrogen-enriched CH_4–air turbulent premixed flames. *Proc. Combust. Inst.* 31:1345–52.

Colin, O., Ducros, F., Veynante, D., and T. Poinsot. 2000. A thickened flame model for large eddy simulations of turbulent premixed combustion. *Phys. Fluids* 12:1843–63.

Cuenot, B. and T. Poinsot. 1994. Effects of curvature and unsteadiness in diffusion flames. Implications for turbulent diffusion combustion. *Proc. Combust. Inst.* 25:1383–90.

Damköhler, G. 1940. Der einfuss der turbulenz auf die flammengeschwindigkeit in gasgemischen. *Zs. Electrochemie* 6:601–26.

Daniele, S., Jansohn, P., Mantzaras, J., and K. Boulouchos. 2011. Turbulent flame speed for syngas at gas turbine relevant conditions. *Proc. Combust. Inst.* 33:2937–44.

Darabiha, N. 1992. Transient behaviour of laminar counterflow hydrogen–air diffusion flames with complex chemistry. *Combust. Sci. Tech.* 86:163–81.

Darbyshire, O.R., Swaminathan, N., and S. Hochgreb. 2010. The effects of small-scale mixing models on the prediction of turbulent premixed and stratified combustion. *Combust. Sci. Tech.* 182:1141–70.

Darrieus, G. 1938. Propagation d'un front de flamme. Paper presented at *La Technique Moderne*, Paris.

Das, A.K. and R.L. Evans. 1997. An experimental study to determine fractal parameters for lean premixed flames. *Exp. Fluids* 22:312–20.

Davis, S.G. and C.K. Law. 1998. Laminar flame speeds and oxidation kinetics of *iso*-octane–air and *n*-heptane–air flames. *Proc. Combust. Inst.* 27:521–7.

Day, M., Tachibana, S., Bell, J., Lijewski, M., Beckner, V., and R.K. Cheng. 2012. A combined computational and experimental characterization of lean premixed low swirl laboratory flames. I: Methane flames. *Combust. Flame* 159:275–90.

Deschamps, B., Boukhalfa, A., Chauveau, C., Gökalp, I., Shepherd, I.G., and R.K. Cheng. 1992. An experimental estimation of flame surface density and mean reaction rate in turbulent premixed flames. *Proc. Combust. Inst.* 24:469–75.

Dinkelacker, F. 2002. Numerical calculation of turbulent premixed flames with an efficient turbulent flame speed closure model. In *High-Performance Scientific and Engineering Computing*, eds. M. Breuer, F. Durst, and C. Zenger, pp. 81–8. Berlin: Springer.

Dinkelacker F. 2003. Experimental validation of flame regimes for highly turbulent premixed flames. In *European Combustion Meeting*, paper no. 158. CD.

Dinkelacker, F. and S. Hölzler. 2000. Investigation of a turbulent flame speed closure approach for premixed flame calculations. *Combust. Sci. Tech.* 158:321–40.

Dinkelacker, F., Soika, A., Most, D., Hofmann, D., Leipertz, A., Polifke, W., and K. Döbbeling. 1998. Structure of locally quenched highly turbulent lean premixed flames. *Proc. Combust. Inst.* 27:857–65.

Dixon-Lewis, G. 1990. Structure of laminar flames. *Proc. Combust. Inst.* 23:305–24.

Domingo, P., Vervisch, L., and K.N.C. Bray. 2002. Partially premixed flamelets in LES of nonpremixed turbulent combustion. *Combust. Theory Model.* 6:529–51.

Domingo, P., Vervisch, L., and J. Réveillon. 2005a. DNS analysis of partially premixed combustion in spray and gaseous turbulent flame-bases stabilized in hot air. *Combust. Flame* 140:172–95.

Domingo, P., Vervisch, L., Payet, S., and R. Hauguel. 2005b. DNS of a premixed turbulent V flame and LES of a ducted flame using a FSD-PDF subgrid scale closure with FPI-tabulated chemistry. *Combust. Flame* 143:566–86.

Dopazo, C. 1994. Recent developments in PDF methods. In *Turbulent Reacting Flows*, eds. P.A. Libby and F.A. Williams, pp. 375–474. London: Academic Press.

Driscoll, J.F. 2008. Turbulent premixed combustion: Flamelet structure and its effect on turbulent burning velocities. *Prog. Energy Combust. Sci.* 34:91–134.

Driscoll, J.F. and A. Gulati. 1988. Measurement of various terms in the turbulent kinetic energy balance within a flame and comparison with theory. *Combust. Flame* 72:131–52.

Duclos, J.M., Veynante, D., and T. Poinsot. 1993. A comparison of flamelet models for premixed turbulent combustion. *Combust. Flame* 95:101–17.

Dunn, M.J., Masri, A.R., and R.W. Bilger. 2007. A new piloted premixed jet burner to study strong finite-rate chemistry effects. *Combust. Flame* 151:46–60.

Dunn, M.J., Masri, A.R., Bilger, R.W., Barlow, R.S., and G.-H. Wang. 2009. The compositional structure of highly turbulent piloted premixed flames issuing into a hot coflow. *Proc. Combust. Inst.* 32:1779–86.

Dunn, M.J., Masri, A.R., Bilger, R.W., and R.S. Barlow. 2010. Finite rate chemistry effects in highly sheared turbulent premixed flames. *Flow Turbul. Combust.* 85:621–48.

Duwig, C. and L. Fuchs. 2005. Study of flame stabilization in a swirling combustor using a new flamelet formulation. *Combust. Sci. Tech.* 177:1584–610.

van Dyke, M. 1982. *An Album of Fluid Motion.* Stanford: Parabolic Press.

Ebert, U. and W. van Saarloos. 2000. Front propagation into unstable states: Universal algebraic convergence towards uniformly translating pulled fronts. *Physica D* 146:1–99.

Egolfopoulos, F.N., Zhu, D.L., and C.K. Law. 1990. Experimental and numerical determination of laminar flame speeds: Mixtures of C_2–hydrocarbons with oxygen and nitrogen. *Proc. Combust. Inst.* 23:471–8.

Erard, V., Boukhalfa, A., Puechberty, D., and M. Trinité. 1996. A statistical study on surface properties of freely-propagating premixed turbulent flames. *Combust. Sci. Tech.* 113–114:313–27.

Filatyev, S.A., Driscoll, J.F., Carter, C.D., and J.M. Donbar. 2005. Measured properties of turbulent premixed flames for model assessment, including burning velocities, stretch rates, and surface densities. *Combust. Flame* 141:1–21.

Flohr, P. and H. Pitsch. 2000. A turbulent flame speed closure model for LES of industrial burner flows. In *Proceeding of the 2000 Summer Program*, pp. 169–79, Center for Turbulence Research, Stanford, CA: Stanford University.

Foucher, F. and C. Mounaïm-Rousselle. 2005. Fractal approach to the evaluation of burning rates in the vicinity of the piston in a spark ignition engine. *Combust. Flame* 143:323–32.

Fox, M.D. and F.J. Weinberg. 1962. An experimental study of burner stabilized turbulent flames in premixed reactants. *Proc. R. Soc. Lond. A* 268:222–39.

Frank, J.H., Kalt, P.A.M., and R.W. Bilger. 1999. Measurements of conditional velocities in turbulent premixed flames by simultaneous OH PLIF and PIV. *Combust. Flame* 116:220–32.

Frankel, M.L. and G.J. Sivashinsky. 1982. The effect of viscosity on hydrodynamic stability of a plane flame front. *Combust. Sci. Tech.* 29:207–24.

Frankel, M.L. and G.J. Sivashinsky. 1983. On effects due to thermal expansion and Lewis number in spherical flame propagation. *Combust. Sci. Tech.* 31:131–8.

Frisch, U. 1995. *Turbulence. The Legacy of A.N. Kolmogorov.* Cambridge: Cambridge University Press.

Fureby, C. 2005. A fractal flame-wrinkling large eddy simulation model for premixed turbulent combustion. *Proc. Combust. Inst.* 30:593–601.

Garrido-López, D. and S. Sarkar. 2005. Effects of imperfect premixing coupled with hydrodynamic instability on flame propagation. *Proc. Combust. Inst.* 30:621–8.

Ghirelli, F. 2011. Turbulent premixed flame model based on a recent dispersion model. *Comput. Fluids* 44:369–76.

Gicquel, O., Darabiha, N., and D. Thévenin. 2000. Laminar premixed hydrogen/air counterflow flame simulations using flame prolongation of ILDM with differential diffusion. *Proc. Combust. Inst.* 28:1901–8.

Giovangigli, V. 1999. *Multicomponent Flow Modeling*. Berlin: Springer.

Girimaji, S.S. and S.B. Pope. 1992. Propagating surfaces in isotropic turbulence. *J. Fluid Mech.* 234:247–77.

Givi, P. 1994. Spectral and random vortex methods in turbulent reacting flows. In *Turbulent Reacting Flows*, eds. P.A. Libby and F.A. Williams, pp. 475–572. London: Academic Press.

de Goey, L.P.H. and J.H.M. ten Thije Boonkkamp. 1999. A flamelet description of premixed laminar flame and the relation with flame stretch. *Combust. Flame* 119:253–71.

de Goey, L.P.H., Plessing, T., Hermanns, R.T.E., and N. Peters. 2005. Analysis of the flame thickness of turbulent flamelets in the thin reaction zone regime. *Proc. Combust. Inst.* 30:859–66.

de Goey, L.P.H., van Oijen, J.A., Kornilov, V.N., and J.H.M. ten Thije Boonkkamp. 2011. Propagation, dynamics and control of laminar premixed flames. *Proc. Combust. Inst.* 33:863–86.

Goix, P. and I.G. Shepherd. 1993. Lewis number effects in turbulent premixed flame structure. *Combust. Sci. Tech.* 91:191–206.

Goix, P., Paranthoen, P., and M. Trinité. 1990. A tomographic study of measurements in a V-shaped H_2–air flame and a Lagrangian interpretation of the turbulent flame brush thickness. *Combust. Flame* 81:229–41.

Goix, P., Shepherd, I.G., and M. Trinité. 1989. A fractal study of a premixed V-shaped H_2/air flame. *Combust. Sci. Tech.* 63:275–86.

Goldstein, S. 1951. On diffusion by discontinuous movements, and on the telegraph equation. *Quart. J. Mech. Appl. Math.* 4:129–56.

Gouldin, F.C. 1987. An application of fractals to modelling premixed turbulent flames. *Combust. Flame* 68:249–66.

Gouldin, F.C. 1996. Combustion intensity and burning rate integral of premixed flames. *Proc. Combust. Inst.* 26:381–8.

Gouldin, F.C. and K.V. Dandekar. 1984. Time-resolved density measurements in premixed turbulent flames. *AIAA J.* 22:655–63.

Gouldin, F.C. and P.C. Miles. 1995. Chemical closure and burning rates in premixed turbulent flames. *Combust. Flame* 100:202–10.

Griebel, P., Siewert, P., and P. Jansohn. 2007. Flame characteristics of turbulent lean premixed methane/air flames at high-pressure: Turbulent flame speed and flame brush thickness. *Proc. Combust. Inst.* 31:3083–90.

Groff, E.G. 1982. The cellular nature of confined spherical propane–air flames. *Combust. Flame* 48:51–62.

Groff, E.G. 1987. An experimental evaluation of an entrainment flame propagation model. *Combust. Flame* 67:153–62.

Grout, R.W., Swaminathan, N., and R.S. Cant. 2009. Effects of compositional fluctuations on premixed flames. *Combust. Theory Model.* 13:823–52.

Gulati, A. and J.F. Driscoll. 1986a. Velocity–density correlations and Favre averages measured in a premixed turbulent flame. *Combust. Sci. Tech.* 48:285–307.

Gulati, A. and J.F. Driscoll. 1986b. Flame-generated turbulence and mass fluxes: Effect of varying heat release. *Proc. Combust. Inst.* 21:1367–75.

Gülder, Ö.L. 1982. Laminar burning velocities of methanol, ethanol, and isooctane–air mixtures. *Proc. Combust. Inst.* 19:275–81.

Gülder, Ö.L. 1990a. Turbulent premixed flame propagation models for different combustion regimes. *Proc. Combust. Inst.* 23:743–50.

Gülder, Ö.L. 1990b. Turbulent premixed combustion modelling using fractal geometry. *Proc. Combust. Inst.* 23:835–41.

Gülder, Ö.L. 2007. Contribution of small-scale turbulence to burning velocity of flamelets in the thin reaction zone regime. *Proc. Combust. Inst.* 31:1369–75.

Gülder, Ö.L. and G.J. Smallwood. 1995. Inner cutoff scale of flame surface wrinkling in turbulent premixed flames. *Combust. Flame* 103:107–14.

Gülder, Ö.L. and G.J. Smallwood. 2007. Flame surface densities in premixed combustion at medium to high turbulence intensities. *Combust. Sci. Tech.* 179:191–206.

Gülder, Ö.L, Smallwood, G.J., Wong, R., Snelling, D.R., Smith, R., Deschamps, B.M., and J.C. Sautet. 2000. Flame front surface characteristics in turbulent premixed propane/air combustion. *Combust. Flame* 120:407–16.

Hainsworth, E. 1985. Study of free turbulent premixed flames. Master thesis, Massachusetts Institute of Technology, Cambridge.

Hakberg, B. and A.D. Gosman. 1984. Analytical determination of turbulent flame speed from combustion models. *Proc. Combust. Inst.* 20:225–32.

Hartung, G., Hult, J., Kaminski, C.F. Rogerson, J.W., and N. Swaminathan. 2008. Effect of heat release on turbulence and scalar–turbulence interaction in premixed combustion. *Phys. Fluids* 20:035110.

Haworth, D.C. 2010. Progress in probability density function methods for turbulent reacting flows. *Prog. Energy Combust. Sci.* 36:168–259.

Haworth, D.C., Drake, M.C., Pope, S.B., and R.J. Blint. 1988. The importance of time-dependent flame structures in stretched laminar flamelet models for turbulent jet diffusion flames. *Proc. Combust. Inst.* 22:589–97.

Haworth, D.C., Blint, R.J., Cuenot, B., and T. Poinsot. 2000. Numerical simulation of turbulent propane–air combustion with nonhomogeneous reactants. *Combust. Flame* 121:395–417.

Hélie, J. and A. Trouvé. 1998. Turbulent flame propagation in partially premixed combustion. *Proc. Combust. Inst.* 27:891–8.

Hentschel, H.G.E. and I. Procaccia. 1984. Relative diffusion in turbulent media: The fractal dimension of clouds. *Phys. Rev.* A 29:1461–70.

Herrmann, M. 2006. Numerical simulation of turbulent Bunsen flames with a level set flamelet model. *Combust. Flame* 145:357–75.

Hinze, J.O. 1975. *Turbulence*, 2nd edn. New York: McGraw Hill.

Hirschfelder, J.O., Curtiss, C.F., and R.B. Bird. 1954. *Molecular Theory of Gases and Liquids*. New York: Wiley.

Ho, C.M., Jakus, K., and K.H. Parker. 1976. Temperature fluctuations in a turbulent flame. *Combust. Flame* 27:113–23.

Huang, C. and A.N. Lipatnikov. 2012. Comparison of presumed PDF models of turbulent flames. *J. Combust.* (in press).

Huang, C., Golovitchev, V., and A.N. Lipatnikov. 2010. Chemical model of gasoline–ethanol blends for internal combustion engine applications. *SAE Paper* 2010-01-0543.

Huang, Y., Sung, C.J., and J.A. Eng. 2004. Laminar flame speeds of primary reference fuels and reformer gas mixtures. *Combust. Flame* 139:239–51.

Ichikawa, Y., Otawara, Y., Kobayashi, H., Ogami, Y., Kudo, T., Okuyama, M., and S. Kadowaki. 2011. Flame structure and radiation characteristics of $CO/H_2/CO_2$/air turbulent premixed flames at high pressure. *Proc. Combust. Inst.* 33:1543–50.

Im, Y.H., Huh, K.Y., Nishiki, S., and T. Hasegawa. 2004. Zone conditional assessment of flame-generated turbulence with DNS database of a turbulent premixed flame. *Combust. Flame* 137:478–88.

Ishizuka, S. 1990. On the flame propagation in a rotating flow field. *Combust. Flame* 82:176–90.

Ishizuka, S. 2002. Flame propagation along a vortex axis. *Prog. Energy Combust. Sci.* 28:477–542.

Ishizuka, S., Murakami, T., Hamazaki, T., Koumura, K., and R. Hasegawa. 1998. Flame speeds in combustible vortex rings. *Combust. Flame* 113:542–53.

Ishizuka, S., Ikeda, M., and K. Kameda. 2002. Vortex ring combustion in an atmosphere of the same mixture as the combustible. *Proc. Combust. Inst.* 29:1705–12.

Janicka, J. and W. Kollmann. 1979. A two-variables formalism for the treatment of chemical reactions in turbulent H₂-air diffusion flames. *Proc. Combust. Inst.* 17:421–30.

Janicka, J. and N. Peters. 1982. Prediction of turbulent jet diffusion flame lift-off using a pdf transport equation. *Proc. Combust. Inst.* 19:367–74.

Janicka, J. and S. Sadiki. 2005. Large eddy simulation of turbulent combustion systems. *Proc. Combust. Inst.* 30:537–47.

Jarosinski, J. 1986. A survey of recent studies on flame extinction. *Prog. Energy Combust. Sci.* 12:81–116.

Jarosinski, J., Podfilipski, J., and T. Fodemski. 2002a. Properties of flames propagating in propane–air mixtures near flammability and quenching limits. *Combust. Sci. Tech.* 174:167–87.

Jarosinski, J., Podfilipski, J., Gorczakowski, A., and B. Veyssiere. 2002b. Experimental study of flame propagation in propane–air mixture near rich flammability limits in microgravity. *Combust. Sci. Tech.* 174:21–48.

Jerzembeck, S., Peters, N., Pepiot-Desjardins, P., and H. Pitsch. 2009. Laminar burning velocities at high pressure for primary reference fuels and gasoline: Experimental and numerical investigation. *Combust. Flame* 156:292–301.

Jiménez, C., Cuenot, B., Poinsot, T., and D. Haworth. 2002. Numerical simulation and modeling for lean stratified propane–air flames. *Combust. Flame* 128:1–21.

Jiménez, J. and A.A. Wray. 1998. On the characteristics of vortex filaments in isotropic turbulence. *J. Fluid Mech.* 373:255–85.

Jin, B., Grout, R., and W.K. Bushe. 2008. Conditional source-term estimation as a method for chemical closure in premixed turbulent reacting flow. *Flow Turbul. Combust.* 81:563–82.

Jones, W.P. 1994. Turbulence modelling and numerical solution methods for variable density and combusting flows. In *Turbulent Reacting Flows*, eds. P.A. Libby and F.A. Williams, pp. 309–74. London: Academic Press.

Jones, W.P. and B.E. Launder. 1972. The prediction of laminarization with a two-equation model of turbulence. *Int. J. Heat Mass Transfer* 15:301–14.

Kadowaki, S. and T. Hasegawa. 2005. Numerical simulation of dynamics of premixed flames: Flame instability and vortex–flame interaction. *Prog. Energy Combust. Sci.* 31:193–241.

Kalt, P.A.M., Frank, J.H., and R.W. Bilger. 1998. Laser imaging of conditional velocities in premixed propane–air flames by simultaneous OH PLIF and PIV. *Proc. Combust. Inst.* 27:751–8.

Kalt, P.A.M., Chen, Y.-C., and R.W. Bilger. 2002. Experimental investigation of turbulent scalar flux in premixed stagnation-type flames. *Combust. Flame* 129:401–15.

Kang, T. and D.C. Kyritsis. 2007. Departure from quasi-homogeneity during laminar flame propagation in lean, compositionally stratified methane–air mixtures. *Proc. Combust. Inst.* 31:1075–83.

Kang, T. and D.C. Kyritsis. 2009. Phenomenology of methane flame propagation into compositionally stratified, gradually richer mixtures. *Proc. Combust. Inst.* 32:979–85.

Karlovitz, B., Denniston, D.W., and F.E. Wells. 1951. Investigation of turbulent flames. *J. Chem. Phys.* 19:541–7.

Karpov, V.P. 1965. Cellular flame structure under conditions of a constant-volume bomb and its relationship with vibratory combustion. *Combust. Explos. Shock Waves* 1(3):39–42.

Karpov, V.P. and A.N. Lipatnikov. 1994. Premixed turbulent combustion and some thermodiffusional effects in laminar flames. In *Combustion, Detonation, Shock Waves*, Proceedings of the Zel'dovich Memorial, eds. A.G. Merzhanov and S.M. Frolov, vol. 1, pp. 168–80. Moscow: ENAS.

Karpov, V.P. and A.N. Lipatnikov. 1995. An effect of molecular thermal conductivity and diffusion on premixed combustion. *Dokl. Phys. Chem.* 341:83–5.

Karpov, V.P. and E.S. Severin. 1977. Turbulent burning velocities of gas mixtures for describing combustion in engines. In *Chemical Physics of Combustion and Explosion Processes. Combustion of Heterogeneous and Gas Systems*. Proceedings of the Fourth All-Union Combustion Symposium, pp. 74–6. Chernogolovka: Institute of Chemical Physics (in Russian).

Karpov, V.P. and E.S. Severin. 1978. Turbulent burn-up rates of propane–air flames determined in a bomb with agitators. *Combust. Explos. Shock Waves* 14:158–63.

Karpov, V.P. and E.S. Severin. 1980. Effects of molecular-transport coefficients on the rate of turbulent combustion. *Combust. Explos. Shock Waves* 16:41–6.

Karpov, V.P., Lipatnikov, A.N., and V.L. Zimont. 1994. A model of premixed turbulent combustion and its validation. *Arch. Combust.* 14(3–4):125–41.

Karpov, V.P., Lipatnikov, A.N., and V.L. Zimont. 1996a. A test of an engineering model of premixed turbulent combustion. *Proc. Combust. Inst.* 26:249–57.

Karpov, V.P., Lipatnikov, A.N., and V.L. Zimont. 1996b. Influence of molecular heat and mass transfer processes on premixed turbulent combustion. In *Transport Phenomena in Combustion*, ed. C.H. Chan, vol. 1, pp. 629–40. New York: Taylor & Francis.

Karpov, V.P., Lipatnikov, A.N., and V.L. Zimont. 1997. Flame curvature as a determinant of preferential diffusion effects in premixed turbulent combustion. In *Advances in Combustion Science: In Honor of Ya.B. Zel'dovich*. eds. W.A. Sirignano, A.G. Merzhanov, and L. De Luca, *Progress in Astronautics and Aeronautics*, vol. 173, chapter 14, pp. 235–50. Reston: AIAA.

Kelley, A.P., Liu, W., Xin, Y.X., Smallbone, A.J., and C.K. Law. 2011. Laminar flame speeds, non-premixed stagnation ignition, and reduced mechanisms in the oxidation of *iso*-octane. *Proc. Combust. Inst.* 33:501–8.

Kent, J.H. and R.W. Bilger. 1973. Turbulent diffusion flames. *Proc. Combust. Inst.* 14:615–25.

Kerstein, A.R. 1988. Linear-eddy model of turbulent transport and mixing. *Combust. Sci. Tech.* 60:391–421.

Kerstein, A.R. 1992a. Linear-eddy modelling of turbulent transport. Part 4: Structure of diffusion flames. *Combust. Sci. Tech.* 81:75–96.

Kerstein, A.R. 1992b. Linear-eddy modelling of turbulent transport. Part 7: Finite-rate chemistry and multi-stream mixing. *J. Fluid Mech.* 240:289–313.

Kerstein, A.R. 2002. Turbulence in combustion processes: Modeling challenges. *Proc. Combust. Inst.* 29:1763–73.

Kido, H., Kitagawa, T., Nakashima, K., and K. Kato. 1989. An improved model of turbulent mass burning velocity. *Memoirs Fac. Eng. Kyushu Univ.* 49:229–47.

Klimenko, A.Y. 1990. Multicomponent diffusion of various scalars in turbulent flows. *Fluid Dyn.* 25:327–34.

Klimenko, A.Y. 2001. On the relation between the conditional moment closure and unsteady flamelets. *Combust. Theory Model.* 5:275–94.

Klimenko, A.Y. and R.W. Bilger. 1999. Conditional moment closure for turbulent combustion. *Prog. Energy Combust. Sci.* 25:595–687.

Klimenko, A.Y. and S.B. Pope. 2003. The modeling of turbulent reactive flows based on multiple mapping conditioning. *Phys. Fluids* 15:1907–25.

Klimov, A.M. 1963. Laminar flame in a turbulent flow. *Zh. Prikladnoi Mekh. Tekh. Fiz.* 3:49–58.

Kobayashi, H., Tamura, T., Maruta, K., Niioka, T., and F.A. Williams. 1996. Burning velocity of turbulent premixed flames in a high-pressure environment. *Proc. Combust. Inst.* 26:389–96.

Kobayashi, H., Nakashima, T., Tamura, T., Maruta K., and T. Niioka. 1997. Turbulence measurements and observations of turbulent premixed flames at elevated pressures up to 3.0 MPa. *Combust. Flame* 108:104–17.

Kobayashi, H., Kawabata, Y., and K. Maruta. 1998. Experimental study on general correlation of turbulent burning velocity at high pressure. *Proc. Combust. Inst.* 27:941–8.

Kobayashi, H., Kawahata, T., Seyama, K., Fujimari, T., and J.-S. Kim. 2002. Relationship between the smallest scale of flame wrinkles and turbulence characteristics of high-pressure, high-temperature turbulent premixed flames. *Proc. Combust. Inst.* 29:1793–800.

Kobayashi, H., Seyama, K., Hagiwara, H., and Y. Ogami. 2005. Burning velocity correlation of methane/air turbulent premixed flames at high-pressure and high-temperature. *Proc. Combust. Inst.* 30:827–34.

Kolla, H. and N. Swaminathan. 2010a. Strained flamelets for turbulent premixed flames. I: Formulation and planar flame results. *Combust. Flame* 157:943–54.

Kolla, H. and N. Swaminathan. 2010b. Strained flamelets for turbulent premixed flames. II: Laboratory flame results. *Combust. Flame* 157:1274–89,

Kolla, H., Rogerson, J.W., Chakraborty, N., and N. Swaminathan. 2009. Scalar dissipation rate modeling and its validation. *Combust. Sci. Tech.* 181:518–35.

Kolla, H., Rogerson, J.W., and N. Swaminathan. 2010. Validation of a turbulent flame speed model across combustion regimes. *Combust. Sci. Tech.* 182:284–308.

Kolmogorov, A.N. 1941a. The local structure of turbulence in incompressible viscous fluid for very large Reynolds number. *Dokl. Akad. Nauk SSSR* 30:9–13 (in Russian; English translation in *Proc. R. Soc. Lond. A* 434:9–13, 1991).

Kolmogorov, A.N. 1941b. Dissipation of energy in locally isotropic turbulence. *Dokl. Akad. Nauk SSSR* 32:16–18 (in Russian; English translation in *Proc. R. Soc. Lond. A* 434:15–17, 1991).

Kolmogorov, A.N. 1942. Equations of turbulent motion of an incompressible fluid. *Izv. Akad. Nauk SSSR Fizika* 6(1–2):56–8 (in Russian; English translation in *Proc. R. Soc. Lond. A* 434:214–16, 1991).

Kolmogorov, A.N. 1962. A refinement of previous hypotheses concerning the local structure of turbulence in a viscous incompressible fluid at high Reynolds number. *J. Fluid Mech.* 13:82–5.

Kolmogorov, A.N., Petrovsky, E.G., and N.S. Piskounov. 1937. A study of the diffusion equation with a source term and its application to a biological problem. *Bjulleten' MGU, Moscow State Univ.* A1(6):1–26 (in Russian; English translation in *Dynamics of Curved Fronts*, ed. P. Pelcé, pp. 105–30. San Diego: Academic Press, 1988).

Kortschik, C., Plessing, T., and N. Peters. 2004. Laser optical investigation of turbulent transport of temperature ahead of the preheat zone in a premixed flame. *Combust. Flame* 136:43–50.

Kostiuk, L.W. and I.G. Shepherd. 1996. Measuring the burning rate of premixed turbulent flames in stagnation flows. *Combust. Sci. Tech.* 112:359–68.

Kostiuk, L.W., Shepherd, I.G., and K.N.C. Bray. 1999. Experimental study of premixed turbulent combustion in opposed streams. Part III: Spatial structure of flames. *Combust. Flame* 118:129–39.

Kronenburg, A. 2004. Double conditioning of reactive scalar transport equations in turbulent nonpremixed flames. *Phys. Fluids* 16:2640–8.

Kronenburg, A. and A.E. Papoutsakis. 2005. Conditional moment closure modelling of extinction and reignition in turbulent non-premixed flames. *Proc. Combust. Inst.* 30:757–64.

Kronenburg, A., Bilger, R.W., and J.H. Kent. 1998. Second order conditional moment closure for turbulent jet diffusion flames. *Proc. Combust. Inst.* 27:1097–104,

Kumar, K., Freeth, J.E., Sung, C.J., and Y. Huang. 2007. Laminar flame speeds of preheated *iso*-octane/O_2/N_2 and *n*-heptane/O_2/N_2 mixtures. *J. Propulsion Power* 16:513–22.

Kuznetsov, V.R. 1972. Passive contaminant concentration probability in turbulent flows with transverse shear. *Izv. AN SSSR MZhG* 3:86–91.

Kuznetsov, V.R. 1977. Mixing up to a molecular level and the development of a chemical reaction in a turbulent flow. *Izv. AN SSSR MZhG.* 3:32–41.

Kuznetsov, V.R. 1979. Estimate of the correlation between pressure pulsations and the divergence of the velocity in subsonic flows of variable density. *Fluid Dynamics* 14:328–34.

Kuznetsov, V.R. 1982. Effects of turbulence on the formation of large superequilibrium concentration of atoms and free radicals in diffusion flames. *Mekhan. Zhidkosti Gasa* 6:3–9.

Kuznetsov, V.R. and V.A. Sabel'nikov. 1977. Combustion characteristics of mixed gases in a strongly turbulent flow. *Combust. Explos. Shock Waves* 13:425–34.

Kuznetsov, V.R. and V.A. Sabel'nikov. 1990. *Turbulence and Combustion.* New York: Hemisphere.

Kwon, O.C., Hassan, M.I., and G.M. Faeth. 2000. Flame/stretch interactions of premixed fuel-vapor/O_2/N_2 flames. *J. Propulsion Power* 16:513–22.

Kwon, S. and G.M. Faeth. 2001. Flame/stretch interactions of premixed hydrogen-fueled flames: Measurements and predictions. *Combust. Flame* 124:590–610.

Kwon, S., Wu, M.S., Driscoll, J.F., and G.M. Faeth. 1992. Flame surface properties of premixed flames in isotropic turbulence: Measurements and numerical simulations. *Combust. Flame* 88:221–38.

Lam, S.H. and D.A. Goussis. 1991. Conventional asymptotics and computational singular perturbation for simplified kinetic modeling. In *Reduced Kinetic Mechanisms and Asymptotic Approximations for Methane–Air Flames*, ed. M.D. Smooke, pp. 227–42. Berlin: Springer.

Landau, L.D. 1944. On the theory of slow combustion. *Acta Physicochem. URSS* 19:77–85.

Landau, L.D. and E.M. Lifshitz. 1987. *Fluid Mechanics.* Oxford: Pergamon Press.

Launder, B.E. and D.B. Spalding. 1972. *Mathematical Models of Turbulence.* London: Academic Press.

Law, C.K. and C.J. Sung. 2000. Structure, aerodynamics, and geometry of premixed flames. *Prog. Energy Combust. Sci.* 26:459–505.

Lawn, C.J. 2009. Lifted flames on fuel jets on co-flowing air. *Prog. Energy Combust. Sci.* 35:1–30.

Lee, B., Choi, C.R., and K.Y. Huh. 1998. Application of the coherent flamelet model to counterflow turbulent premixed combustion and extinction. *Combust. Sci. Tech.* 138:1–25.

Lee, E. and K.Y. Huh. 2004. Zone conditional modeling of premixed turbulent flames at a high Damköhler number. *Combust. Flame* 138:211–24.

Lewis, B. and G. von Elbe. 1961. *Combustion, Flames and Explosions of Gases*, 2nd edn. New York: Academic Press.

Li, B., Baudoin, E., Yu, R., Sun, Z.W., Li, Z.S., Bai, X.S., Aldén, M., and M.S. Mansour. 2009. Experimental and numerical study of a conical turbulent partially premixed flame. *Proc. Combust. Inst.* 32:1811–18.

Li, S.C., Libby, P.A., and F.A. Williams 1994. Experimental investigation of a premixed flame in an impinging turbulent stream. *Proc. Combust. Inst.* 25:1207–14.

Li, Z.S., Li, B., Sun, Z.W., Bai, X.S., and M. Aldén. 2010. Turbulence and combustion interaction: High resolution local flame front structure visualization with simultaneous single-shot PLIF of CH and CH_2O in a piloted premixed jet flame. *Combust. Flame* 157:1087–96.

Libby, P.A. 1975. On the prediction of intermittent turbulent flows. *J. Fluid Mech.* 68:273–95.

Libby, P.A. 1985. Theory of normal premixed turbulent flames revisited. *Prog. Energy Combust. Sci.* 11:83–96.

Libby, P.A. and K.N.C. Bray. 1977. Variable density effects in premixed turbulent flames. *AIAA J.* 15:1186–93.

Libby, P.A. and K.N.C. Bray. 1980. Implications of the laminar flamelet model in premixed turbulent combustion. *Combust. Flame* 39:33–41.

Libby, P.A. and K.N.C. Bray. 1981. Countergradient diffusion in premixed turbulent flames. *AIAA J.* 19:205–13.

Libby, P.A. and F.A. Williams. 1982. Structure of laminar flamelets in premixed turbulent flames. *Combust. Flame* 44:287–303.

Libby, P.A. and F.A. Williams. 1984. Strained premixed laminar flames with two reaction zones. *Combust. Sci. Tech.* 37:221–52.

Libby, P.A. and F.A. Williams. 1994. Fundamental aspects and a review. In *Turbulent Reactive Flows*, eds. P.A. Libby and F.A. Williams, pp. 1–61. London: Academic Press.

Libby, P.A. and F.A. Williams. 2000. Presumed PDF analysis of partially premixed turbulent combustion. *Combust. Sci. Tech.* 161:359–90.

Libby, P.A., Linán, A., and F.A. Williams. 1983. Strained premixed laminar flames with non-unity Lewis numbers. *Combust. Sci. Tech.* 34:257–93.

Liew, S.K., Bray, K.N.C., and J.B. Moss. 1981. A flamelet model of turbulent non-premixed combustion. *Combust. Sci. Tech.* 27:69–73.

Liew, S.K., Bray, K.N.C., and J.B. Moss. 1984. A stretched laminar flamelet model of turbulent nonpremixed combustion. *Combust. Flame* 56:199–213.

Lindstedt, R.P., Milosavljevic, V.D., and M. Persson. 2011. Turbulent burning velocity predictions using transported PDF methods. *Proc. Combust. Inst.* 33:1277–84.

Lipatnikov, A.N. 1997. Modeling of the influence of mixture properties on premixed turbulent combustion. In *Advanced Computation and Analysis of Combustion*, eds. G.D. Roy, S.M. Frolov, and P. Givi, pp. 335–59. Moscow: ENAS.

Lipatnikov, A.N. 2002. Comments on the paper "Premixed flames in stagnating turbulence. Part V: Evaluation of models for the chemical source term" by K.N.C. Bray, M. Champion, and P.A. Libby. *Combust. Flame* 131:219–21.

Lipatnikov, A.N. 2007a. Scalar transport in self-similar, developing, premixed, turbulent flames. *Combust. Sci. Tech.* 179:91–115.

Lipatnikov, A.N. 2007b. Premixed turbulent flame as a developing front with a self-similar structure. In *Focus on Combustion Research*, ed. S.Z. Jaing, pp. 89–141. New York: Nova Science.

Lipatnikov, A.N. 2008. Conditionally averaged balance equations for modeling premixed turbulent combustion in flamelet regime. *Combust. Flame* 152:529–47.

Lipatnikov, A.N. 2009a. Can we characterize turbulence in premixed flames? *Combust. Flame* 156:1242–7.

Lipatnikov, A.N. 2009b. Testing premixed turbulent combustion models by studying flame dynamics. *Int. J. Spray Combust. Dyn.* 1:39–66.

Lipatnikov, A.N. 2011a. Conditioned moments in premixed turbulent reacting flows. *Proc. Combust. Inst.* 33:1489–96.

Lipatnikov, A.N. 2011b. A test of conditioned balance equation approach. *Proc. Combust. Inst.* 33:1497–504.

Lipatnikov, A.N. 2011c. Transient behavior of turbulent scalar transport in premixed flames. *Flow Turbul. Combust.* 86:609–37.

Lipatnikov, A.N. 2011d. Burning rate in impinging jet flames. *J. Combust.* 2:737914.

Lipatnikov, A.N. and J. Chomiak. 1997. A simple model of unsteady turbulent flame propagation. *SAE Paper* 972993.

Lipatnikov, A.N. and J. Chomiak. 1998. Lewis number effects in premixed turbulent combustion and highly perturbed laminar flames. *Combust. Sci. Tech.* 137:277–98.

Lipatnikov, A.N. and J. Chomiak. 1999. A quasi-fractal subgrid flame speed closure for large eddy simulations of premixed flames. In *17th International Colloquium on the Dynamics of Explosion and Reactive Systems*, Heidelberg.

Lipatnikov, A.N. and J. Chomiak. 2000. Transient and geometrical effects in expanding turbulent flames. *Combust. Sci. Tech.* 154:75–117.

Lipatnikov, A.N. and J. Chomiak. 2002a. Turbulent flame speed and thickness: Phenomenology, evaluation, and application in multi-dimensional simulations. *Prog. Energy Combust. Sci.* 28:1–74.

Lipatnikov, A.N. and J. Chomiak. 2002b. Are premixed turbulent stagnation flames equivalent to fully developed ones? A computational study. *Combust. Sci. Tech.* 174(11&12):3–26.

Lipatnikov, A.N. and J. Chomiak. 2002c. Turbulent burning velocity and speed of developing, curved, and strained flames. *Proc. Combust. Inst.* 29:2113–21.

Lipatnikov, A.N. and J. Chomiak. 2003. A numerical study of weakly turbulent premixed combustion with Flame Speed Closure model. *SAE Paper* 2003-01-1839.

Lipatnikov, A.N. and J. Chomiak. 2004a. Application of the Markstein number concept to curved turbulent flames. *Combust. Sci. Tech.* 176:331–58.

Lipatnikov, A.N. and J. Chomiak. 2004b. Flame Speed Closure model of premixed and partially premixed turbulent combustion: Further development and validation. In *Proceedings of the Sixth International Symposium on Diagnostics and Modeling of Combustion in Internal Combustion Engines—COMODIA2004*, pp. 583–90. Yokohama: JSME.

Lipatnikov, A.N. and J. Chomiak. 2004c. Comment on "Turbulent burning velocity, burned gas distribution, and associated flame surface definition," D. Bradley, M.Z. Haq, R.A. Hicks, T. Kitagawa, M. Lawes, C.G.W. Sheppard, and R. Woolley. *Combust. Flame* 133:415, 2003; *Combust. Flame* 137:261–3, 2004.

Lipatnikov, A.N. and J. Chomiak. 2005a. Molecular transport effects on turbulent flame propagation and structure. *Prog. Energy Combust. Sci.* 31:1–73.

Lipatnikov, A.N. and J. Chomiak, J. 2005b. Self-similarly developing, premixed, turbulent flames: A theoretical study. *Phys. Fluids* 17:065105.

Lipatnikov, A.N. and J. Chomiak. 2007. Global stretch effects in premixed turbulent combustion. *Proc. Combust. Inst.* 31:1361–8.

Lipatnikov, A.N. and J. Chomiak. 2010. Effects of premixed flames on turbulence and turbulent scalar transport. *Prog. Energy Combust. Sci.* 36:1–102.

Lipatnikov, A.N. and V.A. Sabel'nikov. 2008. Some basic issues of the averaged G-equation approach to premixed turbulent combustion modeling. *Open Thermodyn. J.* 2:53–8.

Lipatnikov, A.N. and V.A. Sabel'nikov. 2011. Transition from countergradient to gradient turbulent scalar transport in developing premixed turbulent flames. In *Seventh Mediterranean Combustion Symposium*, Sardinia.

Lipatnikov, A.N., Wallesten, J., and J. Nisbet. 1998. Testing of a model for multi-dimensional computations of turbulent combustion in spark ignition engines. In *Proceedings of the Fourth International Symposium on Diagnostics and Modeling of Combustion in Internal Combustion Engines—COMODIA98*, pp. 239–44. Kyoto: JSME.

Lipzig, J.P.J., Nilsson, E.J.K., de Goey, L.P.H., and A.A. Konnov. 2011. Laminar burning velocities of *n*-heptane, *iso*-octane, ethanol and their binary and tertiary mixtures. *Fuel* 90:2773–81.

Liu, J.-K. and P.D. Ronney. 1999. Premixed edge flames in spatially varying straining flows. *Combust. Sci. Tech.* 144:21–46.

Lockwood, F.G. and A.S. Naguib. 1975. The prediction of fluctuations in the properties of free, round-jet, turbulent, diffusion flames. *Combust. Flame* 24:109–24.

Lu, T. and C.K. Law. 2009. Toward accommodating realistic fuel chemistry in large-scale computations. *Prog. Energy Combust. Sci.* 35:192–215.

Lyons, K.M. 2007. Toward an understanding of the stabilization mechanisms of lifted turbulent jet flames: Experiments. *Prog. Energy Combust. Sci.* 33:211–31.

Maas, U. and S.B. Pope. 1992. Simplifying chemical kinetics: Intrinsic low-dimensional manifold in composition space. *Combust. Flame* 88:239–64.

Maas, U. and J. Warnatz. 1988. Ignition processes in carbon-monoixide–hydrogen–oxygen mixtures. *Proc. Combust. Inst.* 22:1695–704.

Magnussen, B.F. and B.H. Hjertager. 1976. On mathematical modeling of turbulent combustion with special emphasis on soot formation and combustion. *Proc. Combust. Inst.* 16:719–29.

Malkeson, S.P. and N. Chakraborty. 2010. A priori direct numerical simulation assessment of algebraic models of variances and dissipation rates in the context of Reynolds-averaged Navier-Stokes simulations of low Damköhler number partially premixed combustion. *Combust. Sci. Tech.* 182:960–99.

Mandelbrot, B.B. 1975. On the geometry of homogeneous turbulence with stress on the fractal dimension of iso-surfaces of scalars. *J. Fluid Mech.* 72:401–16.

Mandelbrot, B.B. 1983. *Fractal Geometry of Nature*. San Francisco: W.H. Freeman.

Mantel, T. and R. Borghi. 1994. A new model of premixed wrinkled flame propagation based on a scalar dissipation equation. *Combust. Flame* 96:443–57.

Mantel, T. and J.M. Samaniego. 1999. Fundamental mechanisms in premixed turbulent flame propagation via flame–vortex interactions. Part II: Numerical simulation. *Combust. Flame* 118:557–82.

Mantzaras, J., Felton, P.G., and F.V. Bracco. 1989. Fractals and turbulent premixed engine flames. *Combust. Flame* 77:295–310.

Marshall, S.P., Taylor, S., Stone, C.R., Davies, T.J., and R.F. Cracknell. 2011. Laminar burning velocity measurements of liquid fuels at elevated pressures and temperature with combustion residuals. *Combust. Flame* 158:1920–32.

Markstein, G.H. 1951. Experimental and theoretical studies of flame front stability. *J. Aeronaut. Sci.* 18:199–220.

Markstein, G.H. 1964. *Nonsteady Flame Propagation*. New York: MacMillan.

Marzouk, Y.M., Ghoniem, A.F., and H.N. Najm. 2000. Dynamic response of stratified premixed flames to equivalence ratio gradients. *Proc. Combust. Inst.* 28:1859–66.

Mason, H.B. and D.B. Spalding. 1973. Prediction of reaction rates in turbulent premixed boundary layer flows. In *The Combustion Institute European Symposium*, ed. F.J. Weinberg, pp. 601–6. New York: Academic Press.

Masuya, G. 1986. Influence of laminar flame speed on turbulent premixed combustion. *Combust. Flame* 64:353–67.

Matalon, M. 2009. Flame dynamics. *Proc. Combust. Inst.* 32:57–82.

Matalon, M. and B.J. Matkowsky. 1982. Flames as gas dynamic discontinuities. *J. Fluid Mech.* 124:239–60.

Matalon, M., Cui, C., and J.K. Bechtold. 2003. Hydrodynamic theory of premixed flames: Effects of stoichiometry, variable transport coefficients and arbitrary reaction orders. *J. Fluid Mech.* 487:179–200.

Mauss, F., Keller, D., and N. Peters. 1990. A Lagrangian simulation of flamelet extinction and re-ignition in turbulent jet diffusion flames. *Proc. Combust. Inst.* 23:693–8.

McCormack, P.D., Scheller, K., Mueller, G., and R. Tisher. 1972. Flame propagation in a vortex core. *Combust. Flame* 19:297–303.

Meneveau, C. and T. Poinsot. 1991. Stretching and quenching of flamelets in premixed turbulent combustion. *Combust. Flame* 86:311–32.

Metghalchi, M. and J.C. Keck. 1982. Burning velocities of mixtures of air with methanol, isooctane, and indolene at high pressure and temperature. *Combust. Flame* 48:191–210.

Mikolaitis, D.W. 1984. The interaction of flame curvature and stretch. Part 1: The concave premixed flame. *Combust. Flame* 57:25–31.

Miller, J.A. and C.T. Bowman. 1989. Mechanism and modeling of nitrogen chemistry in combustion. *Prog. Energy Combust. Sci.* 15:287–338.

Miller, J.A., Pilling, M.J., and J. Troe. 2005. Unravelling combustion mechanisms through a quantitative understanding of elementary reactions. *Proc. Combust. Inst.* 30:43–88.

Monin, A.S. and A.M. Yaglom. 1971. *Statistical Fluid Mechanics: Mechanics of Turbulence*, vol. 1. Cambridge, MA: MIT Press.

Monin, A.S. and A.M. Yaglom. 1975. *Statistical Fluid Mechanics: Mechanics of Turbulence*, vol. 2. Cambridge, MA: MIT Press.

Moreau, P. 1977. Turbulent flame development in a high velocity premixed flow. *AIAA Paper* 77/49.

Moreau, V. 2009. A self-similar premixed turbulent flame model. *Appl. Math. Model.* 33:835–51.

Moss, J.B. 1980. Simultaneous measurements of concentration and velocity in an open premixed turbulent flame. *Combust. Sci. Tech.* 22:119–29.

Mouqallid, M., Lecordier, B., and M. Trinité. 1994. High speed laser tomography analysis of flame propagation in a simulated internal combustion engine—Applications to nonuniform mixture. *SAE Paper* 941990.

Mueller, C.J., Driscoll, J.F., Reuss, D.L., and M.C. Drake. 1996. Effects of unsteady stretch on the strength of a freely-propagating flame wrinkled by a vortex. *Proc. Combust. Inst.* 26:347–55.

Müller, C.M., Breitbach, H., and N. Peters. 1994. Partially premixed turbulent flame propagation in jet flames. *Proc. Combust. Inst.* 25:1099–106.

Müller, U.C., Bollig, M., and N. Peters. 1997. Approximations for burning velocities and Markstein numbers for lean hydrocarbon and methanol flames. *Combust. Flame* 108:349–56.

Muppala, S.R.P. and F. Dinkelacker. 2004. Numerical modelling of the pressure dependent reaction source term for turbulent premixed methane–air flames. *Prog. Comput. Fluid Dyn.* 4:328–33.

Mura, A. and R. Borghi. 2003. Towards an extended scalar dissipation equation for turbulent premixed combustion. *Combust. Flame* 133:193–6.

Mura, A. and M. Champion. 2009. Relevance of the Bray number in the small-scale modeling of turbulent premixed flames. *Combust. Flame* 156:729–33.

Mura, A., Tsuboi, K., and T. Hasegawa. 2008. Modelling of the correlation between velocity and reactive scalar gradients in turbulent premixed flames based on DNS data. *Combust. Theory Model.* 12:671–98.

Mura, A., Robin, V., Champion, M., and T. Hasegawa. 2009. Small scale features of velocity and scalar fields in turbulent premixed flames. *Flow Turbul. Combust.* 82:339–58.

Murayama, M. and T. Takeno. 1988. Fractal-like character of flamelets in turbulent premixed combustion. *Proc. Combust. Inst.* 22:551–9.

Namazian, M., Shepherd, I.G., and L. Talbot. 1986. Characterization of the density fluctuations in turbulent V-shaped premixed flames. *Combust. Flame* 64:299–308.

Niemeyer, J.C. and A.R. Kerstein. 1997. Numerical investigation of scaling properties of turbulent premixed flames. *Combust. Sci. Tech.* 128:343–58.

Nilsson, P. and X.S. Bai. 2002. Effects of flame stretch and wrinkling on CO formation in turbulent premixed combustion. *Proc. Combust. Inst.* 29:1873–9.

Nishiki, S. 2003. DNS and modeling of turbulent premixed combustion. Ph.D. dissertation, Nagoya Institute of Technology, Nagoya.

Nishiki, S., Hasegawa, T., Borghi, R., and R. Himeno. 2002. Modeling of flame-generated turbulence based on direct numerical simulation databases. *Proc. Combust. Inst.* 29:2017–22.

Nishiki, S., Hasegawa, T., Borghi, R., and R. Himeno. 2006. Modelling of turbulent scalar flux in turbulent premixed flames based on DNS databases. *Combust. Theory Model.* 10:39–55.

Nishioka, M. and R. Ogura. 2005. A numerical study of the high-speed flame propagation in a vortex tube. In *Proceedings of the 19th International Colloquium on the Dynamics of Explosions and Reactive Systems*, Hakone.

Nogenmyr, K.J., Petersson, P., Bai, X.S., Nauert, A., Olofsson, J., Brackman, C., et al. 2007. Large eddy simulation and experiments of stratified lean premixed methane/air turbulent flames. *Proc. Combust. Inst.* 31:1467–75.

Nogenmyr, K.J., Fureby, C., Bai, X.S., Petersson, P., Collin, R., and M. Linne. 2008. Large eddy simulation and laser diagnostic studies on a low swirl stratified premixed flame. *Combust. Flame* 155:357–68.

Nogenmyr, K.J., Kiefer, J., Li, Z.S., Bai, X.S., and M. Aldén. 2010. Numerical computations and optical diagnostics of unsteady partially premixed methane/air flames. *Combust. Flame* 157:915–24.

North, G.L. and D.A. Santavicca. 1990. The fractal nature of premixed turbulent flames. *Combust. Sci. Tech.* 72:215–32.

Nwagwe, I.K., Weller, H.G., Tabor, G.R., Gosman, A.D., Lawes, M., Sheppard, C.G.W., and R. Woolley. 2000. Measurements and large eddy simulations of turbulent premixed flame kernel growth. *Proc. Combust. Inst.* 28:59–65.

Oberlack, M., Wenzel, H., and N. Peters. 2001. On symmetries and averaging of the *G*-equation for premixed combustion. *Combust. Theory Model.* 5:363–83.

O'Brien, E.E. 1980. The probability density function (PDF) approach to reacting turbulent flows. In *Turbulent Reacting Flows*, eds. P.A. Libby and F.A. Williams, pp. 185–207. Berlin: Springer.

O'Rourke, P.J. and F.V. Bracco. 1979. Two scaling transformations for the numerical computation of multidimensional unsteady laminar flames. *J. Comput. Phys.* 33:185–203.

van Oijen, J.A., Lammers, F.A., and L.P.H. de Goey. 2001. Modeling of complex premixed burner systems by using flamelet-generated manifolds. *Combust. Flame* 127:2124–34.

Pagnini, G. and F. Bonomi. 2011. Lagrangian formulation of turbulent premixed combustion. *Phys. Rev. Lett.* 107:044503.

Pasquier, N., Lecordier, B., Trinité, M., and A. Cessou. 2007. An experimental investigation of flame propagation through a turbulent stratified mixture. *Proc. Combust. Inst.* 31:1567–74.

Paul, R.N. and K.N.C. Bray. 1996. Study of premixed turbulent combustion including Landau-Darrieus instability effects. *Proc. Combust. Inst.* 26:259–66.

Peters, N. 1982. The premixed turbulent flame in the limit of a large activation energy. *J. Non-Equilib. Thermodyn.* 7:25–38.

Peters, N. 1983. Local quenching due to flame stretch and non-premixed turbulent combustion. *Combust. Sci. Tech.* 30:1–17.

Peters, N. 1984. Laminar diffusion flamelet models in non-premixed turbulent combustion. *Prog. Energy Combust. Sci.* 10:319–39.

Peters, N. 1986. Laminar flamelet concepts in turbulent combustion. *Proc. Combust. Inst.* 21:1231–49.

Peters, N. 1992. A spectral closure for premixed turbulent combustion in the flamelet regime. *J. Fluid Mech.* 242:611–29.

Peters, N. 1994. Kinetic foundation of thermal flame theory. In *Advances in Combustion Science: In Honor of Ya.B. Zel'dovich*, eds. W.A. Sirignano, A.G. Merzhanov, and L. De Luca, *Progress in Astronautics and Aeronautics*, vol. 173, chapter 5, pp. 73–91. Reston: AIAA.

Peters, N. 1999. The turbulent burning velocity for large-scale and small-scale turbulence. *J. Fluid Mech.* 384:107–32.

Peters, N. 2000. *Turbulent Combustion*. Cambridge: Cambridge University Press.

Peters, N. and F.A. Williams. 1983. Lift-off characteristics of turbulent jet diffusion flames. *AIAA J.* 21:423–9.

Peters, N. and F.A. Williams. 1987. The asymptotic structure of stoichiometric methane–air flames. *Combust. Flame* 68:185–207.

Peters, N., Wenzel, H., and F.A. Williams. 2000. Modification of the turbulent burning velocity by gas expansion. *Proc. Combust. Inst.* 28:235–43.

Pfadler, S., Leipertz, A., Dinkelacker, F., Wäsle, J., Winkler, A., and T. Sattelmayer. 2007. Two-dimensional direct measurement of the turbulent flux in turbulent premixed swirl flames. *Proc. Combust. Inst.* 31:1337–44.

Philips, H. 1965. Flame in a buoyant methane air. *Proc. Combust. Inst.* 10:1277–83.

Pirez da Cruz, A., Dean, A.M., and J.M. Grenda. 2000. A numerical study of the laminar flame speed of stratified methane/air flames. *Proc. Combust. Inst.* 28:1925–32.

Pitsch, H. 2005. A consistent level set formulation for large-eddy simulation of premixed turbulent combustion. *Combust. Flame* 143:587–98.

Pitsch, H. 2006. Large-eddy simulation of turbulent combustion. *Ann. Rev. Fluid Mech.* 38:453–82.

Pitsch, H., Peters, N., and K. Seshadri. 1996. Numerical and asymptotic studies of the structure of premixed iso-octane flames. *Proc. Combust. Inst.* 26:763–71.

Pitsch, H., Chen, M., and N. Peters. 1998. Unsteady flamelet modeling of turbulent hydrogen–air diffusion flames. *Proc. Combust. Inst.* 27:1057–64.

Poinsot, T. 1996. Using direct numerical simulations to understand premixed turbulent combustion. *Proc. Combust. Inst.* 26:219–32.

Poinsot, T. and D. Veynante. 2005. *Theoretical and Numerical Combustion*, 2nd edn. Philadelphia, PA: Edwards.

Poinsot, T., Veynante, D., and S. Candel. 1990. Diagrams of premixed turbulent combustion based on direct simulation. *Proc. Combust. Inst.* 23:613–19.

Poinsot, T., Veynante, D., and S. Candel. 1991. Quenching processes and premixed turbulent combustion diagrams. *J. Fluid Mech.* 228:561–606.

Poinsot, T., Candel, S., and A. Trouvé. 1995. Applications of direct numerical simulations to premixed turbulent combustion. *Prog. Energy Combust. Sci.* 21:531–76.

Polifke, W., Flohr, P., and M. Brandt. 2002. Modeling of inhomogeneously premixed combustion with an extended TFC model. *ASME J. Eng. Gas Turbines Power* 124:58–65.

Poludnenko, A.Y. and E.S. Oran. 2010. The interaction of high-speed turbulence with flames: Global properties and internal flame structure. *Combust. Flame* 157:995–1011.

Poludnenko, A.Y. and E.S. Oran. 2011. The interaction of high-speed turbulence with flames: Turbulent flame speed. *Combust. Flame* 158:301–26.

Pope, S.B. 1985. PDF methods for turbulent reacting flows. *Prog. Energy Combust. Sci.* 11:119–92.

Pope, S.B. 1988. The evolution of surface in turbulence. *Int. J. Eng. Sci.* 26:445–69.

Pope, S.B. 1994. Lagrangian PDF methods for turbulent flows. *Ann. Rev. Fluid Mech.* 26:23–63.

Pope, S.B. 2000. *Turbulent Flows*. Cambridge: Cambridge University Press.

Prasad, R.O.S. and J.P. Gore. 1999. An evaluation of flame surface density models for turbulent premixed jet flames. *Combust. Flame* 116:1–14.

Procaccia, I. 1984. Fractal structures in turbulence. *J. Stat. Phys.* 5:649–63.

Prudnikov, A.G. 1960. Hydrodynamics equations in turbulent flames. In *Combustion in a Turbulent Flow*, ed. A.G. Prudnikov, pp. 7–29. Moscow: Oborongiz (in Russian).

Prudnikov, A.G. 1967. Combustion of homogeneous fuel–air mixtures in turbulent flows. In *Physical Principles of the Working Process in Combustion Chambers of Jet Engines*, ed. B.V. Raushenbakh, pp. 244–336. Springfield: Clearing House for Federal Scientific & Technical Information.

Ra, Y. and W.K. Chen. 2001. Laminar flame propagation through a step-stratified charge. In *Proceedings of the Fifth International Symposium on Diagnostics and Modeling of Combustion in Internal Combustion Engines—COMODIA2001*, pp. 251–7. Nagoya, Japan: JSME.

Renard, P.-H., Thévenin, D., Rolon, J.C., and S. Candel. 2000. Dynamics of flame/vortex interactions. *Prog. Energy Combust. Sci.* 26:225–82.

Renou, B., Mura, A., Samson, E., and A. Boukhalfa. 2002. Characterization of the local flame structure and flame surface density for freely-propagating premixed flames at various Lewis numbers. *Combust. Sci. Tech.* 174:143–79.

Ribert, G., Champion, M., and P. Plion. 2004. Modeling a turbulent reactive flow with variable equivalence ratio: Application to the calculation of a reactive shear layer. *Combust. Sci. Tech.* 176:907–23.

Ribert, G., Gicquel, O., Darabiha, N., and D. Veynante. 2006. Tabulation of complex chemistry based on self-similar behavior of laminar premixed flames. *Combust. Flame* 146:649–64.

Richardson, E.S., Granet, V.E., Eyssartier, A., and J.H. Chen. 2010. Effects of equivalence ratio variation on lean, stratified methane–air laminar counterflow flames. *Combust. Theory Model.* 14:775–92.

Richardson, L.F. 1926. Atmospheric diffusion shown on a distance–neighbour graph. *Proc. R. Soc. Lond. A* 110:709–37.

Roberts, W.L., Driscoll, J.F., Drake, M.C., and L.P. Goss. 1993. Images of the quenching of a flame by vortex to quantify regimes of turbulent combustion. *Combust. Flame* 94:58–69.

Robin, V., Mura, A., Champion, M., and P. Plion. 2006. A multi-Dirac presumed PDF model for turbulent reactive flows with variable equivalence ratio. *Combust. Sci. Tech.* 178:1843–70.

Robin, V., Mura, A., Champion, M., Degardin, O., Renou, B., and M. Doulthalfa 2008. Experimental and numerical analysis of stratified turbulent V-shaped flames. *Combust. Flame* 153:288–315.

Ronney, P.D. 1998. Understanding combustion processes through microgravity research. *Proc. Combust. Inst.* 27:2485–506.

Ronney, P.D. and V. Yakhot. 1992. Flame broadening effects on premixed turbulent flame speed. *Combust. Sci. Tech.* 86:31–43.

Ruetsch, G.R., Vervisch, L., and A. Liñán. 1995. Effects of heat release on triple flames. *Phys. Fluids* 7:1447–54.

Rutland, C.J. and A. Trouvé. 1993. Direct simulations of premixed turbulent flames with non-unity Lewis numbers. *Combust. Flame* 94:41–57.

Sabel'nikov, V.A. and A.N. Lipatnikov. 2010. Rigorous derivation of an unclosed mean G-equation for statistically 1D premixed turbulent flame. *Int. J. Spray Combust. Dyn.* 2:301–24.

Sabel'nikov, V.A. and A.N. Lipatnikov. 2011a. A simple model for evaluating conditioned velocities in premixed turbulent flames. *Combust. Sci. Tech.* 183:588–613.

Sabel'nikov, V.A. and A.N. Lipatnikov. 2011b. Averaging of flamelet-conditioned kinematic equation in turbulent reacting flows. In *Turbulence: Theory, Types and Simulations*, ed. R.J. Marcuso, pp. 321–64. New York: Nova Science.

Sakai, Y. and S. Ishizuka. 1996. The phenomena of flame propagation in a rotating tube. *Proc. Combust. Inst.* 26:847–53.

Samaniego, J.M. and T. Mantel. 1999. Fundamental mechanisms in premixed turbulent flame propagation via flame–vortex interactions. Part I: Experiment. *Combust. Flame* 118:537–56.

Sankaran, R. and H.G. Im. 2002. Dynamic flammability limits of methane/air premixed flames with mixture composition fluctuations. *Proc. Combust. Inst.* 29:77–84.

Sathiah, P. and A.N. Lipatnikov. 2007. Effects of flame development on stationary premixed turbulent combustion. *Proc. Combust. Inst.* 31:3115–22.

Savarianandam, V.R. and C.J. Lawn. 2006. Burning velocity of premixed turbulent flames in the weakly wrinkled regime. *Combust Flame* 146:1–18.

Schneider, E., Maltsev, A., Sadiki, A., and J. Janicka. 2008. Study of the potential of BML-approach and G-equation concept-based models for predicting swirling partially premixed combustion systems: URANS computations. *Combust. Flame* 152:548–72.

Schumacher, J., Sreenivasan, K.R., and V. Yakhot. 2007. Asymptotic exponents from low-Reynolds-number flows. *New J. Phys.* 9:89.

Scurlock, A.C. 1948. Flame stabilization and propagation in high-velocity gas streams. MIT Press.

Seiser, R., Truett, L., Trees, D., and K. Seshadri. 1998. Structure and extinction of non-premixed n-heptane flames. *Proc. Combust. Inst.* 27:649–57.

Seshadri, K. 1996. Multistep asymptotic analyses of flame structure. *Proc. Combust. Inst.* 26:831–46.

Seshadri, K. and N. Peters. 1990. The inner structure of methane–air flames. *Combust. Flame* 81:96–118.

Seshadri, K., Bollig, M., and N. Peters. 1997. Numerical and asymptotic studies of the structure of stoichiometric and lean premixed heptane flames. *Combust. Flame* 108:518–36.

Shchelkin, K.I. 1943. Combustion in turbulent flow. *Zh. Tekh. Fiz.* 13:520–30 (in Russian; English translation NACA TM 1112, 1947).

She, Z.-S., Jackson, E., and A. Orszag. 1991. Structure and dynamics of homogeneous turbulence: Models and simulations. *Proc. R. Soc. Lond. A* 434:101–24.

Shepherd, I.G. 1995. Heat release and induced strain in premixed flames. *Combust. Flame* 103:1–10.

Shepherd, I.G. 1996. Flame surface density and burning rate in premixed turbulent flames. *Proc. Combust. Inst.* 26:373–9.

Shepherd, I.G. and L.W. Kostiuk. 1994. The burning rate of premixed turbulent flames in divergent flows. *Combust. Flame* 96:371–80.

Shepherd, I.G., Cheng, R.K., and P.J. Goix. 1990. The spatial scalar structure of premixed turbulent stagnation point flames. *Proc. Combust. Inst.* 23:781–7.

Shepherd, I.G., Cheng, R.K., and L. Talbot. 1992. Experimental criteria for the determination of fractal parameters of premixed turbulent flames. *Exp. Fluids* 13:386–92.

Shepherd, I.G., Cheng, R.K., Plessing, T., Kortschik, C., and N. Peters. 2002. Premixed flame front structure in intense turbulence. *Proc. Combust. Inst.* 29:1833–40.

Shim, Y., Tanaka, S., Tanahashi, M., and T. Miyauchi. 2011. Local structure and fractal characteristics of H_2–air turbulent premixed flame. *Proc. Combust. Inst.* 33:1455–62.

Shy, S.S., Lin, W.J., and J.C. Wei. 2000a. An experimental correlation of turbulent burning velocities for premixed turbulent methane–air combustion. *Proc. R. Soc. Lond. A* 456:1997–2019.

Shy, S.S., Lin, W.J., and K.Z. Peng. 2000b. Highly intensity turbulent premixed combustion: General correlations of turbulent burning velocities in a new cruciform burner. *Proc. Combust. Inst.* 28:561–8.

Sinibaldi, J.O., Driscoll, J.F., Mueller, C.J., Donbar, J.M., and C.D. Carter. 2003. Propagation speeds and stretch rates measured along wrinkled flames to assess the theory of flame stretch. *Combust. Flame* 133:323–34.

Sivashinsky, G.I. 2002. Some developments in premixed combustion modeling. *Proc. Combust. Inst.* 29:1737–61.

Sivashinsky, G.I., Law, C.K., and G. Joulin. 1982. On stability of premixed flames in stagnation-point flow. *Combust. Sci. Tech.* 28:155–9.

Sjunnesson, A., Henriksson, P., and C. Löfström. 1992. CARS measurements and visualization of reacting flows in a bluff body stabilized flame. *AIAA Paper* 92/3650.

Smallwood, G.J., Gülder, Ö.L., Snelling, D.R., Deschamps, B.M., and I. Gökalp. 1995. Characterization of flame front surfaces in turbulent premixed methane/air combustion. *Combust. Flame* 101:461–70.

Smith, K.G. and F.G. Gouldin. 1979. Turbulence effects on flame speed and flame structure. *AIAA J.* 17:1243–50.

Soika, A., Dinkelacker, F., and A. Leipertz. 2003. Pressure influence on the flame front curvature of turbulent premixed flames: Comparison between experiment and theory. *Combust. Flame* 132:451–62.

Spalding, D.B. 1971. Mixing and chemical reaction in steady confined turbulent flame. *Proc. Combust. Inst.* 13:649–57.

Spalding, D.B. 1976. Development of the eddy-break-up model of turbulent combustion. *Proc. Combust. Inst.* 16:1657–63.

Sreenivasan, K.R. 1991. Fractals and multifractals in fluid turbulence. *Ann. Rev. Fluid Mech.* 23:539–600.

Sreenivasan, K.R. and C. Meneveau. 1986. The fractal facets of turbulence. *J. Fluid Mech.* 17:357–86.

Sreenivasan, K.R. and R.R. Prasad. 1988. Fractal dimension of scalar interfaces in turbulent jets. *Polish Acad. Sci.* 14:1–14.

Steinberg, A.M., Driscoll, J.F., and S.L. Ceccio. 2008. Measurements of turbulent premixed flame dynamics using cinema stereoscopic PIV. *Exp. Fluids* 44:985–99.

Stevens, E.J., Bray, K.N.C., and B. Lecordier. 1998. Velocity and scalar statistics for premixed turbulent stagnation flames using PIV. *Proc. Combust. Inst.* 27:949–55.

Summerfield, M., Reiter, S.H., Kebely, V., and R.W. Mascolo. 1955. The structure and propagation mechanism of turbulent flames in high speed flow. *Jet Propulsion* 25:377–84.

Sun, C.J., Sung, C.J., He, L., and C.K. Law. 1999. Dynamics of weakly stretched flames. Quantitative description and extraction of global flame parameters. *Combust. Flame* 118:108–28.

Swaminathan, N. and R.W. Bilger. 1998. Conditional variance equation and its analysis. *Proc. Combust. Inst.* 27:1191–8.

Swaminathan, N. and K.N.C. Bray. 2005. Effect of dilatation on scalar dissipation in turbulent premixed flames. *Combust. Flame* 143:549–65.

Swaminathan, N. and R.W. Grout. 2006. Interaction of turbulence and scalar fields in premixed flames. *Phys. Fluids* 18:045102.

Swaminathan, N., Bilger, R.W., and G.R. Ruetsch. 1997. Interdependence of the instantaneous flame front structure and the overall scalar flux in turbulent premixed flames. *Combust. Sci. Tech.* 128:73–97.

Sweeney, M.S., Hochgreb, S., and R.S. Barlow. 2011. The structure of premixed and stratified low turbulence flames. *Combust. Flame* 158:935–48.

Tahtouh, T., Halter, F., and C. Mounaïm-Rousselle. 2009. Measurement of laminar burning speeds and Markstein lengths using a novel methodology. *Combust. Flame* 156:1735–43.

Takeno, T., Murayama, M., and Y. Tanida. 1990. Fractal analysis of turbulent premixed surface. *Exp. Fluids* 10:61–70.

Talantov, A.V. 1978. *Combustion in Flows*. Moscow: Mashinostroenie (in Russian).

Tanahashi, M., Fujimura, M., and T. Miyachii. 2002. Coherent fine-scale eddies in turbulent premixed flames. *Proc. Combust. Inst.* 28:529–35.

Tanahashi, M., Sato, M., Shimura, M., and T. Miyachii. 2008. DNS and combined laser diagnostics of turbulent combustion. *J. Thermal Sci. Tech.* 3:391–409.

Tanaka, H. and T. Yanagi. 1981. Velocity–temperature correlation in premixed flame. *Proc. Combust. Inst.* 18:1031–9.

Taylor, G.I. 1921. Diffusion by continuous movements. *Proc. Lond. Math. Soc. Ser. 2*, 20:196–211.

Taylor, G.I. 1935. Statistical theory of turbulence. IV: Diffusion in a turbulent air stream. *Proc. R. Soc. Lond. A* 151:421–78.

Tennekes, H. and J.L. Lumley. 1972. *A First Course in Turbulence*. Cambridge, MA: MIT Press.

Townsend, A.A. 1976. *The Structure of Turbulent Shear Flow*, 2nd edn. Cambridge: Cambridge University Press.

TRANSPORT. 2000. *A Software Package for the Evaluation of Gas-Phase, Multicomponent Transport Properties*. San Diego: Reaction Design.

Treurniet, T.C., Nieuwstadt, F.T.M., and B.J. Boersma. 2006. Direct numerical simulation of homogeneous turbulence in combination with premixed combustion at low Mach number modelled by the *G*-equation. *J. Fluid Mech.* 565:25–62.

Troiani, G., Marrocco, M., Giammartini, S., and C.M. Casciola. 2009. Counter-gradient transport in the combustion of a premixed CH_4/air annular jet by combined PIV/OH-LIF. *Combust. Flame* 156:608–20.

Trouvé, A. and T. Poinsot. 1994. Evolution equation for flame surface density in turbulent premixed combustion. *J. Fluid Mech.* 278:1–31.

Truffaut, J.-M. and G. Searby. 1999. Experimental study of the Darrieus-Landau instability on an inverted-V flame, and measurement of the Markstein number. *Combust. Sci. Tech.* 149:35–52.

Uberoi, M.S., Kuethe, A.M., and H.R. Menkes. 1958. Flow field of a Bunsen flame. *Phys. Fluids* 1:150–8.

Üngüt, A., Gorgeon, A., and I. Gökalp. 1993. A planar laser induced fluorescence study of turbulent flame kernel growth and fractal characteristics. *Combust. Sci. Tech.* 92:265–90.

Upatnieks, A., Driscoll, J.F., Rasmussen, C.C., and S.L. Ceccio. 2004. Liftoff of turbulent jet flames—Assessment of edge flame and other concepts using cinema PIV. *Combust. Flame* 138:259–72.

Vagelopoulos, C.M. and F.N. Egolfopoulos. 1998. Direct experimental determination of laminar flame speeds. *Proc. Combust. Inst.* 27:513–19.

Varea, E., Modica, V., Vandel, A., and B. Renou. 2012. Measurement of laminar burning velocity and Markstein length relative to fresh gas using a new postprocessing procedure: Application to laminar spherical flames for methane, ethanol and isooctane/air mixtures. *Combust. Flame* 159:577–90.

Vena, P.C., Deschamps, B., Smallwood, G.J., and M.R. Johnson. 2011. Equivalence ratio gradient effects on flame front topology in a stratified iso-octane/air turbulent V-flame. *Proc. Combust. Inst.* 33:1551–8.

Venkateswaran, P., Marshall, A., Shin, D.H., Noble, D., Seitzman, J., and T. Lieuwen. 2011. Measurements and analysis of turbulent consumption speeds of H_2/CO mixtures. *Combust. Flame* 158:1602–14.

Veynante, D. and T. Poinsot. 1997. Effects of pressure gradients on turbulent premixed flames. *J. Fluid Mech.* 353:83–114.

Veynante, D. and L. Vervisch. 2002. Turbulent combustion modeling. *Prog. Energy Combust. Sci.* 28:193–266.

Veynante, D., Duclos, J.M., and J. Piana. 1994. Experimental analysis of flamelet models for premixed turbulent combustion. *Proc. Combust. Inst.* 25:1249–56.

Veynante, D., Piana, J., Duclos, J.M., and C. Martel. 1996. Experimental analysis of flame surface density models for premixed turbulent combustion. *Proc. Combust. Inst.* 26:413–20.

Veynante, D., Trouvé, A., Bray, K.N.C., and T. Mantel. 1997. Gradient and counter-gradient scalar transport in turbulent premixed flames. *J. Fluid Mech.* 332:263–93.

Villermaux, E., Sixou, B., and Y. Gagne. 1995. Intense vortical structures in grid-generated turbulence. *Phys. Fluids* 7:2008–13.

Vincent, A. and M. Meneguzzi. 1991. The spatial structure and statistical properties of homogeneous turbulence. *J. Fluid Mech.* 225:1–25.

Vreman, A.W., van Oijen, J.A., de Goey, L.P.H, and R.J.M. Bastiaans. 2009. Subgrid scale modeling in large-eddy simulation of turbulent combustion using premixed flamelet chemistry. *Flow Turbul. Combust.* 82:511–35.

Wallesten, J., Lipatnikov, A.N., and J. Chomiak. 2002a. Simulations of fuel/air mixing, combustion, and pollutant formation in a direct injection gasoline engine. *SAE Paper* 2002-01-0835.

Wallesten, J., Lipatnikov, A.N., and J. Chomiak. 2002b. Modeling of stratified combustion in a DI SI engine using detailed chemistry pre-processing. *Proc. Combust. Inst.* 29:703–9.

Warnatz, J., Maas, U., and R.W. Dibble. 1996. *Combustion: Physical and Chemical Fundamentals, Modeling and Simulation, Experiments, Pollutant Formation.* Berlin: Springer.

Weller, H.G. 1993. The development of a new flame area combustion model using conditional averaging. Imperial College of Science, Technology and Medicine, London.

Weller, H.G., Uslu, S., Gosman, A.D., Maly, R.R., Herweg, R., and B. Heel. 1994. Prediction of combustion in homogeneous-charge spark-ignition engines. In *Proceedings of the Third International Symposium on Diagnostics and Modeling of Combustion in Internal Combustion Engines—COMODIA 94*, pp. 163–69. Yokohama: JSME.

Westbrook, C.K. and F.L. Dryer. 1984. Chemical kinetic modeling of hydrocarbon combustion. *Prog. Energy Combust. Sci.* 10:1–57.

Williams, F.A. 1975. Recent advances in theoretical descriptions of turbulent diffusion flames. In *Turbulent Mixing in Nonreactive and Reactive Flows*, ed. S.N.B. Murthy, pp. 189–208. New York: Plenum.

Williams, F.A. 1985a. *Combustion Theory*, 2nd edn. Menlo Park, CA: Benjamin/Cummings.

Williams, F.A. 1985b. Turbulent combustion In *The Mathematics of Combustion*, ed. J. D. Buckmaster, pp. 97–131. Philadelphia, PA: SIAM.

Williams, F.A. 2000. Progress in knowledge of flamelet structure and extinction. *Prog. Energy Combust. Sci.* 26:657–82.

Williams, G.C., Hottel, H.C., and A.C. Scurlock. 1949. Flame stabilization and propagation in high velocity gas streams. *Proc. Combust. Inst.* 3:21–40.

Wu, M.S., Kwon, A., Driscoll, G., and G.M. Faeth. 1991. Preferential diffusion effects on the surface structure of turbulent premixed hydrogen/air flames. *Combust. Sci. Tech.* 78:69–96.

Yakhot, V. 2008. Dissipation-scale fluctuations and mixing transition in turbulent flows. *J. Fluid Mech.* 606:325–37.

Yakhot, V. and J. Wanderer. 2012. Anomalous scaling of structure functions and sub-grid models for large eddy simulations of strong turbulence. Submitted. Available online http://arxiv.org/pdf/1109.6188v1.

Yeung, P.K. 2002. Lagrangian investigations of turbulence. *Ann. Rev. Fluid Mech.* 34:115–42.

Yeung, P.K., Girimaji, S.S., and S.B. Pope. 1990. Straining and scalar dissipation of material surfaces in turbulence: Implications for flamelets. *Combust. Flame* 79:340–65.

Yoshida, A. 1988. Structure of opposed jet premixed flame and transition of turbulent premixed flame structure. *Proc. Combust. Inst.* 22:1471–8.

Yoshida, A., Narisawa, M., and H. Tsuji. 1992. Structure of highly turbulent premixed flames. *Proc. Combust. Inst.* 24:519–25.

Yoshida, A., Kasahara, M., Tsuji, H., and T. Yanagisawa. 1994. Fractal geometry application in estimation of turbulent burning velocity of wrinkled laminar flame. *Combust. Sci. Tech.* 103:207–18.

Yuen, F.T.C. and Ö.L. Gülder. 2009. Premixed turbulent flame front structure investigation by Rayleigh scattering in the thin reaction zone regime. *Proc. Combust. Inst.* 32:1747–54.

Yuen, F.T.C. and Ö.L. Gülder. 2010. Dynamics of lean-premixed turbulent combustion at high turbulence intensities. *Combust. Sci. Tech.* 182:544–58.

Zel'dovich, Ya.B. 1944. *Theory of Gas Combustion and Detonation*. Moscow: AN SSSR.

Zel'dovich, Ya.B. 1949. On the theory of combustion of nonpremixed gases. *Zh. Tekh. Fiz.* 19:1199–210.

Zel'dovich, Ya.B. 1980. Flame propagation in a substance reacting at initial temperature. *Combust. Flame* 19:219–24.

Zel'dovich, Ya.B. and D.A. Frank-Kamenetskii. 1938. A theory of thermal flame propagation. *Acta Physicochem. URSS* 9:341–6.

Zel'dovich, Ya.B. and D.A. Frank-Kamenetskii. 1947. *Turbulent and Heterogeneous Combustion*. Moscow: MMI (in Russian).

Zel'dovich, Ya.B., Barenblatt, G.I., Librovich, V.B., and G.M. Makhviladze. 1985. *The Mathematical Theory of Combustion and Explosions*. New York: Plenum.

Zhao, Z., Conley, J.P., Kazakov, A., and F.L. Dryer. 2003. Burning velocities of real gasoline fuel at 353 K and 500 K. *SAE Paper* 2003-01-3265.

Zhou, J., Yoshizaki, T., Nishida, K., and H. Hiroyasu. 1997. Effects of mixture heterogeneity on flame propagation in a constant volume combustion chamber. *SAE Paper* 972943.

Zhou, J., Nishida, K., Yoshizaki, T., and H. Hiroyasu. 1998. Flame propagation characteristics in a heterogeneous concentration distribution of a fuel–air mixture. *SAE Paper* 982563.

Zimont, V.L. 1977. To computations of turbulent combustion of partially premixed gases. In *Chemical Physics of Combustion and Explosion Processes. Combustion of Multi-phase and Gas Systems*, pp. 77–80. Chernogolovka: OIKhF (in Russian).

Zimont, V.L. 1979. Theory of turbulent combustion of a homogeneous fuel mixture at high Reynolds number. *Combust. Explos. Shock Waves* 15:305–11.

Zimont, V.L. 2000. Gas premixed combustion at high turbulence. Turbulent flame closure combustion model. *Exp. Thermal Fluid Sci.* 21:179–86.

Zimont, V.L. 2006. Kolmogorov's legacy and turbulent premixed combustion modeling. In *New Developments in Combustion Research*, ed. W.J. Carey, pp. 1–93. New York: Nova Science.

Zimont, V.L. and V. Battaglia. 2006. Joint RANS/LES approach to premixed flame modelling in the context of the TFC combustion model. *Flow Turbul. Combust.* 77:305–31.

Zimont, V.L. and A.N. Lipatnikov. 1992. A model of heat release in premixed turbulent combustion. In *Applied Problems of Aeromechanics and Geospase Physics*, ed. T.V. Kondranin, pp. 48–58. Moscow: MIPT (in Russian).

Zimont, V.L. and A.N. Lipatnikov. 1993. To computations of the heat release rate in turbulent flames. *Dokl. Phys. Chem.* 332:592–4.

Zimont, V.L. and A.N. Lipatnikov. 1995. A numerical model of premixed turbulent combustion. *Chem. Phys. Rep.* 14:993–1025.

Zimont, V.L. and E.A. Mesheryakov. 1988. A model of combustion of partially premixed gases. In *Structure of Gas Flames*, Proceedings of International Colloquium, Part II, pp. 35–43. Novosibirsk: ITPM So AN USSR.

Zimont, V.L., Polifke, W., Bettelini, M., and W. Weisenstein. 1998. An efficient computational model for premixed turbulent combustion at high Reynolds number based on a turbulent flame speed closure. *ASME J. Eng. Gas Turbines Power* 120:526–32.

Zimont, V.L., Biagioli, F., and K. Syed. 2001. Modelling turbulent premixed combustion in the intermediate steady propagation regime. *Prog. Comput. Fluid Dyn.* 1:14–28.

Index